REMEDIAL MATHEMATICS

By

Dr. J.K. TYAGI
Professor & Head
Department of Applied Science
Maharaja Agarsain Institute of Technology
Ghaziabad (U.P.)

Dr. S.K. TYAGI
Senior Lecturer
Amity School of Engg. and Technology
Amity University,
NOIDA (U.P.)

KHANNA BOOK PUBLISHING CO. (P) LTD.
DELHI

REMEDIAL MATHEMATICS

By J.K. TYAGI
S.K. TYAGI

Laser Typeset by : **PLUS COMPUTERS**, 696/1, Madhav Puram, Meerut. Ph. 0-9897079785.

Printed and Bound by : **MEHAK OFFSET PRINTERS**

Published by : **Khanna Book Publishing Co. (P) Ltd.**
Regd. Office : 1694-95, Nai Sarak, Delhi-110006
Telephone : 011-23256658, 23283211
Fax : 011-23256658
E-mail : contact@khannaedu.com
Website : www.khannaedu.com

Preface

The primary objective of this book is to provide the basic concepts and applications of mathematics suited to the needs of students appearing in the examinations of **B. Pharm** of various universities. The aim of the book is to present the subject in such a way that even an average student may understand it without any difficulty. The language is simple and easily understandable. Numerous worked out examples have been given which will help the readers in understanding various concepts easily. Thus the approach is to communicate the practical aspects of mathematics in a way that drives out the common fear of learning any mathematical subject. The book contains complete course prescribed by various universities and so it can be safely recommended as a text book as well as a reference book. A problem set is given at the end of main topics so that student may try the problems and assess their confidence of answering the problem after studying the topics of the book. Most of the questions conform to the trend questions appearing in the University examinations.

The authors gratefully acknowledge the inspiration, encouragement and valuable suggestions received from the well wishers during the preparation of this book.

They also would like to thank the Director and the members of the Management Committee for their kind co-operation during this period.

The authors also thankful to many other authors whose books have been consulted with during the preparation of this book.

We would also like to thank our families for their patience, understanding, love and assistance in making this book a reality.

Finally, we are thankful to the publishers and printers for their full co-operation and utmost efficiency with which they have brought out the book in the present nice form.

Suggestions for improvement of the book will be gratefully acknowledged.

AUTHORS
E-mail : jaikishoretyagi@rediffmail.com

SYLLABUS

B.Pharm

(PHAR-III M / PH-III M)

REMEDIAL MATHEMATICS

UNIT I : **Algebra :** Equations reducible to quadratics, simultaneous equations (linear and quadratic), Determinants, properties of determinants, solution of simultaneous equations by Cramer's rule, Matrices, Properties of matrices, solution of simultaneous equations by matrices, pharmaceutical applications of determinants and matrices.

UNIT II : **Measures of Central Value :** Objectives and pre-requisites of an ideal measure, mean, mode and median.

UNIT III : **Trigonometry :** Measurement of angle, T-ratios, • subtraction and transformation formulae, T-ratio of multiple, submultiple, allied and certain angles, application of Logarithms in Pharmaceutical computations.

UNIT IV : **Analytical Plain Geometry :** Certain co-ordinates, distance between two points, area of triangle, locus of a point, straight line, slope and intercept form, normal (perpendicular) form, slope-point and two point form, general equation of first degree.

UNIT V : **Calculus :**

Differential : Limits and functions, definition of differential coefficient, diffrentiation of standard functions, including function of a function (chain rule).

Integral : Integration as inverse of differentiation, indefinite integrals of standard form, integration by parts.

CONTENTS

UNIT I ALGEBRA 1 - 97

UNIT II MEASURES OF CENTRAL VALUE 98 - 152

UNIT III TRIGONOMETRY 153 - 210

UNIT IV ANALYTICAL PLAIN GEOMETRY 211 - 273

UNIT V CALCULUS 274 - 400

UNIT - I

Algebra

1.1 LINEAR AND QUADRATIC EQUATIONS

1.1.1 Equations

If the two algebraic expressions are related by the sign of equality then it is called an equation. For example:

$ax = b$; a, b are constants (real numbers)

$ax^2 + bx + c = 0$; a, b, c are constants (real numbers)

$ax^3 + bx^2 + cx + d = 0$; a, b, c, d are constants (real numbers)

$a_n x^n + a_{n-1}x^{n-1} + \ldots + a_2x^2 + a_1x + a_0 = 0$;

$a_n, a_{n-1} \ldots a_1, a_0$ are constants (real numbers)

Here x is a variable.

1.1.2 Degree of an Equation

The highest power (positive integer) of an unknown variable in an equation is said to be the degree of equation. For example, consider an equation of the form

$$a_nx^n + a_{n-1}\, x^{n-1} + \ldots a_1x + a_0 = 0$$

The degree of this equation is n.

■ **NOTE :** (i) The coefficient of highest power of variable must be non-zero, *i.e.,* in the above equation $a_n \neq 0$.
(ii) It should be carefully noted that the variable is in the numerator only.

Example 1.1

Find the degree of each of the following equations:

(a) $2y^2 + 5y + 4 = 0$ (b) $3x + 4 = 0$

(c) $x^2 + 3 = x - \dfrac{4}{x}$ (d) $x^2 + \dfrac{1}{x} + 5 = \dfrac{1}{x}$

(e) $\dfrac{1}{x+5} = \dfrac{1}{x-1} + \dfrac{1}{x+3}$ (f) $y + 2y^{3/2} + 4y^{1/2} = 3$

SOLUTION :

(a) Since, the highest power of variable y is 2, so this is a second degree equation.

(b) Since, the highest power of variable x is one, so it is a first degree equation.

(c) The given equation can be written as

$$x^2 + 3 = \frac{x^2 - 4}{x} \quad \text{or} \quad x\,(x^2 + 3) = x^2 - 4 \qquad \text{(cross multiplication)}$$

$$\text{or} \qquad x^3 + 3x - x^2 + 4 = 0 \text{ or } x^3 - x^2 + 3x + 4 = 0$$

Clearly, the highest power of variable x is 3, so it is of degree 3.

(d) After some simplifications, the equatin is

$$x^2 + 5 = 0$$

which is of second degree in x

(e)
$$\frac{1}{x+5} = \frac{1}{x-1} + \frac{1}{x+3}$$

$$\frac{1}{x+5} = \frac{x+3+x-1}{(x-1)(x+3)}$$

$$\frac{1}{x+5} = \frac{2x+2}{x^2+2x-3}$$

After cross multiplication

$$x^2 + 2x - 3 = (x+5)(2x+2) = 2x^2 + 12x + 10$$

$$\text{or} \qquad x^2 + 2x - 3 - 2x^2 - 12x - 10 = 0$$

$$\text{or} \qquad -x^2 - 10x - 13 = 0$$

$$\text{or} \qquad x^2 + 10x + 13 = 0$$

which is a second degree equation in x.

(f)
$$y + 2y^{3/2} + 4y^{1/2} = 3$$

In the given equation we have the powers of y in fractions. So first we have to reduce the powers of y as an integer number. For this the given equation can be written as

$$2y^{3/2} + 4y^{1/2} = 3 - y$$

$$2y^{1/2}\,(y + 2) = 3 - y$$

Squaring both sides, we get

$$4y\,(y + 2)^2 = (3 - y)^2$$

$$\text{or} \qquad 4y\,(y^2 + 4y + 4) = 9 + y^2 - 6y$$

$$\text{or} \qquad 4y^3 + 16y^2 + 16y - 9 - y^2 + 6y = 0$$

$$\text{or} \qquad 4y^3 + 15y^2 + 22y - 9 = 0$$

which is of degree three.

1.2 LINEAR EQUATION

■ **1.2.1 DEFENITION** : *An equation in which the highest power of variable is one is called a linear equation.*

We have different type of linear equations depending upon the number of variables such as; equation of one variable, linear equation in two variables, linear equation in three variables, simultaneous linear equations.

1.2.2 Solution of an Equation

The solution of an equation means that we have to find the value of x which satisfies the given equation and also known as the root of given equation.

Consider a linear equation $2x + 5 = 0$. If we replace x by $-\frac{5}{2}$ then the LHS = RHS. Hence, $x = -\frac{5}{2}$ satisfies the equation $2x + 5 = 0$ so $x = -\frac{5}{2}$ is the root of $2x + 5 = 0$

1.2.3 Linear Equation of One Variable

The linear equations of the form $ax + b = 0, a \neq 0$ where a, b are constants and x is variable, are called linear equations of one variable.

Example 1.2

The equations $2x - 3 = 0, x - \frac{5}{2} = 0, 3x + \frac{5}{2} = 0, 5x = 4$ are all linear equations in one variable.

■ **NOTE :** All linear equations in one variable have a unique solution.

Example 1.3

Solve the linear equation $\frac{x}{6} - \frac{x}{8} = \frac{x}{15} + \frac{x}{5} + 8$

SOLUTION : We have $\frac{x}{6} - \frac{x}{8} = \frac{x}{15} + \frac{x}{5} + 8$

$$\frac{4x - 3x}{24} = \frac{x + 3x + 120}{15}$$

$$\frac{x}{24} = \frac{4x + 120}{15}$$

After cross multiplication

$$15x = 24\,(4x + 120)$$

$$15x - 96x = 2880$$

$$-81x = 2880$$

$$x = -\frac{2880}{81}$$

$$x = -\frac{320}{9}$$

Example 1.4

Solve the equation $x + 4 - [2 + \{x - (2 + x)\}] = \frac{1}{2}$

SULUTION : We have

$$x + 4 - [2 + \{x - (2 + x)\}] = \frac{1}{2}$$

or $$x + 4 - [2 + \{x - 2 - x\}] = \frac{1}{2}$$

or $$x + 4 - [2 - 2] = \frac{1}{2}$$

or $$x + 4 = \frac{1}{2}$$

or $$x = \frac{1}{2} - 4$$

or $$x = \frac{1 - 8}{2}$$

$$x = -\frac{7}{2}$$

Example 1.5

Solve the equation

$$\frac{a - x}{a} + \frac{2a - x}{2a} = \frac{3a - x}{3a}$$

SOLUTION : We have

$$\frac{a - x}{a} + \frac{2a - x}{2a} = \frac{3a - x}{3a}$$

$$\frac{2(a - x) + (2a - x)}{2a} = \frac{3a - x}{3a}$$

or $$\frac{2a - 2x + 2a - x}{2a} = \frac{3a - x}{3a}$$

or $$(4a - 3x)\, 3a = (3a - x)\, 2a$$

or $$12a^2 - 9xa = 6a^2 - 2ax$$

or $$-9ax + 2ax = 6a^2 - 12a^2$$

or $$-7ax = -6a^2$$

$$x = \frac{6a^2}{7a}$$

$$x = \frac{6}{7}a$$

Example 1.6

Solve the equation $(\sqrt[3]{3})^{3x + 8} = (\sqrt[5]{81})^{4x + 3}$

SOLUTION : We have

$$(\sqrt[3]{3})^{3x + 8} = (\sqrt[5]{81})^{4x + 3}$$

or $$3^{\frac{3x + 8}{3}} = (81)^{\frac{4x + 3}{5}}$$

$$3^{\frac{3x + 8}{3}} = (3^4)^{\frac{4x + 3}{5}}$$

$$3^{\frac{3x + 8}{3}} = 3^{\frac{16x + 12}{5}}$$

or $$\frac{3x + 8}{3} = \frac{16x + 12}{5}$$

after cross multiplication

$$5\,(3x + 8) = 3\,(16x + 12)$$

$$15x + 40 = 48x + 36$$
$$33x = 4$$
$$x = \frac{4}{33}$$

1.3 QUADRATIC EQUATIONS

An equation of the form $ax^2 + bx + c = 0, a \neq 0$ is called a quadratic equation. Here a, b, c are arbitrary constants and x is a variable.

A quadratic equation is satisfied by two values of x, called the roots of quadratic equation.

If $b = 0$, then given equation reduces to $ax^2 + c = 0$. This equation is known as **pure quadratic** equation. If $b \neq 0$, then the equation is called complete quadratic equation.

1.3.1 Method of Solving Quadratic Equation

There are two main methods of solving a quadratic equation

(i) Factorization method

(ii) Sridharacharya's method

1.3.2 Factorization Method

In this method the quadratic equation $ax^2 + bx + c = 0$ is first decomposed into two factors, each of the first degree, and then equating them separately to zero to find the values of the unkwnon variable x, *i.e.*,

$$ax^2 + bx + c = (x - \alpha)(x - \beta) = 0 \Rightarrow x = \alpha, x = \beta$$

Example 1.7

Solve the equation $x^2 + 3x + 2 = 0$

SOLUTION : We have $x^2 + 3x + 2 = 0$

$$\Rightarrow \quad x^2 + 2x + x + 2 = 0$$
$$\Rightarrow \quad x(x + 2) + 1(x + 2) = 0$$
$$\Rightarrow \quad (x + 2)(x + 1) = 0$$

Thus, either $x + 2 = 0 \Rightarrow x = -2$

or $x + 1 = 0 \Rightarrow x = -1$

Hence, the required roots are $x = -1$ and -2

1.3.3 Sridharacharya's Method

Consider $ax^2 + bx + c = 0$

$$\Rightarrow \quad ax^2 + bx = -c$$
$$\Rightarrow \quad x^2 + \frac{b}{a}x = -\frac{c}{a} \qquad \text{[Divide by } a\text{, as } a \neq 0]$$
$$\Rightarrow \quad x^2 + \frac{b}{a}x + \frac{b^2}{4a^2} = -\frac{c}{a} + \frac{b^2}{4a^2} \qquad \left[\text{Adding } \frac{b^2}{4a^2} \text{ both sides}\right]$$
$$\Rightarrow \quad \left(x + \frac{b}{2a}\right)^2 = \frac{b^2 - 4ac}{4a^2}$$

$$\Rightarrow \quad x + \frac{b}{2a} = \pm\sqrt{\frac{b^2 - 4ac}{4a^2}} \qquad \text{[Taking the square root]}$$

$$\Rightarrow \quad x = -\frac{b}{2a} \pm \frac{\sqrt{b^2 - 4ac}}{2a}$$

$$\text{or} \quad x = \frac{-b \pm \sqrt{b^2 - 4ac}}{2a}$$

If α and β are two roots of $ax^2 + bx + c = 0$ then

$$\alpha = \frac{-b + \sqrt{b^2 - 4ac}}{2a}, \; \beta = \frac{-b - \sqrt{b^2 - 4ac}}{2a}$$

1.3.4 Nature of Roots

The term $b^2 - 4ac$ which appears in the above formula is called **discriminant** because it separates (discriminates) between different types of roots and is denoted by D. Hence, with this notation the roots α, β can be denoted as

$$\alpha = \frac{-b + \sqrt{D}}{2a}, \; \beta = \frac{-b - \sqrt{D}}{2a}, \; a \neq 0$$

The following different cases arise depending upon the value of D:

Case 1 : If $D = b^2 - 4ac > 0$, then the roots α, β are real and unequal.

Case 2 : If $D = b^2 - 4ac = 0$, then the roots α, β are equal, each being equal to $-\frac{b}{2a}$

Case 3 : If $D = b^2 - 4ac < 0$, then the roots α, β are not real, *i.e.*, the roots are imaginary.

Example 1.8

Find the nature of the following equations and hence, solve:

(i) $\sqrt{3}y^2 + 10y - 8\sqrt{3} = 0$

(ii) $16x^2 - 8x + 1 = 0$

(iii) $5x^2 - 4x + 2 = 0$

SOLUTION : (i) Given $\sqrt{3}y^2 + 10y - 8\sqrt{3} = 0$ ...(1)

Here, $a = \sqrt{3}$, $b = 10$, $c = -8\sqrt{3}$

$$\therefore \quad D = b^2 - 4ac = (10)^2 - 4(\sqrt{3})(-8\sqrt{3})$$

$$= 100 + 96 = 196 > 0$$

Hence, the roots of (1) are equal and unequal given by

$$x = \frac{-b \pm \sqrt{b^2 - 4ac}}{2a} = \frac{-10 \pm \sqrt{196}}{2\sqrt{3}} = \frac{-10 \pm 14}{2\sqrt{3}}$$

$$\Rightarrow \quad x = \frac{-10 + 14}{2\sqrt{3}} \text{ or } x = \frac{-10 - 14}{2\sqrt{3}}$$

$$\Rightarrow \quad x = \frac{2}{\sqrt{3}} \text{ or } x = \frac{-12}{\sqrt{3}}$$

(ii) Given $16x^2 - 8x + 1 = 0$

Here $a = 16$, $b = -8$, $c = 1$

$$\therefore \quad D = b^2 - 4ac = (-8)^2 - 4(16)(1) = 64 - 64 = 0$$

Hence, roots are equal and real, given by

$$x = \frac{-b \pm \sqrt{b^2 - 4ac}}{2a} = \frac{-(-8) \pm \sqrt{0}}{2(16)}$$

$$\Rightarrow \quad x = \frac{8+0}{32} \text{ or } x = \frac{8-0}{32}$$

$$\Rightarrow \quad x = \frac{1}{4} \text{ or } x = \frac{1}{4}$$

(iii) Given $5x^2 - 4x + 2 = 0$

Here, $a = 5$, $b = -4$, $c = 2$

$$\therefore \quad D = b^2 - 4ac = (-4)^2 - 4(5)(2) = 16 - 40 = -24 < 0$$

Hence, the roots are imaginary, given by

$$x = \frac{-b \pm \sqrt{b^2 - 4ac}}{2a} = \frac{-(-4) \pm \sqrt{-24}}{2(5)} = \frac{4 \pm i\,2\sqrt{6}}{10}$$

$$\Rightarrow \quad x = \frac{2 + i\sqrt{6}}{5} \text{ or } x = \frac{2 - i\sqrt{6}}{5}$$

Example 1.9

For what values of k will the following equations have real and equal roots.

(i) $9x^2 + kx + 1 = 0$

(ii) $3kx^2 = 4(kx - 1)$

SOLUTION : (i) Given $9x^2 + kx + 1 = 0$

$$\therefore \quad D = b^2 - 4ac = k^2 - 4(9)(1) = k^2 - 36$$

The roots will be equal if $D = 0$

$$\Rightarrow \quad k^2 - 36 = 0 \Rightarrow k = \pm 6$$

(ii) Given, $3kx^2 - 4kx + 4 = 0$

$$\therefore \quad D = b^2 - 4ac = 16k^2 - 48k$$

The roots will be equal if $D = 0 \Rightarrow 16k^2 - 48k = 0 \Rightarrow k = 0, 3$

1.3.5 Relations between Roots and Coefficients

We know that the roots of the equation $ax^2 + bx + c = 0$ are given by

$$\alpha = \frac{-b + \sqrt{b^2 - 4ac}}{2a} \quad \ldots(1)$$

and

$$\beta = \frac{-b - \sqrt{b^2 - 4ac}}{2a} \quad \ldots(2)$$

(i)

$$\alpha + \beta = \left(\frac{-b + \sqrt{b^2 - 4ac}}{2a}\right) + \left(\frac{-b - \sqrt{b^2 - 4ac}}{2a}\right)$$

$$= \frac{-b + \sqrt{b^2 - 4ac} - b - \sqrt{b^2 - 4ac}}{2a} = \frac{-2b}{2a}$$

$\therefore \quad \alpha + \beta = -\frac{b}{a}$

(ii) $$\alpha \cdot \beta = \left(\frac{-b + \sqrt{b^2 - 4ac}}{2a}\right)\left(\frac{-b - \sqrt{b^2 - 4ac}}{2a}\right)$$

$$= \frac{(-b)^2 - (\sqrt{b^2 - 4ac})^2}{4a^2} = \frac{b^2 - (b^2 - 4ac)}{4a^2}$$

$$= \frac{b^2 - b^2 + 4ac}{4a^2} = \frac{4ac}{4a^2}$$

$\therefore \quad \alpha \cdot \beta = \frac{c}{a}$

Hence, the sum of roots $\alpha + \beta = -\frac{b}{a} = -\frac{\text{Coefficient of } x}{\text{Coefficient of } x^2}$

and the product of roots $= \alpha \cdot \beta = \frac{c}{a} = \frac{\text{Constant term}}{\text{Coefficient of } x^2}$

1.3.6 Construction of Quadratic Equations

Let α, β be the given numbers and let it be required to find the equation whose roots are α, β.

Since, α, β are the roots

$\therefore \quad x - \alpha = 0$ and $x - \beta = 0$

$\Rightarrow \quad (x - \alpha)(x - \beta) = 0 \Rightarrow x^2 - (\alpha + \beta)x + \alpha\beta = 0$

which is the required equation.

In other words the equation is

$$x^2 - (\text{sum of roots})\, x + \text{Product of roots} = 0$$

Example 1.10

Obtain a quadratic equation whose roots are 2 and 3.

SOLUTION : Given $\alpha = 2$, $\beta = 3$

$\therefore$ Sum of roots $= 2 + 3 = 5$

Product of roots $= 2 \times 3 = 6$

Required equation is

$$x^2 - (\text{sum of roots})\, x + \text{product of roots} = 0$$

$\therefore \quad x^2 - 5x + 6 = 0$

Example 1.11

If α, β are the roots of $3x^2 - 2x - 6 = 0$, find

(i) $\frac{1}{\alpha} + \frac{1}{\beta}$ (ii) $\alpha^2 + \beta^2$ (iii) $\alpha^3 + \beta^3$

SOLUTION : Since, α, β are roots of $3x^2 - 2x - 6 = 0$

$\therefore \quad \alpha + \beta = -\left(\frac{-2}{3}\right) = \frac{2}{3}$

$$\alpha \cdot \beta = \left(\frac{-6}{3}\right) = -2$$

(i) $\dfrac{1}{\alpha} + \dfrac{1}{\beta} = \dfrac{\alpha + \beta}{\alpha\beta} = \dfrac{\frac{2}{3}}{-2} = -\dfrac{1}{3}$

(ii) $\alpha^2 + \beta^2 = (\alpha + \beta)^2 - 2\alpha\beta$

$$= \left(\frac{2}{3}\right)^2 - 2(-2) = \frac{4}{9} + 4 = \frac{40}{9}$$

(iii) $\alpha^3 + \beta^3 = (\alpha + \beta)^3 - 3\,\alpha\beta\,(\alpha + \beta)$

$$= \left(\frac{2}{3}\right)^3 - 3(-2)\left(\frac{2}{3}\right)$$

$$= \frac{8}{27} + 4 = \frac{116}{27}$$

PROBLEM SET 1.1

1. Form the equation whose roots are $p + \sqrt{q}$ and $p - \sqrt{q}$.

2. Find the equation whose roots are twice the reciprocals of the roots of the equation $ax^2 + 2bx + 4c = 0$.

3. If α, β are the roots of the equation $x^2 + x + 1 = 0$, find the equation whose roots are α^2, β^2.

4. If α, β are the roots of $ax^2 + bx + c = 0$, find the values of :

(i) $\alpha^2 - \beta^2$ (ii) $\dfrac{\alpha + \beta}{\alpha^{-1} + \beta^{-1}}$ (iii) $\sqrt{\dfrac{\alpha}{\beta}} + \sqrt{\dfrac{\beta}{\alpha}}$

5. If α, β are the roots of the equation $x^2 - 2x + 3 = 0$. Find the equations whose roots are

(i) $\dfrac{1}{\alpha^2}, \dfrac{1}{\beta^2}$ (ii) $\dfrac{\alpha}{\beta^2}, \dfrac{\beta}{\alpha^2}$

6. If one root of the quadratic equation $ax^2 + bx + c = 0$ is double the other, prove that $2b^2 = 9ac$.

7. For what value of k are the roots of the equation $3x^2 - 2kx + k = 0$ in the ratio 3 : 1.

8. For what value (s) of m the equation $(m + 1)x^2 + 2(m + 3)x + (m + 8) = 0$ has equal roots?

9. For what values of m will the equation $x^2 - 2mx + 7m - 12 = 0$ have equal roots?

10. For what values of k will each of the following equations have real and equal roots.

(i) $kx^2 - 12x + 9 = 0$

(ii) $3x^2 - kx + k = 0$

ANSWERS

1. $x^2 - 2px + (p^2 - q) = 0$ **2.** $cx^2 + bx + a = 0$ **3.** $x^2 + x + 1 = 0$

4. (i) $\pm \frac{b}{a^2}\sqrt{b^2 - 4ac}$ (ii) $\frac{c}{a}$ (iii) $-\frac{a}{\sqrt{ac}}$ **5.** (i) $9x^2 + 2x + 1 = 0$ (ii) $9x^2 + 10x + 3 = 0$

7. $k = 0$ or $k = 4$ **8.** $m = \frac{1}{3}$ **9.** $m = 3, 4$ **10.** (i) $k = 4$ (ii) $k = 0, 12$

1.4 EQUATIONS REDUCIBLE TO QUADRATIC FORM

Example 1.12

Solve $x^4 - 10x^2 + 9 = 0$

SOLUTION : Put $y = x^2$, we get $y^2 - 10y + 9 = 0$

which is a quadratic equation.

Making factors, we get $y^2 - 9y - y + 9 = 0$

$\Rightarrow \quad y(y - 9) - 1(y - 9) = 0$

$\Rightarrow \quad (y - 9)(y - 1) = 0$

$\Rightarrow \quad y - 9 = 0$ or $y - 1 = 0$

$\Rightarrow \quad y = 9$ or $y = 1$

Put the value of y, we get

$x^2 = 9$ or $x^2 = 1$

$\Rightarrow \quad x = \pm 3$ or $x = \pm 1$

Hence, roots are $x = -1, -3, 1, 3$

Example 1.13

Solve the equation $\sqrt{\frac{x}{1+x}} + 6\sqrt{\frac{1+x}{x}} = 5$

SOLUTION : Let $\sqrt{\frac{x}{1+x}} = y$

The given equation becomes $y + \frac{6}{y} = 5$

$\Rightarrow \quad y^2 - 5y + 6 = 0$

$\Rightarrow \quad (y - 2)(y - 3) = 0$

$\Rightarrow \quad y = 2$ and $y = 3$

$\Rightarrow \quad \sqrt{\frac{x}{1+x}} = 2$ and $\sqrt{\frac{x}{1+x}} = 3$

$\Rightarrow \quad \frac{x}{1+x} = 4$ and $\frac{x}{1+x} = 9$

$\Rightarrow \quad x = 4 + 4x$ and $x = 9 + 9x$

$\Rightarrow \quad -3x = 4 \text{ and } -8x = 9$

$\Rightarrow \quad x = -\frac{4}{3} \text{ and } x = -\frac{9}{8}$

Hence, roots are $x = -\frac{4}{3}$ and $-\frac{9}{8}$

Example 1.14

Solve the equation $\sqrt{3x+10} + \sqrt{9x+7} = 9$

SOLUTION : Given, $\sqrt{3x+10} + \sqrt{9x+7} = 9$

$\Rightarrow \quad \sqrt{3x+10} = 9 - \sqrt{9x+7}$

Squaring both sides, we get

$$3x + 10 = 81 - 18\sqrt{9x+7} + (9x+7)$$

or $\quad 3x + 10 - 81 - 9x - 7 = -18\sqrt{9x+7}$

or $\quad -6x - 78 = -18\sqrt{9x+7}$

or $\quad -6(x+13) = -18\sqrt{9x+7}$

or $\quad x + 13 = 3\sqrt{9x+7}$

Again squaring both sides, we get

$$(x+13)^2 = 9(9x+7)$$

$$x^2 + 26x + 169 - 81x - 63 = 0$$

$\Rightarrow \quad x^2 - 55x + 106 = 0$

$\Rightarrow \quad x^2 - 53x - 2x + 106 = 0$

$\Rightarrow \quad x(x-53) - 2(x-53) = 0$

$\Rightarrow \quad (x-53)(x-2) = 0$

$\Rightarrow \quad x - 53 = 0 \text{ or } x - 2 = 0$

$\Rightarrow \quad x = 53 \text{ or } x = 2$

Hence, roots are $x = 2, 53$

Example 1.15

Solve $(x+1)(x+2)(x+3)(x+4) = 120$ ***[UPTU 2005]***

SOLUTION : To solve such equations, first reduce it to four factors in which two factors have terms containing x^2 and x are the same. Thus, we get

$$[(x+1)(x+4)][(x+2)(x+3)] = 120$$

$\Rightarrow \quad (x^2 + 5x + 4)(x^2 + 5x + 6) = 120$

Let $x^2 + 5x = y$, we get $(y+4)(y+6) = 120$

$\Rightarrow \quad y^2 + 10y + 24 - 120 = 0$

$\Rightarrow \quad y^2 + 10y - 96 = 0$

$\Rightarrow \quad y^2 + 16y - 6y - 96 = 0$

$\Rightarrow \quad y(y+16) - 6(y+16) = 0$

$\Rightarrow \quad (y+16)(y-6) = 0$

$\Rightarrow \quad y = -16 \text{ or } y = 6$

Putting the value of y, we get

$$x^2 + 5x = -16 \text{ or } x^2 + 5x = 6$$

$$\Rightarrow \quad x^2 + 5x + 16 = 0 \text{ or } x^2 + 5x - 6 = 0$$

$$\Rightarrow \quad x = \frac{-5 \pm \sqrt{25 - 64}}{2} \text{ or } x = \frac{-5 \pm \sqrt{25 + 24}}{2}$$

$$x = \frac{-5 \pm \sqrt{-39}}{2} \text{ or } x = \frac{-5 \pm \sqrt{49}}{2}$$

$$x = \frac{-5 \pm \sqrt{39}\, i}{2} \text{ or } x = \frac{-5 \pm 7}{2}$$

Roots are

$$x = \frac{-5 + \sqrt{39}\, i}{2}, \frac{-5 - \sqrt{39}\, i}{2}, 1, -6$$

Example 1.16

Solve the equation $2x^2 - x + 5\sqrt{2x^2 - x + 4} = 10$

SOLUTION : Given $2x^2 - x + 5\sqrt{2x^2 - x + 4} = 10$

$$\Rightarrow \quad (2x^2 - x + 4) - 4 + 5\sqrt{2x^2 - x + 4} = 10$$

Let $\sqrt{2x^2 - x + 4} = y$, we get

$$y^2 - 4 + 5y = 10 \Rightarrow y^2 + 5y - 14 = 0$$

$$\Rightarrow \quad y^2 + 7y - 2y - 14 = 0 \Rightarrow (y + 7)(y - 2) = 0$$

$$\Rightarrow \quad y + 7 = 0 \text{ or } y - 2 = 0$$

$$\Rightarrow \quad y = -7 \text{ or } y = 2$$

Putting the value of y, we get

$$\sqrt{2x^2 - x + 4} = -7 \text{ or } \sqrt{2x^2 - x + 4} = 2$$

$$\Rightarrow \quad 2x^2 - x + 4 = 49 \text{ or } 2x^2 - x + 4 = 4$$

$$\Rightarrow \quad 2x^2 - x - 45 = 0 \text{ or } 2x^2 - x = 0$$

$$\Rightarrow \quad 2x^2 - 10x + 9x - 45 = 0 \text{ or } x(2x - 1) = 0$$

$$\Rightarrow \quad 2x(x - 5) + 9(x - 5) = 0 \text{ or } x(2x - 1) = 0$$

$$\Rightarrow \quad (2x + 9)(x - 5) = 0 \text{ or } x(2x - 1) = 0$$

$$\Rightarrow \quad 2x - 9 = 0, x - 5 = 0 \text{ or } x = 0, 2x - 1 = 0$$

$$\Rightarrow \quad x = -\frac{9}{2}, x = 5 \text{ or } x = 0, x = \frac{1}{2}$$

Hence, roots are $x = -\frac{9}{2}, 5, 0, \frac{1}{2}$.

Example 1.17

Solve the equation: $\sqrt{x^2 - 8x + 15} + \sqrt{x^2 + 2x - 15} = \sqrt{4x^2 - 18x + 18}$

SOLUTION : Given $\sqrt{x^2 - 8x + 15} + \sqrt{x^2 + 2x - 15} = \sqrt{4x^2 - 18x + 18}$

or $\quad \sqrt{(x - 5)(x - 3)} + \sqrt{(x + 5)(x - 3)} = \sqrt{(4x - 6)(x - 3)}$

Taking out the common factor $(x - 3)$, we get

$$\sqrt{x-3}\,[\sqrt{x-5} + \sqrt{x+5} - \sqrt{4x-6}] = 0$$

$\Rightarrow$ $\sqrt{x-3} = 0$ or $\sqrt{x-5} + \sqrt{x+5} = \sqrt{4x-6}$

$\Rightarrow$ $x = 3$

or $(x-5) + (x+5) + 2\sqrt{x^2-25} = 4x - 6$ (squaring both sides)

or $2x + 2\sqrt{x^2-25} = 4x - 6$

or $2x - 4x + 6 = -2\sqrt{x^2-25}$

or $-2x + 6 = -2\sqrt{x^2-25}$

or $-2x + 6 = -2\sqrt{x^2-25}$

or $x - 3 = \sqrt{x^2-25}$

or $(x-3)^2 = x^2 - 25$ [Again squaring]

or $x^2 - 6x + 9 - x^2 + 25 = 0$

or $-6x + 34 = 0$

or $x = \dfrac{34}{6} = \dfrac{17}{3}$

Hence, roots are $x = 3, \dfrac{17}{3}$

Example 1.18

Solve $4^x - 5 \cdot 2^x + 4 = 0$

SOLUTION : Given $4^x - 5 \cdot 2^x + 4 = 0$

$\Rightarrow$ $2^{2x} - 5 \cdot 2^x + 4 = 0$

Put $2^x = y$, we get

$$y^2 - 5y + 4 = 0$$

$\Rightarrow$ $y - 4y - y + 4 = 0 \Rightarrow (y-4)(y-1) = 0$

$\Rightarrow$ $y = 4$ or $y = 1$

Putting the value of y, we get

$$2^x = 4 \text{ or } 2^x = 1$$

$\Rightarrow$ $2^x = 2^2$ or $2^x = 2^0$

$\Rightarrow$ $x = 2$ or $x = 0$

Hence, roots are $x = 0, 2$

Example 1.19

Solve $\left(x + \dfrac{1}{x}\right)^2 - \dfrac{3}{2}\left(x - \dfrac{1}{x}\right) = 4,\ x \neq 0$ ***[UPTU 2003]***

SOLUTION : Given $\left(x + \dfrac{1}{x}\right)^2 - \dfrac{3}{2}\left(x - \dfrac{1}{x}\right) = 4$

$$\Rightarrow \quad x^2 + \frac{1}{x^2} + 2 - \frac{3}{2}\left(x - \frac{1}{x}\right) = 4$$

$$\Rightarrow \quad \left[\left(x - \frac{1}{x}\right)^2 + 2\right] + 2 - \frac{3}{2}\left(x - \frac{1}{x}\right) = 4 \qquad \left[\because \left(x - \frac{1}{x}\right)^2 = x^2 + \frac{1}{x^2} - 2\right]$$

$$\Rightarrow \quad \left(x - \frac{1}{x}\right)^2 + 4 - \frac{3}{2}\left(x - \frac{1}{x}\right) = 4$$

$$\Rightarrow \quad \left(x - \frac{1}{x}\right)^2 - \frac{3}{2}\left(x - \frac{1}{x}\right) = 0 \qquad \text{...(1)}$$

Let $x - \frac{1}{x} = y$

From (1), we get

$$y^2 - \frac{3}{2}y = 0 \Rightarrow y\left(y - \frac{3}{2}\right) = 0 \Rightarrow y = 0 \text{ or } y = \frac{3}{2}$$

Substituting the value of y, we obtain

$$x - \frac{1}{x} = 0 \text{ or } x - \frac{1}{x} = \frac{3}{2}$$

$$\Rightarrow \quad x^2 - 1 = 0 \text{ or } \frac{x^2 - 1}{x} = \frac{3}{2}$$

$$\Rightarrow \quad x^2 = 1 \text{ or } 2x^2 - 2 - 3x = 0$$

$$\Rightarrow \quad x = \pm 1 \text{ or } 2x^2 - 3x - 2 = 0$$

$$\Rightarrow \quad x = \pm 1 \text{ or } (2x - 1)(x - 2) = 0$$

$$\Rightarrow \quad x = \pm 1 \text{ or } x = -\frac{1}{2} \text{ and } x = 2$$

Hence, roots are $x = \pm 1, -\frac{1}{2}, 2$

Example 1.20

Solve the following equation:

$$\left(\frac{2x+1}{x-1}\right)^4 - 10\left(\frac{2x+1}{x-1}\right)^2 + 9 = 0$$

[UPTU 2004]

SOLUTION : Given $\left(\frac{2x+1}{x-1}\right)^4 - 10\left(\frac{2x+1}{x-1}\right)^2 + 9 = 0$ ...(1)

Let $\left(\frac{2x+1}{x-1}\right)^2 = y$

From Eq. (1), we get

$$y^2 - 10y + 9 = 0$$

$$\Rightarrow \quad y^2 - 9y - y + 9 = 0$$

$$\Rightarrow \quad (y - 9)(y - 1) = 0$$

$$\Rightarrow \quad y = 9 \text{ or } y = 1$$

Substituting the value of y, we get

$$\left(\frac{2x+1}{x-1}\right)^2 = 9 \text{ or } \left(\frac{2x+1}{x-1}\right)^2 = 1$$

$$\Rightarrow \quad \frac{2x+1}{x-1} = \pm 3 \text{ or } \frac{2x+1}{x-1} = \pm 1$$

(i) $\frac{2x+1}{x-1} = 3 \Rightarrow 2x+1 = 3x-3 \Rightarrow -x = -4 \Rightarrow x = 4$

(ii) $\frac{2x+1}{x-1} = -3 \Rightarrow 2x+1 = -3x+3 \Rightarrow 5x = 2 \Rightarrow x = \frac{2}{5}$

(iii) $\frac{2x+1}{x-1} = 1 \Rightarrow 2x+1 = x-1 \Rightarrow x = -2$

(iv) $\frac{2x+1}{x-1} = -1 \Rightarrow 2x+1 = -x+1 \Rightarrow 3x = 0 \Rightarrow x = 0$

Hence, roots are $x = 0, -2, \frac{2}{5}, 4$

Example 1.21

Solve the equation:

$$[\sqrt{5+2\sqrt{6}}]^x + [\sqrt{5-2\sqrt{6}}]^x = 10$$ *[UPTU 2002]*

SOLUTION : Given $\quad [\sqrt{5+2\sqrt{6}}]^x + [\sqrt{5-2\sqrt{6}}]^x = 10 \quad$ ...(1)

Let $\quad \sqrt{5+2\sqrt{6}} = y$

$$\Rightarrow \quad \frac{1}{\sqrt{5+2\sqrt{6}}} = \frac{1}{y}$$

$$\Rightarrow \quad \frac{\sqrt{5-2\sqrt{6}}}{(\sqrt{5+2\sqrt{6}})(\sqrt{5-2\sqrt{6}})} = \frac{1}{y}$$

$$\Rightarrow \quad \frac{\sqrt{5-2\sqrt{6}}}{\sqrt{(5+2\sqrt{6})(5-2\sqrt{6})}} = \frac{1}{y} \Rightarrow \frac{\sqrt{5-2\sqrt{6}}}{\sqrt{25-24}} = \frac{1}{y}$$

$$\Rightarrow \quad \sqrt{5-2\sqrt{6}} = \frac{1}{y}$$

From Eq. (1), we get

$$y^x + \left(\frac{1}{y}\right)^x = 10$$

or $\quad y^{2x} - 10y^x + 1 = 0 \quad$ ...(2)

Let $\quad y^x = t \quad$ ...(3)

From Eq. (2), we get

$$t^2 - 10t + 1 = 0$$

$$\Rightarrow \quad t = \frac{10 \pm \sqrt{100-4}}{2} = \frac{10 \pm \sqrt{96}}{2}$$

$$\Rightarrow \quad t = 5 \pm 2\sqrt{6}$$

From Eq. (3),

$$y^x = 5 \pm 2\sqrt{6}$$

$\Rightarrow$ $y^x = 5 + 2\sqrt{6}$ or $y^x = 5 - 2\sqrt{6}$

$\Rightarrow$ $y^x = y$ or $y^x = \frac{1}{y}$

$\Rightarrow$ $x = 1$ or $x = -1$

Example 1.22.

Solve :

$$(x^2 - 5x)^2 - 30(x^2 - 5x) - 216 = 0. \quad \textit{[UPTU (C.O.) 2006]}$$

SOLUTION : We have

$$(x^2 - 5x)^2 - 30(x^2 - 5x) - 216 = 0 \quad ...(1)$$

Let $x^2 - 5x = y$ ...(2)

(1) becomes

$$y^2 - 30y - 216 = 0 \Rightarrow y^2 - 36y + 6y - 216 = 0$$

$\Rightarrow$ $(y - 36)(y + 6) = 0 \Rightarrow y = 36$ and $y = -6$

From (2),

	$x^2 - 5x = 36$		$x^2 - 5x = -6$
$\Rightarrow$	$x^2 - 5x - 36 = 0$	$\Rightarrow$	$x^2 - 5x + 6 = 0$
$\Rightarrow$	$x^2 - 9x + 4x - 36 = 0$	$\Rightarrow$	$x^2 - 3x - 2x + 6 = 0$
$\Rightarrow$	$(x - 9)(x + 4) = 0$	$\Rightarrow$	$(x - 3)(x - 2) = 0$
$\Rightarrow$	$x = 9, -4$	$\Rightarrow$	$x = 2, 3$

Hence, $x = -4, 2, 3, 9$.

Example 1.23

Solve

$$7^{x+1} + 7^{1-x} = 50 \quad \textit{[UPTU 2007]}$$

SOLUTION : $7^{x+1} + 7^{1-x} = 50$

$\Rightarrow$ $7 \cdot 7^x + \frac{7}{7^x} = 50$

$\Rightarrow$ $7 \cdot 7^{2x} - 50 \cdot 7^x + 7 + 0$ ...(1)

Let $7^x = y$ ...(2)

From (1), we get

$$7y^2 - 50y + 7 = 0 \Rightarrow 7y^2 - 49y - y + 7 = 0$$

$\Rightarrow$ $7y(y - 7) - 1(y - 7) = 0 \Rightarrow (y - 7)(7y - 1) = 0$

$\Rightarrow$ $y = 7$ or $y = \frac{1}{7}$

From (2),

$$7^x = 7 \Rightarrow x = 1$$

and $7^x = \frac{1}{7} \Rightarrow x = -1$

Thus, $x = 1, -1$

PROBLEM SET 1.2

Solve the following equations :

1. $x + \frac{1}{x} = \frac{5}{2}$

2. $\frac{4x+5}{x} - \frac{3x}{4x+5} = 2$

3. $\frac{x-1}{x+1} + \frac{x+1}{x-1} = \frac{5}{2}$

4. $\left(x - \frac{1}{x}\right)^2 + 7\left(x - \frac{1}{x}\right) = \frac{51}{4}$

5. $\frac{a-x}{a} + \frac{2a-x}{2a} = \frac{3a-x}{3a}$

6. If $(5 \cdot 3)^x = (0.53)^x = 10000$. Show that $\frac{1}{x} - \frac{1}{y} = \frac{1}{4}$

Solve the following equations :

7. $(x-1)(x-2)(3x-2)(3x+1) = 21$

8. $(x-1)(x-2)(x-3)(x-4) = 24$

9. $\sqrt{3x^2+1} + \frac{4}{\sqrt{3x^2+1}} = 5$

10. $x^2 - 6x - \sqrt{x^2 - 6x - 3} = 5$

11. $\sqrt{17x^2 - 40x + 16} = 4x - 5$

12. $\frac{x - \sqrt{2x+1}}{x + \sqrt{2x+1}} = \frac{1}{7}$

13. $\sqrt{4x+1} + \sqrt{3x+9} = 7$

14. $\sqrt{3x^2 - 5x - 6} + \sqrt{3x^2 - 5x + 3} = 9$

15. $\sqrt{5x^2 + 7x + 2} + \sqrt{4x^2 + 7x + 18} = x + 4$

16. $\left(x - \frac{1}{x}\right)^2 + 4 = \frac{5}{2}\left(x + \frac{1}{x}\right)$

17. $x^{2/3} + x^{1/3} - 2 = 0$

18. $4x^4 - 33x^2 + 8 = 0$

19. $\sqrt{x+4} + \sqrt{x-4} = 2\sqrt{x-1}$

20. $5^{2x} - 126.5^x + 125 = 0$

ANSWERS

1. $x = 2, \frac{1}{2}$ **2.** $x = -1, -5$ **3.** $x = 3, -3$ **4.** $x = 2, -\frac{1}{2}, x = \frac{-17 \pm \sqrt{305}}{4}$

5. $x = \frac{6a}{7}$ **7.** $x = \frac{5 \pm \sqrt{85}}{6}$ $x = \frac{5 \pm \sqrt{-55}}{6}$ **8.** $x = 0, 5, x = \frac{5 \pm \sqrt{-15}}{2}$ **9.** $x = 0$

10. $x = 2, -1, 7$ **11.** $x = 3, -3$ **12.** $x = 4, -\frac{4}{9}$ **13.** $x = 4, 480$

14. $x = -2, \frac{11}{3}$ **15.** $x = \frac{-7 \pm \sqrt{17}}{8}$ **16.** $x = \pm 1, 2, \frac{1}{2}$ **17.** $x = 1, 8$

18. $x = \pm 2\sqrt{2}, x = \pm \frac{1}{2}$ **19.** $x = 5$ **20.** $x = 0, 3.$

1.5 SIMULTANEOUS EQUATIONS

1.5.1 Linear Simultaneous Equations with Two Variables

The sets of equations containing two or more variables are said to be simultaneous equations, if they are satisfied simultaneously by the same values of the variables.

A general set of equations is given by

$$a_1x + b_1y + c_1 = 0$$
$$a_2x + b_2y + c_2 = 0$$

(i) This system of equations is **consistent** if $\frac{a_1}{a_2} \neq \frac{b_1}{b_2}$ and in this case the system have unique solution.

(ii) This system of equations is **inconsistent** if $\frac{a_1}{a_2} = \frac{b_1}{b_2} \neq \frac{c_1}{c_2}$ and in this case the system have no solution.

1.5.2 Methods to Solve Linear Simultaneous Equations

There are three methods to solve linear simultaneous equations.

(1) Substituion method

(2) Elimination method

(3) Cross multiplication method.

1.5.3 Substitution method

In this method we substitute the value of first variable from first equation to the second equation and get a linear equation of second variable. We can solve this easily. After finding the value of second variable, we can easily find the value of first variable.

Example 1.24

Solve the equations $x + 3y - 4 = 0$, $3x - y - 3 = 0$

SOLUTION : We have

$$x + 3y - 4 = 0 \quad \text{...(1)}$$
$$3x - y - 3 = 0 \quad \text{...(2)}$$

From Eq. (1),

$$x = 4 - 3y \quad \text{...(3)}$$

Substitute this value of x in Eq. (2), we get

$$3(4 - 3y) - y - 3 = 0$$
$$\Rightarrow \quad -10y + 9 = 0$$

which is linear equation in y. Solving this, we get

$$y = \frac{9}{10}$$

Substitute this value of y in Eq. (3), we get

$$x = 4 - 3\left(\frac{9}{10}\right) = 4 - \frac{27}{10} = \frac{40 - 27}{10} = \frac{13}{10}$$

Hence, the solution of given simultaneous equations is

$$x = \frac{13}{10}, y = \frac{9}{10}$$

1.5.4 Elimination Method

Consider $a_1x + b_1y + c_1 = 0$ ...(1)

$a_2x + b_2y + c_2 = 0$ ...(2)

To eliminate x, multiply (1) by a_2 and (2) by a_1, we get

$a_1a_2x + a_2b_1y + a_2c_1 = 0$ ...(3)

$a_1a_2x + a_1b_2y + a_1c_2 = 0$ ...(4)

Subtracting (4) from (3), we get

$$(a_2b_1 - a_1b_2)y + (a_2c_1 - a_1c_2) = 0$$

$$\Rightarrow \quad y = -\frac{(a_2c_1 - a_1c_2)}{a_2b_1 - a_1b_2}$$

Substituting this value of y in either (1) or (2), we get the value of x.

Example 1.25

Using method of elimination, solve the system fo equations

$$3x + 4y = 5, \ 4x + 2y = 9$$

SOLUTION : We have

$3x + 4y = 5$ ...(1)

$4x + 2y = 9$ ...(2)

To eliminate x, multiply (1) by 4 and (2) by 3, we get

$12x + 16y = 20$ ...(3)

$12x + 6y = 27$ ...(4)

Subtracting (4) from (3), we obtain

$$10y = -7 \text{ or } y = -\frac{7}{10}$$

Substituting this value of y in Eq. (1), we get

$$3x + 4\left(-\frac{7}{10}\right) = 5 \Rightarrow 3x - \frac{14}{5} = 5$$

$$\Rightarrow \quad 3x = 5 + \frac{14}{5} \Rightarrow 3x = \frac{25 + 14}{5} \Rightarrow 3x = \frac{39}{5} \Rightarrow x = \frac{13}{5}$$

Hence, the solution of this system of equations is

$$x = \frac{13}{5}, y = -\frac{7}{10}$$

1.5.5 Cross Multiplication method

Consider $a_1x + b_1y + c_1 = 0$ ...(1)

$a_2x + b_2y + c_2 = 0$ ...(2)

Write

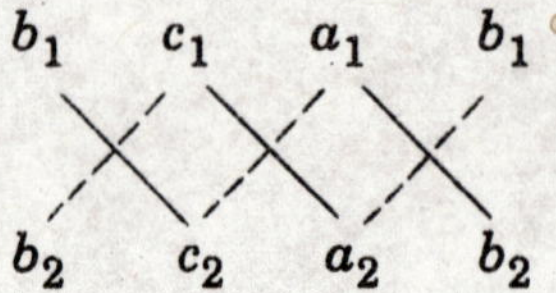

Now use the formula

$$\frac{x}{b_1c_2 - b_2c_1} = \frac{y}{a_2c_1 - a_1c_2} = \frac{1}{a_1b_2 - a_2b_2}$$

Example 1.26

Using method of cross multiplication find the solution of system of equations:

$$2x + 5y = 4, \; x - 2y = 6$$

SOLUTION : Given $2x + 5y - 4 = 0$ ...(1)

$x - 2y - 6 = 0$ ...(2)

Here, $a_1 = 2, \; b_1 = 5, \; c_1 = -4$

$a_2 = 1, b_2 = -2, c_2 - 6$

Using method of cross multiplication, we have

$$\frac{x}{b_1c_2 - b_2c_1} = \frac{y}{a_2c_1 - a_1c_2} = \frac{1}{a_1b_2 - a_2b_1}$$

$$\frac{x}{(5)(-6) - (-2)(-4)} = \frac{y}{(1)(-4) - (2)(-6)} = \frac{1}{(2)(-2) - (1)(5)}$$

or $$\frac{x}{-30 - 8} = \frac{y}{-4 + 12} = \frac{1}{-4 - 5}$$

or $$\frac{x}{-38} = \frac{y}{8} = \frac{1}{-9}$$

or $$\frac{x}{-38} = \frac{1}{-9}$$

and $$\frac{y}{8} = \frac{1}{-9}$$

$\Rightarrow$ $$x = \frac{38}{9}$$

and $$y = -\frac{8}{9}$$

1.6 QUADRATIC SIMULTANEOUS EQUATIONS

1.6.1 One Linear and One Quadratic Equation

Example 1.27

Solve the equations $x^2 + y^2 = 5$ and $x + 2y = 4$

SOLUTION : We have $x^2 + y^2 = 5$ ...(1)

$x + 2y = 4$ ...(2)

Here (2) is a linear equation. From (2)

$x = 4 - 2y$ ...(3)

Substituting this value of x in (1), we get

$$(4 - 2y)^2 + y^2 = 5$$

$$\Rightarrow \quad 16 - 16y + 4y^2 + y^2 = 5$$

$$\Rightarrow \quad 5y^2 - 16y + 11 = 0$$

$$\Rightarrow \quad 5y^2 - 5y - 11y + 11 = 0$$

$$\Rightarrow \quad 5y(y - 1) - 11(y - 1) = 0$$

$$\Rightarrow \quad (y - 1)(5y - 11) = 0$$

$$\Rightarrow \quad y - 1 = 0 \text{ or } 5y - 11 = 0$$

$$\Rightarrow \quad y = 1 \text{ or } y = \frac{11}{5}$$

If $y = 1,\ x = 4 - 2(1) = 2$

If $y = \frac{11}{5},\ x = 4 - 2\left(\frac{11}{5}\right) = 4 - \frac{22}{5} = \frac{20 - 22}{5} = -\frac{2}{5}$

Hence, the solutions of this system are:

$$(2, 1), \left(-\frac{2}{5}, \frac{11}{5}\right)$$

Example 1.28

Solve the equations:

$$\sqrt{\frac{y}{x}} + \sqrt{\frac{x}{y}} = \frac{13}{6} \text{ and } xy = 36$$

SOLUTION : Given $\sqrt{\frac{y}{x}} + \sqrt{\frac{x}{y}} = \frac{13}{6}$ ...(1)

and $xy = 36$ ...(2)

Solving (1), we get

$$\frac{y + x}{\sqrt{xy}} = \frac{13}{6}$$

$$\frac{x + y}{\sqrt{36}} = \frac{13}{6} \Rightarrow \frac{x + y}{6} = \frac{13}{6}$$

$\Rightarrow \quad x + y = 13 \Rightarrow y = 13 - x$ ...(3)

Substituting the value of y from (3) in (2), we get

$$x(13 - x) = 36 \Rightarrow 13x - x^2 - 36 = 0$$

$$\Rightarrow \quad x^2 - 13x + 36 = 0$$

$$\Rightarrow \quad x^2 - 9x - 4x + 36 = 0$$

$$\Rightarrow \quad x(x - 9) - 4(x - 9) = 0$$

$$\Rightarrow \quad (x - 9)(x - 4) = 0$$

$$\Rightarrow \quad x - 9 = 0 \text{ or } x - 4 = 0$$

$$\Rightarrow \quad x = 9 \text{ or } x = 4$$

If $x = 9$, from Eq. (3), $y = 13 - 9 = 4$

If $x = 4$, from Eq. (3), $y = 13 - 4 = 9$

Hence, the solutions of the system are

$$(9, 4) \text{ and } (4, 9)$$

1.6.2 Two Quadratic Simultaneous Equations

Example 1.29

Solve the equations $2x^2 + 3xy = 26$, $3y^2 + 2xy = 39$

SOLUTION : Put $y = mx$ in both the equations, we get

$$2x^2 + 3mx^2 = 26 \Rightarrow x^2(2 + 3m) = 26 \quad \text{...(1)}$$

$$3m^2x^2 + 2mx^2 = 39 \Rightarrow x^2(3m^2 + 2m) = 39 \quad \text{...(2)}$$

Divide (2) and (1), we obtain

$$\frac{2 + 3m}{3m^2 + 2m} = \frac{26}{39}$$

By cross multiplication, we get

$$39(2 + 3m) = 26(3m^2 + 2m)$$

$$\Rightarrow \quad 78 + 117m = 78m^2 + 52m$$

$$\Rightarrow \quad 78m^2 - 117m + 52m - 78 = 0$$

$$\Rightarrow \quad 78m^2 - 65m - 78 = 0$$

$$\Rightarrow \quad m = \frac{65 \pm \sqrt{(-65)^2 - 4(78)(-78)}}{2 \times 78} = \frac{65 \pm \sqrt{4225 + 24336}}{156}$$

$$= \frac{65 \pm \sqrt{28561}}{156} = \frac{65 \pm 169}{156}$$

$$\Rightarrow \quad m = \frac{65 + 169}{156}, m = \frac{65 - 169}{156}$$

$$m = \frac{234}{156}, m = -\frac{104}{156}$$

From (1), if $\quad m = \frac{234}{156}, x^2\left(2 + \frac{234}{52}\right) = 26 \Rightarrow x^2 = 4 \Rightarrow x = \pm 2$

If $\quad m = \frac{-104}{156}, x^2\left(2 - \frac{104}{52}\right) = 26$

In this case x^2 does not exist

So, if $x = 2,\ y = 2\left(\frac{234}{156}\right) = \frac{234}{78} = 3$

$$\text{if } x = -2,\ y = -2\left(\frac{234}{156}\right) = -3$$

Hence, the solutions of the given set of equations are **$x = 2, y = 3$ and $x = -2,\ y = -3$**

Example 1.30

Solve : $4^x \cdot 2^y = 128,\ 3^{3x+2y} = 9^{xy}$

SOLUTION : Given $4^x \cdot 2^y = 128$

$\Rightarrow \quad 2^{2x} \cdot 2^y = 2^7 \Rightarrow 2^{2x+y} = 2^7$

$\Rightarrow \quad 2x + y = 7$ ...(1)

Also $\quad 3^{3x+2y} = 9^{xy} \Rightarrow 3^{3x+2y} = 3^{2xy} \Rightarrow 3x + 2y = 2xy$ ...(2)

From (1), we get

$y = 7 - 2x$ ...(3)

From (2), we obtain

$$3x + 2(7 - 2x) = 2x(7 - 2x)$$

$$\Rightarrow \quad 3x + 14 - 4x = 14x - 4x^2$$

$$\Rightarrow \quad 4x^2 - 15x + 14 = 0$$

$$\Rightarrow \quad 4x^2 - 8x - 7x + 14 = 0$$

$$\Rightarrow \quad 4x(x-2) - 7(x-2) = 0$$

$$\Rightarrow \quad (x-2)(4x-7) = 0$$

$$\Rightarrow \quad x = 2 \text{ or } x = -\frac{7}{4}$$

From (3), if $x = 2,\ y = 7 - 4 = 3$

If $x = -\frac{7}{4},\ y = 7 - 2\left(\frac{-7}{4}\right) = \frac{21}{2}$

Hence, the solution is $x = 2, y = 3$ and $x = \frac{-7}{4},\ y = \frac{21}{2}$

PROBLEMS SET 1.3

Solve the following equations :

1. $x^2 + y^2 = 29,\ x - y = 3$
2. $x - 2y = 3,\ 7y - 2x = 3$
3. $x + \frac{3}{y} = 2,\ y + \frac{3}{x} = -2$

4. $\sqrt{\frac{x}{y}} + \sqrt{\frac{y}{x}} = 1 + \frac{7}{\sqrt{xy}}$; $\sqrt{xy}\,(x + y) = 78$

5. $3x^2 + 11xy - 4y^2 = 0$; $x^2 - xy + y^2 = 7$

6. $\frac{1}{x^2} + \frac{1}{y^2} = 13$; $\frac{1}{x} - \frac{1}{y} = 1$

7. $\frac{x + y}{1 - xy} = 3$; $\frac{x - y}{1 + xy} = \frac{1}{3}$

8. $x + 3y - 4 = 0$; $3x - y - 3 = 0$

ANSWERS

1. $x = -2, y = -5$; $x = 5, y = 2$ **2.** $x = 9, y = 3$ **3.** $x = 3, -1, y = -3, 1$

4. $x = 9, y = 4; x = 4, y = 9$ **5.** $x = \pm 1, \pm\frac{4}{\sqrt{3}}$, $y = \pm 3, \pm\frac{1}{\sqrt{3}}$ **6.** $x = -\frac{1}{2}, \frac{1}{3}, y = -\frac{1}{3}, \frac{1}{2}$

7. $x = \pm 1$, $y = \frac{1}{2}, -2$ **8.** $x = \frac{13}{10}, y = \frac{9}{10}$

1.7 APPLICATIONS OF QUADRATIC AND LINEAR EQUATIONS

Example 1.31

A number exceeds its positive square root by 12. Find the number.

SOLUTION : Let the number be x. By given condition

$$x = \sqrt{x} + 12$$

$$\Rightarrow \quad x - 12 = \sqrt{x}$$

Squaring both sides, we get

$$(x - 12)^2 = (\sqrt{x})^2 \Rightarrow x^2 - 24x + 144 = x$$

$$\Rightarrow \quad x^2 - 25x + 144 = 0$$

$$\Rightarrow \quad x = \frac{25 \pm \sqrt{625 - 576}}{2} = \frac{25 \pm \sqrt{49}}{2} = \frac{25 \pm 7}{2}$$

$$\Rightarrow \quad x = 16, 9$$

But $x = 9$ does not satisfy given condition

$\therefore x = 16$. Hence, the required number is 16.

Example 1.32

Divide 33 into two part whose product is 242. ***[UPTU 2000]***

SOLUTION : Let one part of 33 is x. Then the other part will be $33 - x$.

By given condition, $x\,(33 - x) = 242$

$$\Rightarrow \quad 33x - x^2 = 242$$

$$\Rightarrow \quad x^2 - 33x + 242 = 0$$

$\Rightarrow \quad x^2 - 11x - 22x + 242 = 0$

$\Rightarrow \quad (x - 11)(x - 22) = 0$

$\Rightarrow \quad x = 11 \text{ or } x = 22$

Hence, the two parts are 11 and 22.

Example 1.33

The sum of the present ages of Kamal and his mother is 89 years. After 11 years, mother's age will be twice Kamal's age. Find their present ages. *[UPTU 2000]*

SOLUTION : Let the present age of Kamal = x years

and the present age of his mother = y years.

then $\quad x + y = 89 \quad$ ...(1)

After 11 years, Kamal's age = $x + 11$ years

Mother's age = $y + 11$ years.

Then $\quad y + 11 = 2(x + 11) \quad$ ...(2)

From (1) and (2), we have

$$x + y = 89$$

$$2x - y = -11$$

On adding, we get

$$3x = 78 \Rightarrow x = 26$$

From (1),

$$y = 89 - 26 = 53$$

Hence, the present age of Kamal = 26 years

and the present age of his mother = 53 years.

Example 1.34

In a given fraction, if the numerator is multiplied by 3 and the denominator is reduced by 1, we get $\frac{3}{2}$. But if the numerator is increased by 12 and the denominator is multiplied by 3, we get $\frac{1}{3}$. Find the fraction. *[UPTU 2007]*

SOLUTION : Let the given fraction is $\frac{x}{y}$

By condition (1), we get

$$\frac{3x}{y-1} = \frac{3}{2}$$

$\Rightarrow \quad 6x - 3y = -3 \quad$ ...(1)

By condition (2), we get

$$\frac{x+12}{3y} = \frac{1}{3}$$

$\Rightarrow \quad 3x - 3y = -36 \quad$ ...(2)

On subtracting (1) and (2), we get

$$3x - 33 \Rightarrow x = 11$$

From (1),

$$6(11) - 3y = -3$$

$$\Rightarrow \quad 66 - 3y = -3 \Rightarrow y = 23$$

Thus, the required fraction is $\frac{11}{23}$.

PROBLEMS SET 1.4

1. The sum of the squares of two numbers is 233 and one of the numbers is 3 less than twice the other. Find the numbers.
2. The length of a rectangle is three times its breadth. If the length is diminshed by 1 cm and the breadth is increased by 3 cm, the area of rectangle becomes 72 cm^2. Find the sides of the rectangle.
3. Find two numbers whose sum is 64 and difference is 16.
4. Total salary of A and B is equal. If A gets 65% allowance of his basic salary and B gets 80% of his basic salary, what is the basic salary of A if the basic salary of B is Rs. 1100.
5. A father's age is four times that of his son. Before eight years the father's age was sixteen times that of the son. Find father's age.
6. A man is five times as old as his son and the sum of the square of their ages is equal to 2106. Find their ages.

ANSWERS

1. $-\frac{71}{5}, -\frac{28}{5}$ and 8, 13 **2.** 5, 15 **3.** 24, 40 **4.** Basic salary of A is Rs. 1200.

5. 40 years **6.** Father's age = 45 years, Son's age = 9 years.

1.8 MATRICES

1.8.1 Introduction

A French mathematician **'Cayley'** discovered the method of matrices in the year 1860. A matrix is an ordered system of numbers arranged in formation of rows and columns, describing various aspects of a phenomenon inter related to each other. Matrices have been found to be a powerful and useful tool for solving many problems involving electrical circuits, linear equations, mechanics, system of ordinary and partial differential equations, astronomy, game theory, LPP etc. It is also used in the study of dominance within a group by sociologists and in the study of births and survivals, marriage and descent, class structure and mobility by Demographers. So matrices have wide applications in modern mathematics, engineering and technology. Therefore, it is necessary for us to be familiar with the concept of matrix algebra.

■ **1.8.2 DEFINITION :** *A matrix of order p by q is defined as a rectangular arrangement of scalars (real or complex) in p rows and q columns subject to certain rules of operations, denoted by* $p \times q$.

A general form of the matrix of order $p \times q$ is usually written as

$$\begin{bmatrix} a_{11} & a_{12} \,\ldots\ldots\, a_{1j} \,\ldots\ldots\, a_{1q} \\ a_{21} & a_{22} \,\ldots\ldots\, a_{2j} \,\ldots\ldots\, a_{2q} \\ \vdots & \vdots \qquad \vdots \\ a_{i1} & a_{i2} \,\ldots\ldots\, a_{ij} \,\ldots\ldots\, a_{iq} \\ \vdots & \\ a_{p1} & a_{p2} \,\ldots\ldots\, a_{pj} \,\ldots\ldots\, a_{pq} \end{bmatrix} \rightarrow \text{Row}$$

$\uparrow$ column

The objects a_{ij} $(i = 1, 2, ..., p$ and $j = 1, 2, ..., q)$ in this arrangement (rectangular or square) of pq scalars is said to be an element of the matrix. Each element of the matrix can be a real or a complex number or a function of one or more variables or any other objects.

Here from the above matrix we can observe that each element has two suffixes. The first suffix indicates the row and the second suffix indicates the column in which the element lies. Thus, a_{13} is the element lying in the first row and third column, a_{ij} is the element liying in the ith row and j^{th} column.

In compact form the matrix of order $p \times q$ is also written as

$$A = [a_{ij}]_{p \times q},\ i = 1, 2, ..., p; j = 1,2, \ldots q.$$

■ **NOTE :** A matrix is not a number. It has got no numerical value. For example, 7 is simply a number, but [7] is a matrix of order 1×1.

Notations : There are different notations of enclosing the elements constituting a matrix i.e. [ ], ( ), || || but square bracket is generally used. Matrices are usually denoted by a single capital letter such as A, B, C, X, Y, etc., and their elements by corresponding small letters. For example, the following matrix

$$\begin{bmatrix} 1 & 3 & 9 \\ 0 & 2 & 4 \end{bmatrix} \text{ or } \begin{pmatrix} 1 & 3 & 9 \\ 0 & 2 & 4 \end{pmatrix} \text{ or } \begin{Vmatrix} 1 & 3 & 9 \\ 0 & 2 & 4 \end{Vmatrix}$$

has two rows and three columns. Thus, the order of the matrix is 2×3.

1.9 TYPES OF MATRICES

1.9.1 Real Matrix

A matrix is called real if all its elements are real numbers.

e.g., $$\begin{bmatrix} 2 & \sqrt{3} & 0 & 1 \\ -1 & 3 & 4 & -2 \end{bmatrix}$$

1.9.2 Row Matrix

A matrix in which there is exactly one row and any number of columns is called row matrix. A matrix $[a_{ij}]_{1 \times q}$ is a row matrix. *e.g.*, $[1, 0, 9, 4]_{1 \times 4}$ is a row matrix.

1.9.3 Column Matrix

A matrix in which there is exactly one column and any number of rows is called column matrix. A matrix $[a_{ij}]_{p \times 1}$ is a column matrix. *e.g.*, $\begin{bmatrix} 1 \\ 0 \\ 9 \\ 4 \end{bmatrix}_{4 \times 1}$ is a column matrix.

1.9.4 Null Matrix

A matrix in which each element is zero is said to be a null matrix or a zero matrix. A null matrix of order $p \times q$ is denoted by $O_{p \times q}$ *e.g.*, $O_{2 \times 3} = \begin{bmatrix} 0 & 0 & 0 \\ 0 & 0 & 0 \end{bmatrix}$, $O_{3 \times 3} = \begin{bmatrix} 0 & 0 & 0 \\ 0 & 0 & 0 \\ 0 & 0 & 0 \end{bmatrix}$

1.9.5 Equal Matrix

Two matices are said to be equal if and only if (i) they have the same number of rows and columns *i.e.*, the order of two matrices are same (ii) every element of one is equal to the corresponding element of the other.

For example : Given the matrices

$$A = \begin{bmatrix} x & y \\ z & u \end{bmatrix}, B = \begin{bmatrix} 0 & 2 \\ -1 & 5 \end{bmatrix}$$

clearly the order of two matrices is same *i.e.*, 2×2. So A and B are equal if $x = 0, y = 2, z = -1, u = 5$.

1.9.6 Sub Matrix

A matrix obtainad from a given matrix A by deleting some of its rows or columns or both is called a sub-matrix of matrix A.

Consider $A = \begin{bmatrix} 1 & 2 & 5 & 6 \\ 7 & 9 & 3 & 0 \\ 4 & 7 & 2 & -1 \end{bmatrix}$ and $B = \begin{bmatrix} 1 & 2 \\ 7 & 9 \end{bmatrix}$

Then B is a submatrix of matrix A obtained by deleting third row, third and fourth columns of matrix A.

1.9.7 Square Matrix

An $p \times q$ matrix A is said to be a square matrix if $p = q$, *i.e.*, the number of rows is equal to number of columns.

A square matrix having p rows and p columns is said to be 'a square matrix of order p. Thus, the matrix $A = [a_{ij}]_{p \times p}$; $i = 1, 2, ..., p, j = 1, 2, ..., p$ is a $p \times p$ square matrix.

For example

$$\begin{bmatrix} a_{11} & a_{12} \\ a_{21} & a_{22} \end{bmatrix}_{2 \times 2} ; \begin{bmatrix} a_{11} & a_{12} & a_{13} \\ a_{21} & a_{22} & a_{23} \\ a_{31} & a_{32} & a_{33} \end{bmatrix}_{3 \times 3} ; \begin{bmatrix} a_{11} & a_{12} & \cdots & a_{1p} \\ a_{21} & a_{22} & \cdots & a_{2p} \\ \cdots & \cdots & \cdots & \cdots \\ a_{p1} & a_{p2} & \cdots & a_{pp} \end{bmatrix}_{p \times p}$$

are square matrices of order 2×2, 3×3 and $p \times p$

In a square matrix of order p, elements a_{ij} for which $i = j$ are called *diagonal elements i.e.*, $a_{11}, a_{22}, ..., a_{pp}$ are diagonal elements. The line along which the diagonal elements lie is called the *principal diagonal* or *leading diagonal* or simply *'diagonal'* of the matrix.

For example, the matrix $\begin{bmatrix} 5 & 6 & 0 \\ 9 & -1 & 2 \\ 0 & 0 & 4 \end{bmatrix}_{3 \times 3}$

is a square matrix of order 3, in which the diagonal elements are 5, – 1, and 4.

■ **NOTE :** The sum of the diagonal elements of a square matrix is called trace of the matrix *i.e.*, *trace* of matrix A of order p is $a_{11} + a_{22} + ... + a_{pp}$

1.9.8 Scalar Matrix

A diagonal matrix in which all the diagonal elements are equal and all non-diagonal (remaining) elements are zero is called a *scalar matrix*.

Thus, $A = [a_{ij}]_{p \times p}$ is a scalar matrix

If $a_{ij} = \begin{bmatrix} 0 & \text{if} & i \neq j \\ m & \text{if} & i = j \end{bmatrix}$

For example, $\begin{bmatrix} 3 & 0 \\ 0 & 3 \end{bmatrix}, \begin{bmatrix} 5 & 0 & 0 \\ 0 & 5 & 0 \\ 0 & 0 & 5 \end{bmatrix}$ are scalar matrices.

1.9.9 Unit (or Identity) Matrix

A scalar matrix in which every diagonal element is one (or unity) is said to be a *unit* or *identity* matrix.

Thus $A = [a_{ij}]_{p \times p}$ is a unit matrix if $a_{ij} = \begin{bmatrix} 0 & \text{if} & i \neq j \\ 1 & \text{if} & i = j \end{bmatrix}$

Such a matrix is denoted by I_p.

For example, $I_2 = \begin{bmatrix} 1 & 0 \\ 0 & 1 \end{bmatrix}, I_3 = \begin{bmatrix} 1 & 0 & 0 \\ 0 & 1 & 0 \\ 0 & 0 & 1 \end{bmatrix}$

are unit matrices of order two and three respectively.

1.9.10 Upper Triangular Matrix

A square matrix is said to be an *upper triangular matrix* if all the elements below the principal diagonal are zero.

Thus, $A = [a_{ij}]_{p \times p}$ is an upper triangular matrix if $a_{ij} = 0$ for $i > j$.

For example, $\begin{bmatrix} 5 & 7 & 0 \\ 0 & -2 & 1 \\ 0 & 0 & 0 \end{bmatrix}$ is an upper triangular matrix.

1.9.11 Lower Triangular Matrix

A square matrix is said to be a *lower triangular matrix* if all the elements above the principal diagonal are zero.

Thus $A = [a_{ij}]_{p \times p}$ is a lower triangular matrix if $a_{ij} = 0$ for $i < j$.

For example, $\begin{bmatrix} 0 & 0 & 0 \\ 5 & -2 & 0 \\ 1 & 2 & -1 \end{bmatrix}$ is a lower triangular matrix.

1.9.12 Triangular Matrix

A square matrix $A = [a_{ij}]_{p \times p}$ is said to be a *triangular matrix* if it is either an upper triangular or a lower triangular matrix.

1.9.13 Strictly Triangular Matrix

A triangular matrix $A = [a_{ij}]_{p \times p}$ is said to be strictly triangular if $a_{ii} = 0$ for all $i = 1, 2, \ldots, p$.

For example, $A = \begin{bmatrix} 0 & 0 & 0 \\ 1 & 0 & 0 \\ 0 & 2 & 0 \end{bmatrix}$ is strictly triangular matrix of order 3.

1.10 ALGEBRA OF MATRICES

In the algebra of matrices, we shall define compositions on matrices by virtue of which we shall be able to combine matrices so as to produce new matrices. The main compositions are :

(1) Addition (Sum) and Subtraction of matrices.

(2) Multiplication of a matrix by a scalar *i.e.*, scalar multiplication.

(3) Multiplication of matrices.

1.10.1 Addition of Matrices

Two matrices are conformable for addition if they are of the same order.

If A and B are two matrices of the same order $p \times q$, then their sum $A + B$ is a matrix of the same order $p \times q$ and obtained by adding the corresponding elements of A and B.

If $A = [a_{ij}]_{p \times q}$ and $B = [b_{ij}]_{p \times q}$ are two matrices of the same order $p \times q$, their sum $A + B$ is another matrix $C = [c_{ij}]_{p \times q}$, where $c_{ij} = a_{ij} + b_{ij} \ \forall\ i, j$

Example 1.35

If $A = \begin{bmatrix} 2 & 0 & 5 \\ -1 & 2 & 4 \end{bmatrix}, B = \begin{bmatrix} 4 & -1 & 7 \\ 0 & 2 & 3 \end{bmatrix}$

then $A + B = \begin{bmatrix} 2+4 & 0-1 & 5+7 \\ -1+0 & 2+2 & 4+3 \end{bmatrix} = \begin{bmatrix} 6 & -1 & 12 \\ -1 & 4 & 7 \end{bmatrix}$

1.10.2 Subtraction of Matrices

If A and B are two matrices of the same order $p \times q$, than their difference $A - B$ is a matrix of the same order $p \times q$ and obtained by subtracting the corresponding elements of A and B.

Let $A = [a_{ij}]_{p \times q}$ and $B = [b_{ij}]_{p \times q}$ are two matrices of the same order $p \times q$, their difference $A - B$ is another matrix $C = [c_{ij}]_{p \times q}$.

Example 1.36

If $A = \begin{bmatrix} 3 & 2 & 2 \\ 2 & 4 & -1 \end{bmatrix}, B = \begin{bmatrix} 1 & 0 & 1 \\ 2 & 0 & 3 \end{bmatrix}$, are two matrices of same order. Than

$$A - B = \begin{bmatrix} 3-1 & 2-0 & 2-1 \\ 2-2 & 4-0 & -1-3 \end{bmatrix} = \begin{bmatrix} 2 & 2 & 1 \\ 0 & 4 & -4 \end{bmatrix}$$

■ **NOTE :** Two matrices of different orders cannot be added or subtracted.

1.10.3 Properties of Matrix Addition

1. **Commutativity :** If A and B are two matrices of the same order $p \times q$, then
$$A + B = B + A$$
2. **Associativity :** If A, B and C are three martices of the same order $p \times q$, then
$$(A + B) + C = A + (B + C)$$
3. **Existence of Additive Identity :** For any matrix A of order $p \times q$, there exists a null (zero) matrix O of the same order $p \times q$, such that
$$A + O = O + A = A$$
4. **Existence of Additive Inverese :** For any given matrix A of order $p \times q$, there exists a matrix $-A$ (negative of A) of the same order $p \times q$, such that
$$A + (-A) = O = (-A) + A.$$
5. **Cancellation Law :** If A, B and C are matrices of the same order, then
$$A + C = B + C \Rightarrow A = B.$$

Example 1.37

Let $A = \begin{bmatrix} 4 & 7 & -1 \\ 2 & 4 & 2 \end{bmatrix}; B = \begin{bmatrix} 3 & 9 & 0 \\ 0 & 5 & -1 \end{bmatrix}; C = \begin{bmatrix} 0 & 4 & 2 \\ -1 & -1 & 5 \end{bmatrix}$

and $O = \begin{bmatrix} 0 & 0 & 0 \\ 0 & 0 & 0 \end{bmatrix}$

1. **Commutativity :**

$$A + B = \begin{bmatrix} 4+3 & 7+9 & -1+0 \\ 2+0 & 4+5 & 2-1 \end{bmatrix} = \begin{bmatrix} 7 & 16 & -1 \\ 2 & 9 & 1 \end{bmatrix}$$

$$B + A = \begin{bmatrix} 3+4 & 9+7 & 0-1 \\ 0+2 & 5+4 & -1+2 \end{bmatrix} = \begin{bmatrix} 7 & 16 & -1 \\ 2 & 9 & 1 \end{bmatrix}$$

Thus, $A + B = B + A$

2. **Associativity :**

$$(A + B) + C = \begin{bmatrix} 7 & 16 & -1 \\ 2 & 9 & 1 \end{bmatrix} + \begin{bmatrix} 0 & 4 & 2 \\ -1 & -1 & 5 \end{bmatrix} = \begin{bmatrix} 7 & 20 & 1 \\ 1 & 8 & 6 \end{bmatrix}$$

$$A + (B + C) = \begin{bmatrix} 4 & 7 & -1 \\ 2 & 4 & 2 \end{bmatrix} + \begin{bmatrix} 3 & 13 & 2 \\ -1 & 4 & 4 \end{bmatrix} = \begin{bmatrix} 7 & 20 & 1 \\ 1 & 8 & 6 \end{bmatrix}$$

3. **Additive Identity :**

$$A + O = \begin{bmatrix} 4 & 7 & -1 \\ 2 & 4 & 2 \end{bmatrix} + \begin{bmatrix} 0 & 0 & 0 \\ 0 & 0 & 0 \end{bmatrix}$$

$$= \begin{bmatrix} 4 & 7 & -1 \\ 2 & 4 & 2 \end{bmatrix} = A$$

4. Additive Inverse :

$$A + (-A) = \begin{bmatrix} 4 & 7 & -1 \\ 2 & 4 & 2 \end{bmatrix} + \begin{bmatrix} -4 & -7 & 1 \\ -2 & -4 & -2 \end{bmatrix}$$

$$= \begin{bmatrix} 0 & 0 & 0 \\ 0 & 0 & 0 \end{bmatrix} = O$$

1.10.4. Scalar Multiplication of A Matrix

If A is a matrix of order $p \times q$ and k is a scalar, than the matrix obtained by multiplying each element of matrix A by k is said to be the *scalar-multiple* of A by k and denoled by kA *i.e.*,

$$kA = [ka_{ij}]_{p \times q} \quad \text{if} \quad A = [a_{ij}]_{p \times q}$$

Example 1.38

$$A = \begin{bmatrix} 3 & 9 & 5 \\ 4 & 7 & 0 \end{bmatrix}, \text{ than } 2A = \begin{bmatrix} 2\times 3 & 2\times 9 & 2\times 5 \\ 2\times 4 & 2\times 7 & 2\times 0 \end{bmatrix} = \begin{bmatrix} 6 & 18 & 10 \\ 8 & 14 & 0 \end{bmatrix}$$

1.10.5. Properties of Scalar Multiplication

If A, B and O are three matrices of same order, then for any scalar k, k_1 and k_2, the following important properties hold.

(i) $k(A + B) = kA + kB$

(ii) $(k_1 + k_2)A = k_1 A + k_2 A$

(iii) $k_1 (k_2 A) = (k_1 k_2) A$

(iv) $OA = O$

(v) $kO = O$

(vi) $(-k)A = -(kA) = k(-A)$

Example 1.39

If $A = \begin{bmatrix} 2 & 0 & 0 \\ 5 & 4 & 7 \\ 10 & 7 & 2 \end{bmatrix}$ and $B = \begin{bmatrix} 5 & 9 & -1 \\ 0 & 2 & 7 \\ 0 & 0 & 4 \end{bmatrix}$, find

(i) $3A$ (ii) $4B$

(iii) $3A + 4B$ (iv) $3A - 4B$

SOLUTION : (i) $3A = 3\begin{bmatrix} 2 & 0 & 0 \\ 5 & 4 & 7 \\ 10 & 7 & 2 \end{bmatrix} = \begin{bmatrix} 3\times 2 & 3\times 0 & 3\times 0 \\ 3\times 5 & 3\times 4 & 3\times 7 \\ 3\times 10 & 3\times 7 & 3\times 2 \end{bmatrix} = \begin{bmatrix} 6 & 0 & 0 \\ 15 & 12 & 21 \\ 30 & 21 & 6 \end{bmatrix}$

(ii) $4B = 4\begin{bmatrix} 5 & 9 & -1 \\ 0 & 2 & 7 \\ 0 & 0 & 4 \end{bmatrix} = \begin{bmatrix} 4\times 5 & 4\times 9 & 4\times -1 \\ 4\times 0 & 4\times 2 & 4\times 7 \\ 4\times 0 & 4\times 0 & 4\times 4 \end{bmatrix} = \begin{bmatrix} 20 & 36 & -4 \\ 0 & 8 & 28 \\ 0 & 0 & 16 \end{bmatrix}$

(iii) $3A + 4B = \begin{bmatrix} 6 & 0 & 0 \\ 15 & 12 & 21 \\ 30 & 21 & 6 \end{bmatrix} + \begin{bmatrix} 20 & 36 & -4 \\ 0 & 8 & 28 \\ 0 & 0 & 16 \end{bmatrix}$

$$= \begin{bmatrix} 6+20 & 0+36 & 0-4 \\ 15+0 & 12+8 & 21+28 \\ 30+0 & 21+0 & 6+16 \end{bmatrix}$$

$$= \begin{bmatrix} 26 & 36 & -4 \\ 15 & 20 & 49 \\ 30 & 21 & 22 \end{bmatrix}$$

(iv) $$3A - 4B = \begin{bmatrix} 6 & 0 & 0 \\ 15 & 12 & 21 \\ 30 & 21 & 6 \end{bmatrix} - \begin{bmatrix} 20 & 36 & -4 \\ 0 & 8 & 28 \\ 0 & 0 & 16 \end{bmatrix} = \begin{bmatrix} 6-20 & 0-36 & 0+4 \\ 15-0 & 12-8 & 21-28 \\ 30-0 & 21-0 & 6-16 \end{bmatrix}$$

$$= \begin{bmatrix} -14 & -36 & 4 \\ 15 & 4 & -7 \\ 30 & 21 & -10 \end{bmatrix}$$

Example 1.40

If $A = \begin{bmatrix} 2 & -1 & 3 \\ 4 & 2 & 0 \\ -2 & 7 & 3 \end{bmatrix}$ and $B = \begin{bmatrix} 6 & -2 & 5 \\ 0 & 4 & 9 \\ 3 & 1 & -5 \end{bmatrix}$ find a matrix C such that $3A + 4C = 2B$.

SOLUTION : Given that $3A + 4C = 2B$. Thus $4C = 2B - 3A$

$$4C = 2\begin{bmatrix} 6 & -2 & 5 \\ 0 & 4 & 9 \\ 3 & 1 & -5 \end{bmatrix} - 3\begin{bmatrix} 2 & -1 & 3 \\ 4 & 2 & 0 \\ -2 & 7 & 3 \end{bmatrix}$$

$$= \begin{bmatrix} 12 & -4 & 10 \\ 0 & 8 & 18 \\ 6 & 2 & -10 \end{bmatrix} - \begin{bmatrix} 6 & -3 & 9 \\ 12 & 6 & 0 \\ -6 & 21 & 9 \end{bmatrix}$$

$$= \begin{bmatrix} 6 & -1 & 1 \\ -12 & 2 & 18 \\ 12 & -19 & -19 \end{bmatrix}$$

or $$C = \frac{1}{4}\begin{bmatrix} 6 & -1 & 1 \\ -12 & 2 & 18 \\ 12 & -19 & -19 \end{bmatrix} = \begin{bmatrix} 6/4 & -1/4 & 1/4 \\ -12/4 & 2/4 & 18/4 \\ 12/4 & -19/4 & -19/4 \end{bmatrix}$$

$$= \begin{bmatrix} 3/2 & -1/4 & 1/4 \\ -3 & 1/2 & 9/2 \\ 3 & -19/4 & -19/4 \end{bmatrix}$$

Example 1.41

Find a, b, c and d if

$$3\begin{bmatrix} a & b \\ c & d \end{bmatrix} = \begin{bmatrix} a & 6 \\ -1 & 2d \end{bmatrix} + \begin{bmatrix} 4 & a+b \\ c+d & 3 \end{bmatrix}$$

SOLUTION : Given

$$3\begin{bmatrix} a & b \\ c & d \end{bmatrix} = \begin{bmatrix} a & 6 \\ -1 & 2d \end{bmatrix} + \begin{bmatrix} 4 & a+b \\ c+d & 3 \end{bmatrix}$$

or $$\begin{bmatrix} 3a & 3b \\ 3c & 3d \end{bmatrix} = \begin{bmatrix} a+4 & 6+a+b \\ -1+c+d & 2d+3 \end{bmatrix}$$

By equating the corresponding elements of the two matrices, we get

$$3a = a + 4,\ 3b = 6 + a + b,\ 3c = -1 + c + d,\ 3d = 2d + 3.$$

$\Rightarrow$ $2a = 4,\ 2b = 6 + a,\ 2c = -1 + d,\ d = 3$

$\Rightarrow \quad a = 2, b = \frac{6+2}{2}, c = \frac{-1+3}{2}, d = 3$

$\Rightarrow \quad a = 2, b = 4, c = 1, d = 3$

Example 1.42

Find A and B if $A + B = \begin{bmatrix} 3 & 0 \\ 2 & 5 \end{bmatrix}$ and $A - B = \begin{bmatrix} 7 & 2 \\ 0 & 5 \end{bmatrix}$.

SOLUTION : We have $A + B = \begin{bmatrix} 3 & 0 \\ 2 & 5 \end{bmatrix}$ and $A - B = \begin{bmatrix} 7 & 2 \\ 0 & 5 \end{bmatrix}$

Thus on adding,

$$(A + B) + (A - B) = \begin{bmatrix} 3 & 0 \\ 2 & 5 \end{bmatrix} + \begin{bmatrix} 7 & 2 \\ 0 & 5 \end{bmatrix}$$

$$2A = \begin{bmatrix} 10 & 2 \\ 2 & 10 \end{bmatrix}$$

or

$$A = \frac{1}{2}\begin{bmatrix} 10 & 2 \\ 2 & 10 \end{bmatrix} = \begin{bmatrix} 5 & 1 \\ 1 & 5 \end{bmatrix}$$

on subtracting,

$$(A + B) - (A - B) = \begin{bmatrix} 3 & 0 \\ 2 & 5 \end{bmatrix} - \begin{bmatrix} 7 & 2 \\ 0 & 5 \end{bmatrix}$$

$$2B = \begin{bmatrix} -4 & -2 \\ 2 & 0 \end{bmatrix}$$

or

$$B = \frac{1}{2}\begin{bmatrix} -4 & -2 \\ 2 & 0 \end{bmatrix} = \begin{bmatrix} -2 & -1 \\ 1 & 0 \end{bmatrix}$$

Thus, $A = \begin{bmatrix} 5 & 1 \\ 1 & 5 \end{bmatrix}$ and $B = \begin{bmatrix} -2 & -1 \\ 1 & 0 \end{bmatrix}$

Example 1.43

A company produces three types of products X, Y and Z. The total annual sales in thousands of units of these products for the years 1999 and 2000 in the four regions is given below :

For the year 1999 :

Product/Region	Eastern (*E*)	Western (*W*)	Southern (*S*)	Northern (*N*)
X	10	8	6	10
Y	5	3	20	32
Z	9	7	30	10

For the year 2000 :

Product/Region	Eastern (*E*)	Western (*W*)	Southern (*S*)	Northern (*N*)
X	12	7	4	9
Y	6	5	25	15
Z	9	10	22	7

Find the total sales of three products for two years.

SOLUTION : The matrix for the year 1999 and 2000 are obtaind as :

$$A = \begin{matrix} \\ X \\ Y \\ Z \end{matrix}\begin{bmatrix} E & W & S & N \\ 10 & 8 & 6 & 10 \\ 5 & 3 & 20 & 32 \\ 9 & 7 & 30 & 10 \end{bmatrix}$$

$$B = \begin{matrix} \\ X \\ Y \\ Z \end{matrix}\begin{bmatrix} E & W & S & N \\ 12 & 7 & 4 & 9 \\ 6 & 5 & 25 & 15 \\ 9 & 10 & 22 & 7 \end{bmatrix}$$

Total sales of the products is obtained by adding the two matrices A and B.

Thus, $$C = A + B = \begin{bmatrix} 22 & 15 & 10 & 19 \\ 11 & 8 & 45 & 47 \\ 18 & 17 & 52 & 17 \end{bmatrix}$$

PROBLEM SET 1.5

1. If $A = \begin{bmatrix} 0 & 2 & 3 \\ 2 & 1 & 4 \end{bmatrix}$, $B = \begin{bmatrix} 7 & 6 & 3 \\ 1 & 4 & 5 \end{bmatrix}$. Find the matrices for (i) $3A$ (ii) $4B$ (iii) $3A + 4B$ (iv) $3A - 4B$.

2. If $A = \begin{bmatrix} 2 & 4 & 7 \\ 0 & 1 & 2 \\ -1 & 0 & 4 \end{bmatrix}$, $B = \begin{bmatrix} 3 & -7 & 5 \\ 4 & -2 & 10 \\ -2 & 7 & 3 \end{bmatrix}$ and $C = \begin{bmatrix} 2 & -3 & 4 \\ 0 & 0 & 1 \\ -1 & -3 & 2 \end{bmatrix}$ find the matrix X such that $5A + 4B + 3X = 2C$.

3. Find the values of a, b, c and d which satisfy the matrix equation

$$\begin{bmatrix} a+3 & 2b+a \\ c-1 & 4d-6 \end{bmatrix} = \begin{bmatrix} 0 & -7 \\ 3 & 2d \end{bmatrix}.$$

4. If $A = \begin{bmatrix} 2 & 0 & 9 \\ 7 & 9 & 4 \end{bmatrix}$, $B = \begin{bmatrix} -1 & -5 & 0 \\ 2 & 7 & 5 \end{bmatrix}$, find the matrix X such that $A + B + X = 0$.

5. If $A = \begin{bmatrix} 9 & 1 \\ 4 & 3 \end{bmatrix}$, $B = \begin{bmatrix} 1 & 5 \\ 7 & 12 \end{bmatrix}$, find matrix X such that $3A + 5B + 2X = 0$.

6. Find a, b, c and d if :

(i) $\begin{bmatrix} 2a-3b & c-d & 3 \\ 1 & a+4b & 3c+4d \end{bmatrix} = \begin{bmatrix} 1 & -2 & 3 \\ 1 & 6 & 29 \end{bmatrix}$

(ii) $\begin{bmatrix} 3a+2b & 2 & a-2b \\ c+d & 2c-d & -1 \end{bmatrix} = \begin{bmatrix} 2 & 2 & 4 \\ 5 & -5 & -1 \end{bmatrix}$.

7. Solve the matrix equation :

(i) $3\begin{bmatrix} a & b \\ c & d \end{bmatrix} - 4\begin{bmatrix} 2 & 5 \\ 7 & 9 \end{bmatrix} = 3\begin{bmatrix} 0 & 2 \\ 7 & 9 \end{bmatrix}$

(ii) $2\begin{bmatrix} x & y \\ z & u \end{bmatrix} = 7\begin{bmatrix} 2 & 4 \\ z & 2 \end{bmatrix} + 3\begin{bmatrix} x & x+y \\ z-3u & u \end{bmatrix}$.

8. A Bakery produces basic pastry mix types X, Y and Z. In the past the mix of ingredients has been as given in the table 1.

Pastry Mix	Ingredients		
	Flour	Fat	Sugar
X	5	1	1
Y	5·5	2·5	0·5
Z	4·5	2	4

Due to changes in the consumer's tastes it has been decided to change the mixes using the following amendment matrix :

Pastry Mix	Ingredients		
	Flour	Fat	Sugar
X	0	1	0
Y	− 0·5	0·5	0·5
Z	0·5	0	0

obtain the matrix for the new mix.
[**HINT :** The matrix of the new mix is obtained by the addition of the original mix matrix and the amendment matrix.]

9. If $A = \begin{bmatrix} 1 & -1 \\ 2 & -1 \end{bmatrix}$, $B = \begin{bmatrix} a & -1 \\ b & -1 \end{bmatrix}$ and $(A+B)^2 = A^2 + B^2$, find a and b.

10. If $A = \begin{bmatrix} 2 & 3 & 4 \\ -3 & 0 & 2 \end{bmatrix}$, $B = \begin{bmatrix} 3 & -4 & -5 \\ 1 & 2 & 1 \end{bmatrix}$, $C = \begin{bmatrix} 5 & -1 & 0 \\ 7 & 0 & 3 \end{bmatrix}$, find X such that $2A + 2B - X = C$

11. Determine the matrices A and B when $A + 2B = \begin{bmatrix} 1 & 2 & 0 \\ 6 & -3 & 3 \\ -5 & 3 & 1 \end{bmatrix}$ and $2A - B = \begin{bmatrix} 2 & -1 & 5 \\ 2 & -1 & 6 \\ 0 & 1 & 2 \end{bmatrix}$

12. If matrices $\begin{bmatrix} a+b & 2 \\ 5 & ab \end{bmatrix} = \begin{bmatrix} 6 & 2 \\ 5 & 8 \end{bmatrix}$, find the values of a and b.

13. Find matrices X and Y, if $X + Y = \begin{bmatrix} 5 & 2 \\ 0 & 9 \end{bmatrix}$ and $X - Y = \begin{bmatrix} 3 & 6 \\ 0 & -1 \end{bmatrix}$.

ANSWERS

1. (i) $\begin{bmatrix} 0 & 6 & 9 \\ 6 & 3 & 12 \end{bmatrix}$ (ii) $\begin{bmatrix} 28 & 24 & 12 \\ 4 & 16 & 20 \end{bmatrix}$ (iii) $\begin{bmatrix} 28 & 30 & 21 \\ 10 & 19 & 32 \end{bmatrix}$ (iv) $\begin{bmatrix} -28 & -18 & -3 \\ 2 & -13 & -8 \end{bmatrix}$

2. $X = \frac{1}{3}\begin{bmatrix} -18 & 2 & -47 \\ -16 & 3 & -48 \\ 11 & -34 & -28 \end{bmatrix}$ 3. $a = -3,\ b = -2,\ c = 4,\ d = 3$

4. $x = \begin{bmatrix} -1 & 5 & -9 \\ -9 & 16 & -9 \end{bmatrix}$ 5. $X = \frac{1}{2}\begin{bmatrix} -32 & -28 \\ -47 & -69 \end{bmatrix}$

6. (i) $a = 2,\ b = 1,\ c = 3,\ d = 5$ (ii) $a = \frac{3}{2},\ b = -\frac{5}{4},\ c = 0,\ d = 5$

7. (i) $a = \frac{8}{3},\ b = \frac{26}{3},\ c = \frac{49}{3},\ d = 21$ (ii) $x = -14,\ y = 14,\ u = -14,\ z = \frac{63}{4}$

8. $\begin{bmatrix} 5 & 2 & 1 \\ 5 & 3 & 1 \\ 5 & 2 & 4 \end{bmatrix}$ 9. $a = 1,\ b = 4$ 10. $\begin{bmatrix} 5 & -1 & -2 \\ -11 & 4 & 3 \end{bmatrix}$

11. $A = \begin{bmatrix} 1 & 0 & 2 \\ 2 & -1 & 3 \\ -1 & 1 & 1 \end{bmatrix},\ B = \begin{bmatrix} 0 & 1 & -1 \\ 2 & -1 & 0 \\ -2 & 1 & 0 \end{bmatrix}$. 12. $a = 2, 4$ and $b = 4, 2$

13. $Y = \begin{bmatrix} 1 & -2 \\ 0 & 5 \end{bmatrix},\ X = \begin{bmatrix} 4 & 4 \\ 0 & 4 \end{bmatrix}$

1.11 MULTIPLICATION OF MATRICES

The multiplication (Product) of two matrices A and B denoted by AB, can be evaluated if and only if the number of columns in A (pre-factor of AB) are equal to the number of rows in B (Post-factor of AB).

Thus, if the order of A is $p \times q$ and order of B is $m \times n$, then

(i) AB exists if and only if $q = m$

(ii) BA exists if and only if $n = p$

If $A = [a_{ij}]_{p \times q}$ and $B = [b_{jk}]_{q \times r}$, then the product of A and B is defined as $C = [c_{ik}]_{p \times r}$

where, $c_{ik} = a_{i1}\, b_{1k} + a_{i2}\, b_{2k} + \ldots + a_{iq}\, b_{qk} = \sum_{j=1}^{q} a_{ij}\, b_{jk}$

= sum of the products of the elements of ith row and j^{th} column of A with the element of jth row and k^{th} column of B.

Example 1.44

Let $A = \begin{bmatrix} a_{11} & a_{12} \\ a_{21} & a_{22} \\ a_{31} & a_{32} \end{bmatrix}_{3 \times 2}$,

$B = \begin{bmatrix} b_{11} & b_{12} & b_{13} \\ b_{21} & b_{22} & b_{23} \end{bmatrix}_{2 \times 3}$

than $$AB = \begin{bmatrix} a_{11} b_{11} + a_{12} b_{21} & a_{11} b_{12} + a_{12} b_{22} & a_{11} b_{13} + a_{12} b_{23} \\ a_{21} b_{11} + a_{22} b_{21} & a_{21} b_{12} + a_{22} b_{22} & a_{21} b_{13} + a_{22} b_{23} \\ a_{31} b_{11} + a_{32} b_{21} & a_{31} b_{12} + a_{32} b_{22} & a_{31} b_{13} + a_{32} b_{23} \end{bmatrix}_{3 \times 3}$$

Example 1.45

Let $$A = \begin{bmatrix} 3 & 4 & 2 \\ -1 & 0 & 2 \end{bmatrix}_{2 \times 3}, B = \begin{bmatrix} 4 & 5 & 7 \\ 0 & 0 & 2 \\ -1 & 4 & 0 \end{bmatrix}_{3 \times 3}$$

then $$AB = \begin{bmatrix} 3{\cdot}4 + 4{\cdot}0 + 2{\cdot}(-1) & 3{\cdot}5 + 4{\cdot}0 + 2{\cdot}4 & 3{\cdot}7 + 4{\cdot}2 + 2{\cdot}0 \\ -1{\cdot}4 + 0{\cdot}0 + 2{\cdot}(-1) & -1{\cdot}5 + 0{\cdot}0 + 2{\cdot}4 & -1{\cdot}7 + 0{\cdot}2 + 2{\cdot}0 \end{bmatrix} = \begin{bmatrix} 10 & 23 & 29 \\ -6 & 3 & -7 \end{bmatrix}_{2 \times 3}$$

Here BA is not defined because number of columns in $B \neq$ number of rows in A.

■ **NOTE :** (1) The product AB can be evaluated if and only if the number of columns of A are equal to the number of rows in B.
(2) In the matrix product AB, the matrix A is called premultplier with matrix B and matrix B is called post-multiplier with matrix A.

1.11.1 Properties of Matrix Multplication

(i) Matrix Multiplication is Associative : It A, B and$C$ are three matrices of order $p \times q$, $q \times r$ and $r \times s$ respectively, then $(AB)\,C = A\,(BC)$.

Example 1.46

Let $$A = \begin{bmatrix} 2 & -1 \\ 0 & 1 \end{bmatrix}_{2 \times 2}, B = \begin{bmatrix} 3 & 4 & 0 \\ 5 & 7 & 9 \end{bmatrix}_{2 \times 3} \text{ and } C = \begin{bmatrix} 2 \\ 7 \\ 5 \end{bmatrix}_{3 \times 1}$$

than, $$(AB)\,C = \begin{bmatrix} -36 \\ 104 \end{bmatrix}_{2 \times 1} \text{ and } A(BC) = \begin{bmatrix} -36 \\ 104 \end{bmatrix}_{2 \times 1}$$

So $(AB)\,C = A(BC)$

(ii) Matrix Multiplication is Distributive with Respect to Matrix Addition : If A, B and C are three matrices of order $p \times q$, $q \times r$ and $q \times r$ respectively, then

$$A\,(B + C) = AB + AC$$

Example 1.47

Let $$A = \begin{bmatrix} 2 & 3 & 1 \\ 3 & 1 & 0 \end{bmatrix}_{2 \times 3}, B = \begin{bmatrix} 2 & 0 \\ 0 & 1 \\ 1 & -1 \end{bmatrix}_{3 \times 2} \text{ and } C = \begin{bmatrix} 2 & 0 \\ 2 & 2 \\ 1 & 5 \end{bmatrix}_{3 \times 2}$$

then, $$A \cdot (B + C) = \begin{bmatrix} 16 & 13 \\ 14 & 3 \end{bmatrix} \text{ and } AB + AC = \begin{bmatrix} 16 & 13 \\ 14 & 3 \end{bmatrix}$$

Thus, $A \cdot (B + C) = AB + AC$

(iii) Matrix Multiplication is not Commutative in General : There are three different cases :

(a) If A is $p \times q$ and B is $q \times r$ matrix, than AB exists whereas BA does not exists, because we know that AB can be evaluated if and only if the number of columns in A is equal to the number of rows in B.

(b) If A is $p \times q$ and B is $q \times p$ matrix, than AB and BA both exist, but AB is $p \times p$ matrix while BA is $q \times q$ matrix. Thus, $AB \neq BA$ because the order of two products are not same.

(c) If A and B are $p \times q$ (same order) matrices, then AB as well as BA exists but are not necessarily equal.

Example 1.48

If $$A = \begin{bmatrix} 2 & 1 \\ 3 & 0 \end{bmatrix}, B = \begin{bmatrix} 4 & -1 \\ 5 & 2 \end{bmatrix}$$

then, $AB = \begin{bmatrix} 13 & 0 \\ 12 & -3 \end{bmatrix}$ and $BA = \begin{bmatrix} 5 & 4 \\ 16 & 5 \end{bmatrix}$ clearly, $AB \neq BA$

Example 1.49

If $$A = \begin{bmatrix} 2 & 0 \\ 0 & 4 \end{bmatrix}, B = \begin{bmatrix} 1 & 0 \\ 0 & 7 \end{bmatrix}$$

then $AB = \begin{bmatrix} 2 & 0 \\ 0 & 28 \end{bmatrix}$ and $BA = \begin{bmatrix} 2 & 0 \\ 0 & 28 \end{bmatrix}$

Here, $AB = BA$

(iv) Multiplication of a Matrix by a Unit Matrix : If A and I are square matrices of the same order, then $AI = A = IA$.

Example 1.50

If $$A = \begin{bmatrix} 2 & 0 & 3 \\ 0 & 4 & 5 \\ 1 & 2 & 0 \end{bmatrix}_{3 \times 3} \text{ and } I = \begin{bmatrix} 1 & 0 & 0 \\ 0 & 1 & 0 \\ 0 & 0 & 1 \end{bmatrix}_{3 \times 3}$$

then, $AI = \begin{bmatrix} 2 & 0 & 3 \\ 0 & 4 & 5 \\ 1 & 2 & 0 \end{bmatrix}$ and $IA = \begin{bmatrix} 2 & 0 & 3 \\ 0 & 4 & 5 \\ 1 & 2 & 0 \end{bmatrix}$

Thus, $AI = A = IA$

(v) $AB = O$ does not means that $A = O$ or $B = O$ or both are zero.

Example 1.51

If $$A = \begin{bmatrix} 2 & 0 \\ 0 & 0 \end{bmatrix} \neq O, \quad B = \begin{bmatrix} 0 & 0 \\ 1 & 0 \end{bmatrix} \neq O$$

Then, $$AB = \begin{bmatrix} 0 & 0 \\ 0 & 0 \end{bmatrix} = O$$

Thus, the product of two non-zero matrices may be a null matrix.

(vi) Multiplication of a Matrix by Itself : The product AA exists only if A is a square matrix. Thus, if A is a square matrix of order p, then $A{\cdot}A = A^2, A{\cdot}A{\cdot}A = A^3, \ldots, A{\cdot}A\ldots, m$ times $= A^m$ are all square matrices of order p.

■ **NOTE :** $A^\circ = I$ and $I^n = I$

(vii) If A is a square matrix, than $A^n A^m = A^m{\cdot}A^n = A^{m+n}$

Example 1.52

If $A = \begin{bmatrix} 1 & 3 & 0 \\ -1 & 2 & 1 \\ 0 & 0 & 2 \end{bmatrix}$ and $B = \begin{bmatrix} 2 & 3 & 4 \\ 1 & 2 & 3 \\ -1 & 1 & 2 \end{bmatrix}$,

find AB and BA and show that $AB \neq BA$.

SOLUTION : $AB = \begin{bmatrix} 1 & 3 & 0 \\ -1 & 2 & 1 \\ 0 & 0 & 2 \end{bmatrix} \begin{bmatrix} 2 & 3 & 4 \\ 1 & 2 & 3 \\ -1 & 1 & 2 \end{bmatrix}$

$$= \begin{bmatrix} 2+3+0 & 3+6+0 & 4+9+0 \\ -2+2-1 & -3+4+1 & -4+6+2 \\ 0+0-2 & 0+0+2 & 0+0+4 \end{bmatrix}$$

$$= \begin{bmatrix} 5 & 9 & 13 \\ -1 & 2 & 4 \\ -2 & 2 & 4 \end{bmatrix}$$

$$BA = \begin{bmatrix} 2 & 3 & 4 \\ 1 & 2 & 3 \\ -1 & 1 & 2 \end{bmatrix} \begin{bmatrix} 1 & 3 & 0 \\ -1 & 2 & 1 \\ 0 & 0 & 2 \end{bmatrix}$$

$$= \begin{bmatrix} 2-3+0 & 6+6+0 & 0+3+8 \\ 1-2+0 & 3+4+0 & 0+2+6 \\ -1-1+0 & -3+2+0 & 0+1+4 \end{bmatrix} = \begin{bmatrix} -1 & 12 & 11 \\ -1 & 7 & 8 \\ -2 & -1 & 5 \end{bmatrix}$$

clearly, $AB \neq BA$.

Example 1.53

Prove by an example that $AB = AC$ does not imply that $B = C$.

SOLUTION : Let $A = \begin{bmatrix} 1 & 4 \\ 0 & 0 \end{bmatrix}$, $B = \begin{bmatrix} 1 & 5 \\ 2 & 1 \end{bmatrix}$ and $C = \begin{bmatrix} 1 & 1 \\ 2 & 2 \end{bmatrix}$

Then $AB = \begin{bmatrix} 1 & 4 \\ 0 & 0 \end{bmatrix} \begin{bmatrix} 1 & 5 \\ 2 & 1 \end{bmatrix} = \begin{bmatrix} 1+8 & 5+4 \\ 0+0 & 0+0 \end{bmatrix} = \begin{bmatrix} 9 & 9 \\ 0 & 0 \end{bmatrix}$

$$AC = \begin{bmatrix} 1 & 4 \\ 0 & 0 \end{bmatrix} \begin{bmatrix} 1 & 1 \\ 2 & 2 \end{bmatrix} = \begin{bmatrix} 1+8 & 1+8 \\ 0+0 & 0+0 \end{bmatrix} = \begin{bmatrix} 9 & 9 \\ 0 & 0 \end{bmatrix}$$

clearly, $AB = AC$ but $B \neq C$.

Example 1.54

Let $f(x) = x^2 - 5x + 6$. Find $F(A)$ where $A = \begin{bmatrix} 2 & 0 & 1 \\ 2 & 1 & 3 \\ 0 & -1 & 0 \end{bmatrix}$.

SOLUTION : To obtain $f(A)$, replace x by A and multiply the constant term by the unit matrix of order equal to the order of given matrix A. Now $f(A) = A^2 - 5A + 6 \cdot I$, where I is a unit matrix of order 3×3.

Now $A^2 = A \cdot A = \begin{bmatrix} 2 & 0 & 1 \\ 2 & 1 & 3 \\ 1 & -1 & 0 \end{bmatrix} \begin{bmatrix} 2 & 0 & 1 \\ 2 & 1 & 3 \\ 1 & -1 & 0 \end{bmatrix} = \begin{bmatrix} 5 & -1 & 2 \\ 9 & -2 & 5 \\ 0 & -1 & -2 \end{bmatrix}$

$$5A = \begin{bmatrix} 10 & 0 & 5 \\ 10 & 5 & 15 \\ 5 & -5 & 0 \end{bmatrix}$$

$$6I = \begin{bmatrix} 6 & 0 & 0 \\ 0 & 6 & 0 \\ 0 & 0 & 6 \end{bmatrix}$$

Now, $$A^2 - 5A + 6I = \begin{bmatrix} 5 & -1 & 2 \\ 9 & -2 & 5 \\ 0 & -1 & -2 \end{bmatrix} - \begin{bmatrix} 10 & 0 & 5 \\ 10 & 5 & 15 \\ 5 & -5 & 0 \end{bmatrix} + \begin{bmatrix} 6 & 0 & 0 \\ 0 & 6 & 0 \\ 0 & 0 & 6 \end{bmatrix}$$

$$= \begin{bmatrix} 1 & -1 & -3 \\ -1 & -1 & -10 \\ -5 & 4 & 4 \end{bmatrix}$$

Example 1.55

If a is a non-zero real or complex number and $A = \begin{bmatrix} a & 1 \\ 0 & a \end{bmatrix}$, then use the principle of mathematical induction to prove that $A^n = \begin{bmatrix} a^n & na^{n-1} \\ 0 & a^n \end{bmatrix}$ for every positive integer n.

SOLUTION : We will prove the result by using mathematical induction. For $n = 1$,

$$A^1 = A = \begin{bmatrix} a & 1 \\ 0 & a \end{bmatrix} = \begin{bmatrix} a^1 & 1 \cdot a^{1-1} \\ 0 & a^1 \end{bmatrix}$$

So, the result is true for $n = 1$

Assume that the result is true for $n = k$. Then,

$$A^k = \begin{bmatrix} a^k & ka^{k-1} \\ 0 & a^k \end{bmatrix}$$

Now, we have to show that it is true for $n = k + 1$ consider $A^{k+1} = A^k \cdot A^1$

$$= \begin{bmatrix} a^k & ka^{k-1} \\ 0 & a^k \end{bmatrix} \begin{bmatrix} a & 1 \\ 0 & a \end{bmatrix}$$

$$= \begin{bmatrix} a^k\, a + 0 \cdot ka^{k-1} & 1 \cdot a^k + ka^{k-1}\, a \\ 0 \cdot a + 0 \cdot a^k & 0 \cdot 1 + a^k \cdot a \end{bmatrix}$$

$$= \begin{bmatrix} a^{k+1} & (k+1)\, a^k \\ 0 & a^{k+1} \end{bmatrix}$$

Thus, the result is true for $n = k + 1$

Hence, by mathematical induction, the result is true for all positive integer values of n.

■ **NOTE : Mathematical induction** is a very important method for proving results for all positive integers. If the result to be proved having n, a positive integer, then the proof by mathematical induction having the following steps :

(1) Verify the result for $n = 1$.

(2) Assume the result to be true for $n = k$ and then prove that it is true for $n = k + 1$.

Example 1.56

By mathematical induction, prove that it :

$$A = \begin{bmatrix} 3 & -4 \\ 1 & -1 \end{bmatrix} \text{ then } A^n = \begin{bmatrix} 1 + 2n & -4n \\ n & 1 - 2n \end{bmatrix}$$

where n is any positive integer.

SOLUTION : We will prove the result by using mathematical induction for $n = 1$,

$$A^1 = \begin{bmatrix} 1+2\,(1) & -4\,(1) \\ 1 & 1-2\,(1) \end{bmatrix} = \begin{bmatrix} 3 & -4 \\ 1 & -1 \end{bmatrix} = A$$

Thus, the result is true for $n = 1$

Assume that it is true for $n = k$ *i.e.*,

$$A^k = \begin{bmatrix} 1+2k & -4k \\ k & 1-2k \end{bmatrix}$$

We have to show that it is true for $n = k + 1$

Now, $A^{k+1} = A^k \cdot A = \begin{bmatrix} 1+2k & -4k \\ k & 1-2k \end{bmatrix} \begin{bmatrix} 3 & -4 \\ 1 & -1 \end{bmatrix}$

$$= \begin{bmatrix} 3+6k-4k & -4-8k+4k \\ 3k+1-2k & -4k-1+2k \end{bmatrix} = \begin{bmatrix} 3+2k & -4-4k \\ k+1 & -1-2k \end{bmatrix}$$

$$= \begin{bmatrix} 1+2\,(k+1) & -4\,(k+1) \\ k+1 & 1-2\,(k+1) \end{bmatrix}$$

The result is true for $n = k + 1$

Hence, by mathematical induction, the result is true for all positive integers n.

Example 1.57

A man buys 8 dozens of mangoes, 10 dozens of apples and 4 dozens of bananas. The cost of mangoes is Rs. 18 per dozen, apples is Rs. 9 per dozen and bananas is Rs. 6 per dozen. Represent the quantities bought by a row matrix and the prices by a column matrix and hence obtain the total cost.

SOLUTION : Let A be a row matrix representing the quantities bought, *i.e.*, 8 dozens of mangoes, 10 dozens of apples and 4 dozens of bananas, then $A = [8\ 10\ 4]$ and B be a column matrix representing the prices *i.e.*,

$$B = \begin{bmatrix} 18 \\ 9 \\ 6 \end{bmatrix}$$

The total cost is obtained by multiplying matrix A and B.

Thus, $AB = [8\ 10\ 4] \times \begin{bmatrix} 18 \\ 9 \\ 6 \end{bmatrix}$

$$= [8 \cdot 18 + 10 \cdot 9 + 4 \cdot 6] = [258]$$

So, the required total cost is Rs. 258.

Example 1.58

If $A = \begin{bmatrix} 0 & -\tan\dfrac{\theta}{2} \\ \tan\dfrac{\theta}{2} & 0 \end{bmatrix}$

Show that $\qquad I + A = (I - A)\begin{bmatrix} \cos\theta & -\sin\theta \\ \sin\theta & \cos\theta \end{bmatrix}$.

SOLUTION : L.H.S. $= I + A = \begin{bmatrix} 1+0 & 0-\tan\theta/2 \\ 0+\tan\theta/2 & 1+0 \end{bmatrix}$,

$$= \begin{bmatrix} 1 & -\tan\theta/2 \\ \tan\theta/2 & 1 \end{bmatrix}$$

$$\text{R.H.S.} = (I - A) \begin{bmatrix} \cos\theta & -\sin\theta \\ \sin\theta & \cos\theta \end{bmatrix}$$

$$= \begin{bmatrix} 1 & \tan\theta/2 \\ -\tan\theta/2 & 1 \end{bmatrix} \times \begin{bmatrix} \cos\theta & -\sin\theta \\ \sin\theta & \cos\theta \end{bmatrix}$$

$$= \begin{bmatrix} \cos\theta + \sin\theta\tan\theta/2 & -\sin\theta + \cos\theta\tan\theta/2 \\ -\tan\theta/2\cos\theta + \sin\theta & \sin\theta\tan\theta/2 + \cos\theta \end{bmatrix}$$

$$= \begin{bmatrix} \dfrac{\cos\theta\cos\theta/2 + \sin\theta\sin\theta/2}{\cos\theta/2} & \dfrac{-\sin\theta\cos\theta/2 + \cos\theta\sin\theta/2}{\cos\theta/2} \\ \dfrac{-\sin\theta/2\cos\theta + \cos\theta/2\sin\theta}{\cos\theta/2} & \dfrac{\sin\theta\sin\theta/2 + \cos\theta\cos\theta/2}{\cos\theta/2} \end{bmatrix}$$

$$= \begin{bmatrix} \dfrac{\cos(\theta - \theta/2)}{\cos\theta/2} & \dfrac{-\sin(\theta - \theta/2)}{\cos\theta/2} \\ \dfrac{\sin(\theta - \theta/2}{\cos\theta/2} & \dfrac{\cos(\theta - \theta/2)}{\cos\theta/2} \end{bmatrix}$$

$$= \begin{bmatrix} \dfrac{\cos\theta/2}{\cos\theta/2} & \dfrac{-\sin\theta/2}{\cos\theta/2} \\ \dfrac{\sin\theta/2}{\cos\theta/2} & \dfrac{\cos\theta/2}{\cos\theta/2} \end{bmatrix}$$

$$= \begin{bmatrix} 1 & -\tan\theta/2 \\ \tan\theta/2 & 1 \end{bmatrix}$$

Thus L.H.S. = R.H.S.

Example 1.59

If $[1 \quad 1 \quad x] \begin{bmatrix} 1 & 0 & 2 \\ 0 & 2 & 1 \\ 2 & 1 & 0 \end{bmatrix} \begin{bmatrix} 1 \\ 1 \\ 1 \end{bmatrix} = 0$. Find x

[UPTU 2003, 05]

SOLUTION : Given $[1 \quad 1 \quad x] \begin{bmatrix} 1 & 0 & 2 \\ 0 & 2 & 1 \\ 2 & 1 & 0 \end{bmatrix} \begin{bmatrix} 1 \\ 1 \\ 1 \end{bmatrix} = 0$

$$\Rightarrow \quad [1 \quad 1 \quad x] \begin{bmatrix} 1+0+2 \\ 0+2+1 \\ 2+1+0 \end{bmatrix} = 0$$

$$\Rightarrow \quad [1 \quad 1 \quad x] \begin{bmatrix} 3 \\ 3 \\ 3 \end{bmatrix} = 0$$

$$\Rightarrow \quad [3 + 3 + 3x] = 0$$

$$\Rightarrow \quad [6 + 3x] = 0$$

$$\Rightarrow \quad 6 + 3x = 0 \Rightarrow 3x = -6 \Rightarrow x = -2$$

Example 1.60

If $A = \begin{bmatrix} 4 & 1 \\ 3 & 2 \end{bmatrix}$, then prove that

$$A^2 - 6A + 5I = \begin{bmatrix} 0 & 0 \\ 0 & 0 \end{bmatrix}, \text{ when } I = \begin{bmatrix} 1 & 0 \\ 0 & 1 \end{bmatrix}$$ ***[UPTU 2001]***

SOLUTION : Given $A = \begin{bmatrix} 4 & 1 \\ 3 & 2 \end{bmatrix}$

$$\therefore \quad A^2 = A \cdot A = \begin{bmatrix} 4 & 1 \\ 3 & 2 \end{bmatrix} \begin{bmatrix} 4 & 1 \\ 3 & 2 \end{bmatrix} = \begin{bmatrix} 19 & 6 \\ 18 & 7 \end{bmatrix}$$

Now
$$A^2 - 6A + 5I = \begin{bmatrix} 19 & 6 \\ 18 & 7 \end{bmatrix} - 6\begin{bmatrix} 4 & 1 \\ 3 & 2 \end{bmatrix} + 5\begin{bmatrix} 1 & 0 \\ 0 & 1 \end{bmatrix}$$

$$= \begin{bmatrix} 19 & 6 \\ 18 & 7 \end{bmatrix} - \begin{bmatrix} 24 & 6 \\ 18 & 12 \end{bmatrix} + \begin{bmatrix} 5 & 0 \\ 0 & 5 \end{bmatrix}$$

$$= \begin{bmatrix} 19 - 24 + 5 & 6 - 6 + 0 \\ 18 - 18 + 0 & 7 - 12 + 5 \end{bmatrix}$$

$$= \begin{bmatrix} 0 & 0 \\ 0 & 0 \end{bmatrix}$$

PROBLEM SET 1.6

1. If $A = \begin{bmatrix} 1 & 3 & 0 \\ -1 & 2 & 1 \\ 0 & 0 & 2 \end{bmatrix}$ and $B = \begin{bmatrix} 2 & 3 & 4 \\ 1 & 2 & 3 \\ -1 & 1 & 2 \end{bmatrix}$. Find AB and BA and show that $AB \neq BA$.

2. If $A = \begin{bmatrix} 0 & 1 & 2 \\ 1 & 2 & 3 \\ 2 & 3 & 4 \end{bmatrix}$ and $B = \begin{bmatrix} 1 & -2 \\ -1 & 0 \\ 2 & -1 \end{bmatrix}$. Find AB and explain why BA is not defined.

3. If $A = \begin{bmatrix} 1 & 2 \\ 3 & 4 \end{bmatrix}$, $B = \begin{bmatrix} 4 & 3 \\ 0 & 1 \end{bmatrix}$ and $C = \begin{bmatrix} -1 & 2 \\ 0 & 7 \end{bmatrix}$.

Verify $(AB)\,C = A\,(BC)$ and $A\,(B + C) = AB + AC$.

4. If $A = \begin{bmatrix} 2 & 3 \\ 3 & 10 \end{bmatrix}$. Show that $(2I - A)\,(10\,I - A) = 9I$.

5. If $A = \begin{bmatrix} 3 & 1 \\ -1 & 2 \end{bmatrix}$, show that $A^2 - 5A + 7I = 0$.

6. If $A = \begin{bmatrix} 0 & 1 \\ 1 & 0 \end{bmatrix}$, $B = \begin{bmatrix} 0 & -i \\ i & 0 \end{bmatrix}$, where $i = \sqrt{-1}$. Show that $(A + B)^2 = A^2 + B^2$.

7. If $A = \begin{bmatrix} \cos\theta & \sin\theta \\ -\sin\theta & \cos\theta \end{bmatrix}$, show that $A^n = \begin{bmatrix} \cos n\theta & \sin n\theta \\ -\sin n\theta & \cos n\theta \end{bmatrix}$, where n is a positive integer.

8. If $A = \begin{bmatrix} 1 & -2 & 3 \\ 2 & 3 & -1 \\ -3 & 1 & 2 \end{bmatrix}$, then evaluate $A^2 - 3A + 9I$.

9. If A and B are square matrices of order p, prove that

$$(A + B)^2 = A^2 + AB + BA + B^2 \neq A^2 + B^2 + 2AB.$$

10. If $A = \begin{bmatrix} i & 0 \\ 0 & -i \end{bmatrix}$, $B = \begin{bmatrix} 0 & -1 \\ 1 & 0 \end{bmatrix}$, $C = \begin{bmatrix} 0 & i \\ i & 0 \end{bmatrix}$, verify

(i) $A^2 = B^2 = C^2 = -I$ (ii) $AB = -BA = -C$.

11. Let $A = \begin{bmatrix} 2 & 3 \\ -1 & 2 \end{bmatrix}$. If $f(x) = x^2 - 4x + 7$ find $f(A)$.

12. A firm produces three products A, B and C which it sells in two markets. Annual sales in units are given below :

Markets	Units sold		
	A	B	C
I	8,000	4,000	16,000
II	7,000	18,000	9,000

If the price per unit of A, B and C is Rs. 2.50, Rs. 1·25 and Rs. 1·50 respectively and the costs per unit are Rs. 1·70, Rs. 1·20 and Rs. 0·80 respectively. Find the total profit in each market by using matrix algebra.

13. If $A = \begin{bmatrix} 0 & 1 \\ 0 & 0 \end{bmatrix}$, prove that $(aI + bA)^n = a^nI + na^{-1}bA$ where I is the unit matrix of order 2, n is a positive integer and a, b are arbitrary saclars.

14. If $A = \begin{bmatrix} 1 & 2 \\ 2 & 1 \end{bmatrix}$, show that $A^2 - 3I = 2A$.

15. If $A = \begin{bmatrix} 1 & 0 & -2 \\ 2 & 2 & 4 \\ 0 & 0 & 2 \end{bmatrix}$, verify that $A^2 - 3A + 2I = 0$.

16. Find ABC if $A = [x \quad y \quad z]$, $B = \begin{bmatrix} a & h & g \\ h & b & f \\ g & f & c \end{bmatrix}$, $C = \begin{bmatrix} x \\ y \\ z \end{bmatrix}$.

17. Prove that $A = \begin{bmatrix} a & b \\ c & d \end{bmatrix}$ satisfies the relation $A^2 - (a + d)A - (bc - ad)I = 0$.

I is the unit matrix of order 2.

ANSWERS

1. $AB = \begin{bmatrix} 5 & 9 & 13 \\ -1 & 2 & 4 \\ -2 & 2 & 4 \end{bmatrix}$, $BA = \begin{bmatrix} -1 & 12 & 11 \\ -1 & 7 & 8 \\ -2 & -1 & 5 \end{bmatrix}$

2. $AB = \begin{bmatrix} 3 & -2 \\ 5 & -5 \\ 7 & -8 \end{bmatrix}$, BA does not exist because the order of 3 is 3×2 and the order of A is 3×3 so the multiplication does not exist.

8. $\begin{bmatrix} -6 & 1 & 1 \\ 5 & 4 & 2 \\ 2 & 8 & -6 \end{bmatrix}$ **11.** $\begin{bmatrix} 0 & 0 \\ 0 & 0 \end{bmatrix}$ **12.** $A = \begin{bmatrix} 1 & 0 \\ 0 & 0 \end{bmatrix}, B = \begin{bmatrix} 0 & 0 \\ 0 & 1 \end{bmatrix}$

16. $ax^2 + by^2 + cz^2 + 2hxy + 2fyz + 2gzx$.

1.12 TRANSPOSE OF A MATRIX

Consider a matrix A of order $p \times q$ *i.e.*, (p rows and q columns). If we interchange the rows and columns of given matrix A, then we obtain a new matrix of order $q \times p$ which is known as *transpose* of matrix A and is usually denoted by $\overline{A}$ or A^T or A'.

Thus, if $A = [a_{ij}]_{p \times q}$, then $A' = [a_{ji}]_{q \times p}$

Example 1.61

$$A = \begin{bmatrix} 2 & 0 & 3 \\ 4 & 5 & 0 \end{bmatrix}_{2 \times 3} \text{ then } A' = \begin{bmatrix} 2 & 4 \\ 0 & 5 \\ 3 & 0 \end{bmatrix}_{3 \times 2}$$

1.12.1 Properties of Transpose of A Matrix

(i) $(A')' = A$, *i.e.*, the transpose of the transpose of a matrix A is equal to the matrix A itself.

(ii) $(A + B)' = A' + B'$, *i.e.*, the transpose of the sum (or difference) of matrices is equal to the sum (or difference) of their individual transposes.

(iii) If k is any number and A is any matrix of order $p \times q$, then $(kA)' = kA'$.

(iv) $(AB)' = B'A'$, *i.e.*, the transpose of the product of conformable matrices is equal to the product of the transpose of the individual matrices taken in the reverse order.

Example 1.62

If $A = \begin{bmatrix} 3 & 4 \\ 0 & 2 \end{bmatrix}$ and $B = \begin{bmatrix} -1 & 2 \\ 3 & 5 \end{bmatrix}$, then show that

(i) $(A + B)' = A' + B'$ (ii) $(AB)' = B'A'$

(iii) $(3A)' = 3A'$ (iv) $(A')' = A$ and $(B')' = B$.

SOLUTION : Here $A = \begin{bmatrix} 3 & 4 \\ 0 & 2 \end{bmatrix} \Rightarrow A' = \begin{bmatrix} 3 & 0 \\ 4 & 2 \end{bmatrix}$

and $B = \begin{bmatrix} -1 & 2 \\ 3 & 5 \end{bmatrix} \Rightarrow B' = \begin{bmatrix} -1 & 3 \\ 2 & 5 \end{bmatrix}$

(i) $A + B = \begin{bmatrix} 2 & 6 \\ 3 & 7 \end{bmatrix} \Rightarrow (A + B)' = \begin{bmatrix} 2 & 3 \\ 6 & 7 \end{bmatrix}$

Again, $A' + B' = \begin{bmatrix} 2 & 3 \\ 6 & 7 \end{bmatrix}$

So, $(A+B)' = A' + B'$

(ii) Now, $AB = \begin{bmatrix} -3+12 & 6+20 \\ 0+6 & 0+10 \end{bmatrix} = \begin{bmatrix} 9 & 26 \\ 6 & 10 \end{bmatrix}$

$$(AB)' = \begin{bmatrix} 9 & 6 \\ 26 & 10 \end{bmatrix}$$

$$B'A' = \begin{bmatrix} -3+12 & 0+6 \\ 6+20 & 0+10 \end{bmatrix} = \begin{bmatrix} 9 & 6 \\ 26 & 10 \end{bmatrix}$$

Thus, $(AB)' = B'A'$

(iii) Here, $3A = \begin{bmatrix} 9 & 12 \\ 0 & 6 \end{bmatrix} \Rightarrow (3A)' = \begin{bmatrix} 9 & 0 \\ 12 & 6 \end{bmatrix}$

Again, $3A' = \begin{bmatrix} 9 & 0 \\ 12 & 6 \end{bmatrix}$

Hence, $(3A)' = 3A'$

(iv) Here $(A')' = \begin{bmatrix} 3 & 4 \\ 0 & 2 \end{bmatrix} = A$

and $(B')' = \begin{bmatrix} -1 & 2 \\ 3 & 5 \end{bmatrix} = B$

1.13 SOME IMPORTENT MATRICES

1.13.1 Symmetric Matrix

A square matrix A is said to be a symmetric matrix if $A' = A$, *i.e.*, the transpose of matrix A is equal to the matrix itself.

Example 1.63

For $A = \begin{bmatrix} 3 & 5 & 7 \\ 5 & 2 & 9 \\ 7 & 9 & 1 \end{bmatrix}$, $A' = \begin{bmatrix} 3 & 5 & 7 \\ 5 & 2 & 9 \\ 7 & 9 & 1 \end{bmatrix} = A$. Thus, A is a symmetric matrix.

1.13.2 Skew-Symmetric Matrix

A square matrix A is said to be a skew-symmetric matrix if $A' = -A$.

Example 1.64

For $A = \begin{bmatrix} 0 & 2 & 9 \\ -2 & 0 & 7 \\ -9 & -7 & 0 \end{bmatrix}$, $A' = \begin{bmatrix} 0 & -2 & -9 \\ 2 & 0 & -7 \\ 9 & 7 & 0 \end{bmatrix} = -A$

Thus, A is skew-symmetric matrix.

■ **NOTE :** In a skew-symmetric matrix all the diagonal elements must be zero.

1.13.3 Orthogonal Matrix

A square matrix A is said to be orthogonal matrix if $AA' = I = A'A$; in other words a square matrix A is orthogonal if the product of matrix A and its transpose is equal to the identify (unit) matrix.

Example 1.65

Show that the matrix $A = \begin{bmatrix} \cos\theta & 0 & \sin\theta \\ 0 & 1 & 0 \\ -\sin\theta & 0 & \cos\theta \end{bmatrix}$ is orthogonal.

SOLUTION : Here, $A = \begin{bmatrix} \cos\theta & 0 & \sin\theta \\ 0 & 1 & 0 \\ -\sin\theta & 0 & \cos\theta \end{bmatrix}$

$\Rightarrow$ $A' = \begin{bmatrix} \cos\theta & 0 & -\sin\theta \\ 0 & 1 & 0 \\ \sin\theta & 0 & \cos\theta \end{bmatrix}$

Now, $AA' = \begin{bmatrix} \cos^2\theta + \sin^2\theta & 0 & -\cos\theta\sin\theta + \sin\theta\cos\theta \\ 0 & 1 & 0 \\ -\sin\theta\cos\theta + \sin\theta\cos\theta & 0 & \sin^2\theta + \cos^2\theta \end{bmatrix}$

$$= \begin{bmatrix} 1 & 0 & 0 \\ 0 & 1 & 0 \\ 0 & 0 & 1 \end{bmatrix} = I$$

Hence, A is orthogonal matrix.

1.13.4 Nilpotent Matrix

A square matrix is *Nilpotent* if $A^2 = O$.

Example 1.66

Show that martix

$A = \begin{bmatrix} ab & b^2 \\ -a^2 & -ab \end{bmatrix}$ is nilpotent of order 2.

SOLUTION : We know that matrix A is nilpotent if $A^2 = O$.

Now, $A^2 = \begin{bmatrix} ab & b^2 \\ -a^2 & -ab \end{bmatrix}\begin{bmatrix} ab & b^2 \\ -a^2 & -ab \end{bmatrix}$

$$= \begin{bmatrix} a^2b^2 - a^2b^2 & ab^3 - ab^3 \\ -a^3b + a^3b & -a^2b^2 + a^2b^2 \end{bmatrix}$$

$$= \begin{bmatrix} 0 & 0 \\ 0 & 0 \end{bmatrix}$$

Hence, A is nilpotent of order 2.

1.13.5 Involutary Matrix

A square matrix A is said to be *involutary* matrix if $A^2 = I$.

Example 1.67

Show that matrix

$A = \begin{bmatrix} -5 & -8 & 0 \\ 3 & 5 & 0 \\ 1 & 2 & -1 \end{bmatrix}$ is an involutary matrix.

SOLUTION : We know that matrix A is involutary if $A^2 = I$.

Now, $A^2 = \begin{bmatrix} -5 & -8 & 0 \\ 3 & 5 & 0 \\ 1 & 2 & -1 \end{bmatrix} \begin{bmatrix} -5 & -8 & 0 \\ 3 & 5 & 0 \\ 1 & 2 & -1 \end{bmatrix}$

$$= \begin{bmatrix} 25-24+0 & 40-40+0 & 0+0+0 \\ -15+15+0 & -24+25+0 & 0+0+0 \\ -5+6-1 & -8+10-2 & 0+0+1 \end{bmatrix}$$

$$= \begin{bmatrix} 1 & 0 & 0 \\ 0 & 1 & 0 \\ 0 & 0 & 1 \end{bmatrix} = I$$

So, A is an involutary matrix.

1.13.6 Idempotent Matrix

A square matrix A is idempotent matrix if $A^2 = A$.

Example 1.68

Show that the matrix $A = \begin{bmatrix} 2 & -2 & -4 \\ -1 & 3 & 4 \\ 1 & -2 & -3 \end{bmatrix}$ is an idempotent matrix.

SOLUTION : We know that matrix A is idempotent if $A^2 = A$.

Now, $A^2 = \begin{bmatrix} 2 & -2 & -4 \\ -1 & 3 & 4 \\ 1 & -2 & -3 \end{bmatrix} \begin{bmatrix} 2 & -2 & -4 \\ -1 & 3 & 4 \\ 1 & -2 & -3 \end{bmatrix}$

$$= \begin{bmatrix} 4+2-4 & -4-6+8 & -8-8+12 \\ -2-3+4 & 2+9-8 & 4+12-12 \\ 2+2-3 & -2-6+6 & -4-8+9 \end{bmatrix}$$

$$= \begin{bmatrix} 2 & -2 & -4 \\ -1 & 3 & 4 \\ 1 & -2 & -3 \end{bmatrix}$$

$$= A$$

Hence, A is an idempotent matrix.

THEOREM 1.1 Show that every square matrix can be uniquely expressed as the sum of a symmetric and a skew-symmetric matrix.

PROOF : Let A be any square matrix and A' be its transpose, then

$$A = \frac{1}{2}\,[(A + A') + (A - A')] = P + Q \text{ (say)}$$

where $P = \dfrac{A + A'}{2}$ and $Q = \dfrac{A - A'}{2}$

Now, we have to show that P is symmetric matrix and Q is skew-symmetric matrix.

Here, $\quad P = \dfrac{A + A'}{2} \Rightarrow P' = \left(\dfrac{A + A'}{2}\right)' = \dfrac{1}{2}\,[A' + (A')'] = \dfrac{1}{2}\,(A' + A) = P$

Hence, P is symmetric matrix.

Also, $\quad Q = \dfrac{A - A'}{2} \Rightarrow Q' = \dfrac{1}{2}\,(A - A')' = \dfrac{1}{2}\,[A' - (A')']$

$$= \frac{1}{2}\,(A' - A) = \frac{-1}{2}\,(A - A') = -Q$$

Hence, Q is skew-symmetric matrix.

Thus, any square matrix can be expressed as a sum of a symmetric and skew-symmetric matrix.

Example 1.69

Express the following matrix as the sum of a symmetric and skew-symmetic matrix

$$A = \begin{bmatrix} -1 & 7 & 1 \\ 2 & 3 & 4 \\ 5 & 0 & 5 \end{bmatrix}.$$

SOLUTION : Let $A = P + Q$ where, $P = \frac{1}{2}(A + A')$ and $Q = \frac{1}{2}(A - A')$.

Now, we have to show that P is symmetric and Q is skew-symmetric matrix.

Here, $A = \begin{bmatrix} -1 & 7 & 1 \\ 2 & 3 & 4 \\ 5 & 0 & 5 \end{bmatrix} \Rightarrow A' = \begin{bmatrix} -1 & 2 & 5 \\ 7 & 3 & 0 \\ 1 & 4 & 5 \end{bmatrix}$

Then, $P = \dfrac{A + A'}{2} = \dfrac{1}{2}\begin{bmatrix} -2 & 9 & 6 \\ 9 & 6 & 4 \\ 6 & 4 & 10 \end{bmatrix}$ and $= \dfrac{A - A'}{2} = \dfrac{1}{2}\begin{bmatrix} 0 & 5 & -4 \\ -5 & 0 & 4 \\ 4 & -4 & 0 \end{bmatrix}$

Now $P' = \dfrac{1}{2}\begin{bmatrix} -2 & 9 & 6 \\ 9 & 6 & 4 \\ 6 & 4 & 10 \end{bmatrix} = P$ and $Q' = \dfrac{1}{2}\begin{bmatrix} 0 & -5 & 4 \\ 5 & 0 & -4 \\ -4 & 4 & 0 \end{bmatrix} = Q$

Thus, P is symmetric and Q is skew symmetric matrix.

Hence, $A = \dfrac{1}{2}\begin{bmatrix} -2 & 9 & 6 \\ 9 & 6 & 4 \\ 6 & 4 & 10 \end{bmatrix} + \dfrac{1}{2}\begin{bmatrix} 0 & -5 & 4 \\ 5 & 0 & -4 \\ -4 & 4 & 0 \end{bmatrix}$

PROBLEM SET 1.7

1. If $A = \begin{bmatrix} 1 & 5 & 0 \\ 4 & 7 & 9 \\ -1 & 2 & 7 \end{bmatrix}$ and $B = \begin{bmatrix} 3 & 2 & 5 \\ -1 & 2 & 9 \\ -2 & -3 & 4 \end{bmatrix}$ verify that $(AB)' = B'A'$.
2. If A and B are symmetric matrices, prove that $AB + BA$ is symmetric and $AB - BA$ is skew-symmetric matrix.
3. Express the following matrices as a sum of symmetric and skew-symmetric matrix.

   (i) $\begin{bmatrix} 3 & 0 & 4 \\ 4 & 8 & 5 \\ -1 & 2 & 9 \end{bmatrix}$ (ii) $\begin{bmatrix} 2 & 9 & 7 \\ 3 & 4 & 0 \\ 5 & 7 & 9 \end{bmatrix}$ (iii) $\begin{bmatrix} 1 & 2 \\ 7 & 9 \end{bmatrix}$.
4. Show that A is involutary iff $(I + A)(I - A) = O$.
5. Prove that if A is an orthogonal matrix, than A' is also an orthogonal matrix.

   HINT : [A is orthogonal $\Rightarrow AA' = I$. Now $A'(A')' = A'A = I \Rightarrow A$ is orthogonal]

6. Show that the matrix $A = \begin{bmatrix} 1 & 2 & 3 \\ 1 & 2 & 3 \\ -1 & -2 & -3 \end{bmatrix}$ is nilpotent matrix.

7. Show that if A is an $m \times n$ matrix, then AA' is symmetric.

8. If $A = \begin{bmatrix} 3 & 4 \\ 6 & 2 \end{bmatrix}$, $B = \begin{bmatrix} 7 & 8 \\ 4 & 3 \end{bmatrix}$. Show that $(A+B)' = A' + B'$

9. If $A = \begin{bmatrix} 1 & -2 & 3 \\ -4 & 2 & 5 \end{bmatrix}$, $B = \begin{bmatrix} 1 & 3 \\ -1 & 0 \\ 2 & 4 \end{bmatrix}$

Show that $(AB)' = B'A'$

10. If $A = \begin{bmatrix} 1 & 1 & 2 \\ 2 & 1 & 0 \end{bmatrix}$ and $B = \begin{bmatrix} 1 & 2 \\ 1 & 1 \\ 2 & 0 \end{bmatrix}$

Show that $(AB)' = B'A'$.

ANSWERS

3. (i) $P = \frac{1}{2}\begin{bmatrix} 6 & 4 & 3 \\ 4 & 16 & 7 \\ 3 & 7 & 18 \end{bmatrix}$, $Q = \frac{1}{2}\begin{bmatrix} 0 & -4 & 5 \\ 4 & 0 & 3 \\ -5 & -3 & 0 \end{bmatrix}$

(ii) $P = \frac{1}{2}\begin{bmatrix} 4 & 12 & 12 \\ 12 & 8 & 7 \\ 12 & 7 & 18 \end{bmatrix}$, $Q = \frac{1}{2}\begin{bmatrix} 0 & 6 & 2 \\ -6 & 0 & -7 \\ -2 & 7 & 0 \end{bmatrix}$

(iii) $P = \frac{1}{2}\begin{bmatrix} 2 & 9 \\ 9 & 18 \end{bmatrix}$, $Q = \frac{1}{2}\begin{bmatrix} 0 & -5 \\ 5 & 0 \end{bmatrix}$

1.14 COMPLEX MATRIX

1.14.1 Conjugate of A Matrix

Let $A = \begin{bmatrix} 1-3i & 2+4i & i \\ 0 & -1-i & 4 \end{bmatrix}$

Conjugate of matrix A is denoted by $\overline{A}$ and defined by

$$\overline{A} = \begin{bmatrix} 1+3i & 2-4i & -i \\ 0 & -1+i & 4 \end{bmatrix}$$

■ **NOTE :** To obtain the conjugate of given matrix, just replace the complex numbers by their complex conjugate, *i.e.*, replace i by $-i$.

1.14.2 Matrix A^{θ}

Let A be any matrix having at least one element as complex number. The conjugate of the transpose of matrix A or the transpose of the conjugate of matrix A is A^θ. Another notation for A^θ is $(\overline{A})'$ or $\overline{(A')}$.

Example 1.70

Let $A = \begin{bmatrix} 1-i & 2+4i & 5 \\ 0 & 7 & 3-2i \\ i & -i & 4 \end{bmatrix}$

Then $\overline{A} = \begin{bmatrix} 1+i & 2-4i & 5 \\ 0 & 7 & 3+2i \\ -i & i & 4 \end{bmatrix}$

$$(\overline{A})' = \begin{bmatrix} 1+i & 0 & -i \\ 2-4i & 7 & i \\ 5 & 3+2i & 4 \end{bmatrix} = A^\theta$$

or

$$A' = \begin{bmatrix} 1-i & 0 & i \\ 2+4i & 7 & -i \\ 5 & 3-2i & 4 \end{bmatrix}$$

$$\overline{(A')} = \begin{bmatrix} 1+i & 0 & -i \\ 2-4i & 7 & i \\ 5 & 3+2i & 4 \end{bmatrix} = A^\theta$$

1.14.3 Hermitian Matrix

A square matrix A is said to be Hermitian matrix if $A = A^\theta$, *i.e.*, if the conjgate transpose of a square matrix is equal to the matrix itself, then the matrix is Hermitian matrix

Example 1.71

Let $A = \begin{bmatrix} 5 & 3-2i & 4-i \\ 3+2i & 4 & 7 \\ 4+i & 7 & 2 \end{bmatrix}$

then $\overline{A} = \begin{bmatrix} 5 & 3+2i & 4+i \\ 3-2i & 4 & 7 \\ 4-i & 7 & 2 \end{bmatrix}$ and $(\overline{A})' = \begin{bmatrix} 5 & 3-2i & 4-i \\ 3+2i & 4 & 7 \\ 4+i & 7 & 2 \end{bmatrix} = A$

Clearly, $(\overline{A})' = A \Rightarrow A$ is Hermitian matrix.

1.14.4 Skew-Hermitian Matrix

A square matrix A is said to be skew Hermitian matrix if $A^\theta = -A$, *i.e.*, if the conjugate transpose of square matrix A is equal to negative of the matrix itself then matrix is skew Hermitian matrix.

Example 1.72

Let $A = \begin{bmatrix} i & 3-2i & 4-i \\ -(3+2i) & 4i & 2+i \\ -(4+i) & -(2-i) & 0 \end{bmatrix}$

then $\bar{A} = \begin{bmatrix} -i & 3+2i & 4+i \\ -(3-2i) & -4i & 2-i \\ -(4-i) & -(2+i) & 0 \end{bmatrix}$

and $(\bar{A})' = \begin{bmatrix} i & -(3-2i) & -(4-i) \\ 3+2i & -4i & -(2+i) \\ 4+i & 2-i & 0 \end{bmatrix} = -\begin{bmatrix} i & 3-2i & 4-i \\ -(3+2i) & 4i & 2+i \\ -(4+i) & -(2-i) & 0 \end{bmatrix} = -A$

Clearly, $(\bar{A})' = -A \Rightarrow A$ is skew Hermitian matrix.

THEOREM 1.2 **Show that every square matrix can be uniquely expressed as the sum of Hermitian and skew Hermitian matrix.**

SOLUTION : Suppose that A is any square matrix, than

$$A = \frac{1}{2}[(A + A^\theta) + (A - A^\theta)] = P + Q$$

where $$P = \frac{1}{2}(A + A^\theta) \text{ and } Q = \frac{1}{2}(A - A^\theta)$$

Now, $$P^\theta = \frac{1}{2}(A + A^\theta)^\theta = \frac{1}{2}[(A^\theta + (A^\theta)^\theta] = \frac{1}{2}[A^\theta + A] = P$$

$\Rightarrow$ P is Hermitian matrix.

Also $$Q^\theta = \frac{1}{2}[A - A^\theta]^\theta = \frac{1}{2}[A^\theta - (A^\theta)^\theta] = \frac{1}{2}(A^\theta - A) = -\frac{1}{2}(A - A^\theta)$$

$$= -Q$$

$\Rightarrow$ Q is skew-hermition matrix.

Therefore, every square matrix can be expressed as sum of Hermitian and skew Hermitian matrix.

1.14.5 Unitary Matrix

Any square matrix A is said to be unitary if $A^\theta A = I$, *i.e.*, the product of matrix A and its conjugate transpose is equal to identity matrix.

Example 1.73

Let $$A = \begin{bmatrix} \frac{1-i}{2} & \frac{-1-i}{2} \\ \frac{1-i}{2} & \frac{1+i}{2} \end{bmatrix} \Rightarrow A^\theta = \begin{bmatrix} \frac{1+i}{2} & \frac{1+i}{2} \\ \frac{-1+i}{2} & \frac{1-i}{2} \end{bmatrix}$$

$$AA^\theta = \begin{bmatrix} \frac{1-i}{2} & \frac{-1-i}{2} \\ \frac{1-i}{2} & \frac{1+i}{2} \end{bmatrix} \begin{bmatrix} \frac{1+i}{2} & \frac{1+i}{2} \\ \frac{-1+i}{2} & \frac{1-i}{2} \end{bmatrix} = \begin{bmatrix} 1 & 0 \\ 0 & 1 \end{bmatrix} = I$$

$\Rightarrow$ A is unitary matrix.

■ **NOTE :** **(1)** A square matrix is skew Hermitian if all diagonal elements of the matrix are either zero or pure imaginary number.

(2) If all elements of square matrix A are real numbers, then an Hermitian matrix is same as a symmetric matrix and a skew Hermitian matrix is equal to a skew symmetric matrix.

Example 1.74

If A is an Hermitian matrix, then show that iA is skew Hermitian.

SOLUTION : If A is Hermitian matrix, then $A^\theta = A$ (by definition).

Now $(iA)^\theta = (i)^\theta A^\theta = -iA^\theta = -(iA)$

$\Rightarrow$ iA is skew Hermitian.

Example 1.75

If A is skew-Hermitian matrix, then show that iA is Hermitian matrix.

SOLUTION : If A is skew Hermitian matrix, then $A^\theta = -A$

Now, $(iA)^\theta = i^\theta A^\theta = -iA^\theta = (-i)(-A) = iA$

$\Rightarrow$ iA is an Hermitian matrix.

Example 1.76

If $A = \begin{bmatrix} 0 & 1+2i \\ -1+2i & 0 \end{bmatrix}$ obtain the matrix $(I - A)(I + A)^{-1}$ and show that it is unitary.

SOLUTION : We have $A = \begin{bmatrix} 0 & 1+2i \\ -1+2i & 0 \end{bmatrix}$

Now, $$I - A = \begin{bmatrix} 1 & 0 \\ 0 & 1 \end{bmatrix} - \begin{bmatrix} 0 & 1+2i \\ -1+2i & 0 \end{bmatrix} = \begin{bmatrix} 1 & -1-2i \\ 1-2i & 1 \end{bmatrix} \quad \text{...(1)}$$

$$I + A = \begin{bmatrix} 1 & 0 \\ 0 & 1 \end{bmatrix} + \begin{bmatrix} 0 & 1+2i \\ -1+2i & 0 \end{bmatrix} = \begin{bmatrix} 1 & 1+2i \\ -1+2i & 1 \end{bmatrix} \quad \text{...(2)}$$

$$|I + A| = \begin{vmatrix} 1 & 1+2i \\ -1+2i & 1 \end{vmatrix} = 1 - (1+2i)(-1+2i) = 1 - (-1 - 2i + 2i + 4i^2)$$

$$= 1 - (-1 - 4) = 1 + 5 = 6$$

Now, $$\text{Adj. } (I + A) = \begin{bmatrix} 1 & -1-2i \\ 1-2i & 1 \end{bmatrix}$$

$\therefore$ $$(I + A)^{-1} = \frac{\text{Adj. } (I + A)}{|I + A|} = \frac{1}{6}\begin{bmatrix} 1 & -1-2i \\ 1-2i & 1 \end{bmatrix}$$

Now, $$(I - A)(I + A)^{-1} = \frac{1}{6}\begin{bmatrix} 1 & -1-2i \\ 1-2i & 1 \end{bmatrix}\begin{bmatrix} 1 & -1-2i \\ 1-2i & 1 \end{bmatrix}$$

$$= \frac{1}{6}\begin{bmatrix} -4 & -2-4i \\ 2-4i & -4 \end{bmatrix} = T \text{ (let)}$$

Now, $$T' = \frac{1}{6}\begin{bmatrix} -4 & 2-4i \\ -2-4i & -4 \end{bmatrix}$$

$$(\overline{T'}) = T^\theta = \frac{1}{6}\begin{bmatrix} -4 & 2+4i \\ -2+4i & -4 \end{bmatrix}$$

and $$TT^\theta = \frac{1}{36}\begin{bmatrix} -4 & 2+4i \\ -2+4i & -4 \end{bmatrix}\begin{bmatrix} -4 & -2-4i \\ 2-4i & -4 \end{bmatrix} = \begin{bmatrix} 1 & 0 \\ 0 & 1 \end{bmatrix} = I$$

Thus, $TT^\theta = I$ $\Rightarrow$ T is unitary matrix.

PROBLEM SET 1.8

1. Show that the matrix $\begin{bmatrix} 4 & 2-4i & 3+2i \\ 2+4i & 0 & i \\ 3-2i & -i & 1 \end{bmatrix}$ is Hermitian.

2. Express $\begin{bmatrix} i & 2-3i & 4+5i \\ 6+i & 0 & 4-5i \\ -i & 2-i & 2+i \end{bmatrix}$ as the sum of Hermitian and skew Hermitian matrix.

3. Show that the matrix $\begin{bmatrix} 0 & 2+5i & -3-2i \\ 2-5i & i & 2 \\ -3+2i & 2 & -i \end{bmatrix}$ is skew Hermitain.

4. If A is any square matrix, prove that $A + A^{\theta}$, AA^{θ} are all Hermitian and $A - A^{\theta}$ is skew-Hermitian.

5. If $A = \begin{bmatrix} 3 & 5+2i & -3 \\ 5-2i & 7 & 4i \\ -3 & -4i & 5 \end{bmatrix}$, show that A is Hermitian matrix. Verify that iA is skew Hermitian.

6. If A is a unitary matrix, then show that $\overline{A}$ and A^{θ} are also unitary.

7. If $A = \begin{bmatrix} 2+i & 3 & -1+3i \\ -5 & i & 4-2i \end{bmatrix}$, verify that $A^{\theta}A$ is Hermitian matrix.

8. Define unitary matrix. Show that the matrix

$$A = \begin{bmatrix} \frac{1}{\sqrt{3}} & \frac{1+i}{\sqrt{3}} \\ \frac{1-i}{\sqrt{3}} & \frac{-1}{\sqrt{3}} \end{bmatrix} \text{ is unitary.}$$

9. Show that the matrix $\begin{bmatrix} a+ic & -b+id \\ b+id & a-ic \end{bmatrix}$ is unitary iff $a^2 + b^2 + c^2 + d^2 = 1$.

ANSWERS

2. $P = \begin{bmatrix} 0 & 4-2i & 2+3i \\ 4+2i & 0 & 3-2i \\ 2-3i & 3+2i & 2 \end{bmatrix}$,

$Q = \begin{bmatrix} i & -2-i & 2+2i \\ 2-i & 0 & 1-3i \\ -2+2i & -1-3i & i \end{bmatrix}$

1.15 DETERMINANTS

■ **1.15.1 DEFINITION :** *The determinant of any square matrix A is denoted by |A| or det. A and is defined as a scalar quantity obtained by expanding the elements of A with respect to the elements of any row or column.*

1.15.2 Minors

The minor of any element a_{ij} of a square matrix A is the determinent obtained by removing (deleting) the row and the column in which that element is lying and is denoted by M_{ij}.

For example, if $A = \begin{bmatrix} a_{11} & a_{12} & a_{13} \\ a_{21} & a_{22} & a_{23} \\ a_{31} & a_{32} & a_{33} \end{bmatrix}$

Then M_{11} = minor of a_{11} = determinent obtained by deleting first row and first column from A.

$$= \begin{vmatrix} a_{22} & a_{23} \\ a_{32} & a_{33} \end{vmatrix}.$$

Similarly, M_{23} = minor of a_{23}, obtained by deleting second row and third column from

$$A = \begin{vmatrix} a_{11} & a_{12} \\ a_{31} & a_{32} \end{vmatrix} \text{ and so on.}$$

Example 1.77

If $A = \begin{bmatrix} 2 & 3 & 4 \\ 5 & 6 & 7 \\ 8 & 9 & 10 \end{bmatrix}$, obtain (i) M_{33} (ii) M_{23}.

SOLUTION : (i) $M_{33} = \begin{vmatrix} 2 & 3 \\ 5 & 6 \end{vmatrix}$ (ii) $M_{23} = \begin{vmatrix} 2 & 3 \\ 8 & 9 \end{vmatrix}$

1.15.3 Co-Factors

The co-factor of an element a_{ij} of any square matrix A is denoted by C_{ij} and is obtained by the relation

$$C_{ij} = (-1)^{i+j} M_{ij}$$

or

$$C_{ij} = \begin{cases} -M_{ij} & \text{if } i+j \text{ is odd} \\ M_{ij} & \text{if } i+j \text{ is even} \end{cases}$$

For example, if $A = \begin{bmatrix} a_{11} & a_{12} & a_{13} \\ a_{21} & a_{22} & a_{23} \\ a_{31} & a_{32} & a_{33} \end{bmatrix}$

then, $C_{11} = (-1)^{1+1} M_{11} = M_{11} = \begin{vmatrix} a_{22} & a_{23} \\ a_{32} & a_{33} \end{vmatrix}$

1.15.4 Rule for Expansion of Determinant

(1) To expand determinant of square matrix A with respect to elements of row (or column) we multiply each element of the first row (or column) by that determinant which is obtained after deleting the row and column to which that element belongs. Starting from the first element, the sign of the products being taken alternately positive and negative. We can find the value of a determinant of order 3 by expanding it along any of its row or column.

(2) The value of the determinant of square matrix A is also obtained by the sum of the products of the elements of any row or column and their corresponding co-factors.

For example,

(i) If $A = \begin{bmatrix} a_{11} & a_{12} \\ a_{21} & a_{22} \end{bmatrix}$, then $|A| = \begin{vmatrix} a_{11} & a_{12} \\ a_{21} & a_{22} \end{vmatrix} = a_{11}\,a_{22} - a_{21}\,a_{12}$

(ii) If $A = \begin{bmatrix} a_{11} & a_{12} & a_{13} \\ a_{21} & a_{22} & a_{23} \\ a_{31} & a_{32} & a_{33} \end{bmatrix}$, then the expansion of the determinant along the first row is as follows :

$$|A| = a_{11}\begin{vmatrix} a_{22} & a_{23} \\ a_{32} & a_{33} \end{vmatrix} - a_{12}\begin{vmatrix} a_{21} & a_{23} \\ a_{31} & a_{33} \end{vmatrix} + a_{13}\begin{vmatrix} a_{21} & a_{22} \\ a_{31} & a_{32} \end{vmatrix}$$

$$= a_{11}\,(a_{22}\,a_{33} - a_{32}\,a_{23}) - a_{12}\,(a_{21}\,a_{33} - a_{23}\,a_{31}) + a_{13}\,(a_{21}\,a_{32} - a_{31}\,a_{22})$$

Example 1.78

Find the determinant of the following matrix $A = \begin{bmatrix} 3 & 5 & 7 \\ 9 & 0 & 2 \\ 3 & 4 & 5 \end{bmatrix}$.

SOLUTION : $|A| = 3\begin{vmatrix} 0 & 2 \\ 4 & 5 \end{vmatrix} - 5\begin{vmatrix} 9 & 2 \\ 3 & 5 \end{vmatrix} + 7\begin{vmatrix} 9 & 0 \\ 3 & 4 \end{vmatrix}$

$= 3\,(0 - 8) - 5\,(45 - 6) + 7\,(36 - 0)$

$= -\,24 - 195 + 252$

$= 33$

1.15.5 Difference between Matrix and Determinant

(i) A matrix has no numerical value while a determinant has a numerical value.

(ii) A matrix can be square as well as rectangular while a determinant can not be rectangular.

(iii) If we interchange rows (or columns) of a matrix, we obtain a different matrix while an interchange of rows (or columns) in a determinant gives the same determinant but with positive or negative sign.

Example 1.79

Find the co-factors of the elements of matrix $A = \begin{bmatrix} 9 & 2 & 7 \\ 5 & 7 & 6 \\ 2 & 4 & 3 \end{bmatrix}$.

SOLUTION : C_{11} = co-factor of 9

$$= (-1)^{1+1}\begin{vmatrix}7 & 6\\4 & 3\end{vmatrix} = 21 - 24 = -3$$

$$C_{12} = \text{co-factor of } 2 = (-1)^{1+2}\begin{vmatrix}5 & 6\\2 & 3\end{vmatrix} = -3$$

$$C_{13} = \text{co-factor of } 7 = (-1)^{1+3}\begin{vmatrix}5 & 7\\2 & 4\end{vmatrix} = 6$$

$$C_{21} = \text{co-factor of } 5 = (-1)^{2+1}\begin{vmatrix}2 & 7\\4 & 3\end{vmatrix} = 22$$

$$C_{22} = \text{co-factor of } 7 = (-1)^{2+2}\begin{vmatrix}9 & 7\\2 & 3\end{vmatrix} = 13$$

$$C_{23} = \text{co-factor of } 6 = (-1)^{2+3}\begin{vmatrix}9 & 2\\2 & 4\end{vmatrix} = -32$$

$$C_{31} = \text{co-factor of } 2 = (-1)^{3+1}\begin{vmatrix}2 & 7\\7 & 6\end{vmatrix} = -37$$

$$C_{32} = \text{co-factor of } 4 = (-1)^{3+2}\begin{vmatrix}9 & 7\\5 & 6\end{vmatrix} = -19$$

$$C_{33} = \text{co-factor of } 3 = (-1)^{3+3}\begin{vmatrix}9 & 2\\5 & 7\end{vmatrix} = 53$$

1.15.6 Properties of Determinants

(I) **If two adjacent rows or columns of a determinant are interchanged, the sign of the determinant is changed, but the numerical value remains same.**

For example,

$$\begin{vmatrix}a_{11} & a_{12} & a_{13}\\a_{21} & a_{22} & a_{23}\\a_{31} & a_{32} & a_{33}\end{vmatrix} = -\begin{vmatrix}a_{21} & a_{22} & a_{23}\\a_{11} & a_{12} & a_{13}\\a_{31} & a_{32} & a_{33}\end{vmatrix} \; R_1 \leftrightarrow R_2$$

■ **NOTE :** By repeated application of the above property, it is clear that sign of a determinant will or will not change if there are an odd or even number of interchanges in adjacent rows (or columns) respectively.

(II) **If A' is the transpose of square matrix A, then $|A| = |A'|$**

For example,

$$\begin{vmatrix}a_{11} & a_{12} & a_{13}\\a_{21} & a_{22} & a_{23}\\a_{31} & a_{32} & a_{33}\end{vmatrix} = -\begin{vmatrix}a_{11} & a_{21} & a_{31}\\a_{12} & a_{22} & a_{32}\\a_{13} & a_{23} & a_{33}\end{vmatrix}$$

(III) **If two rows or columns of a determinant are identical, then the value of the determinant is zero.**

For example,

$$\begin{vmatrix}a_{11} & a_{12} & a_{13}\\a_{11} & a_{12} & a_{13}\\a_{31} & a_{32} & a_{33}\end{vmatrix} = 0 \quad \text{or} \quad \begin{vmatrix}a_{11} & a_{11} & a_{13}\\a_{21} & a_{21} & a_{23}\\a_{31} & a_{31} & a_{33}\end{vmatrix} = 0$$

(IV) If every element in any row or column be multiplied by the same factor k, the determinant is multiplied by that factor. In other words, the value of new determinant is k times the value of original determinant.

For example,

$$\begin{vmatrix} ka_{11} & ka_{12} & ka_{13} \\ a_{21} & a_{22} & a_{23} \\ a_{31} & a_{32} & a_{33} \end{vmatrix} = k \begin{vmatrix} a_{11} & a_{12} & a_{13} \\ a_{21} & a_{22} & a_{23} \\ a_{31} & a_{32} & a_{33} \end{vmatrix}$$

or
$$\begin{vmatrix} ka_{11} & a_{12} & a_{13} \\ ka_{21} & a_{22} & a_{23} \\ ka_{31} & a_{32} & a_{33} \end{vmatrix} = k \begin{vmatrix} a_{11} & a_{12} & a_{13} \\ a_{21} & a_{22} & a_{23} \\ a_{31} & a_{32} & a_{33} \end{vmatrix}$$

■ **NOTE :** $\begin{vmatrix} k_1 a_{11} & k_2 a_{12} & k_3 a_{13} \\ k_1 a_{21} & k_2 a_{22} & k_3 a_{23} \\ k_1 a_{31} & k_2 a_{32} & k_3 a_{33} \end{vmatrix} = k_1 k_2 k_3 \begin{vmatrix} a_{11} & a_{12} & a_{13} \\ a_{21} & a_{22} & a_{23} \\ a_{31} & a_{32} & a_{33} \end{vmatrix}$

(V) If each element of any row or column is the sum (difference) of two quantities, then the determinant can be expressed as the sum (difference) of two determinants of the same order.

For example,

$$\begin{vmatrix} a_{11}+b_{11} & a_{12} & a_{13} \\ a_{21}+b_{21} & a_{22} & a_{23} \\ a_{31}+b_{31} & a_{32} & a_{33} \end{vmatrix} = \begin{vmatrix} a_{11} & a_{12} & a_{13} \\ a_{21} & a_{22} & a_{23} \\ a_{31} & a_{32} & a_{33} \end{vmatrix} + \begin{vmatrix} b_{11} & a_{12} & a_{13} \\ b_{21} & a_{22} & a_{23} \\ b_{31} & a_{32} & a_{33} \end{vmatrix}$$

(VI) The value of a determinant remains same if all the elements of one row (or column) are multiplied by a scalar and added to the corresponding elements of a different row (column).

For example,

$$\begin{vmatrix} a_{11} & a_{12} & a_{13} \\ a_{21} & a_{22} & a_{23} \\ a_{31} & a_{32} & a_{33} \end{vmatrix} = \begin{vmatrix} a_{11}-\lambda a_{12}+m\, a_{13} & a_{12} & a_{13} \\ a_{21}-\lambda a_{22}+m\, a_{23} & a_{22} & a_{23} \\ a_{31}-\lambda a_{32}+m\, a_{33} & a_{32} & a_{33} \end{vmatrix}$$

(VII) If X and Y are square matrices of same order, then $|XY| = |X|\,|Y|$.

Example 1.80

Show that $\begin{vmatrix} p & q & r \\ q & t & u \\ r & u & x \end{vmatrix} = ptx + 2urq - pu^2 - tr^2 - xq^2.$

SOLUTION : Expanding the determinant in terms of the elements of first row, we get

$$\begin{vmatrix} p & q & r \\ q & t & u \\ r & u & x \end{vmatrix} = p \begin{vmatrix} t & u \\ u & x \end{vmatrix} - q \begin{vmatrix} q & u \\ r & x \end{vmatrix} + r \begin{vmatrix} q & t \\ r & u \end{vmatrix}$$

$$= p\,(tx - u^2) - q\,(qx - ru) + r\,(qu - rt)$$

$$= ptx - pu^2 - q^2x + qru + rqu - r^2t$$

$$= ptx + 2urq - pu^2 - tr^2 - xq^2$$

Example 1.81

Prove that $\begin{vmatrix} 1 & 1 & 1 \\ a & b & c \\ bc & ca & ab \end{vmatrix} = (a-b)(b-c)(c-a)$

SOLUTION : The given determinant

$$= \begin{vmatrix} 1 & 1 & 1 \\ a & b & c \\ bc & ca & ab \end{vmatrix}$$

$$= \begin{vmatrix} 0 & 0 & 1 \\ a-c & b-c & c \\ b(c-a) & a(c-b) & ab \end{vmatrix} \begin{matrix} C_1 \to C_1 - C_3 \\ C_2 \to C_2 - C_3 \end{matrix}$$

taking out $(a-c)$ common from C_1 and $(b-c)$ from C_2

$$= (a-c)(b-c) \begin{vmatrix} 0 & 0 & 1 \\ 1 & 1 & c \\ -b & -a & ab \end{vmatrix}$$

$$= (a-c)(b-c)\, 1 \begin{vmatrix} 1 & 1 \\ -b & -a \end{vmatrix} \quad \text{[expanded by } R_1\text{]}$$

$$= (a-c)(b-c)(-a+b)$$
$$= (b-a)(b-c)(a-c)$$
$$= (a-b)(b-c)(c-a)$$

Example 1.82

Prove that $\begin{vmatrix} 1 & 1 & 1 \\ a & b & c \\ a^3 & b^3 & c^3 \end{vmatrix} = (a-b)(b-c)(c-a)(a+b+c).$

SOLUTION : The given determinant

$$= \begin{vmatrix} 1 & 1 & 1 \\ a & b & c \\ a^3 & b^3 & c^3 \end{vmatrix}$$

$$= \begin{vmatrix} 1 & 0 & 0 \\ a & b-a & c-a \\ a^3 & b^3-a^3 & c^3-a^3 \end{vmatrix} \begin{matrix} C_2 \to C_2 - C_1 \\ C_3 \to C_3 - C_1 \end{matrix}$$

$$= \begin{vmatrix} 1 & 0 & 0 \\ a & b-a & c-a \\ a^3 & (b-a)(a^2+ab+b^2) & (c-a)(c^2+ca+a^2) \end{vmatrix}$$

$$= (b-a)(c-a) = \begin{vmatrix} 1 & 0 & 0 \\ a & 1 & 1 \\ a^3 & a^2+ab+b^2 & c^2+ca+a^2 \end{vmatrix}$$

$$= (b-a)(c-a)\, 1 \begin{vmatrix} 1 & 1 \\ a^2+ab+b^2 & c^2+ca+a^2 \end{vmatrix}$$

$$= (b-a)(c-a)[c^2+ca+a^2-ab-b^2-a^2]$$
$$= (b-a)(c-a)(c^2+ca-ab-b^2)$$

$$= (b-a)(c-a)[(c^2-b^2)+a(c-b)]$$
$$= (b-a)(c-a)[(c-b)(c+b)+a(c-b)]$$
$$= (a-b)(b-c)(c-a)(a+b+c)$$

Example 1.83

Prove that $= \begin{vmatrix} 1 & bc & a(b+c) \\ 1 & ca & b(c+a) \\ 1 & ab & c(a+b) \end{vmatrix} = 0$

SOLUTION : Applying the following operations :

$R_2 \to R_2 - R_1, R_3 \to R_3 - R_1$, we get

$$\begin{vmatrix} 1 & bc & a(b+c) \\ 0 & c(a-b) & c(b-a) \\ 0 & b(a-c) & b(c-a) \end{vmatrix}$$

Taking out $c(a-b)$ and $b(a-c)$ common from R_2 and R_3, we get

$$c(a-b)\cdot b(a-c)\begin{vmatrix} 1 & bc & a(b+c) \\ 0 & 1 & -1 \\ 0 & 1 & -1 \end{vmatrix} = 0 \qquad [\because R_2 = R_3]$$

Example 1.84

Show that $\begin{vmatrix} b+c & c+a & a+b \\ q+r & r+p & p+q \\ y+z & z+x & x+y \end{vmatrix} = 2\begin{vmatrix} a & b & c \\ p & q & r \\ x & y & z \end{vmatrix}$

SOLUTION : $C_1 \to C_1 + C_2 + C_3$

$$= \begin{vmatrix} 2(a+b+c) & c+a & a+b \\ 2(p+q+r) & r+p & p+q \\ 2(x+y+z) & z+x & x+y \end{vmatrix} = 2\begin{vmatrix} a+b+c & c+a & a+b \\ p+q+r & r+p & p+q \\ x+y+z & z+x & x+y \end{vmatrix}$$

$C_2 \to C_1 - C_2, \ C_3 \to C_1 - C_3$

$$= 2\begin{vmatrix} a+b+c & b & c \\ p+q+r & q & r \\ x+y+z & y & z \end{vmatrix}$$

$C_1 \to C_1 - (C_2 + C_3)$

$$= 2\begin{vmatrix} a & b & c \\ p & q & r \\ x & y & z \end{vmatrix}$$

Example 1.85

Show that $\begin{vmatrix} a & a^2 & 1+a^3 \\ b & b^2 & 1+b^3 \\ c & c^2 & 1+c^3 \end{vmatrix} = (a-b)(b-c)(c-a)(1+abc).$

SOLUTION : Applying elementary row and column transformation as follows :

$$\begin{vmatrix} a & a^2 & 1+a^3 \\ b & b^2 & 1+b^3 \\ c & c^2 & 1+c^3 \end{vmatrix} = \begin{vmatrix} a & a^2 & 1 \\ b & b^2 & 1 \\ c & c^2 & 1 \end{vmatrix} + \begin{vmatrix} a & a^2 & a^3 \\ b & b^2 & b^3 \\ c & c^2 & c^3 \end{vmatrix} \quad \text{(by property)}$$

$$= (-1)\begin{vmatrix} 1 & a^2 & a \\ 1 & b^2 & b \\ 1 & c^2 & c \end{vmatrix} + \begin{vmatrix} a & a^2 & a^3 \\ b & b^2 & b^3 \\ c & c^2 & c^3 \end{vmatrix}$$

[($C_1 \Leftrightarrow C_3$) in Ist determinant]

$$= (-1)^2\begin{vmatrix} 1 & a & a^2 \\ 1 & b & b^2 \\ 1 & c & c^2 \end{vmatrix} + abc\begin{vmatrix} 1 & a & a^2 \\ 1 & b & b^2 \\ 1 & c & c^2 \end{vmatrix}$$

[($C_2 \Leftrightarrow C_3$) in Ist determinant]

$$= (1 + abc)\begin{vmatrix} 1 & a & a^2 \\ 1 & b & b^2 \\ 1 & c & c^2 \end{vmatrix}$$

$$= (1 + abc)\begin{vmatrix} 1 & a & a^2 \\ 0 & b-a & b^2-a^2 \\ 0 & c-a & c^2-a^2 \end{vmatrix} \quad \begin{matrix} R_2 \to R_2 - R_1 \\ R_3 \to R_3 - R_1 \end{matrix}$$

$$= (1 + abc)\,(b-a)\,(c-a)\begin{vmatrix} 1 & a & a^2 \\ 0 & 1 & b+a \\ 0 & 1 & c+a \end{vmatrix}$$

$$= (1 + abc)\,(b-a)\,(c-a) \cdot \{1 \cdot (c+a-b-a)\}$$

$$= (1 + abc)\,(b-a)\,(c-a)\,(c-b)$$

$$= (1 + abc)\,(a-b)\,(b-c)\,(c-a)$$

Example 1.86

Prove that $\begin{vmatrix} 1 & 1 & 1 \\ a & b & c \\ a^2 & b^2 & c^2 \end{vmatrix} = (a-b)\,(b-c)\,(c-a)$

[UPTU 2000]

SOLUTION : The given determinant

$$= \begin{vmatrix} 1 & 1 & 1 \\ a & b & c \\ a^2 & b^2 & c^2 \end{vmatrix}$$

$$= \begin{vmatrix} 0 & 0 & 1 \\ a-c & b-c & c \\ a^2-c^2 & b^2-c^2 & c^2 \end{vmatrix} \quad \begin{matrix} c_1 \longrightarrow c_1 - c_3 \\ c_2 \longrightarrow c_2 - c_3 \end{matrix}$$

$$= \begin{vmatrix} 0 & 0 & 1 \\ a-c & b-c & c \\ (a-c)\,(a+c) & (b-c)\,(b+c) & c^2 \end{vmatrix}$$

$$= (a-c)\,(b-c)\begin{vmatrix} 0 & 0 & 1 \\ 1 & 1 & c \\ a+c & b+c & c^2 \end{vmatrix}$$

$$= (a-c)\,(b-c)\begin{vmatrix} 1 & 1 \\ a+c & b+c \end{vmatrix}$$ [expanded by R_1]

$$= (a-c)\,(b-c)\,[b+c-a-c]$$

$= (a-c)(b-c)(b-a)$
$= (a-b)(b-c)(c-a)$

Example 1.87

Without expanding the determinant, show that

$$= \begin{vmatrix} 0 & b & -c \\ -b & 0 & a \\ c & -a & 0 \end{vmatrix} = 0$$

[UPTU 2001]

SOLUTION : Given $\begin{vmatrix} 0 & b & -c \\ -b & 0 & a \\ c & -a & 0 \end{vmatrix}$

$$= \begin{vmatrix} 0 & b & -c \\ -b+c & -a+b & a-c \\ -b+c & -a+b & a-c \end{vmatrix} \quad \begin{matrix} R_2 \longrightarrow R_1 + R_2 + R_3 \\ R_3 \longrightarrow R_1 + R_2 + R_3 \end{matrix}$$

$= 0$ [$\because$ R_2 and R_3 are identical]

Example 1.88

Prove that $\begin{vmatrix} a & b & c \\ b & c & a \\ c & a & b \end{vmatrix} = (a+b+c)(ab+bc+ca-a^2-b^2-c^2)$

[UPTU 2002]

SOLUTION : Given $\begin{vmatrix} a & b & c \\ b & c & a \\ c & a & b \end{vmatrix}$

$$= \begin{vmatrix} a+b+c & a+b+c & a+b+c \\ b & c & a \\ c & a & b \end{vmatrix} \quad R_1 \longrightarrow R_1 + R_2 + R_3$$

$$= (a+b+c) \begin{vmatrix} 1 & 1 & 1 \\ b & c & a \\ c & a & b \end{vmatrix}$$

$= (a+b+c)\,[1\,(bc-a^2) - 1\,(b^2-ac) + 1\,(ab-c^2)]$

$= (a+b+c)(bc-a^2-b^2+ac+ab-c^2)$

$= (a+b+c)(ab+ac+bc-a^2-b^2-c^2)$

Example 1.89

Show that $\begin{vmatrix} a & a+b & a+b+c \\ 2a & 3a+2b & 4a+3b+2c \\ 3a & 6a+3b & 10a+6b+3c \end{vmatrix} = a^3$

[UPTU (C.O.) 2006]

SOLUTION : We have

$$\begin{vmatrix} a & a+b & a+b+c \\ 2a & 3a+2b & 4a+3b+2c \\ 3a & 6a+3b & 10a+6b+3c \end{vmatrix} = a \begin{vmatrix} 1 & a+b & a+b+c \\ 2 & 3a+2b & 4a+3b+2c \\ 3 & 6a+3b & 10a+6b+3c \end{vmatrix}$$

$R_2 \to R_2 - 2R_1, R_3 \to R_3 - 3R_1$

$$= a \begin{vmatrix} 1 & a+b & a+b+c \\ 0 & a & 2a+b \\ 0 & 3a & 7a+3b \end{vmatrix}$$

$$= a \begin{vmatrix} a & 2a+b \\ 3a & 7a+3b \end{vmatrix}$$

$$= a\,[7a^2 + 3ab - 6a^2 - 3ab]$$

$$= a^3$$

PROBLEM SET 1.9

1. Evaluate $\begin{vmatrix} x & y & z \\ y+z & z+x & x+y \\ x^2 & y^2 & z^2 \end{vmatrix}$.

2. If $p \neq q \neq r$ and $\begin{vmatrix} p & p^2 & p^3-1 \\ q & q^2 & q^3-1 \\ r & r^2 & r^3-1 \end{vmatrix} = 0$, then show that $pqr = 1$

3. Evaluate $\begin{vmatrix} a-b-c & 2b & 2c \\ 2a & b-c-a & 2c \\ 2a & 2b & c-a-b \end{vmatrix}$

4. Show that $\begin{vmatrix} a & b & ax+by \\ b & c & bx+cy \\ ax+by & bx+cy & 0 \end{vmatrix} = (b^2 - ac)\,(ax^2 + 2bxy + cy^2)$

5. Show that $\begin{vmatrix} yz-x^2 & xz-y^2 & xy-z^2 \\ zx-y^2 & xy-z^2 & yz-x^2 \\ xy-z^2 & yz-x^2 & zx-y^2 \end{vmatrix} = \begin{vmatrix} x & y & z \\ y & z & x \\ z & x & y \end{vmatrix}^2$

6. Show that $\begin{vmatrix} 1+a & 1 & 1 \\ 1 & 1+b & 1 \\ 1 & 1 & 1+c \end{vmatrix} = abc\left(1 + \frac{1}{a} + \frac{1}{b} + \frac{1}{c}\right)$ **[UPTU 2003, 05]**

7. Find the value of $\begin{vmatrix} 1 & 1 & 1 \\ 1 & 1+x & 1 \\ 1 & 1 & 1+y \end{vmatrix}$

8. If $a + b + c = 0$, prove that $\begin{vmatrix} a-x & c & b \\ c & b-x & a \\ b & a & c-x \end{vmatrix} = 0$

9. Prove that $\begin{vmatrix} x & y & z \\ x^2 & y^2 & z^2 \\ yz & zx & xy \end{vmatrix} = \begin{vmatrix} 1 & 1 & 1 \\ x^2 & y^2 & z^2 \\ x^3 & y^3 & z^3 \end{vmatrix}$

10. Show that $\begin{vmatrix} (y+z)^2 & x^2 & x^2 \\ y^2 & (z+x)^2 & y^2 \\ z^2 & z^2 & (x+y)^2 \end{vmatrix} = 2xyz\,(x+y+z)^3$

11. Show that $\begin{vmatrix} x & x+y & y+z \\ y & y+z & z+x \\ z & z+y & x+y \end{vmatrix} = x^3+y^3+z^3-3xyz$

12. Prove that $\begin{vmatrix} x+a & b & c \\ a & x+b & c \\ a & b & x+c \end{vmatrix} = x^2\,(x+a+b+c)$

13. Prove that $\begin{vmatrix} x^2+1 & xy & xz \\ xy & y^2+1 & yz \\ xz & yz & z^2+1 \end{vmatrix} = 1+x^2+y^2+z^2$

14. Prove that $\begin{vmatrix} -a^2 & ab & ac \\ ab & -b^2 & bc \\ ab & bc & -c^2 \end{vmatrix} = 4a^2b^2c^2$

15. Show that

$$\begin{vmatrix} a+b+2c & a & b \\ c & b+c+2a & b \\ c & a & c+a+2b \end{vmatrix} = 2\,(a+b+c)^2$$

16. Show that $x = 2$ is the root of the equation

$$\begin{vmatrix} x & -6 & -1 \\ 2 & -3x & x-3 \\ -3 & 2x & x+2 \end{vmatrix} = 0$$

17. Prove that

$$\begin{vmatrix} a^2 & a^2-(b-c)^2 & bc \\ b^2 & b^2-(c-a)^2 & ca \\ c^2 & c^2-(a-b)^2 & ab \end{vmatrix}$$

$$= (a-b)\,(b-c)\,(c-a)\,(a+b+c)\,(a^2+b^2+c^2)$$

18. Show that

$$\begin{vmatrix} b+c & c+a & a+b \\ q+r & r+p & p+q \\ y+z & z+x & x+y \end{vmatrix} = 2\begin{vmatrix} a & b & c \\ p & q & r \\ x & y & z \end{vmatrix}$$

ANSWERS

1. $-(x-y)\,(y-z)\,(z-x)\,(x+y+z)$ **3.** $(a+b+c)^3$ **7.** $-xy$

1.16 ADJOINT OF A MATRIX

Let $A = [a_{ij}]_{n \times n}$ be a square matrix of order $n \times n$, then the corresponding determinant is:

$$|A| = \begin{vmatrix} a_{11} & a_{12} & a_{1n} \\ a_{21} & a_{22} & a_{2n} \\ \dots & \dots & \dots \\ \dots & \dots & \dots \\ a_{n1} & a_{n2} & a_{nn} \end{vmatrix}$$

Let A_{ij} be the co-factor of a_{ij} in the determinant A, then the transpose of the co-factor matrix :

$$\begin{bmatrix} A_{11} & A_{12} & \dots & A_{1n} \\ A_{21} & A_{22} & \dots & A_{2n} \\ \dots & \dots & \dots & \dots \\ A_{n1} & A_{n2} & \dots & A_{nn} \end{bmatrix} \text{ is given by } \begin{bmatrix} A_{11} & A_{21} & \dots & A_{n1} \\ A_{12} & A_{22} & \dots & A_{n2} \\ \dots & \dots & \dots & \dots \\ A_{1n} & A_{2n} & \dots & A_{nn} \end{bmatrix}$$

is called the *adjoint* of A and denoted by $Adj\ (A)$.

Therefore the adjoint of any square matrix is equal to the transpose of its co-factor matrix.

Example 1.90

Let $A = \begin{bmatrix} 3 & 2 & 4 \\ 5 & 6 & 0 \\ 2 & 3 & 5 \end{bmatrix}$

Here $A_{11} = \begin{vmatrix} 6 & 0 \\ 3 & 5 \end{vmatrix} = 30$ $\qquad A_{21} = -\begin{vmatrix} 2 & 4 \\ 3 & 5 \end{vmatrix} = 2$

$A_{12} = -\begin{vmatrix} 5 & 0 \\ 2 & 5 \end{vmatrix} = -25$ $\qquad A_{22} = \begin{vmatrix} 3 & 4 \\ 2 & 5 \end{vmatrix} = 7$

$A_{13} = \begin{vmatrix} 5 & 6 \\ 2 & 3 \end{vmatrix} = 3$ $\qquad A_{23} = -\begin{vmatrix} 3 & 2 \\ 2 & 3 \end{vmatrix} = -5$

$A_{31} = \begin{vmatrix} 2 & 4 \\ 6 & 0 \end{vmatrix} = -24,$ $\qquad A_{32} = \begin{vmatrix} 3 & 4 \\ 5 & 0 \end{vmatrix} = 20$

$A_{33} = \begin{vmatrix} 3 & 2 \\ 5 & 6 \end{vmatrix} = 8$

$$\therefore \quad Adj\ (A) = \begin{bmatrix} A_{21} & A_{21} & A_{31} \\ A_{12} & A_{22} & A_{32} \\ A_{13} & A_{23} & A_{33} \end{bmatrix} = \begin{bmatrix} 30 & 2 & -24 \\ -21 & 7 & 20 \\ 3 & -5 & 8 \end{bmatrix}$$

1.17 INVERSE OF A MATRIX

1.17.1 Singular and Non-singular Matrices

Those matrices whose determinant is equal to zero are called singular matrices, otherwise the matrix is non-singular, *i.e.*, if $|A| = 0$ then matrix A is singulare and if $|A| \neq 0$ the matrix A is non-singular.

■ **DEFINITION :** *Let A be a non-singular square matrix of order $n \times n$ and B be another matrix of same order. If $AB = BA = I$ where I is identity matrix of order $n \times n$, then B is called inverse of A and is denoted by A^{-1} i.e., $B = A^{-1}$ so that $AA^{-1} = A^{-1}A = I$*

THEOREM 1.3. To show that $A^{-1} = \dfrac{Adj\,(A)}{|A|}$.

PROOF : Let $A = [a_{ij}]$ be a square matrix of order $n \times n$ and I be identity matrix of same order then by the properties of adjoint of A we know that

$$A \cdot Adj\,(A) = |A|\, I$$

$$Adj\,(A) = \frac{|A|}{A} \qquad [\because |A| \cdot I = |A|]$$

$$= |A| \cdot A^{-1}$$

$$\Rightarrow \quad A^{-1} = \frac{Adj\,(A)}{|A|}$$

THEOREM 1.4. Show that $(AB)^{-1} = B^{-1}A^{-1}$ **[Reversal Rule]**.

PROOF : (i) Consider $(AB)(B^{-1}A^{-1}) = A(BB^{-1})A^{-1}$

$$= A \cdot I \cdot A^{-1} \qquad [\because BB^{-1} = I]$$

$$= AA^{-1}$$

$$= I$$

(ii) Also $(B^{-1}A^{-1})(AB) = B^{-1}(A^{-1}A)B$

$$= B^{-1} I B \qquad [\because A^{-1}A = I]$$

$$= B^{-1}B = I$$

We know that if $A \cdot B = I$ then $B = A^{-1}$

Thus, $(AB)(B^{-1}A^{-1}) = I \quad \Rightarrow \quad (AB)^{-1} = B^{-1}A^{-1}$

1.17.2 Some Useful Properties of Inverse of a Matrix

1. The inverse of a matrix which is non-singular is unique.
   PROOF : Let A be a square matrix of order n such that $|A| \neq 0$. Let if possible B and C are two inverses of A *i.e.*, $AB = BA = I$ and $AC = CA = I$, where I is the identify matrix.
   Consider $B = BI = B(AC) = (BA)C = IC = C$. Thus the inverse of A is unique.
2. The inverse of inverse of matrix A is matrix A itself *i.e.*, $(A^{-1})^{-1} = A$.
   PROOF : Let A be a square matrix and A^{-1} be the inverse of A then $AA^{-1} = I$
   Post multiply by $(A^{-1})^{-1}$
   $$(AA^{-1})(A^{-1})^{-1} = I(A^{-1})^{-1} \Rightarrow A[A^{-1}(A^{-1})^{-1}] = (A^{-1})^{-1}$$
   $$\Rightarrow A \cdot I = (A^{-1})^{-1} \Rightarrow A = (A^{-1})^{-1}$$
3. The inverse of transpose of a matrix is the transpase of its inverse, *i.e.*, $(A^T)^{-1} = (A^{-1})^T$.
4. If A is non-singular and $AB = O$, then $B = O$.
5. $I^{-1} = I$, *i.e.*, the identity matrix is its own inverse.

Example 1.91

Find the inverse of matrix $A = \begin{bmatrix} 2 & 4 & 3 \\ 0 & 1 & 2 \\ 2 & 2 & 3 \end{bmatrix}$.

SOLUTION : $|A| = 2\begin{vmatrix} 1 & 2 \\ 2 & 3 \end{vmatrix} - 4\begin{vmatrix} 0 & 2 \\ 2 & 3 \end{vmatrix} + 3\begin{vmatrix} 0 & 1 \\ 2 & 2 \end{vmatrix}$

$$= 2(3-4) - 4(0-4) + 3(0-2)$$
$$= 2(-1) - 4(-4) + 3(-2)$$
$$= -2 + 16 - 6$$
$$= 8$$

$\therefore$ $|A| = 8 \neq 0$. Hence inverse of the given matrix exists. The cofactors of A are :

$$A_{11} = \begin{vmatrix} 1 & 2 \\ 2 & 3 \end{vmatrix} = -1, \quad A_{21} = -\begin{vmatrix} 4 & 3 \\ 2 & 3 \end{vmatrix} = -6, \quad A_{31} = \begin{vmatrix} 4 & 3 \\ 1 & 2 \end{vmatrix} = 5$$

$$A_{12} = -\begin{vmatrix} 0 & 2 \\ 2 & 3 \end{vmatrix} = 4, \quad A_{22} = -\begin{vmatrix} 2 & 3 \\ 2 & 3 \end{vmatrix} = 0, \quad A_{32} = -\begin{vmatrix} 2 & 3 \\ 0 & 2 \end{vmatrix} = -4$$

$$A_{13} = \begin{vmatrix} 0 & 1 \\ 2 & 2 \end{vmatrix} = -2, \quad A_{23} = -\begin{vmatrix} 2 & 4 \\ 2 & 2 \end{vmatrix} = 4, \quad A_{33} = \begin{vmatrix} 2 & 4 \\ 0 & 1 \end{vmatrix} = 2$$

$$Adj\,(A) = \begin{bmatrix} A_{11} & A_{21} & A_{31} \\ A_{12} & A_{22} & A_{32} \\ A_{13} & A_{23} & A_{33} \end{bmatrix} = \begin{bmatrix} -1 & -6 & 5 \\ 4 & 0 & -4 \\ -2 & 4 & 2 \end{bmatrix}$$

Now $A^{-1} = \dfrac{Adj\,(A)}{|A|} = \dfrac{1}{8}\begin{bmatrix} -1 & -6 & 5 \\ 4 & 0 & -4 \\ -2 & 4 & 2 \end{bmatrix}$

■ **NOTE :** Verification, $AA^{-1} = I$

Example 1.92

Let $A = \begin{bmatrix} 1 & 2 & 2 \\ 2 & 1 & 2 \\ 2 & 2 & 1 \end{bmatrix}$. Prove that $A^2 - 4A - 5I = O$. Hence obtain A^{-1}.

SOLUTION : $A^2 = A \cdot A = \begin{bmatrix} 1 & 2 & 2 \\ 2 & 1 & 2 \\ 2 & 2 & 1 \end{bmatrix}\begin{bmatrix} 1 & 2 & 2 \\ 2 & 1 & 2 \\ 2 & 2 & 1 \end{bmatrix} = \begin{bmatrix} 9 & 8 & 8 \\ 8 & 9 & 8 \\ 8 & 8 & 9 \end{bmatrix}$

Now $A^2 - 4A - 5I = \begin{bmatrix} 9 & 8 & 8 \\ 8 & 9 & 8 \\ 8 & 8 & 9 \end{bmatrix} - 4\begin{bmatrix} 1 & 2 & 2 \\ 2 & 1 & 2 \\ 2 & 2 & 1 \end{bmatrix} - 5\begin{bmatrix} 1 & 0 & 0 \\ 0 & 1 & 0 \\ 0 & 0 & 1 \end{bmatrix}$

$$= \begin{bmatrix} 9 & 8 & 8 \\ 8 & 9 & 8 \\ 8 & 8 & 9 \end{bmatrix} - \begin{bmatrix} 4 & 8 & 8 \\ 8 & 4 & 8 \\ 8 & 8 & 4 \end{bmatrix} - \begin{bmatrix} 5 & 0 & 0 \\ 0 & 5 & 0 \\ 0 & 0 & 5 \end{bmatrix}$$

$$= \begin{bmatrix} 9-4-5 & 8-8-0 & 8-8-0 \\ 8-8-0 & 9-4-5 & 8-8-0 \\ 8-8-0 & 8-8-0 & 9-4-5 \end{bmatrix}$$

$$= \begin{bmatrix} 0 & 0 & 0 \\ 0 & 0 & 0 \\ 0 & 0 & 0 \end{bmatrix} = O$$

Since $\quad A^2 - 4A - 5I = O$

$\therefore \quad A^2 - 4A = 5I$

Post multiply by A^{-1}

$$A^2 A^{-1} - 4AA^{-1} = 5IA^{-1}$$

$$A - 4I = 5A^{-1}$$

or $\quad A^{-1} = \frac{1}{5}(A - 4I)$

$$A^{-1} = \frac{1}{5}\begin{bmatrix} -3 & 2 & 2 \\ 2 & -3 & 2 \\ 2 & 2 & -3 \end{bmatrix}$$

Example 1.93

Find matrix X such that $\begin{bmatrix} 1 & -4 \\ 3 & -2 \end{bmatrix} X = \begin{bmatrix} 2 & 8 \\ 9 & 4 \end{bmatrix}$.

SOLUTION : Let $\quad AX = B \quad$ ...(i)

where $A = \begin{bmatrix} 1 & -4 \\ 3 & -2 \end{bmatrix}$, $B = \begin{bmatrix} 2 & 8 \\ 9 & 4 \end{bmatrix}$

Pre-multply (i) by A^{-1}

$$X = A^{-1} B \quad \text{...(ii)}$$

To find A^{-1}, $\quad |A| = -2 + 12 = 10 \neq 0$

$$Adj\,(A) = \begin{bmatrix} -2 & 4 \\ -3 & 1 \end{bmatrix}$$

$$A^{-1} = \frac{1}{10}\begin{bmatrix} -2 & 4 \\ -3 & 1 \end{bmatrix}$$

From (ii) $\quad X = \frac{1}{10}\begin{bmatrix} -2 & 4 \\ -3 & 1 \end{bmatrix}\begin{bmatrix} 2 & 8 \\ 9 & 4 \end{bmatrix}$

$$X = \frac{1}{10}\begin{bmatrix} 32 & 0 \\ 3 & -20 \end{bmatrix}.$$

Example 1.94

Find A^{-1}, if $A = \begin{bmatrix} 1 & 2 & 5 \\ 1 & -1 & -1 \\ 2 & 3 & 1 \end{bmatrix}$ *[UPTU 2000]*

SOLUTION : Here, $A = \begin{bmatrix} 1 & 2 & 5 \\ 1 & -1 & -1 \\ 2 & 3 & 1 \end{bmatrix}$

Then $\quad |A| = \begin{vmatrix} 1 & 2 & 5 \\ 1 & -1 & -1 \\ 2 & 3 & 1 \end{vmatrix} = 1\begin{vmatrix} -1 & -1 \\ 3 & 1 \end{vmatrix} - 2\begin{vmatrix} 1 & -1 \\ 2 & 1 \end{vmatrix} + 5\begin{vmatrix} 1 & -1 \\ 2 & 3 \end{vmatrix}$

$$= 1(-1+3) - 2(1+2) + 5(3+2)$$
$$= 21$$

$\therefore$ $|A| = 21 \neq 0$. Hence, inverse of the given matrix exists.

The cofactors of A are :

$$A_{11} = \begin{vmatrix} -1 & -1 \\ 3 & 1 \end{vmatrix} = 2, A_{12} = -\begin{vmatrix} 1 & -1 \\ 2 & 1 \end{vmatrix} = -3$$

$$A_{13} = \begin{vmatrix} 1 & -1 \\ 2 & 3 \end{vmatrix} = 5, A_{21} = -\begin{vmatrix} 2 & 5 \\ 3 & 1 \end{vmatrix} = 13$$

$$A_{22} = \begin{vmatrix} 1 & 5 \\ 2 & 1 \end{vmatrix} = -9, A_{23} = -\begin{vmatrix} 1 & 2 \\ 2 & 3 \end{vmatrix} = 1$$

$$A_{31} = \begin{vmatrix} 2 & 5 \\ -1 & -1 \end{vmatrix} = 3, A_{32} = -\begin{vmatrix} 1 & 5 \\ 1 & -1 \end{vmatrix} = 6$$

$$A_{33} = \begin{vmatrix} 1 & 2 \\ 1 & -1 \end{vmatrix} = -3$$

$$\therefore \quad \text{Adj. }(A) = \begin{bmatrix} A_{11} & A_{21} & A_{31} \\ A_{12} & A_{22} & A_{32} \\ A_{13} & A_{23} & A_{33} \end{bmatrix} = \begin{bmatrix} 2 & 13 & 3 \\ -3 & -9 & 6 \\ 5 & 1 & -3 \end{bmatrix}$$

Now, $$A^{-1} = \frac{\text{Adj. }(A)}{|A|} = \frac{1}{21}\begin{bmatrix} 2 & 13 & 3 \\ -3 & -9 & 6 \\ 5 & 1 & -3 \end{bmatrix}$$

Example 1.95

Find the inverse of the matrix $A = \begin{bmatrix} 2 & 3 & 4 \\ 4 & 3 & 1 \\ 1 & 2 & 4 \end{bmatrix}$ ***[UPTU 2001, 05]***

SOLUTION : Given $A = \begin{bmatrix} 2 & 3 & 4 \\ 4 & 3 & 1 \\ 1 & 2 & 4 \end{bmatrix}$

$$\therefore \quad |A| = \begin{vmatrix} 2 & 3 & 4 \\ 4 & 3 & 1 \\ 1 & 2 & 4 \end{vmatrix} = 2\begin{vmatrix} 3 & 1 \\ 2 & 4 \end{vmatrix} - 3\begin{vmatrix} 4 & 1 \\ 1 & 4 \end{vmatrix} + 4\begin{vmatrix} 4 & 3 \\ 1 & 2 \end{vmatrix}$$
$$= 2(12-2) - 3(16-1) + 4(8-3)$$
$$= -5$$

Here, $|A| \neq 0$. Inverse of the given matrix exists.

The co-factors of A are

$$A_{11} = \begin{vmatrix} 3 & 1 \\ 2 & 4 \end{vmatrix} = 10, A_{12} = -\begin{vmatrix} 4 & 1 \\ 1 & 4 \end{vmatrix} = -15, A_{13} = \begin{vmatrix} 4 & 3 \\ 1 & 2 \end{vmatrix} = 5$$

$$A_{21} = -\begin{vmatrix} 3 & 4 \\ 2 & 4 \end{vmatrix} = -4, A_{22} = \begin{vmatrix} 2 & 4 \\ 1 & 4 \end{vmatrix} = 4, A_{23} = -\begin{vmatrix} 2 & 3 \\ 1 & 2 \end{vmatrix} = -1$$

$$A_{31} = \begin{vmatrix} 3 & 4 \\ 3 & 1 \end{vmatrix} = -9,\ A_{32} = -\begin{vmatrix} 2 & 4 \\ 4 & 1 \end{vmatrix} = 14,\ A_{33} = \begin{vmatrix} 2 & 3 \\ 4 & 3 \end{vmatrix} = -6$$

$$\therefore \quad \text{Adj. }(A) = \begin{bmatrix} A_{11} & A_{21} & A_{31} \\ A_{12} & A_{22} & A_{32} \\ A_{13} & A_{23} & A_{33} \end{bmatrix} = \begin{bmatrix} 10 & -4 & -9 \\ -15 & 4 & 14 \\ 5 & -1 & -6 \end{bmatrix}$$

and $$A^{-1} = \frac{\text{Adj. }(A)}{|A|} = -\frac{1}{5}\begin{bmatrix} 10 & -4 & -9 \\ -15 & 4 & 14 \\ 5 & -1 & -6 \end{bmatrix}$$

Example 1.96

Find the inverse of matrix A, if

$$A = \begin{bmatrix} 2 & -1 & 1 \\ -1 & 2 & -1 \\ 1 & -1 & 2 \end{bmatrix}$$

[UPTU 2002]

SOLUTION : Given $A = \begin{bmatrix} 2 & -1 & 1 \\ -1 & 2 & -1 \\ 1 & -1 & 2 \end{bmatrix}$

$$\therefore \quad |A| = \begin{vmatrix} 2 & -1 & 1 \\ -1 & 2 & -1 \\ 1 & -1 & 2 \end{vmatrix} = 2\begin{vmatrix} 2 & -1 \\ -1 & 2 \end{vmatrix} + 1\begin{vmatrix} -1 & -1 \\ 1 & 2 \end{vmatrix} + 1\begin{vmatrix} -1 & 2 \\ 1 & -1 \end{vmatrix}$$

$$= 2(4-1) + 1(-2+1) + 1(1-2)$$
$$= 4$$

Here $|A| \neq 0$. So inverse of the matrix exists.

The co-factors of matrix A are

$$A_{11} = \begin{vmatrix} 2 & -1 \\ -1 & 2 \end{vmatrix} = 3,\quad A_{12} = -\begin{vmatrix} -1 & -1 \\ 1 & 2 \end{vmatrix} = 1,\quad A_{13} = \begin{vmatrix} -1 & 2 \\ 1 & -1 \end{vmatrix} = -1$$

$$A_{21} = -\begin{vmatrix} -1 & 1 \\ -1 & 2 \end{vmatrix} = 1,\quad A_{22} = \begin{vmatrix} 2 & 1 \\ 1 & 2 \end{vmatrix} = 3,\quad A_{23} = -\begin{vmatrix} 2 & -1 \\ 1 & -1 \end{vmatrix} = 1$$

$$A_{31} = \begin{vmatrix} -1 & 1 \\ 2 & -1 \end{vmatrix} = -1,\quad A_{32} = -\begin{vmatrix} 2 & 1 \\ -1 & -1 \end{vmatrix} = 1,\quad A_{33} = \begin{vmatrix} 2 & -1 \\ -1 & 2 \end{vmatrix} = 3$$

$$\therefore \quad \text{Adj. }(A) = \begin{bmatrix} A_{11} & A_{21} & A_{31} \\ A_{12} & A_{22} & A_{32} \\ A_{13} & A_{23} & A_{33} \end{bmatrix} = \begin{bmatrix} 3 & 1 & -1 \\ 1 & 3 & 1 \\ -1 & 1 & 3 \end{bmatrix}$$

and $$A^{-1} = \frac{\text{Adj. }(A)}{|A|} = \frac{1}{4}\begin{bmatrix} 3 & 1 & -1 \\ 1 & 3 & 1 \\ -1 & 1 & 3 \end{bmatrix}$$

PROBLEM SET 1.10

1. Determine the adjoint of matrix $A = \begin{bmatrix} 5 & -2 & 4 \\ -2 & 1 & 1 \\ 4 & 1 & 0 \end{bmatrix}$ and hence obtain A^{-1}.

2. If $A = \begin{bmatrix} 3 & 4 \\ 0 & 5 \end{bmatrix}$, $B = \begin{bmatrix} 7 & 2 \\ 4 & 9 \end{bmatrix}$, show that

   (i) $(AB)^{-1} = B^{-1} A^{-1}$

   (ii) $(A^{-1})^{-1} = A$

3. If $A = \begin{bmatrix} 1 & 3 & 4 \\ 3 & -1 & 6 \\ -1 & 5 & 1 \end{bmatrix}$, prove that $(A')^{-1} = (A^{-1})'$

4. If $A = \begin{bmatrix} 1 & 2 & 0 \\ 2 & -1 & 0 \\ 0 & 0 & -1 \end{bmatrix}$ show that $A^3 + A^2 - 5A - 5I = 0$

5. Solve for matrix A the equations

$$\begin{bmatrix} 2 & 1 \\ 3 & 2 \end{bmatrix} A = \begin{bmatrix} 3 & -1 \\ 4 & 2 \end{bmatrix}$$

6. Given that $A = \begin{bmatrix} 2 & 1 & 1 \\ 3 & 1 & 2 \\ 1 & -1 & 0 \end{bmatrix}$, evaluate $A^2 - A - I$

7. Compute $f(A)$ if $f(x) = \dfrac{1+x}{1-x}$ and $A = \begin{bmatrix} 1 & 2 \\ 2 & 1 \end{bmatrix}$

8. Find X from the equation

$$X \begin{bmatrix} 1 & 1 & -1 \\ 2 & 1 & 0 \\ 1 & -1 & 1 \end{bmatrix} = \begin{bmatrix} 1 & -1 & 3 \\ 4 & 3 & 2 \\ 1 & -2 & 5 \end{bmatrix}$$

9. Given the matrix $A = \begin{bmatrix} 0 & 1 & 0 \\ 0 & 0 & 1 \\ 1 & 2 & -1 \end{bmatrix}$ determine constants p, q, r such that

$$A^3 = pA^2 + qA + rI$$

10. If $A = \begin{bmatrix} 2 & 2 & 0 \\ 2 & 1 & 1 \\ -7 & 2 & -3 \end{bmatrix}$, find A^{-1}

11. Is the inverse of a non-singular symmetric matrix is symmetric matrix. Give an example.

12. Find A^{-1}, if $A = \begin{bmatrix} a+ib & c+id \\ -c+id & a-ib \end{bmatrix}$. Given $a^2 + b^2 + c^2 + d^2 = 1$.

13. Find the adjoint of the matrix $A = \begin{bmatrix} 1 & 4 & 0 \\ -1 & 2 & 2 \\ 0 & 0 & 2 \end{bmatrix}$. Hence, obtain A^{-1}.

14. If $A = \begin{bmatrix} 3 & 2 \\ 7 & 5 \end{bmatrix}$, $B = \begin{bmatrix} 6 & 7 \\ 8 & 9 \end{bmatrix}$.

Verify that $(AB)^{-1} = B^{-1} \cdot A^{-1}$

15. If $A = \begin{bmatrix} 2 & 2 & 1 \\ -2 & 1 & 2 \\ 1 & -2 & 2 \end{bmatrix}$ and $B = \begin{bmatrix} 1 & 3 & 2 \\ 1 & 1 & 1 \\ 2 & -3 & -1 \end{bmatrix}$.

Verify that $(AB)^{-1} = B^{-1} A^{-1}$.

16. Find the inverse of matrix $A = \begin{bmatrix} 3 & -3 & 4 \\ 2 & -3 & 4 \\ 0 & -1 & 1 \end{bmatrix}$.

Also verify that $A^{-1} = A^3$.

17. Find the inverse of the matrix

$$A = \begin{bmatrix} a & b \\ b & \dfrac{1+bc}{a} \end{bmatrix}$$

and show that $aA^{-1} = (a^2 + bc + 1) I - aA$.

18. If $A = \begin{bmatrix} 2 & 2 & 0 \\ 2 & 1 & 1 \\ -7 & 2 & -3 \end{bmatrix}$, verify that $A^3 - 13A + 12I = 0$, where I is the unit matrix of order 3. Hence, obtain A^{-1}.

ANSWERS

1. Adj. $A = \begin{bmatrix} -1 & 4 & -6 \\ 4 & -16 & -9 \\ -6 & -13 & 1 \end{bmatrix}$, $A^{-1} = \dfrac{-1}{37}\begin{bmatrix} -1 & 4 & -6 \\ 4 & -16 & -9 \\ -6 & -13 & 1 \end{bmatrix}$ **5.** $A = \begin{bmatrix} 2 & -4 \\ -1 & 7 \end{bmatrix}$.

6. $\begin{bmatrix} 5 & 1 & 3 \\ 8 & 0 & 2 \\ -2 & 1 & -2 \end{bmatrix}$ **7.** $\begin{bmatrix} -1 & -1 \\ -1 & -1 \end{bmatrix}$ **8.** $\begin{bmatrix} -3 & 2 & 0 \\ -4 & 5 & -2 \\ -5 & 3 & 0 \end{bmatrix}$ **9.** $p = -1, q = 2, r = 1$

10. $A^{-1} = -\dfrac{1}{12}\begin{bmatrix} -5 & 6 & 2 \\ -1 & -6 & -2 \\ 11 & -18 & -2 \end{bmatrix}$ **11.** Yes **12.** $A^{-1} = \begin{bmatrix} a-ib & -c-id \\ c-id & a+ib \end{bmatrix}$

13. Adj. $(A) = \begin{bmatrix} 4 & -8 & 8 \\ 2 & 2 & -2 \\ 0 & 0 & 6 \end{bmatrix}$, $A^{-1} = \dfrac{1}{12}\begin{bmatrix} 4 & -8 & 8 \\ 2 & 2 & -2 \\ 0 & 0 & 6 \end{bmatrix}$ **16.** $A^{-1} = \begin{bmatrix} 1 & -1 & 0 \\ -2 & 3 & -4 \\ -2 & 3 & -3 \end{bmatrix}$

17. $A^{-1} = \begin{bmatrix} \dfrac{1+bc}{a} & -b \\ -c & a \end{bmatrix}$ **18.** $A^{-1} = \dfrac{1}{2}\begin{bmatrix} 5 & -6 & -2 \\ 1 & 6 & 2 \\ -11 & 18 & 2 \end{bmatrix}$

1.18 SOLUTION OF SYSTEM OF LINEAR EQUATIONS

Consider a system of n linear simultaneous equations in n variables $x_1, x_2, ..., x_n$ such that

$$a_{11} x_1 + a_{12} x_2 + ... + a_{1n} x_n = b_1$$
$$b_{21} x_1 + a_{22} x_2 + ... + a_{2n} x_n = b_2$$
$$\ldots\ldots\ldots\ldots\ldots\ldots\ldots\ldots\ldots$$
$$a_{n1} x_1 + a_{n2} x_2 + ... + a_{nn} x_n = b_n$$

where $b_1, b_2, ..., b_n$ are constants.

This system of equations is called system of linear simultaneous equations.

The system can be written in matrix form as

$$AX = B \quad ...(1)$$

where $A = \begin{bmatrix} a_{11} & a_{12} & \ldots & a_{1n} \\ a_{21} & a_{22} & \ldots & a_{2n} \\ \vdots & \vdots & \ldots & \vdots \\ a_{n1} & a_{n2} & \ldots & a_{nn} \end{bmatrix}$ is called coefficient matrix

$X = \begin{bmatrix} x_1 \\ x_2 \\ \vdots \\ x_n \end{bmatrix}$ is called the column matrix of unknowns (variables)

$B = \begin{bmatrix} b_1 \\ b_2 \\ \vdots \\ b_n \end{bmatrix}$ is called the column matrix of the constants appear in R.H.S.

1.18.1 Consistency of Linear Equations

■ **DEFINITION :** *A system of equations is said to be consistent if the system has one or more solutions otherwise it is inconsistent. In other words if the system (1) has solution than it is consistent otherwise inconsistent.*

- If the matrix $B = 0$ *i.e.,* $b_1 = b_2 = ... = b_n = 0$ then (1) is called *system of homogeneous* equations.
- A system of linear homogeneous equations has a non-zero solution if the number of equations is less then the number of unknowns.
- If $|A| \neq 0$, then the system is consistent and has unique solution given by $X = A^{-1} \cdot B$ and obtained by the following method.

If $|A| \neq 0$, then A is non-singular and hence A^{-1} exists which is unique.

Premultiplying (1) by A^{-1}, we get

$$A^{-1}(AX) = A^{-1} B$$
$$(A^{-1} A) X = A^{-1} B \quad [\because A^{-1} A = I]$$

$$IX = A^{-1} B$$
$$X = A^{-1} B$$

This method is called **Matrix Method.**

- If $|A| = 0$, then the system of equations has either no solution or infinite number of solutions.
- If $Adj\,(A) \cdot B \neq 0$, the system is inconsistent and has no solution.
- If $Adj\,(A) \cdot B = 0$, the system is consistent and has infinite number of solutions.

  Here, in this section we have to discuss only those problems in which $|A| \neq 0$. To solve such type of problems we have three methods such as matrix inversion method, Gauss elimination method and Cramer's rule.

1.18.2 Matrix Inversion Method

In this method we use the formula

$$X = A^{-1} \cdot B$$

Example 1.97

Solve by matrix method

$$2x + 3y + 4z = 11$$
$$x + 5y + 7z = 15$$
$$3x + 11y + 13z = 25$$

SOLUTION : The given system can be written as $AX = B$

where $A = \begin{bmatrix} 2 & 3 & 4 \\ 1 & 5 & 7 \\ 3 & 11 & 13 \end{bmatrix}$, $X = \begin{bmatrix} x \\ y \\ z \end{bmatrix}$, $B = \begin{bmatrix} 11 \\ 15 \\ 25 \end{bmatrix}$

Now $|A| = \begin{vmatrix} 2 & 3 & 4 \\ 1 & 5 & 7 \\ 3 & 11 & 13 \end{vmatrix} = -16 \neq 0$ *i.e.*, A is non-singular.

The system is consistent and has unique solution given by

$$X = A^{-1} B \qquad \text{...(1)}$$

Co-factors of each element of matrix A are

$A_{11} = \begin{vmatrix} 5 & 7 \\ 11 & 13 \end{vmatrix} = -12,$ $\quad A_{12} = -\begin{vmatrix} 1 & 7 \\ 3 & 13 \end{vmatrix} = 8,$ $\quad A_{31} = \begin{vmatrix} 3 & 4 \\ 5 & 7 \end{vmatrix} = 1$

$A_{21} = -\begin{vmatrix} 3 & 4 \\ 11 & 13 \end{vmatrix} = 5,$ $\quad A_{22} = \begin{vmatrix} 2 & 4 \\ 3 & 13 \end{vmatrix} = 14,$ $\quad A_{32} = -\begin{vmatrix} 2 & 4 \\ 1 & 7 \end{vmatrix} = -10$

$A_{13} = \begin{vmatrix} 1 & 5 \\ 3 & 11 \end{vmatrix} = -4,$ $\quad A_{23} = -\begin{vmatrix} 2 & 3 \\ 3 & 11 \end{vmatrix} = -13,$ $\quad A_{33} = \begin{vmatrix} 2 & 3 \\ 1 & 5 \end{vmatrix} = 7$

$$Adj\,(A) = \begin{bmatrix} A_{11} & A_{21} & A_{31} \\ A_{12} & A_{22} & A_{32} \\ A_{13} & A_{23} & A_{33} \end{bmatrix} = \begin{bmatrix} -12 & 5 & 1 \\ 8 & 14 & -10 \\ -4 & -13 & 7 \end{bmatrix}$$

$$\text{Now, } A^{-1} = \frac{Adj\,(A)}{|A|} = -\frac{1}{16}\begin{bmatrix} -12 & 5 & 1 \\ 8 & 14 & -10 \\ -4 & -13 & 7 \end{bmatrix} \qquad \text{...(2)}$$

From (1) and (2)

$$\begin{bmatrix} x \\ y \\ z \end{bmatrix} = -\frac{1}{16}\begin{bmatrix} -12 & 5 & 1 \\ 8 & 14 & -10 \\ -4 & -13 & 7 \end{bmatrix}\begin{bmatrix} 11 \\ 15 \\ 25 \end{bmatrix}$$

$$= -\frac{1}{16}\begin{bmatrix} -32 \\ 48 \\ -64 \end{bmatrix} = \begin{bmatrix} 2 \\ -3 \\ 4 \end{bmatrix}$$

$\therefore \quad x = 2, y = -3, z = 4$

Example 1.98

Test for consistency of the following set of equations :
$2x - 3y + 5z = 1$, $3x + y - z = 2$, $x + 4y - 6z = 1$. If consistent, solve the system.

SOLUTION : The system of equations can be written as $AX = B$

where $\quad A = \begin{bmatrix} 2 & -3 & 5 \\ 3 & 1 & -1 \\ 1 & 4 & -6 \end{bmatrix}$, $X = \begin{bmatrix} x \\ y \\ z \end{bmatrix}$, $B = \begin{bmatrix} 1 \\ 2 \\ 1 \end{bmatrix}$

Now $|A| = \begin{vmatrix} 2 & -3 & 5 \\ 3 & 1 & -1 \\ 1 & 4 & -6 \end{vmatrix} = 0$ *i.e.*, A is singular and hence A^{-1} does not exists. Now the system of equations is either consistent or inconsistent if $Adj(A) \cdot B = 0$ or $Adj(A) \cdot B \neq 0$ respectively.

Now, co-factor of each element of A are

$$A_{11} = -2, \quad A_{21} = 2, \quad A_{31} = -2$$
$$A_{12} = 17, \quad A_{22} = -17, \quad A_{32} = 17$$
$$A_{13} = 11, \quad A_{23} = -11, \quad A_{33} = 11$$

$$Adj(A) = \begin{bmatrix} A_{11} & A_{21} & A_{31} \\ A_{12} & A_{22} & A_{32} \\ A_{13} & A_{23} & A_{33} \end{bmatrix} = \begin{bmatrix} -2 & 2 & -2 \\ 17 & -17 & 17 \\ 11 & -11 & 11 \end{bmatrix}$$

Now $\quad Adj(A) \cdot B = \begin{bmatrix} -2 & 2 & -2 \\ 17 & -17 & 17 \\ 11 & -11 & 11 \end{bmatrix} \cdot \begin{bmatrix} 1 \\ 2 \\ 1 \end{bmatrix}$

$$= 2(17)(11)\begin{bmatrix} -1 & 1 & -1 \\ 1 & -1 & 1 \\ 1 & -1 & 1 \end{bmatrix}\begin{bmatrix} 1 \\ 2 \\ 1 \end{bmatrix}$$

$$= 374\begin{bmatrix} -1+2-1 \\ 1-2+1 \\ 1-2+1 \end{bmatrix} = 374\begin{bmatrix} 0 \\ 0 \\ 0 \end{bmatrix}$$

$$= O$$

Thus, the given system of equations is consistent and has infinitely many solutions and obtained as follows:

Take z = some constant say m in first two equations

$$\left.\begin{aligned} 2x - 3y + 5m &= 1 \\ 3x + y - m &= 2 \end{aligned}\right]$$

i.e.,
$$2x - 3y = 1 - 5m$$
$$3x + y = 2 + m$$
$$\begin{bmatrix} 2 & -3 \\ 3 & 1 \end{bmatrix}\begin{bmatrix} x \\ y \end{bmatrix} = \begin{bmatrix} 1 - 5m \\ 2 + m \end{bmatrix}$$

Hence the new $A = \begin{bmatrix} 2 & -3 \\ 3 & 1 \end{bmatrix}, X = \begin{bmatrix} x \\ y \end{bmatrix}, B = \begin{bmatrix} 1 - 5m \\ 2 + m \end{bmatrix}$

$$|A| = \begin{vmatrix} 2 & -3 \\ 3 & 1 \end{vmatrix} = 2 + 9 = 11 \neq 0.$$

A is singular $\Rightarrow$ A^{-1} exists.

Now, $A^{-1} = \dfrac{Adj\,(A)}{|A|} = \dfrac{1}{11}\begin{bmatrix} 1 & 3 \\ -3 & 2 \end{bmatrix}$

$$X = A^{-1} \cdot B = \frac{1}{11}\begin{bmatrix} 1 & 3 \\ -3 & 2 \end{bmatrix} \cdot \begin{bmatrix} 1 - 5m \\ 2 + m \end{bmatrix}$$

$$\begin{bmatrix} x \\ y \end{bmatrix} = \frac{1}{11}\begin{bmatrix} 7 - 2m \\ 1 + 17m \end{bmatrix}$$

$$\therefore \quad x = \frac{7 - 2m}{11}, y = \frac{1 + 17m}{m}.$$

The solution of the given system is

$$x = \frac{7 - 2m}{11}, y = \frac{1 + 17m}{11}, z = m$$

Example 1.99

Solve the following equations

$$2x - y + 3z = 1, \; x + 2y - z = 2, \; 5y - 5z = 3$$

using matrix method. ***[UPTU 2002, 05]***

SOLUTION : The given system can be written as

$$AX = B$$

where $A = \begin{bmatrix} 2 & -1 & 3 \\ 1 & 2 & -1 \\ 0 & 5 & -5 \end{bmatrix}$, $X = \begin{bmatrix} x \\ y \\ z \end{bmatrix}$ and $B = \begin{bmatrix} 1 \\ 2 \\ 3 \end{bmatrix}$

Now $\quad |A| = \begin{vmatrix} 2 & -1 & 3 \\ 1 & 2 & -1 \\ 0 & 5 & -5 \end{vmatrix} = 0$

i.e., A is singular and hence, A^{-1} does not exists. Now the system of equations is either consistent or inconsistent if Adj. $(A) \cdot B = 0$ or Adj. $(A) \cdot B \neq 0$ respectively.

Now, co-factor of each element of A are

$$A_{11} = -5, \; A_{12} = 5, \; A_{13} = 5$$
$$A_{21} = 10, \; A_{22} = -10, \; A_{23} = -10$$
$$A_{31} = -5, \; A_{32} = 5, \; A_{33} = 5$$

$$\therefore \quad \text{Adj. } A = \begin{bmatrix} A_{11} & A_{21} & A_{31} \\ A_{12} & A_{22} & A_{32} \\ A_{13} & A_{23} & A_{33} \end{bmatrix} = \begin{bmatrix} -5 & 10 & -5 \\ 5 & -10 & 5 \\ 5 & -10 & 5 \end{bmatrix}$$

Now,
$$\text{Adj. } A \cdot B = \begin{bmatrix} -5 & 10 & -5 \\ 5 & -10 & 5 \\ 5 & -10 & 5 \end{bmatrix} \begin{bmatrix} 1 \\ 2 \\ 3 \end{bmatrix} = \begin{bmatrix} 0 \\ 0 \\ 0 \end{bmatrix} = 0$$

Thus, the given system is consistent and has infinitely many solutions and obtained as follows:

Take z = some constant say m in first two equations

$$2x - y + 3m = 1$$
$$x + 2y - m = 2$$

or
$$2x - y = 1 - 3m$$
$$x + 2y = 2 + m$$

or
$$\begin{bmatrix} 2 & -1 \\ 1 & 2 \end{bmatrix} \begin{bmatrix} x \\ y \end{bmatrix} = \begin{bmatrix} 1 - 3m \\ 2 + m \end{bmatrix}$$

Hence, the new
$$A = \begin{bmatrix} 2 & -1 \\ 1 & 2 \end{bmatrix}, \; X = \begin{bmatrix} x \\ y \end{bmatrix}, \; B = \begin{bmatrix} 1 - 3m \\ 2 + m \end{bmatrix}$$

$$\therefore \quad |A| = \begin{vmatrix} 2 & -1 \\ 1 & 2 \end{vmatrix} = 4 + 1 = 5 \neq 0$$

$\Rightarrow A$ is singular. Hence, A^{-1} exists.

Now
$$A^{-1} = \frac{\text{Adj. } A}{|A|} = \frac{1}{5} \begin{bmatrix} 2 & -1 \\ 1 & 2 \end{bmatrix}$$

Also
$$X = A^{-1} \cdot B = \frac{1}{5} \begin{bmatrix} 2 & -1 \\ 1 & 2 \end{bmatrix} \begin{bmatrix} 1 - 3m \\ 2 + m \end{bmatrix}$$

$$\Rightarrow \quad \begin{bmatrix} x \\ y \end{bmatrix} = \frac{1}{5} \begin{bmatrix} -7m \\ 5 - m \end{bmatrix}$$

$$\therefore \quad x = \frac{-7m}{5}, \; y = \frac{5 - m}{5}$$

The solution of the given system is

$$x = \frac{-7m}{5}, \; y = \frac{5 - m}{5}, \; z = m$$

Example 1.100

Solve using matrix method, the following equations:

$$x + y = 0, \; y + z = 1, \; z + x = 3$$ ***[UPTU 2004]***

SOLUTION : The given system can be written as

$$AX = B$$

where
$$A = \begin{bmatrix} 1 & 1 & 0 \\ 0 & 1 & 1 \\ 1 & 0 & 1 \end{bmatrix}, \; X = \begin{bmatrix} x \\ y \\ z \end{bmatrix} \text{ and } B = \begin{bmatrix} 0 \\ 1 \\ 3 \end{bmatrix}$$

Now $|A| = \begin{vmatrix} 1 & 1 & 0 \\ 0 & 1 & 1 \\ 1 & 0 & 1 \end{vmatrix} = 2 \neq 0$. A is non-singular.

The system is consistent and has unique solution given by

$$X = A^{-1}B \qquad ...(1)$$

Co -factors of each element of matrix A are:

$$A_{11} = 1,\ A_{12} = 1,\ A_{13} = -1$$

$$A_{21} = -1,\ A_{22} = 1,\ A_{23} = 1$$

$$A_{31} = 1,\ A_{32} = -1,\ A_{33} = 1$$

$$\therefore \quad \text{Adj. }(A) = \begin{bmatrix} A_{11} & A_{21} & A_{31} \\ A_{12} & A_{22} & A_{32} \\ A_{13} & A_{23} & A_{33} \end{bmatrix} = \begin{bmatrix} 1 & -1 & 1 \\ 1 & 1 & -1 \\ -1 & 1 & 1 \end{bmatrix}$$

Now $$A^{-1} = \frac{\text{Adj. }(A)}{|A|} = \frac{1}{2}\begin{bmatrix} 1 & -1 & 1 \\ 1 & 1 & -1 \\ -1 & 1 & 1 \end{bmatrix} \qquad ...(2)$$

From (1) and (2), we obtain

$$\begin{bmatrix} x \\ y \\ z \end{bmatrix} = \frac{1}{2}\begin{bmatrix} 1 & -1 & 1 \\ 1 & 1 & -1 \\ -1 & 1 & 1 \end{bmatrix}\begin{bmatrix} 0 \\ 1 \\ 3 \end{bmatrix} = \frac{1}{2}\begin{bmatrix} 0-1+3 \\ 0+1-3 \\ 0+1+3 \end{bmatrix} = \frac{1}{2}\begin{bmatrix} 2 \\ -2 \\ 4 \end{bmatrix}$$

$$\therefore \quad \begin{bmatrix} x \\ y \\ z \end{bmatrix} = \begin{bmatrix} 1 \\ -1 \\ 2 \end{bmatrix} \Rightarrow x = 1,\ y = -1, z = 2.$$

Example 1.101

Solve the following system of equations by matrix method

$3x - 2y + 3z = 8,\ 2x + y - z = 1,\ 4x - 3y + 2z = 4$ ***[UPTU (C.O.) 2006]***

SOLUTION : The given system can be written as

$$AX = B$$

where $A = \begin{bmatrix} 3 & -2 & 3 \\ 2 & 1 & -1 \\ 4 & -3 & 2 \end{bmatrix}$, $X = \begin{bmatrix} x \\ y \\ z \end{bmatrix}$ and $B = \begin{bmatrix} 8 \\ 1 \\ 4 \end{bmatrix}$

Now $$|A| = \begin{vmatrix} 3 & -2 & 3 \\ 2 & 1 & -1 \\ 4 & -3 & 2 \end{vmatrix} = -17 \neq 0. \Rightarrow A \text{ is non-singular.}$$

The system is consistent and has unique solution given by

$$X = A^{-1}B \qquad ...(1)$$

Here $$\text{Adj } A = \begin{bmatrix} A_{11} & A_{21} & A_{31} \\ A_{12} & A_{22} & A_{32} \\ A_{13} & A_{23} & A_{33} \end{bmatrix}$$

$$= \begin{bmatrix} -1 & -5 & -1 \\ -8 & -6 & 9 \\ -10 & 1 & 7 \end{bmatrix}$$

Now $$A^{-1} = \frac{\text{Adj }(A)}{|A|} = -\frac{1}{17}\begin{bmatrix} -1 & -5 & -1 \\ -8 & -6 & 9 \\ -10 & 1 & 7 \end{bmatrix}$$

From (1)

$$\begin{bmatrix} x \\ y \\ z \end{bmatrix} = -\frac{1}{17}\begin{bmatrix} -1 & -5 & -1 \\ -8 & -6 & 9 \\ -10 & 1 & 7 \end{bmatrix}\begin{bmatrix} 8 \\ 1 \\ 4 \end{bmatrix}$$

$$= -\frac{1}{17}\begin{bmatrix} -17 \\ -34 \\ -51 \end{bmatrix} = \begin{bmatrix} 1 \\ 2 \\ 3 \end{bmatrix}$$

$\Rightarrow x = 1, y = 2, z = 3.$

Example 1.102

If $A = \begin{bmatrix} 1 & 1 & 1 \\ 1 & -1 & 1 \\ 2 & 1 & -1 \end{bmatrix}$, find A^{-1}. Hence solve the equations

$$x + y + z = 6, x - y + z = 2, 2x + y - z = 1$$ *[UPTU 2007]*

SOLUTION : We have

$$A = \begin{bmatrix} 1 & 1 & 1 \\ 1 & -1 & 1 \\ 2 & 1 & -1 \end{bmatrix}$$

$\therefore$ $$|A| = \begin{vmatrix} 1 & -1 & 1 \\ 1 & -1 & 1 \\ 2 & 1 & -1 \end{vmatrix} = 1(1-1) - 1(-1-2) + 1(1+2)$$

$$= 6$$

Cofactors of A are

$$A_{11} = 0, \quad A_{12} = 3, \quad A_{13} = 3$$
$$A_{21} = 2, \quad A_{22} = -3, \quad A_{23} = 1$$
$$A_{31} = 2, \quad A_{32} = 0, \quad A_{33} = -2$$

$$\text{Adj }(A) = \begin{bmatrix} A_{11} & A_{21} & A_{31} \\ A_{12} & A_{22} & A_{32} \\ A_{13} & A_{23} & A_{33} \end{bmatrix} = \begin{bmatrix} 0 & 2 & 2 \\ 3 & -3 & 0 \\ 3 & 1 & -2 \end{bmatrix}$$

$\therefore$ $$A^{-1} = \frac{\text{Adj }(A)}{|A|} = \frac{1}{6}\begin{bmatrix} 0 & 2 & 2 \\ 3 & -3 & 0 \\ 3 & 1 & -2 \end{bmatrix}$$

The system of equations can be written as

$$AX = B$$

where $A = \begin{bmatrix} 1 & 1 & 1 \\ 2 & -1 & 1 \\ 2 & 1 & -1 \end{bmatrix}, X = \begin{bmatrix} x \\ y \\ z \end{bmatrix}, B = \begin{bmatrix} 6 \\ 2 \\ 1 \end{bmatrix}$

By matrix methid, the solution of given system of equatons is

$$X = A^{-1}B$$

$$\Rightarrow \qquad \begin{bmatrix} x \\ y \\ z \end{bmatrix} = \frac{1}{6}\begin{bmatrix} 0 & 2 & 2 \\ 3 & -3 & 0 \\ 3 & 1 & -2 \end{bmatrix}\begin{bmatrix} 6 \\ 2 \\ 1 \end{bmatrix} = \begin{bmatrix} 1 \\ 2 \\ 3 \end{bmatrix}$$

$$\Rightarrow \qquad x = 1, y = 2, z = 3$$

1.18.3 Gauss-Elimination Method

In this method the given coefficient matrix of the system of simultaneous equations is reduced to an equivalent simpler form in such a way that $a_{mn} = 0$ for $n > m$ by using row transformations only.

By this reduction the coefficient matrix reduced to the upper triangular form.

The method is applicable in both type of systems *i.e.*, either homogeneous or non-homogeneous. This method can also be used even if the coefficient matrix is singular or non-singular.

Example 1.103

Solve the system of equations by using Gauss elimination method

$$x + y + z = 6, x - y + z = 2, 2x + y - z = 1$$

SOLUTION : The system of equations can be written as $AX = B$

i.e., $$\begin{bmatrix} 1 & 1 & 1 \\ 1 & -1 & 1 \\ 2 & 1 & -1 \end{bmatrix}\begin{bmatrix} x \\ y \\ z \end{bmatrix} = \begin{bmatrix} 6 \\ 2 \\ 1 \end{bmatrix}$$

$$\begin{bmatrix} 1 & 1 & 1 \\ 0 & -2 & 0 \\ 0 & -1 & -3 \end{bmatrix}\begin{bmatrix} x \\ y \\ z \end{bmatrix} = \begin{bmatrix} 6 \\ -4 \\ -11 \end{bmatrix} \quad \begin{matrix} R_2 \to R_2 - R_1 \\ R_3 \to R_3 - 2R_1 \end{matrix}$$

$$\begin{bmatrix} 1 & 1 & 1 \\ 0 & -2 & 0 \\ 0 & 0 & -6 \end{bmatrix}\begin{bmatrix} x \\ y \\ z \end{bmatrix} = \begin{bmatrix} 6 \\ -4 \\ -18 \end{bmatrix} \quad R_3 \to 2R_3 - R_2$$

$$x + y + z = 6$$
$$-2y = -4$$
$$-6z = -18$$

Solving the three equations, we get

$$z = 3, y = 2, x = 1$$

Example 1.104

Solve $2x - y + 3z = 9, -x + 2y + z = 6, 3x + y - 4z = -7$ by Gauss elimination method.

SOLUTION : The system can be written as

$$AX = B$$

$$\begin{bmatrix} 2 & -1 & 3 \\ -1 & 2 & 1 \\ 3 & 1 & -4 \end{bmatrix}\begin{bmatrix} x \\ y \\ z \end{bmatrix} = \begin{bmatrix} 9 \\ 6 \\ -7 \end{bmatrix}$$

$$\begin{bmatrix} 2 & -1 & 3 \\ 0 & 3 & 5 \\ 0 & 5 & -17 \end{bmatrix}\begin{bmatrix} x \\ y \\ z \end{bmatrix} = \begin{bmatrix} 9 \\ 21 \\ -41 \end{bmatrix} \quad \begin{matrix} R_2 \to 2R_2 + R_1 \\ R_3 \to 2R_3 - 3R_1 \end{matrix}$$

$$\begin{bmatrix} 2 & -1 & 3 \\ 0 & 3 & 5 \\ 0 & 0 & -76 \end{bmatrix}\begin{bmatrix} x \\ y \\ z \end{bmatrix} = \begin{bmatrix} 9 \\ 21 \\ -228 \end{bmatrix} \quad R_3 \to 3R_3 - 5R_2$$

$$2x - y + 3z = 9$$
$$3y + 5z = 21$$
$$-76z = -228$$

Solving these three equations, we get $z = 3, y = 2, x = 1$

1.18.4 Cramer's Rule

By this rule we can find the solution of those system of linear equations in which the number of unknowns are equal to the number of equations and the coefficient matrix of the system is non-singular.

Consider the system of equations

$$a_1x + b_1y + c_1z = d_1$$
$$a_2x + b_2y + c_2z = d_2$$
$$a_3x + b_3y + c_3z = d_3$$

The system of equation can be written as $AX = B$

$$\begin{bmatrix} a_1 & b_1 & c_1 \\ a_2 & b_2 & c_2 \\ a_3 & b_3 & c_3 \end{bmatrix}\begin{bmatrix} x \\ y \\ z \end{bmatrix} = \begin{bmatrix} d_1 \\ d_2 \\ d_3 \end{bmatrix}$$

Let
$$D = \begin{vmatrix} a_1 & b_1 & c_1 \\ a_2 & b_2 & c_2 \\ a_3 & b_3 & c_3 \end{vmatrix}$$

Replace the first column of D by the elements d_1, d_2, d_3 respectively, we get D_x

i.e.,
$$D_x = \begin{vmatrix} d_1 & b_1 & c_1 \\ d_2 & b_2 & c_2 \\ d_3 & b_3 & c_3 \end{vmatrix}$$

Similarly, we can obtain D_y and D_z by replacing second column of D by the elements d_1, d_2, d_3 and third column of D by the elements d_1, d_2 and d_3 respectively, we obtain

$$D_y = \begin{vmatrix} a_1 & d_1 & c_1 \\ a_2 & d_2 & c_2 \\ a_3 & d_3 & c_3 \end{vmatrix}, D_z = \begin{vmatrix} a_1 & b_1 & d_1 \\ a_2 & b_2 & d_2 \\ a_3 & b_3 & d_3 \end{vmatrix}$$

The values of x, y and z can be obtained by the formula

$$\frac{x}{D_x} = \frac{y}{D_y} = \frac{z}{D_z} = \frac{1}{D}, D \neq 0.$$

Example 1.105

Solve the following system of equations using cramer's rule.

$$x - y + 2z = 9,\ 2x - 5y + 3z = 18,\ 6x + 7y + 10z = 35$$

SOLUTION : The given system of equations can be written as $AX = B$,

where, $A = \begin{bmatrix} 1 & -1 & 2 \\ 2 & -5 & 3 \\ 6 & 7 & 10 \end{bmatrix}$, $X = \begin{bmatrix} x \\ y \\ z \end{bmatrix}$ and $B = \begin{bmatrix} 9 \\ 18 \\ 35 \end{bmatrix}$

Let $D = \begin{vmatrix} 1 & -1 & 2 \\ 2 & -5 & 3 \\ 6 & 7 & 10 \end{vmatrix} = 19$

Then, $D_x = \begin{vmatrix} 9 & -1 & 2 \\ 18 & -5 & 3 \\ 35 & 7 & 10 \end{vmatrix} = 38$

$D_y = \begin{vmatrix} 1 & 9 & 2 \\ 2 & 18 & 3 \\ 6 & 35 & 10 \end{vmatrix} = -19$

$D_z = \begin{vmatrix} 1 & -1 & 9 \\ 2 & -5 & 18 \\ 6 & 7 & 35 \end{vmatrix} = 57$

Now by Cramer's rule

$$\frac{x}{D_x} = \frac{y}{D_y} = \frac{z}{D_z} = \frac{1}{D} \quad \Rightarrow \quad x = \frac{D_x}{D}, y = \frac{D_y}{D}, z = \frac{D_z}{D}$$

$\Rightarrow x = 2, y = -1, z = 3$

Example 1.106

Monthly expenditure in an office for three months is given below. Assuming that the salary in all three months, of different categories of staff did not vary, calculate the salary of each type of staff :

Month	Number of employees			Total monthly salary (Rs.)
	Clerks	Typists	Peons	
April	3	4	3	4000
May	5	4	2	4800
June	2	3	4	5000

SOLUTION : Let x, y and z be the salary of Clerk, Typist and Peon respectively, then

$$3x + 4y + 3z = 4000$$
$$5x + 4y + 2z = 4800$$
$$2x + 3y + 4z = 5000$$

This system of equations can solved by any of the three method given above. Let us use matrix method.

$$\begin{bmatrix} 3 & 4 & 3 \\ 5 & 4 & 2 \\ 2 & 3 & 4 \end{bmatrix}\begin{bmatrix} x \\ y \\ z \end{bmatrix} = \begin{bmatrix} 4000 \\ 4800 \\ 5000 \end{bmatrix}$$

Here $A = \begin{bmatrix} 3 & 4 & 3 \\ 5 & 4 & 2 \\ 2 & 3 & 4 \end{bmatrix}$

Then $A = \begin{vmatrix} 3 & 4 & 3 \\ 5 & 4 & 2 \\ 2 & 3 & 4 \end{vmatrix} = -13 \neq 0$

Thus, matrix A is non-singular. The co-factors of each element of matrix A are

$$A_{11} = 10, \quad A_{21} = -7, \quad A_{31} = -4$$
$$A_{12} = -16, \quad A_{22} = 6, \quad A_{32} = 9$$
$$A_{13} = 7, \quad A_{23} = -1, \quad A_{33} = -8$$

$$Adj\,(A) = \begin{bmatrix} A_{11} & A_{21} & A_{31} \\ A_{12} & A_{22} & A_{32} \\ A_{13} & A_{23} & A_{33} \end{bmatrix} = \begin{bmatrix} 10 & -7 & -4 \\ -16 & 6 & 9 \\ 7 & -1 & -8 \end{bmatrix}$$

$$A^{-1} = \frac{Adj\,(A)}{|A|} = -\frac{1}{13}\begin{bmatrix} 10 & -7 & -4 \\ -16 & 6 & 9 \\ 7 & -1 & -8 \end{bmatrix}$$

By matrix inversion method

$$X = A^{-1} B$$

$$\begin{bmatrix} x \\ y \\ z \end{bmatrix} = -\frac{1}{13}\begin{bmatrix} 10 & -7 & -4 \\ -16 & 6 & 9 \\ 7 & -1 & -8 \end{bmatrix}\begin{bmatrix} 4000 \\ 4800 \\ 5000 \end{bmatrix}$$

$$= -\frac{1}{13}\begin{bmatrix} 40000 - 33600 - 20000 \\ -64000 + 28800 + 45000 \\ 28000 - 4800 - 40000 \end{bmatrix}$$

$$= -\frac{1}{13}\begin{bmatrix} -13600 \\ 9800 \\ -16800 \end{bmatrix}$$

$$x = 1046{\cdot}15,\ y = -753{\cdot}85,\ z = 1292{\cdot}31$$

Example 1.107

Solve the following system of equations by Cramer's rule :

$$x - 2y = 4, \quad -3x + 5y = -7$$ ***[UPTU 2000]***

SOLUTION : The given system of equations can be written as $AX = B$

where $A = \begin{bmatrix} 1 & -2 \\ -3 & 5 \end{bmatrix}$, $X = \begin{bmatrix} x \\ y \end{bmatrix}$, $B = \begin{bmatrix} 4 \\ -7 \end{bmatrix}$

Let $D = \begin{vmatrix} 1 & -2 \\ -3 & 5 \end{vmatrix} = 5 - 6 = -1$

Then $D_x = \begin{vmatrix} 4 & -2 \\ -7 & 5 \end{vmatrix} = 20 - 14 = 6$

$$D_y = \begin{vmatrix} 1 & 4 \\ -3 & -7 \end{vmatrix} = -7 + 12 = 5$$

Now by Cramer's rule

$$\frac{x}{D_x} = \frac{y}{D_y} = \frac{1}{D} \Rightarrow x = \frac{D_x}{D} = \frac{6}{-1} = -6$$

and $$y = \frac{D_y}{D} = \frac{5}{-1} = -5$$

Hence, $x = -6, y = -5$

Example 1.108

Solve the following by using Cramer's rule:

$$x - 2y + z = 2,\ 2x - 3z = 3,\ x + y + z = 6$$ ***[UPTU 2001]***

SOLUTION : The given system of equations can be written as

$$AX = B$$

where $A = \begin{bmatrix} 1 & -2 & 1 \\ 2 & 0 & -3 \\ 1 & 1 & 1 \end{bmatrix}$, $X = \begin{bmatrix} x \\ y \\ z \end{bmatrix}$ and $B = \begin{bmatrix} 2 \\ 3 \\ 6 \end{bmatrix}$

Let $D = \begin{vmatrix} 1 & -2 & 1 \\ 2 & 0 & -3 \\ 1 & 1 & 1 \end{vmatrix} = 15$

then $D_x = \begin{vmatrix} 2 & -2 & 1 \\ 3 & 0 & -3 \\ 6 & 1 & 1 \end{vmatrix} = 51$

$$D_y = \begin{vmatrix} 1 & 2 & 1 \\ 2 & 3 & -3 \\ 1 & 6 & 1 \end{vmatrix} = 20$$

$$D_z = \begin{vmatrix} 1 & -2 & 2 \\ 2 & 0 & 3 \\ 1 & 1 & 6 \end{vmatrix} = 19$$

Now by Cramer's rule

$$\frac{x}{D_x} = \frac{y}{D_y} = \frac{z}{D_z} = \frac{1}{D}$$

$$\Rightarrow \quad x = \frac{D_x}{D} = \frac{51}{15}$$

$$y = \frac{D_y}{D} = \frac{20}{15}$$

$$z = \frac{D_z}{D} = \frac{19}{15}$$

Hence, $x = \frac{51}{15},\ y = \frac{20}{15},\ z = \frac{19}{15}$

Example 1.109

By using Cramer's rule solve the following system of equations:

$3x - 4y + 5z = -6,\ x + y - 2z = -1,\ 2x + 3y + z = 5$ *[UPTU 2003]*

SOLUTION : The given system of equations can be written as

$$AX = B$$

where, $A = \begin{bmatrix} 3 & -4 & 5 \\ 1 & 1 & -2 \\ 2 & 3 & 1 \end{bmatrix},\ X = \begin{bmatrix} x \\ y \\ z \end{bmatrix}$ and $B = \begin{bmatrix} -6 \\ -1 \\ 5 \end{bmatrix}$

Let $D = \begin{vmatrix} 3 & -4 & 5 \\ 1 & 1 & -2 \\ 2 & 3 & 1 \end{vmatrix} = 46$

then $D_x = \begin{vmatrix} -6 & -4 & 5 \\ -1 & 1 & -2 \\ 5 & 3 & 1 \end{vmatrix} = -46$

$$D_y = \begin{vmatrix} 3 & -6 & 5 \\ 1 & -1 & -2 \\ 2 & 5 & 1 \end{vmatrix} = 92$$

$$D_z = \begin{vmatrix} 3 & -4 & -6 \\ 1 & 1 & -1 \\ 2 & 3 & 5 \end{vmatrix} = 46$$

Now, by Cramer's rule

$$\frac{x}{D_x} = \frac{y}{D_y} = \frac{z}{D_z} = \frac{1}{D}$$

$$\Rightarrow \quad x = \frac{D_x}{D} = \frac{-46}{46} = -1$$

$$y = \frac{D_y}{D} = \frac{92}{46} = 2$$

$$z = \frac{D_z}{D} = \frac{46}{46} = 1$$

Hence, $x = -1,\ y = 2,\ z = 1$

Example 1.110

Using Cramer's rule, solve the system of equations

$5x - y + 4z = 5,\ 2x + 3y + 5z = 2,\ 5x - 2y + 6z = -1.$ *[UPTU (C.O.) 2006]*

SOLUTION : We have

$$5x - y + 4z = 5$$
$$2x + 3y + 5z = 2$$

$$5x - 2y + 6z = -1$$

The given system of equations can be written as

$$AX = B$$

where $A = \begin{bmatrix} 5 & -1 & 4 \\ 2 & 3 & 5 \\ 5 & -2 & 6 \end{bmatrix}, X = \begin{bmatrix} x \\ y \\ z \end{bmatrix}$ and $B = \begin{bmatrix} 5 \\ 2 \\ -1 \end{bmatrix}$

Let $D = \begin{vmatrix} 5 & -1 & 4 \\ 2 & 3 & 5 \\ 5 & -2 & 6 \end{vmatrix} = 51$

then $D_x = \begin{vmatrix} 5 & -1 & 4 \\ 2 & -2 & 5 \\ -1 & -2 & 6 \end{vmatrix} = 153$

$$D_y = \begin{vmatrix} 5 & 5 & 4 \\ 2 & 2 & 5 \\ 5 & -1 & 6 \end{vmatrix} = 102$$

$$D_z = \begin{vmatrix} 5 & -1 & 5 \\ 2 & 3 & 2 \\ 5 & -2 & -1 \end{vmatrix} = -102$$

Now by Cramer's rule $\dfrac{x}{D_x} = \dfrac{y}{D_y} = \dfrac{z}{D_z} = \dfrac{1}{2}$

$\Rightarrow$ $$x = \frac{D_x}{D} = \frac{153}{51} = 3$$

$$y = \frac{D_y}{D} = \frac{102}{51} = 2$$

$$z = \frac{D_z}{D} = -\frac{102}{51} = -2$$

Hence, $x = 3, y = 2, z = -2$.

Example 1.111

Solve using Cramer's rule $x + y = 5, y + z = 3, z + x = 4$. ***[UPTU 2007]***

SOLUTION : The system of equations can be written as

$$AX = B$$

where $A = \begin{bmatrix} 1 & 1 & 0 \\ 0 & 1 & 1 \\ 1 & 0 & 1 \end{bmatrix}, X = \begin{bmatrix} x \\ y \\ z \end{bmatrix}, B = \begin{bmatrix} 5 \\ 3 \\ 4 \end{bmatrix}$

Let $D = \begin{vmatrix} 1 & 1 & 0 \\ 0 & 1 & 1 \\ 1 & 0 & 1 \end{vmatrix} = 2$

Then $$D_x = \begin{vmatrix} 5 & 1 & 0 \\ 3 & 1 & 1 \\ 4 & 0 & 1 \end{vmatrix} = 6$$

$$D_y = \begin{vmatrix} 1 & 5 & 0 \\ 0 & 3 & 1 \\ 1 & 4 & 1 \end{vmatrix} = 4$$

$$D_z = \begin{vmatrix} 1 & 1 & 5 \\ 0 & 1 & 3 \\ 1 & 0 & 4 \end{vmatrix} = 2$$

By Cramer's rule $x = \frac{D_x}{D}, y = \frac{D_y}{D}, z = \frac{D_z}{D}$

$\Rightarrow$ $$x = \frac{6}{2} = 3, y = \frac{4}{2} = 2, z = \frac{2}{2} = 1$$

PROBLEM SET 1.11

1. Use matrix method to solve the system of equations :
(i) $x + 2y + z = 1,\ x + 3y + 4z = 3,\ x + y - z = 2$
(ii) $2x + 3y = 5,\ -x - y = 4$
(iii) $x - y + 2z = 7,\ 3x - 2y + 4z = 0,\ x - y - z = 3$

2. Find the inverse of the matrix $\begin{bmatrix} 2 & -2 & 0 \\ 0 & 2 & -2 \\ 2 & 0 & 2 \end{bmatrix}$ and hence solve the system of equations
$2x - 2y = 3,\ 2y - 2z = -1,\ 2x + 2z = 5.$

3. Investigate that the following system of equations is consistent or not :
$$x_1 + x_2 + x_3 = 3,\ 2x_1 - x_2 + 3x_3 = 1,\ 4x_1 + x_2 + 5x_3 = 3.$$

4. Solve by matrix method :
$$x + 3y - z = 0,\ 3x + y + 2z = 4,\ 4x - 3y - 4z = -13.$$

5. Test for consistency the following system of the equations.
$$x + y + z = 3,\ x + 2y + 3z = 4,\ x + 4y + 9z = 4.$$

6. Solve the system of equations using Gauss-elimination method :
$2x - y + 3z = 9,\ -x + 2y + z = 6,\ 3x + y - 4z = -7$

7. Show that the equations $5x + 3y + 7z = 4,\ 3x + 26y + 2z = 9,\ 7x + 2y + 10z = 5$ are consistent and solve.

8. Solve by Gauss-elimination method :
(i) $x_1 + 3x_2 - 2x_3 = 0,\ 2x_1 - x_2 + 4x_3 = 0,\ x_1 - 11x_2 + 14x_3 = 0$
(ii) $4x_1 + x_2 = 0,\ -8x_1 + 2x_2 = 0$
(iii) $x_1 - x_2 + 3x_3 = 0,\ 2x_1 + 5x_2 + 6x_3 = 0.$

9. Solve the equations by using Cramer's rule:

(i) $x + y + 2z = 4,\ 2x - y + 3z = 9,\ 3x - y - z = 2$

(ii) $x + y + z = 6,\ -x + 2y + 3z = 14,\ -x + y - z = -2$

(iii) $2x + 4y + z = 5,\ x + y + z = 6,\ 2x + 3y + z = 6$

(iv) $7x - y - z = 0,\ 10x - 2y + z = 8,\ 6x + 3y - 2z = 7$

10. Solve the equations using matrix method.

(i) $2x + 3y - z = 9,\ x + y + z = 9,\ 3x - y - z = -1$

(ii) $2x + 4y - z = 9,\ 3x + y + 2z = 7,\ x + 3y - 3z = 4$

(iii) $2x - 3y = 3,\ 4x - 11y = 11$

(iv) $x + y + z = 3,\ x + 2y + 3z = 4,\ x + 4y + 9z = 6$

11. Solve the following system of equations using Cramer's rule:

(i) $x - y + 2z = 9,\ 2x - 5y + 3z = 18,\ 6x + 7y + 10z = 35$

(ii) $3x + 3y - z = 11,\ 2x - y + 2z = 9,\ 4x + 3y + 2z = 25$

(iii) $x + y - z = -3,\ 2x + 3y + z = 2,\ 8y + 3z = 1$

(iv) $x + y + z + 1 = 0,\ x + 2y + 3z + 4 = 0x + 3y + 4z + 6 = 0$

ANSWERS

1. (i) $x = 12, y = -7, z = 3$ (ii) $x = -17, y = 13$ (iii) $x = \frac{-44}{3},\ y = -19,\ z = \frac{4}{3}$

2. $\frac{1}{4}\begin{bmatrix} 1 & 1 & 1 \\ -1 & 1 & 1 \\ -1 & -1 & 1 \end{bmatrix},\ x = \frac{7}{4},\ y = \frac{1}{4},\ z = \frac{3}{4}$. **3.** Inconsistent, no solution.

4. $x = \frac{-31}{75},\ y = \frac{13}{15},\ z = \frac{164}{15}$. **5.** $x = 1, y = 3, z = -1$ **6.** $x = 1,\ y = 2,\ z = 3$

8. (i) $x_1 = \frac{6k}{7},\ x_2 = \frac{8k}{7},\ x_3 = k$ (ii) $x = 0,\ x_2 = 0$ (iii) $x_1 = -3k,\ x_2 = 0,\ x_3 = k.$

9. (i) $x = 1, y = -1, z = 2$ (ii) $x = \frac{1}{2},\ y = 2,\ z = \frac{7}{2}$ (iii) $x = 2,\ y = -1,\ z = 5$

(iv) $x = 1,\ y = 2,\ z = 4$

10. (i) $x = 2,\ y = 3,\ z = 4$ (ii) $x = 1,\ y = 2, z = 1$ (iii) $x = 0,\ y = -1$ (iv) $x = 2, y = 1, z = 0.$

MULTIPLE CHOICE QUESTIONS

1. The graph of a quadratic equation is
(a) parabola (b) ellipse
(c) circle (d) None of these

2. A quadratic equation $ax^2 + bx + c = 0$ has real distinct roots if
(a) $b^2 - 4ac > 0$ (b) $b^2 - 4ac < 0$
(c) $b^2 - 4ac = 0$ (d) None of these

3. Sum of roots of a quadratic equation $ax^2 + bx + c = 0$ is
(a) $\frac{c}{a}$ (b) $\frac{a}{c}$
(c) $-\frac{b}{a}$ (d) $\frac{b}{a}$

4. What is the degree of the equation $a_0x^3 + a_1x^2 + a_2x + a_3 = 0$
(a) 1 (b) 3
(c) 4 (d) 2

5. The degree of the equation $\frac{1}{x+5} = \frac{1}{x-1} + \frac{1}{x+3}$ is
(a) 2 (b) 1
(c) 4 (d) None of these

6. The graph of a linear equation is
(a) circle (b) ellipse
(c) straight line (d) hyperbola

7. Product of roots of a quadratic equation $ax^2 + bx + c = 0$ is
(a) $-\frac{b}{a}$ (b) $\frac{b}{a}$
(c) $-\frac{c}{a}$ (d) $\frac{c}{a}$

8. If α and β are the roots of quadratic equation $ax^2 + bx + c = 0$. The equation with roots as $\frac{1}{\alpha}, \frac{1}{\beta}$ is
(a) $ax^2 + cx + b = 0$ (b) $cx^2 + bx + a = 0$
(c) $ax^2 + bx + c = 0$ (d) $bx^2 + cx + a = 0$

9. Which of the following is a quadratic equation
(a) $x^2 + \frac{1}{x^2} = 2$ (b) $x^2 + 2x + 3 = 2x^2$
(c) $x^2 + 2\sqrt{x} + 3 = 0$ (d) $x^2 + 3x + 4 = x^2 + 2x$

10. If one root of the quadratic equation $2x^2 + kx - 6 = 0$ is 2, what is the value of k?
(a) $k = -2$ (b) $k = -4$
(c) $k = 1$ (d) $k = -1$

11. Which of the following equation can be reduced to a quadratic equation

(a) $\frac{x}{x+1} + \frac{x+1}{x} = \frac{10}{15}$ (b) $x^2 + \frac{1}{x^2} = x + \frac{1}{x}$

(c) $x + \frac{1}{x} = x - \frac{1}{x}$ (d) None of these

12. If $3^{2x} + 9 - 10 \cdot 3^x = 0$, what are the values of x?

(a) 0 and 2 (b) 1 and 1

(c) 2 and 1 (d) 0 and 3

13. Roots of the equation $x^{2/3} - 2x^{1/3} = 15$ are

(a) 15, 11 (b) 13, 12

(c) 3, 9 (d) –27, 125

14. The root(s) of the equation $\sqrt{2x+9} + x = 13$ is (are)

(a) 8, 20 (b) 0

(c) 8 (d) None of these

15. The quadratic equation with roots $\sqrt{3}$ and $3\sqrt{3}$ is

(a) $x^2 + 9x - 4\sqrt{3} = 0$ (b) $x^2 - 4\sqrt{3}\,x + 9 = 0$

(c) $x^2 + 4\sqrt{3}x + 9 = 0$ (d) none of these

16. If the sum of roots of the equation $kx^2 + 6x - 3k = 0, k \neq 0$ is equal to the product of roots, what will be the value of k?

(a) $k = 2$ (b) $k = -2$

(c) $k = -1$ (d) $k = 3$

17. The equation $5^{x+1} + 5^{2-x} = 5^3 + 1$ gives the roots as

(a) –1, 1 (b) 1, 2

(c) 2, 2 (d) 2, –1

18. The system of simultaneous equation given by $a_1x + b_1y + c_1 = 0$ and $a_2x + b_2y + c_2 = 0$ is constant if

(a) $\frac{a_1}{a_2} = \frac{b_1}{b_2} = \frac{c_1}{c_2}$ (b) $\frac{a_1}{a_2} \neq \frac{b_1}{b_2}$

(c) $\frac{a_1}{a_2} = \frac{b_1}{b_2} \neq \frac{c_1}{c_2}$ (d) None of these

19. For what value of k the system of equations $5x + 2y = k$ and $10x + 4y = 3$ has infinitely many solution

(a) 1 (b) 3

(c) 2 (d) 6

20. What will be the value of k such that the system of equations $x + 2y = 5$ and $3x + ky + 15 = 0$ has no solution

(a) 1 (b) 3

(c) no value of k (d) 5

21. What are the values of x and y if $\frac{x}{a} + \frac{y}{b} = 2$ and $ax - by = a^2 - b^2$

(a) $x = a, y = b$ (b) $x = b, y = a$

(c) $x = ab, y = 1$ (d) $x = 1, y = ab$

22. If the sum of two numbers is 35 and their difference is 13, then the numbers are

(a) 23 and 12 (b) 22 and 13
(c) 24 and 11 (d) 25 and 10

23. I am three times as old as my son. Five years later, I shall be two and a half times as old as my son. How old am I and my son respectively?

(a) 45, 15 (b) 39, 13
(c) 42, 14 (d) 30, 10

24. A number consists of two digits whose sum is five. When the digits are reversed, the number becomes greater by nine. The number is

(a) 43 (a) 23
(c) 53 (d) 41

25. The sum of the numerator and denominator of a fraction is 4 more than twice the numerator. If the numerator and denominator are increased by 3, they are in the ratio 2 : 3, the fraction is

(a) $\frac{4}{7}$ (b) $\frac{2}{5}$
(c) $\frac{5}{9}$ (d) $\frac{11}{13}$

26. Which of the following is a square matrix?

(a) $\begin{bmatrix} 1 & 3 & 2 & 4 \\ 2 & 3 & 7 & 5 \\ 3 & 9 & 1 & 8 \end{bmatrix}$ (b) $\begin{bmatrix} 1 & 2 & 1 \\ 3 & 7 & 12 \end{bmatrix}$

(c) $\begin{bmatrix} 1 & 2 \\ 1 & 3 \\ 5 & 9 \end{bmatrix}$ (d) $\begin{bmatrix} 1 & 5 & 6 \\ 3 & 1 & 7 \\ 9 & 2 & 11 \end{bmatrix}$

27. Addition of two matrices A and B is possible if

(a) No. of rows and columns in both the matrices are same
(b) No. of rows in one is equal to the no. of columns in other
(c) Always possible
(d) Cannot be added

28. Which of the followins is true

(a) A scalar matrix is always a diagonal matrix
(b) A diagonal matrix is a scalar matrix
(c) Identity matrix is not a scalar matrix
(d) None of these

29. Two matrices $A_{m \times n}$ and $B_{l \times k}$ can be multiplied if

(a) $m = l, n = k$ (b) $n = l, m, k$ arbitrary
(c) $m = k, n, l$ arbitrary (d) none of these

30. For two matrices $A_{m \times n}$ $B_{l \times k}$, AB and BA exists simultaneously if

(a) $n = l$ and $k = m$ (b) $n = l, k \neq m$
(c) $n \neq l, k = m$ (d) always

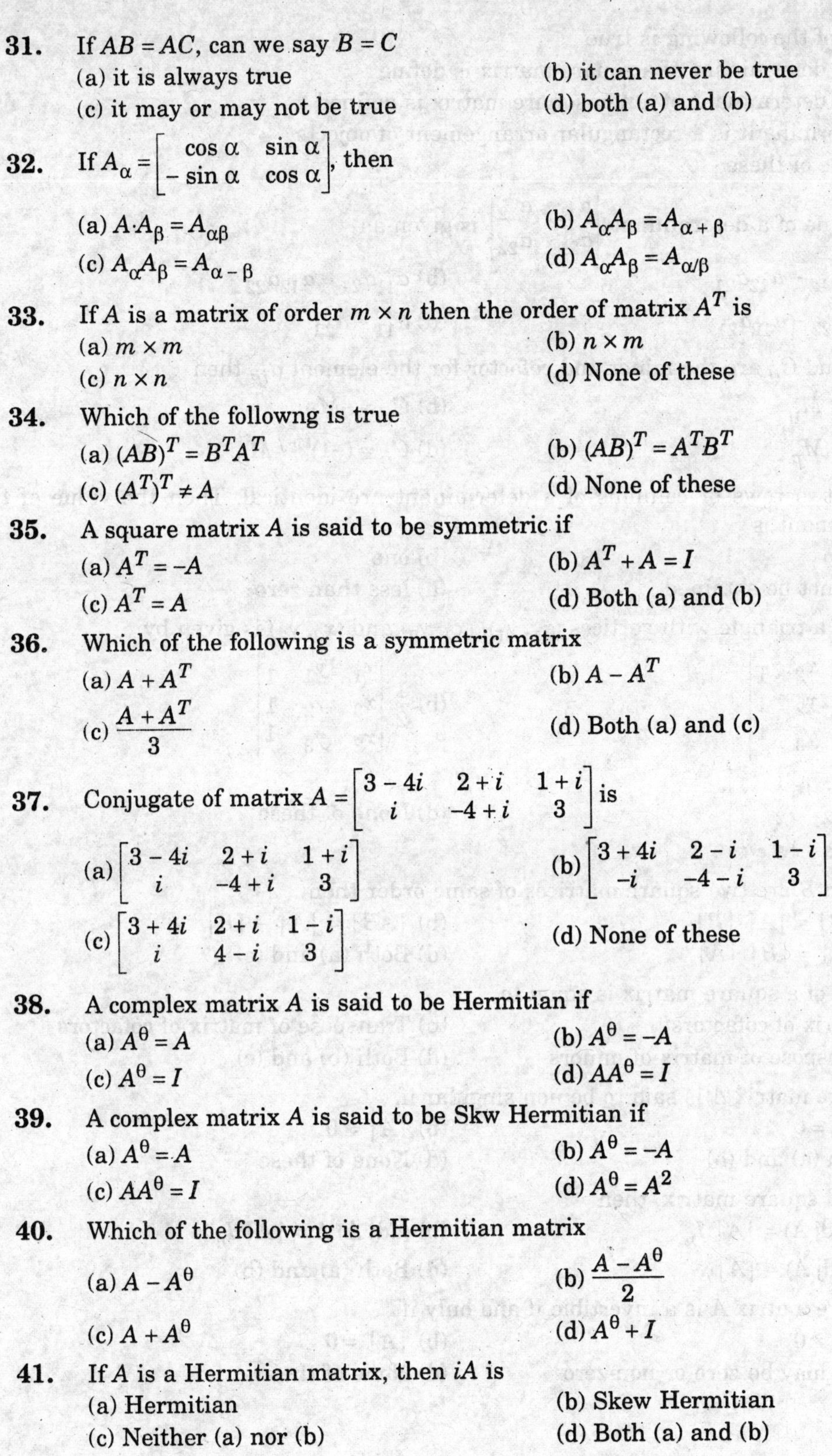

31. If $AB = AC$, can we say $B = C$

(a) it is always true
(b) it can never be true
(c) it may or may not be true
(d) both (a) and (b)

32. If $A_\alpha = \begin{bmatrix} \cos\alpha & \sin\alpha \\ -\sin\alpha & \cos\alpha \end{bmatrix}$, then

(a) $A \cdot A_\beta = A_{\alpha\beta}$
(b) $A_\alpha A_\beta = A_{\alpha+\beta}$
(c) $A_\alpha A_\beta = A_{\alpha-\beta}$
(d) $A_\alpha A_\beta = A_{\alpha/\beta}$

33. If A is a matrix of order $m \times n$ then the order of matrix A^T is

(a) $m \times m$
(b) $n \times m$
(c) $n \times n$
(d) None of these

34. Which of the followng is true

(a) $(AB)^T = B^T A^T$
(b) $(AB)^T = A^T B^T$
(c) $(A^T)^T \neq A$
(d) None of these

35. A square matrix A is said to be symmetric if

(a) $A^T = -A$
(b) $A^T + A = I$
(c) $A^T = A$
(d) Both (a) and (b)

36. Which of the following is a symmetric matrix

(a) $A + A^T$
(b) $A - A^T$
(c) $\dfrac{A + A^T}{3}$
(d) Both (a) and (c)

37. Conjugate of matrix $A = \begin{bmatrix} 3-4i & 2+i & 1+i \\ i & -4+i & 3 \end{bmatrix}$ is

(a) $\begin{bmatrix} 3-4i & 2+i & 1+i \\ i & -4+i & 3 \end{bmatrix}$
(b) $\begin{bmatrix} 3+4i & 2-i & 1-i \\ -i & -4-i & 3 \end{bmatrix}$
(c) $\begin{bmatrix} 3+4i & 2+i & 1-i \\ i & 4-i & 3 \end{bmatrix}$
(d) None of these

38. A complex matrix A is said to be Hermitian if

(a) $A^\theta = A$
(b) $A^\theta = -A$
(c) $A^\theta = I$
(d) $AA^\theta = I$

39. A complex matrix A is said to be Skw Hermitian if

(a) $A^\theta = A$
(b) $A^\theta = -A$
(c) $AA^\theta = I$
(d) $A^\theta = A^2$

40. Which of the following is a Hermitian matrix

(a) $A - A^\theta$
(b) $\dfrac{A - A^\theta}{2}$
(c) $A + A^\theta$
(d) $A^\theta + I$

41. If A is a Hermitian matrix, then iA is

(a) Hermitian
(b) Skew Hermitian
(c) Neither (a) nor (b)
(d) Both (a) and (b)

42. Which of the following is true

(a) The determinant of a square matrix is define

(b) The determinant of a non-square matrix is defined

(c) Determinant is a rectangular arrangement of objects

(d) None of these

43. The value of a determinant $\begin{vmatrix} a_{11} & a_{12} \\ a_{21} & a_{22} \end{vmatrix}$ is given as

(a) $a_{11}a_{12} - a_{12}a_{21}$ (b) $a_{11}a_{21} - a_{12}a_{22}$

(c) $a_{11}a_{22} - a_{12}a_{21}$ (d) $a_{11}^2 - a_{22}^2$

44. If M_{ij} and C_{ij} are the minor and cofactor for the element a_{ij}, then

(a) $C_{ij} = M_{ij}$ (b) $C_{ij} = -M_{ij}$

(c) $C_{ij} = M_{ji}$ (d) $C_{ij} = (-1)^{i+j} M_{ij}$

45. If any two rows or columns of a determinant are identical. Then the value of the determinant is

(a) zero (b) one

(c) can not be obtained (d) less than zero

46. Area of a triangle with vertices (x_1, y_1), (x_2, y_2) and (x_3, y_3) is given by

(a) $\begin{vmatrix} x_1 & x_2 & 1 \\ y_1 & y_2 & 1 \\ x_3 & y_3 & 1 \end{vmatrix}$ (b) $\frac{1}{2}\begin{vmatrix} x_1 & y_1 & 1 \\ x_3 & y_2 & 1 \\ x_2 & y_3 & 1 \end{vmatrix}$

(c) $\frac{1}{2}\begin{vmatrix} x_1 & y_1 & 1 \\ x_2 & y_2 & 1 \\ x_3 & y_3 & 1 \end{vmatrix}$ (d) None of these

47. If A and B are two square matrices of same order then

(a) $|AB| = |A| \cdot |B|$ (b) $|AB| = |A| - |B|$

(c) $|AB| = |B|\ |A|$ (d) Both (a) and (c)

48. Adjoint of a square matrix is equal to

(a) Matrix of cofactors (b) Transpose of matrix of cofactors

(c) Transpose of matrix of minors (d) Both (b) and (c)

49. A square matrix A is said to be non singular if,

(a) $|A| = 0$ (b) $|A| \neq 0$

(c) Both (a) and (b) (d) None of these

50. If A is a square matrix, then

(a) $A \cdot (\text{adj } A) = |A|\ I_n$ (b) $(\text{adj } A)\ A = |A|\ I_n$

(c) $A\ (\text{adj } A) = |A|$ (d) Both (a) and (b)

51. A square matrix A is a invertible if and only if

(a) $|A| \neq 0$ (b) $|A| = 0$

(c) $|A|$ may be zero or non-zero (d) None of these

52. If A is invertible, then

(a) A^T will be invertible (b) A^T can never be invertible

(c) $(A^T)^{-1} = (A^{-1})^T$ (d) Both (a) and (c)

53. If A is an invertible symmetric matrix, then A^{-1} will be

(a) Invertible (b) not invertible

(c) may or may not be invertible (d) none of these

54. Every skew symmetric matrix of odd order is

(a) singualr (b) non singular

(c) invertible (d) None of these

55. If A be an invertible square matrix of order n, then

(a) $|\text{Adj } A| = |A|^{n-1}$ (b) $|\text{Adj } A| = |A|^n$

(c) $|\text{Adj } A| = |A|$ (d) $|\text{Adj } A| = n|A|$

56. What is the condition for a system of equation $AX = B$ of n equations in n unknowns, to be consistent?

(a) $|A| = 0$ and $(\text{adj } A)\, B \neq 0$ (b) $|A| \neq 0$

(c) $|A| = 0$ and $(\text{adj } A)\, B = 0$ (d) Both (b) and (c)

57. The solution of a consistent system of equation $AX = B$ is given by

(a) $X = A^{-1}B$ (b) $X = AB^{-1}$

(c) $X = A^{-1}B^{-1}$ (d) $X = AB$

58. For the matrix $A = \begin{bmatrix} 1 & -1 \\ 2 & 3 \end{bmatrix}$, which of the following is true

(a) $A^2 - 5A + 4I = 0$ (b) $A^2 + 5A - 4I = 0$

(c) $A^2 - 4A + 5I = 0$ (d) None of these

59. Which of the following system of equation is consistent

(a) $x + y + z = 3,\ 2x - y + z = 2,\ x - 2y + 3z = 2$

(b) $8x - 4y + z = 5,\ 10x + 6z = 4,\ 8x + y + 6z = \dfrac{5}{2}$

(c) Both (a) and (b)

(d) None of these

60. Area of the triangle with vertices $A(1, -1)$, $B(2, 4)$, $C(-3, 5)$ is

(a) 12 sq. units (b) 0

(c) 13 sq. units (d) 10 sq. units

True and False :

1. The equation $\left(x^2 + \dfrac{1}{x^2}\right) + 2\left(x + \dfrac{1}{x}\right) + 3 = 0$ can be reduced to a quadratic equaiton.

2. The equation $\dfrac{1}{x+1} + \dfrac{1}{x+5} = \dfrac{1}{x+2} + \dfrac{1}{x+4}$ has two roots.

3. Matrix addition is commutative.

4. Matrix subtraction is associative.

5. Matrix multiplication is always commutative.

6. For matrices $A_{m \times k}$, $B_{k \times l}$, $C_{l \times n}$, $(AB)\,C = A(BC)$.
7. For matrices A of order $m \times n$ and B of order $n \times p$; $\alpha\,(AB) \neq A\,(\alpha\beta)$.
8. If a matrix $A_{2 \times 3}$ is multiplied by an identity matrix of order 3, then $AI = A$.
9. If $A^2 = O$, then does it mean $A = O$?
10. For a square matrix A, $(kA)^T = kA^T$, where k is a factor.
11. For a square matrix A and a non-negative integer γ, $(A^{\gamma})^T = (A^T)^{\gamma}$.
12. If A is unitary matrix then A^{θ} is also unitary.
13. The determinant of a square matrix A and A^T are same.
14. If any two rows or columns of a determinant are interchanged then the determinant retains its absolute value but its sign is changed.
15. If to any row or column of a determinant another row or column is added, the value of determinant is changed.
16. If A is an invertible matrix of order n, then adj (adj A) = $|A|^{n-2} A$.
17. If A and B are invertible square matrices of the order n then (adj AB) = n (adj A) × (adj B).
18. The system of equations $2x + 5y = 7$, $6x + 15y = 13$ is consistent.
19. The system of equations $x - y + z = 3$, $2x + y - z = 2$, $-x - 2y + 2z = 1$ is consistent.
20. Using Cramers rule. The value of $x = \dfrac{D_x}{D}$, $y = \dfrac{D_y}{D}$, $z = \dfrac{D_z}{D}$, where D, D_x, D_y, D_2 have their usual meaning.

Fill in the blanks :

1. The equation $(x + 1)\,(x + 2)\,(x + 3)\,(x + 4) = 24$ has .... roots.
2. If $\sqrt{\dfrac{x}{y}} + \sqrt{\dfrac{y}{x}} = \dfrac{5}{2}$ and $xy = 16$, then the values of x and y are .... respectively.
3. The sum of two upper triangular matrices will be .... matrix.
4. If $AI = IA = A$, and I is an identity matrix of order n. Then A is a .... of order ....
5. If the dimensions of matrices A, B and C are such that the products are defined then $A(B + C)$ = ...
6. If A is a matrix of order $m \times n$, B is a matrix of order $n \times p$, the order of matrix AB is ....
7. If A and B are symmetric matrices, then $AB - BA$ is a ....
8. A square matrix A is said to be idempotent if ....
9. Any square complex matrix A is said to be unitary if ....
10. For a square matrix A, ... is Hermitian part of .... is Skew Hermitian part of A.

1. If for a square matrix A_{11}, A_{12}, A_{13} are cofactors corresponding to the elements a_{11}, a_{12}, a_{13} of first row, then the value of determinant is ....

12. If any two row or column of a determinant are proportional, then its value is ....

13. If A is an invertible square matrix, then $(\text{adj } A)^T$ ....

14. If A and B are invertible matrices of same order, then AB is .... and $(AB)^{-1}$ ....

Answers

Objective type questions :

1. a	**2.** a	**3.** c	**4.** b	**5.** a	**6.** c	**7.** d	**8.** b	**9.** b	**10.** d
11. a	**12.** a	**13.** d	**14.** c	**15.** b	**16.** a	**17.** d	**18.** b	**19.** d	**20.** c
21. a	**22.** c	**23.** a	**24.** b	**25.** c	**26.** d	**27.** a	**28.** a	**29.** b	**30.** a
31. c	**32.** b	**33.** b	**34.** a	**35.** c	**36.** d	**37.** b	**38.** a	**39.** b	**40.** c
41. b	**42.** a	**43.** c	**44.** d	**45.** a	**46.** c	**47.** d	**48.** b	**49.** b	**50.** d
51. a	**52.** d	**53.** a	**54.** a	**55.** a	**56.** d	**57.** a	**58.** c	**59.** c	**60.** c

True and False :

1. T, **2.** F, **3.** Y, **4.** N, **5.** F, **6.** T, **7.** F, **8.** T, **9.** N, **10.** T, **11.** T, **12.** T, **13.** T, **14.** T, **15.** F, **16.** T, **17.** F, **18.** F, **19.** T, **20.** T

Fill in the blanks :

1. 2, **2.** 2 and 8, **3.** Triangular, **4.** Square, n, **5.** $AB + AC$, **6.** $m \times p$,

7. skew symmetric, **8.** $A^2 = A$, **9.** $AA^\theta = I$, **10.** $\frac{A + A^\theta}{2}, \frac{A - A^\theta}{2}$,

11. $a_{11}A_{11} + a_{12}A_{12} + a_{13}A_{13}$, **12.** zero, **13.** $(\text{adj } A^T)$,

14. also invertible, $B^{-1}A^{-1}$

□ ■ □

UNIT - II

Measures of Central Value

2.1 INTRODUCTION

A measure of central tendency is a typical value that divides their number in half, *i.e.*, the value of variate around which they cluster. Plainly speaking, an average of a statistical series is the value of the variable which represents the entire distribution.

2.2 DEFINITION OF CENTRAL TENDENCY

(i) Central tendency is a statistical measure used to find out a single value as a representative of entire distribution. The objective of central tendency is to obtain a single value that is most representative of entire group.

(ii) Clark states that "An average is a figure that represents the whole group".

(iii) In A.E. Wagh's opinion "An average is a single value selected from a group of values to represent them in some ways".

(iv) Croxton and Cowden defines, "An average is a single value within the range of the data that is used to represent all the values in the series". Since, an average is from within the rage of the data, it is sometimes called a measure of central values.

2.3 REQUISITES OF A GOOD AVERAGE

An average should satisfy the following characteristics:

(i) It should be rigorously defined.

(ii) It should be capable of simple interpretation and easy to compute.

(iv) It should be based on all observed values.

(v) It should not unduly be influenced by one or two extremely large or small values.

(vi) It should be fluctuated little from one random sampling to another.

(vii) It should be capable of mathematical manupulation.

2.4 KINDS OF AVERAGE

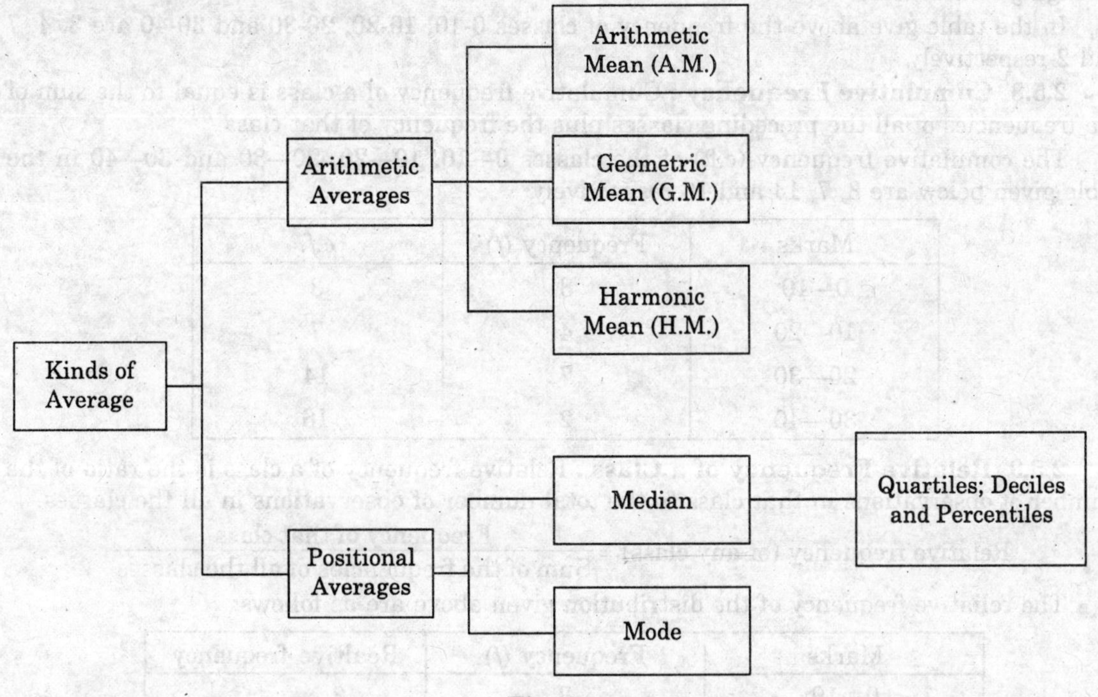

Fig. 2.1

2.5 BASIC DEFINITIONS

2.5.1 Entity : Persons, places and things, which is being observed is called an entity.

2.5.2 Variable : Variable is a quantity that can assume a range of numerical values and each value within the range is called variate. A variable may be discrete or continuous.

2.5.3 Class Interval : Continuous, non over-lapping intervals selected in such a way that they are mutually exclusive and exhaustive are termed as class intervals.

2.5.4 Class Limits : Every class in a frequency distribution is bounded by two numbers. These numbers are called class limits or class boundaries. The lower value of a class is termed as lower limit and upper value as upper limit, *e.g.*, in class 5-10, the lower and upper limits are 5 and 10 respectively.

2.5.5 Middle Value of a Class : Middle-value of a class

$$= \frac{\text{Upper limit + lower limit}}{2}$$

= value which is exactly at the middle of the class.

For example, in class 0-10 the middle-value $= \frac{0 + 10}{2}$

2.5.6 Class Width : The class width of any class in a frequency distribution is equal to the difference between the lower and upper limits of that class.

Marks	0—10	10—20	20—30	30—40
Frequency	3	5	7	2

The width of each class in this table is 10.

2.5.7 Class Frequency : Frequency of a class is the number of items or observations belonging to that class.

In the table give above the frequency of classes 0-10, 10-20, 20-30 and 30-40 are 3, 4, 7 and 2 respectively.

2.5.8 Cumulative Frequency : Cumulative frequency of a class is equal to the sum of the frequencies of all the preceding classes plus the frequency of that class.

The comulative frequency (c.f.) of the classes 0—10, 10—20, 20—30 and 30—40 in the table given below are 3, 7, 14 and 16 respectively.

Marks	Frequency (f)	$c.f.$
0—10	3	3
10—20	4	7
20—30	7	14
30—40	2	16

2.5.9 Relative Frequency of a Class : Relative frequency of a class is the ratio of the number of observations in that class to the total number of observations in all the classes.

$$\text{Relative frequency (of any class)} = \frac{\text{Frequency of that class}}{\text{Sum of the frequencies of all the classes}}$$

The reliative frequency of the distribution given above are as follows:

Marks	Frequency (f)	Realtive frequency
0—10	3	$\frac{3}{16} = 0.1875$
10—20	4	$\frac{4}{16} = 0.2500$
20—30	7	$\frac{7}{16} = 0.4375$
30—40	2	$\frac{2}{16} = 0.1250$

2.6 FREQUENCY DISTRIBUTION

A frequency distribution is a way to display the values assumed by a variable along with the frequency of occurence of these values. In another way frequency distribution is a method to organise and condense data.

The frequency distribution is classified as:

2.6.1 Univariate Frequency Distribution

A frequency distribution that shows the frequency of occurence of different values of a single varaible is termed as univariate frequency distribution. For example :

x	2	4	6	8	10
f	1	7	5	2	3

2.6.2 Bivariate Frequency Distribution

A frequency distribution that deals with two variables is called a bivariate frequency distribution.

For example, if we measure the heights and weights of a certain group of persons, we shall get a bivariable distribution, one varaible for height and other for weight as:

Height in cm (x)	150	155	160	165	170	175	180
Weight in kg (y)	60	62	69	68	72	75	81

2.6.3 Discrete Frequency Distribution

A frequency distribution which is obtained by distinct values of a variable is known as discrete frequency distribution.

For example, let x denotes the marks obtained by the students of a class in a particular subject

x	40	50	60	70	80	90	100
f	2	5	7	8	9	4	1

2.6.4 Grouped Frequency Distribution

A frequnecy distribution formed by dividing the entire range of the observations on a varaible (either discrete or continuous) into groups and distributes the frequencies over these groups is called a grouped frequency distribution. These groups are known as classes.

For example,

Class interval	0—10	10—20	20—30	30—40	40—50
frequency	2	7	11	5	3

2.7 TYPES OF SERIES AND THEIR CONVERSION

2.7.1 Inclusive and Exclusive Series

A series in which the upper limit of a class is included in that class is known as inclusive series, *e.g.*, 0—9, 10—19, etc.

On the other hand exclusive series is a series in which the upper limits are excluded from the respective classes and are included in the immediate next class, *e.g.*, 0—10, 10—20 etc. The conversion of an inclusive series into exclusive series is done by subtracting $\frac{d}{2}$ from lower class limits and adding $\frac{d}{2}$ to upper class limits, where d is equal to the common difference between upper class limit of any class interval and lower class limit of next class interval. In case of exclusive type of classification d is always zero.

For example,

Inclusive series	1—10	11—20	21—30 ...
Exclusive series	0.5—10.5	10.5—20.5	20.5—30.5...

here $d = 1$

2.8 ARITHMETIC MEAN

Arithmetic mean is a most commonly used measures of central tendency which gives the quantitative measure of central tendency for given data. Arithmetic mean of a group of observations is the quotient obtained by dividing the sum of all the observations by their number. Thus, arithmetic mean denoted by A.M. or $\bar{x}$ is

$$\bar{x} = \frac{\text{Sum of all the observations}}{\text{Number of observations}}$$

2.9 METHODS FOR COMPUTATION OF ARITHMETIC MEAN

There are three methods for computation of arithmatic mean:

(i) Direct method
(ii) Short-cut method
(iii) Step deviation method

Using these three methods, the A.M. of various types of series (individuals, discrete, grouped, etc.) can be obtained.

2.10 ARITHMETIC MEAN FOR DISCRETE SERIES OF AN INDIVIDUAL SERIES

In case of individual series, the mean can be obtained by the methods given below:

2.10.1 Direct Method

By this method, the mean can be obtained by dividing the sum of items in a series by the number of items. If $x_1, x_2, \ldots, x_n$ are the n values of a variable x, then the arithmetic mean (A.M.) is given by

$$\text{A.M.} = \frac{x_1 + x_2 + \ldots + x_n}{n} = \frac{\Sigma x}{n} = (\bar{x})$$

Example 2.1

The monthly income of 9 persons in Rs. is given by

4800, 3500, 6500, 4000, 4300, 7000, 4000, 7500, 5200.

Calculate the mean by direct method.

SOLUTION : By direct method

$$\text{A.M.} = (\bar{x}) = \frac{\Sigma x}{n}$$

$$= \frac{4800 + 3500 + 6500 + 4000 + 4300 + 7000 + 4000 + 7500 + 5200}{9}$$

$$= \frac{46800}{9} = 5200 \text{ (in Rs.)}$$

Example 2.2

Calculate the A.M. of the series. 1, 2, 3, 4, 7, 9, 10, 12, 15, 17

SOLUTION : By direct method

$$\text{A.M.} = (\bar{x}) = \frac{\Sigma x}{n} = \frac{1 + 2 + 3 + 4 + 7 + 9 + 10 + 12 + 15 + 17}{10} = \frac{80}{10} = 8$$

2.10.2 Short-Cut Method

Sometimes we have the magnitude of items very large. Then the calculation of A.M. by using short-cut method becomes very easy. In this method let x be a variable and a be assumed number (usually called assumed mean). Then

$$d = x - a = \text{deviation of } x \text{ from } a.$$

Thus, the A.M. is given as:

$$\text{A.M.} = (\bar{x}) = a + \frac{\Sigma d}{n}$$

where $\bar{x}$ is the mean and n be the total number of items.

Example 2.3

By using short-cut method obtain the A.M. of the monthly income of 9 individuals as given in example 1, *i.e.*, Rs. 4800, 3500, 6500, 4000, 4300, 7000, 4000, 7500, 5200.

SOLUTION : Let the assumed mean be 5000.

Monthly Income (x)	Deviation from the assumed mean $d = (x - a)$
4800	– 200
3500	– 1500
6500	1500
4000	– 1000
4300	– 700
7000	2000
4000	– 1000
7500	2500
5200	200
Total	$\Sigma d = 6200 - 4400 = 1800$

Thus, the arithmetic mean (A.M.) $= a + \frac{\Sigma d}{n}$

$$= 5000 + \frac{1800}{9} = 5000 + 200$$

$$= \text{Rs. } 5200$$

2.10.3 Step-Deviation Method

In some cases, where the magnitude of variable is very large and no substantial change occurs on subtracting a number (assumed mean). Then, the size of the variable may be reduced on dividing the deviation by a number (usually called scale). Hence, the A.M. can be obtained by the following expression.

$$\bar{x} = a + h\frac{\Sigma u}{n}$$

where u is obtained on dividing d by h.

Example 2.4

Apply the step deviation method on the data given in example 1 to obtain the arithmetic mean.

SOLUTION : Let x denote the monthly income.

Let the assumed mean $(a) = 5000$ and the scale $h = 100$.

Monthly Income (x)	Deviation $d = (x - a)$	$u = \frac{x - a}{100}$
4800	– 200	– 2
3500	– 1500	– 15
6500	1500	15
4000	– 1000	– 10
4300	– 700	– 7
7000	2000	20
4000	– 1000	– 10
7500	2500	25
5200	200	2
Total		$\Sigma u = 18$

Thus, $\text{A.M.} = (\bar{x}) = a + h\,\frac{\Sigma fu}{n} = 5000 + \frac{18}{9} \times 100$

$= 5000 + 200 = 5200$

2.11 ARITHMETIC MEAN FOR DISCRETE SERIES

In case of discrete series too, there are three methods, to calculate the A.M.

2.11.1 Direct Method

Let the n values $x_1, x_2, \ldots, x_n$ of a variable occurs $f_1, f_2, \ldots, f_n$ times respectively. *i.e.,* $f_1, f_2, \ldots, f_n$ are the respective frequencies of values $x_1, x_2, \ldots, x_n$. Then

$$\text{A.M.} = (\bar{x}) = \frac{f_1x_1 + f_2x_2 + \ldots + f_nx_n}{f_1 + f_2 + \ldots + f_n} = \frac{\Sigma fx}{\Sigma f}$$

Example 2.5

Calculate the arithmetic mean (A.M.) of the following distribution.

x	1	2	3	4	5	6	7
f	2	1	4	5	3	3	6

SOLUTION : $\text{A.M.} = (\bar{x}) = \frac{\Sigma fx}{\Sigma f}$

$$= \frac{1 \times 2 + 2 \times 1 + 3 \times 4 + 4 \times 5 + 5 \times 3 + 6 \times 3 + 7 \times 6}{2 + 1 + 4 + 5 + 3 + 3 + 6}$$

$$= \frac{111}{24}$$

$$= 4.625$$

2.11.2 Short-Cut Method

In case of discrete series, this method is applied almost in a same manner as in an individual series. Let $x_1, x_2, ..., x_n$ be the n different values of a varaible with frequencies $f_1, f_2 ..., f_n$ respectively. Then the A.M. can be obtained by the following expression.

$$\text{A.M.} = (\bar{x}) = a + \frac{\Sigma fd}{\Sigma f}$$

where a is the assumed mean and d denotes the deviation of variable from assumed mean a.

Example 2.6

Calculate the mean of the following.

Heights (in cm)	65	66	67	68	69	70	71	72	73
Number of plants	1	4	5	7	11	10	6	4	2

SOLUTION : Let the value of assumed mean is 68.

Heights (x)	Frequency (f)	Deviation from assumed mean (d)	$f \cdot d$
65	1	– 3	– 3
66	4	– 2	– 8
67	5	– 1	– 5
68	7	0	0
69	11	1	11
70	10	2	20
71	6	3	18
72	4	4	16
73	2	5	10
	$N = \Sigma f = 50$		$\Sigma fd = 59$

Thus, $$\text{A.M.}(\bar{x}) = a + \frac{\Sigma fd}{\Sigma f} = 68 + \frac{59}{50} = 68 + 1.18 = 69.18$$

2.11.3 Step-Deviation Method

In case of discrete series too, this method is used to reduce the size of variable or size of deviation from assumed mean on dividing by a common factor h (called scale).

Let $x_1, x_2, ..., x_n$ be the different values of a variable x with frequencies $f_1, f_2, ..., f_n$ respectively. Then the arithmetic mean (A.M.) can be calculated by the following expression.

$$\text{A.M.} = a + h\frac{\Sigma fu}{\Sigma f}$$

where a is assumed mean, u is the value obtained on dividing deviation d by h and h is a number (called scale).

Example 2.7

Find the value of mean for the following distribution.

x	5	16	27	38	49	60
y	9	17	28	26	15	8

SOLUTION : Let the assumed mean a is 38 and h be 11.

x	f	$d = x - a$	$u = \frac{d}{h}$	fu
5	9	− 33	− 3	− 27
16	17	− 22	− 2	− 34
27	28	− 11	− 1	− 28
38	26	0	0	0
49	15	11	1	15
60	8	22	2	16
	$\Sigma f = 103$			$\Sigma fu = -58$

Thus, $$\text{A.M.} = a + h\frac{\Sigma fu}{\Sigma f} = 38 + 11 \times \frac{(-58)}{103}$$

$$= 38 - \frac{638}{103} = 38 - 6.19 = 31.81$$

Example 2.8

The distribution of 150 patients of ages X or higher, visiting a hospital on a particular day is given below:

x	10	20	30	40	50	60	70	80	90
c.f.	150	133	118	100	75	45	25	09	02

Calculate the mean age. ***[UPTU 2000]***

SOLUTION : First we form the frequency table and then the mean will be obtained by using step deviation method. For this let $a = 50$ and $h = 10$.

x	*c.f.*	f	$d = x - a$	$u = \frac{d}{h}$	fu
10	150	17 (150—133)	− 40	− 4	− 68
20	133	15 (133—118)	− 30	− 3	− 45
30	118	18 (118—100)	− 20	− 2	− 36
40	100	25 (100—75)	− 10	− 1	− 25
50	75	30 (75—45)	0	0	0
60	45	20 (45—25)	10	1	20
70	25	16 (25—09)	20	2	32
80	09	7 (09—02)	30	3	21
90	02	2	40	4	8
		$\Sigma f = 150$			$\Sigma fu = -93$

Thus, $$\text{A.M.} = (\bar{x}) = a + h\frac{\Sigma fu}{\Sigma f}$$

$$= 50 + 10 \times \frac{(-93)}{150}$$

$$= 50 - 6.2 = 43.8$$

Example 2.9

Define and find arithmetic mean and mean deviation of the following data :

12, 3, 18, 17, 4, 9, 17, 19, 20, 15, 8, 17, 2, 3, 16, 11, 3, 1, 0, 5

SOLUTION : Let $a = 12.5$

Class Interval	Mid value x	Frequency f	$d = x - a$	$f.d$	$\mid x - \text{mean} \mid$	$f. \mid x - \text{mean} \mid$
0–5	2.5	7	–10	–70	8	56
5–10	7.5	3	–5	–15	3	9
10–15	12.5	2	0	0	2	4
15–20	17.5	7	5	35	7	49
20–25	22.5	1	10	10	12	12
Total		$\Sigma f = 20$		$\Sigma fd = -40$		$\Sigma f[x - \text{mean}] = 130$

$$\text{Mean} = a + \frac{\Sigma fd}{\Sigma f}$$

$$= 12.5 + \frac{40}{20} = 10.5$$

$$\text{Mean deviation} = \frac{\Sigma f\ |x - \text{Mean}|}{\Sigma f}$$

$$= \frac{130}{20} = 6.5$$

2.12 ARITHMETIC MEAN OF CONTINUOUS SERIES

2.12.1 Equal Intervals

The computation of mean in continuous distribution is similar to that of discrete distribution with only difference that we have obtain a discrete distribution from a given continuous distribution by taking the mid value of all class intervals given to us. The mid points are obtained by taking the mean of lower limit (L_1) and upper limit (L_2), *i.e.*, the mid point for a class $L_1 - L_2$ will be $\frac{L_1 + L_2}{2}$.

Rest of the procedure is same as in discrete series. Hence, the following methods can be used to obtain A.M., *i.e.*,

(a) Direct method
(b) Short cut method
(c) Step deivation method.

Example 2.10

Calculate the arithmetic mean of the following frequency distribution

Class interval	10—20	20—30	30—40	40—50	50—60	60—70
Frequency	3	2	4	9	8	5

SOLUTION : We shall use the short cut method to obtain the mean of the given distribution. For this let, $a = 45$

Class	Mid value (x)	Frequency (f)	$d = x - a$	fd
10—20	15	3	– 30	– 90
20—30	25	2	– 20	– 40
30—40	35	4	– 10	– 40
40—50	45	9	0	0
50—60	55	8	10	80
60—70	65	5	20	100
		$\Sigma f = 31$		$\Sigma fd = 10$

Thus, $\text{A.M.} = (\bar{x}) = a + \dfrac{\Sigma fd}{\Sigma f}$

$$= 45 + \frac{10}{31} = 45 + 0.3226 = 45.3226$$

Example 2.11

Compute the arithmetic mean of the following distribution by short cut method

Class	20—30	30—40	40—50	50—60	60—70
Frequency	8	26	30	20	16

SOLUTION : First we convert the given continuous distribution into discrete distribution and then apply short cut method to obtain the A.M.

Suppose the assumed mean $(a) = 45$

Class	Mid value (x)	Frequency (f)	$d = x - a$	fd
20—30	25	8	– 20	– 160
30—40	35	26	– 10	– 260
40—50	45	30	0	0
50—60	55	20	10	200
60—70	65	16	20	320
		$\Sigma f = 100$		$\Sigma fd = 100$

Thus, $\text{A.M.} = (\bar{x}) = a + \dfrac{\Sigma fd}{\Sigma f}$

$$= 45 + \frac{100}{100} = 45 + 1 = 46$$

Example 2.12

Compute the mean of the following frequency distribution.

Class	0—11	11—22	22—33	33—44	44—55	55—66
Frequency	9	17	28	26	15	8

SOLUTION : First we convert the given distribution into discrete distribution and use step deviation method. Let $a = 38.5$ and $h = 11$

Class	Mid value (x)	Frequency (f)	$d = x - a$	$u = \frac{x-a}{h}$	fu
0—11	5.5	9	− 33	− 3	− 27
11—22	16.5	17	− 22	− 2	− 34
22—33	27.5	28	− 11	− 1	− 28
33—44	38.5	26	0	0	0
44—55	49.5	15	11	1	15
55—66	60.5	8	22	2	16
		$\Sigma f = 103$			$\Sigma fu = -58$

Thus, $\text{A.M.} = (\bar{x}) = a + h\,\frac{\Sigma fu}{\Sigma f}$

$$= 38.5 + 11\,\frac{(-58)}{(103)}$$

$$= 38.5 - \frac{638}{103} = 38.5 - 6.19 = 32.31$$

Example 2.13

Compute the arithmetic mean of the following distribution.

Class	0—10	10—20	20—30	30—40	40—50	50—60	60—70
Frequency	3	7	10	15	20	12	8

SOLUTION : We shall use the short-cut method to obtain the mean of the given distribution. For this let $a = 35$

Class	Mid value (x)	Frequency (f)	$d = x - a$	fu
0—10	5	3	− 30	− 90
10—20	15	7	− 20	− 140
20—30	25	10	− 10	− 100
30—40	35	15	0	0
40—50	45	20	10	200
50—60	55	12	20	240
60—70	65	8	30	240
		$\Sigma f = 75$		$\Sigma fd = 350$

Thus, $\text{A.M.} = (\bar{x}) = a + \frac{\Sigma fd}{\Sigma f}$

$$= 35 + \frac{350}{75}$$

$$= 35 + 4.67 = 39.67$$

2.12.2 Unequal Intervals

If we have a continuous distribution with unequal intervals, then it is not necessary to convert it into an equal class distribution. Rather, we find mid points for each class and then solve the problem in a way similar to discrete distribution.

Example 2.14

Compute the arithmetic mean of the following distribution.

Class	10—20	20—40	40—70	70—120	120—200
Frequency	4	10	26	8	2

SOLUTION : First we convert it into a discrete distribution and then solve by using step deviation method.

Let $a = 55$ and $h = 5$

Class	Mid points (x)	Frequency (f)	$d = x - a$	$u = \frac{x-a}{h}$	$f \cdot u$
10—20	15	4	− 40	− 8	− 32
20—40	30	10	− 25	− 5	− 50
40—70	55	26	0	0	0
70—120	95	8	40	8	64
120—200	160	2	105	21	42
		$\Sigma f = 50$			$\Sigma fu = 24$

Thus, $\quad \text{A.M. } (\bar{x}) = a + h\,\frac{\Sigma fu}{\Sigma f}$

$$= 55 + 5 \times \frac{24}{50} = 55 + 2.4 = 57.4$$

2.13 ARITHEMTIC MEAN FOR INCLUSIVE SERIES

In case of inclusive series, there is no need to convert it into an exclusive series, because mid points are same in both the series.

Example 2.15

Calculate the arithmetic mean of the following distribution:

Class	3—5	6—8	9—11	12—14	15—17	18—20	21—23	24—26
Frequency	2	6	8	11	10	4	3	1

SOLUTION :

Class	Mid points x	Frequency	fx
3—5	4	2	8
6—8	7	6	42
9—11	10	8	80
12—14	13	11	143
15—17	16	10	160
18—20	19	4	76
21—23	22	3	66
24—26	25	1	25
		$\Sigma f = 45$	600

Thus, $$\text{A.M.} = (\bar{x}) = \frac{\Sigma fx}{\Sigma f}$$

$$= \frac{600}{45} = 13.33$$

2.14 ARITHMETIC MEAN FOR OPEN END SERIES

This type of series are first converted to continuous series and then solved by any method available for continuous distribution.

Example 2.16

Compute the mean of the following distribution:

Marks	Below 10	10—20	20—30	30—40	40—50
No. of students	5	4	8	12	16

SOLUTION : First we convert this open series into an ordinary continuous series as below:

Class	Mid points (x)	Frequency (f)	fx
0—10	5	5	25
10—20	15	4	60
20—30	25	8	200
30—40	35	12	300
40—50	4	16	720
		$\Sigma f = 45$	1305

Thus, $$\text{A.M.} = (\bar{x}) = \frac{\Sigma fx}{\Sigma f}$$

$$= \frac{1305}{45} = 29$$

2.15 ARITHMETIC MEAN FOR CUMULATIVE FREQUENCY DISTRIBUTION

In this case, we first convert the given cumulative frequency distribution into an ordinary continuous distribution and calculate the mean by any convenient method.

Example 2.17

Compute the mean for the following distribution.

Value	Less than 10	Less than 20	Less than 30	Less than 40	Less than 50	Less than 60
Frequency	4	16	40	76	96	112

SOLUTION : Here we convert the given distribution into a continuous (ordinary) distribution.

Class	Mid values (x)	Frequency (f)	fx
0—10	5	4	20
10—20	15	12 (16—4)	180
20—30	25	24 (40—16)	600
30—40	35	36 (76—40)	1260
40—50	45	20 (96—76)	900
50—60	55	16 (112—96)	880
		$\Sigma f = 112$	$\Sigma fx = 7240$

Thus, $\text{A.M.} = (\bar{x}) = \dfrac{\Sigma fx}{\Sigma f} = \dfrac{7240}{112} = 64.64$

Example 2.18

Compute the arithmetic mean of the following distribution:

Marks	above 0	above 10	above 20	above 30	above 40	above 50	above 60	above 70	above 80	above 90	above 100
Frequency	88	77	72	65	55	43	28	16	10	8	0

SOLUTION : First we convert the comulative frequency distribution given above into the ordinary continuous distribution and then calculate the A.M. by using step deviation method.

Let $a = 55$ and $h = 10$

Class	Mid value (x)	Frequency (f)	$d = x - a$	$u = \dfrac{x-a}{h}$	fu
0—10	5	3 (80—77)	– 50	– 5	– 15
10—20	15	5 (77—72)	– 40	– 4	– 20
20—30	25	7 (72—65)	– 30	– 3	– 21
30—40	35	10 (65—55)	– 20	– 2	– 20
40—50	45	12 (55—43)	– 10	– 1	– 12
50—60	55	15 (43—28)	0	0	0
60—70	65	12 (28—16)	10	1	12
70—80	75	6 (16—10)	20	2	12
80—90	85	2 (10—8)	30	3	6
90—100	95	8 (8—0)	40	4	32
		$\Sigma f = 80$			$\Sigma fu = -26$

Thus, $\text{A.M.} = (\bar{x}) = a + h\dfrac{\Sigma fu}{\Sigma f}$

$$= 55 + 10 \times \frac{(-26)}{80} = 55 - \frac{26}{8} = 55 - 32.5 = 51.75$$

Example 2.19

The mean marks of 100 students were found to be 40, later on, it was found that a score of 53 was misread as 83. Find the corrected mean corresponding to the corrected score.

SOLUTION : Let n denote the number of students and $\bar{x}$ to be the mean marks of 100 students. Then

$$n = 100 \text{ and } \bar{x} = 40.$$

$\therefore$ Total marks of students $= x = 100 \times 40$

$= 4000$

Total correct score = 4000 – uncorrected score + corrected socre

$= 4000 - 83 + 53$

$= 3970$

Corrected mean $= \dfrac{3970}{100} = 39.7$ marks

Example 2.20

The mean of 200 items is found to be 60. If at the time of calculation, two items are read as 3 and 67 instead of 13 and 17, find the correct mean.

SOLUTION : It is given that, $n = 200$ and $\bar{x} = 100$

Since, $\bar{x} = \dfrac{\sum x}{n}$

$\Rightarrow \quad \sum x = n \cdot \bar{x} = 12{,}000$ (incorrect)

Corrected $\sum x = 12000 - (3 + 67) + (13 + 17)$

$= 12000 - 70 + 30 = 11960$

Thus, corrected mean $= \dfrac{\text{corrected } \sum x}{n} = \dfrac{11960}{200} = 59.8$

Example 2.21

The mean age (in years) of a group of 100 men, was found to be 35. Later it was found that during calculation 77 years was misread as 27. Calculate the arithmatic mean of the given group.

SOLUTION : It is given that $n = 100$ and $\bar{x} = 35$

Since, $\bar{x} = \dfrac{\Sigma x}{n}$

$\Rightarrow \quad \Sigma x = n \cdot \bar{x} = 100 \times 35$

$\Rightarrow \quad \Sigma x = 3500$ (Incorrect)

Corrected $\Sigma x = 3500 - 27 + 77$

$= 3550$

Thus, corrected mean $= (\bar{x}) = \dfrac{\text{Corrected } \Sigma x}{n} = \dfrac{3550}{100} = 35.5$

Example 2.22

Find the missing frequency from the following data, it is being given that 1.46 is the mean of the data and 200 is the sum of frequencies.

size	0	1	2	3	4	5
f	46	–	–	25	10	5

SOLUTION : Let f_1 and f_2 are the missing frequencies corresponding to the size 1 and 2 respectively.

Size (x)	Frequency (f)	fx
0	46	0
1	f_1	f_1
2	f_2	$2f_2$
3	25	75
4	10	40
5	5	25
Total	$86 + f_1 + f_2 = 200$	$f_1 + 2f_2 + 140$

$$86 + f_1 + f_2 = 200 \Rightarrow f_1 + f_2 = 114 \quad \ldots(1)$$

Since, $\bar{x} = \dfrac{\Sigma fx}{\Sigma f} \Rightarrow 1.46 = \dfrac{f_1 + 2f_2 + 140}{200}$

$\Rightarrow f_1 + 2f_2 + 140 = 1.46 \times 200 = 292$

$\Rightarrow f_1 + 2f_2 = 292 - 140 = 152$

$\Rightarrow f_1 + 2f_2 = 152$

On solving equations (1) and (2), we get

$$f_2 = 152 - 114 = 38 \text{ and } f_1 = 114 - 38 = 76$$

Therefore, $f_1 = 76$ and $f_2 = 38$.

Example 2.23

From the following data find the missing frequency when mean is 19.92.

Class	4—8	8—12	12—16	16—20	20—24	24—28	28—32	32.36	36.40
Frequency	11	13	16	14	—	9	17	6	4

SOLUTION : Let the assumed mean is 18, scale h is 4 and the missing frequency be f_1. Then

Class	Mid value (x)	Frequency (f)	$d = x - a$	$u = \dfrac{x-a}{h}$	fu
4—8	6	11	– 12	– 3	– 33
8—12	10	13	– 8	– 2	– 26
12—16	14	16	– 4	– 1	– 16
16—20	18	14	0	0	0
20—24	22	f_1	4	1	f_1
24—28	26	9	8	2	18
28—32	30	17	12	3	51
32—36	34	6	16	4	24
36—40	38	4	20	5	20
		$\Sigma f = 90 + f_1$			$fu = 38 + f_1$

Since, $\text{A.M.} = (\bar{x}) = a + h\dfrac{\Sigma fu}{\Sigma f}$

$$\Rightarrow \quad 19.92 = 18 + 4 \times \frac{38 + f_1}{90 + f_1}$$

$$\Rightarrow \quad 1.92 = \frac{152 + 4f_1}{90 + f_1}$$

$$\Rightarrow \quad 1.92\,(90 + f_1) = 152 + 4f_1$$

$$\Rightarrow \quad 172.8 + 1.92f_1 = 152 + 4f_1$$

$$\Rightarrow \quad 4f_1 - 1.92f_1 = 172.8 - 152$$

$$\Rightarrow \quad 2.08f_1 = 20.8$$

$$\Rightarrow \quad f_1 = \frac{20.8}{2.08} = 10$$

Example 2.24

From the following data find the missing frequency when mean is 15.38 :

Size :	10	12	14	16	18	20
f:	3	7	?	20	8	5

[UPTU 2007]

SOLUTION :

Size (x)	f	$f.x$
10	3	30
12	7	84
14	f_1	$14f_1$
16	20	320
18	8	144
20	5	100
Total	$43 + f_1$	$678 + 14f_1$

$$\text{Mean} = \frac{678 + 14f_1}{43 + f_1} = 15.38 \qquad [\because \text{Mean} = 15.38]$$

$$\Rightarrow \quad 678 + 14f_1 = 661.34 + 15.38f_1$$

$$\Rightarrow \quad 1.38f_1 = 16.66 \Rightarrow f_1 = 12.07$$

2.16 COMBINED (ARITHMETIC) MEAN

If we have two groups G_1 and G_2 of size n_1 and n_2 with mean $\bar{x}_1$ and $\bar{x}_2$ respectively. Then the combined (arithmetic) mean $\bar{x}$ (say) of these groups is given by the following expression.

$$\bar{x} = \frac{n_1\bar{x}_1 + n_2\bar{x}_2}{n_1 + n_2}$$

In general, if we have m groups $G_1, G_2, \ldots, G_m$ of size $n_1, n_2, \ldots, n_m$ with mean $\bar{x}_1, x_2, \ldots, \bar{x}_m$ respectively, then their combined (arithmetic) mean $\bar{x}$ is given as

$$\bar{x} = \frac{n_1 \bar{x}_1 + n_2 \bar{x}_2 + \ldots + n_m \bar{x}_m}{n_1 + n_2 + \ldots + n_m}$$

Example 2.25

Compute the mean weight of 150 students in a class. If the mean weight of 50 boys is 70 and that of 100 girls is 55 kg.

SOLUTION : Let $\bar{x}_1$ and $\bar{x}_2$ are the mean weights of n_1 boys and n_2 girls in a class.

Then, $\bar{x}_1 = 70, \bar{x}_2 = 55, n_1 = 50$ and $n_2 = 100$

Hence, the combined mean $(\bar{x}) = \dfrac{n_1 \bar{x}_1 + n_2 \bar{x}_2}{n_1 + n_2}$

$$= \frac{50 \times 70 + 100 \times 55}{50 + 100}$$

$$= \frac{3500 + 5500}{150} = \frac{9000}{150} = 60$$

Example 2.26

The arithmetic mean of the daily wages of two pharmaceutical concerns A and B are Rs. 30 and Rs. 40 respectively. If the number of workers employed in A and B were 2000 and 4000 respectively, find the average daily wage of all workers.

SOLUTION : Let $\bar{x}_1$ and $\bar{x}_2$ are the mean wage of 2000 workers of firm A and 4000 workers of firm B respectively.

Then, $\bar{x}_1 = 30, n_1 = 2000$ and $\bar{x}_2 = 40, n_2 = 4000$

Hence, the average daily wage of all the workers

$$(\bar{x}) = \frac{n_1 \bar{x}_1 + n_2 \bar{x}_2}{n_1 + n_2}$$

$$= \frac{2000 \times 30 + 4000 \times 40}{2000 + 4000}$$

$$= \frac{60000 + 160000}{6000} = \frac{220000}{6000} = 36.67$$

Example 2.27

From the following particulars relating to employees of a factory determine the number of male and female employees:

Total number of employees = 100
Average salary of all employees = Rs. 5000
Average salary of male employees = Rs.5200
Average salary of female employees = Rs. 4200 ***[UPTU 2002, 03]***

SOLUTION : Let $\bar{x}_1$ and $\bar{x}_2$ denote the average salary of n_1 number of male employees and n_2 number of female employees respectively. Also it is assumed that $\bar{x}$ is the average salary of all the employees.

Thus, $n_1 + n_2 = 100, \bar{x}_1 = 5200, \bar{x}_2 = 4200$ and $\bar{x} = 5000$

Average salary of all employees $(\bar{x}) = \dfrac{n_1 \bar{x}_1 + n_2 \bar{x}_2}{n_1 + n_2}$

$$\Rightarrow \quad 5000 = \frac{5200\,n_1 + 4200\,n_2}{n_1 + n_2}$$

$$\Rightarrow \quad 200\,n_1 = 800\,n_2$$

$$\Rightarrow \quad n_1 = 4n_2$$

But it is given that $n_1 + n_2 = 100$

$$\Rightarrow \quad 4n_2 + n_2 = 100$$

$$\Rightarrow \quad 5n_2 = 100$$

$$\Rightarrow \quad n_2 = 20$$

and $\quad n_1 = 100 - 20 = 80$

Hence, the number of male and female employees are 80 and 20 respectively.

2.17 WEIGHTED (ARITHMETIC) MEAN

There might be various situations, where all the components do not have equal importance. These situations can be handled by weighted arithemetic mean.

If $w_1, w_2, \ldots, w_n$ be the weights assigned to n values $x_1, x_2, \ldots, x_n$ respectively, then the weighted arithmetic mean is given by the following expression

$$\text{Weighted arithmetic mean} = \frac{w_1 x_1 + w_2 x_2 + \ldots + w_n x_n}{w_1 + w_2 + \ldots + w_n}$$

$$= \frac{\Sigma\, wx}{\Sigma\, w}$$

Example 2.28

Calculate the weighted mean for the following series:

x	75	80	85	90	95	100
w	2	4	3	3	5	8

SOLUTION : We form the following table

x	w	wx
75	2	150
80	4	320
85	3	255
90	3	270
95	5	475
100	8	800
	$\Sigma w = 25$	$\Sigma wx = 2270$

$$\text{Thus, weighted arithmetic mean} = \frac{\Sigma\, wx}{\Sigma\, w}$$

$$= \frac{2270}{25} = 90.8$$

Example 2.29

A candidate obtained the following marks in an examination in paper of 100 marks each.

Subject	English	Maths	Physics	Chemistry	Biology
Marks (out of 100)	48	82	70	64	60

It is agreed to give double weight to physics and maths as compared to other subjects; what is the arithmetic mean?

SOLUTION : We form the following table:

Subject	Marks obtained (x)	Weight (w)	wx
English	48	1	48
Maths	82	2	164
Physics	70	2	140
Chemistry	64	1	64
Biology	60	1	60
		$\Sigma w = 7$	$\Sigma wx = 476$

Thus, weighted arithemtic mean $= \dfrac{\Sigma\, wx}{\Sigma\, w}$

$$= \frac{476}{7} = 68$$

Example 2.30

Calculate the weighted arithmetic mean of first n natural numbers, whose weights are equal to the corresponding number.

SOLUTION : We have the following table

x	w	wx
1	1	1^2
2	2	2^2
$\vdots$	$\vdots$	$\vdots$
n	n	n^2
	$\Sigma w = \dfrac{n\,(n+1)}{2}$	$\Sigma wn = \dfrac{n\,(n+1)\,(2n+1)}{6}$

Thus, weighted arithmetic mean $= \dfrac{\Sigma\, wx}{\Sigma\, w}$

$$= \frac{\dfrac{n\,(n+1)\,(2n+1)}{6}}{\dfrac{n\,(n+1)}{2}} = \frac{(2n+1)}{3}$$

Example 2.31

Compute the weighted mean of first n natural numbers whose weights are equal to the squares of the corresponding number.

SOLUTION : We have the following table.

x	w	wx
1	1^2	1^3
2	2^2	2^3
3	3^2	3^3
⋮	⋮	⋮
n	n^2	n^3
	$\Sigma w = \frac{n(n+1)(2n+1)}{6}$	$\Sigma wx = \left[\frac{n(n+1)}{2}\right]^2$

Thus, weighted mean $= \frac{\Sigma wx}{\Sigma w}$

$$= \frac{\left(\frac{n(n+1)}{2}\right)^2}{\frac{n(n+1)(2n+1)}{6}} = \frac{3n(n+1)}{2(2n+1)}$$

2.18 MERITS AND DEMERITS OF ARITHMETIC MEAN

2.18.1 Merits

(i) It is rigidly defined.
(ii) It is easily understood and easy to calculate.
(iii) It is based upon all the observations.
(iv) In comparison of other averages, arithmetic mean is less affected by fluctuations of sampling.
(v) It is the best average to compare two or more series.

2.18.2 Demerits

(i) It cannot be determined by inspection. Also it cannot be located graphically.
(ii) It is an quantitative measure, hence, it cannot be used if we are dealing with qualitative characteristics which cannot be measured quantitatively like, honesty performance, base by etc.
(iii) Arithmetic mean of a distribution may be a value, which may not be present in the distribution.
(iv) Arithmetic mean is affected, very much by extreme values.
(v) Arithmetic mean may give a wrong conclusion if the details of the data are not given.

PROBLEMS SET 2.1

1. Marks obtained by 11 students in statistics are given below :

52, 45, 75, 40, 43, 70, 65, 35, 39, 40, 41.

Calculate the arithmetic mean.

2. Find the mean of first thirty natural numbers.

3. Find the mean of the following distribution

5, 9, 17, 10, 35, 39, 61, 21, 18, 53, 44, 40, 59.

Also find the mean if 3 is deducted from each number.

4. Calculate the mean height for the distribution:

Height in inches (x)	73	72	71	70	69	68	67	66	65
Men (f)	2	4	6	10	11	7	5	4	1

5. Given :

Variable	20	19	18	17	16	15	14	13	12	11
Frequency	1	2	4	8	11	10	7	4	2	1

Find the mean variable by taking 11 as the assumed mean and verify by direct method.

6. Find the mean of the following frequency distribution.

x	10	12	14	16	18	20
f	4	8	12	19	11	6

7. Find out mean of the following frequency distribution.

x	1900	1950	2000	2100	2260	2440	2500
f	26	22	18	21	15	12	7

8. Calculate the mean of the following distribution.

x	100	200	300	400	500	600	700
c.f.	120	110	90	75	65	54	35

9. Calculate the A.M. from the following frequency distribution.

x	400	600	800	1000	1200	1400
c.f.	5	11	21	27	35	42

10. Estimate the A.M. for the following frequency distribution.

Class interval	1–4	4–7	7–10	10–13
f	8	12	14	7

11. Find the mean from the following data:

Class	0–10	10–20	20–30	30–40	40–50	50–60
Frequency	12	18	27	20	17	6

12. Calculate the arithmetic mean of the distrubution given below:

Marks (n)	5–15	15–25	25–35	35–45	45–55	55–65	65–75	75–85	85–95
No. of students (f)	3	4	5	7	9	7	6	5	4

13. The table below gives the distribution of the number of class rooms available in Pharmaceutical Institutes in 1995. Find the mean.

No. of classrooms	1–3	4–6	6–9	10–12	13–15	16–18	19–21	22–24
No. of Institutions	304	682	787	1051	1108	1245	1081	1043

14. The following is the frequency distribution of the number of teachers in the Pharmaceutical Institutions in 1995. Find the average number of teachers per institution for the year 1995.

No. of teachers	6–10	11–15	16–20	21–25	26–30	31–35	36–40	41–45	46–50
No. of Institutions	995	1067	1663	1492	1220	1129	745	637	442

15. Find the mean of the following distribution.

Class	3–5	6–8	9–11	12–14	15–17	18–20	21–23	24–26
Frequency	2	6	8	11	10	4	3	1

16. Determine the mean of the following distribution.

Class	30–39	40–49	50–59	60–69	70–79	80–89
Frequency	2	4	4	20	7	3

17. Find the mean from the following distribution.

Class	less than 20	20–22	22–24	24–26	26–28	28 and more
Frequency	5	18	28	37	24	22

18. Find the mean of the following distribution.

Age	below 20	below 25	below 30	below 35	below 40	below 45	below 50
No. of persons	3	16	37	52	57	61	63

19. Find the mean of the following distribution.

Marks	below 10	below 20	below 30	below 40	below 50	below 60	below 70	below 80	below 90	below 100
No. of students	5	9	17	29	45	60	70	78	83	85

20. Find the mean.

Marks	above 5	above 15	above 25	above 35	above 45	above 55	above 65
No. of students	80	74	63	42	19	5	0

21. Find the mean.

Marks	above 0	above 10	above 20	above 30	above 40	above 50	above 60	above 70	above 80	above 90	above 100
No. of students	80	77	72	65	55	43	28	16	10	8	0

22. The mean of 30 values was 150. It was detected on rechecking that one value 165 was wrongly copied as 135 for the computation of the mean. Find the correct mean.

23. The mean of 200 items was 50. Later on it was discovered that two items were wrongly read as 92 and 8 instead of 192 and 88. Find out the correct mean.

24. The arithmetic mean of a set of 40 values is 65. If each of the 40 values is increased by 5, what will be the mean of the set of new values?

25. The mean of a set of 10 items are known to 17 of the 10 items, one item say 26 was subsequently found to be inaccurate and was left. Find the resulting mean.

26. If the mean of the following data is 20.6. Find the missing frequency f_1.

x	10	15	20	25	35
f	3	10	f_1	7	5

27. The mean of the following frequency table is 50. But the frequencies f_1 and f_2 in classes 20–40 and 60–80 are missing. Find the missing frequencies.

Class	0–20	20–40	40–60	60–80	80–100	Total
Frequency	17	f_1	32	f_2	19	120

28. If the mean of the following distribution is 30.5, find the missing figure.

Marks	10	20	?	40	50
f	8	10	20	15	7

29. The mean age of a combined group of men and women is 30 years. If the mean age for the group of men is 32 and that of women is 27, find out the percentage of the men and women in the group.

30. There are 45 students in a class of which 15 are girls. The average weight of 15 girls is 45 kg and that of the 30 boys is 52 kg. Find the mean weight in kg of the entire class.

31. The mean wage of 500 labourers working in a factory, running two shifts of 360 and 140 labourers respectively in Rs. 70. The mean wage of 360 labours working in first shift is Rs. 40. Find the mean wage of 140 labourers working in second shift.

32. A candidate obtained the following percentage of marks in an examination:

Hindi	Chemistry	Zoology	English	Botany
45%	71%	52%	51%	62%

It is agreed to give double weightage to marks in Zoology and Botany as compared to other. Calculate the weighted mean.

33. A student obtained 60, 75 and 85 marks respectively in quarterly examination in chemistry and 95 marks in final examination. The quarterly examinations are of equal weightage, whereas the final examination is weighted twice as much as a monthly examination. Find his mean marks in chemistry.

ANSWERS

1. 49.54 **2.** 15.5 **3.** 31.61, 28.61 **4.** 84.36 **5.** 15.54 **6.** 15.43 **7.** 2091.57 **8.** 457.5 **9.** 928.5 **10.** 6.96 **11.** 28 **12.** 51.8 **13.** 14.26 **14.** 24.96 **15.** 13.33 **16.** 63.25 **17.** 24.84 **18.** 29.56 **19.** 46.8 **20.** 35.38 **21.** 51.75 **22.** 151 **23.** 50.9 **24.** 70 **25.** 16 **26.** 25 **27.** 28, 24 **28.** 30 **29.** Men : 60%, Women : 40% **30.** 49.76 kg (approx.) **31.** Rs. 57.14 **32.** 56.43 **33.** 82 marks.

2.19 MEDIAN

Median is that value of a variable which divides the group into two equal parts, one part containing all values greater and other all values less than the median.

2.20 COMPUTATION OF MEDIAN

Computation of median depends upon the nature of data (*i.e.*, whether they arranged in the form of an individual series or grouped data)

2.20.1 Median for Individual Series

To estimate the median in ungrouped data, the data are arranged either in ascending order of magnitude or in descending order of magnitude.

After arranging the data in any of the way mentioned above. The middlemost number represents the median.

If the number of data say N is odd, then the value at $\left(\frac{N+1}{2}\right)$th place gives the value of median. On the other hand if the number of item is even, then the mean of the number $\frac{N}{2}$th and $\left(\frac{N}{2}+1\right)$th place is the median value of the given series.

Example 2.32

Find the median of the following series.

2, 11, 9, 7, 5, 12, 15, 19, 21

SOLUTION : After arranging the above series in ascending order, we get

2, 5, 7, 9, 11, 12, 15, 19, 21

The series contains 9 numbers therefore, the number at $\left(\frac{9+1}{2} = 5\right)$th place, *i.e.*, 11 will be the median of the given series.

Example 2.33

Find the median of the following series.

2, 11, 9, 7, 15, 14, 18, 20

SOLUTION : Firstly, we arrange the data in ascending order as :

2, 7, 9, 11, 14, 15, 18, 20

The series contains 8 numbers therefore, the mean of $\frac{8}{2}$ = 4th value and $\left(\frac{8}{2}+1\right)$ = 5th value

$\frac{11+14}{2} = 12.5$ is the median.

2.20.2 Median for Discrete Series

Let we have k observations as $x_1, x_2, \ldots, x_k$ in ascending order with their respective frequencies $f_1, f_2, \ldots, f_k$. Let $F_1, F_2, \ldots, F_k$ be the cumulative frequencies (less than type).

Then median for the above distribution is the lowest value of the variable (x) for which the cumulative frequency (F) exists

(i) $\frac{N+1}{2}$, if N is odd

(ii) The mean of $\left(\frac{N}{2}\right)$ and $\left(\frac{N}{2}+1\right)$, if N is even.

where N is the total frequency.

Example 2.34

Calculate the median of the following distribution.

x	160	150	152	161	156
f	5	8	6	3	7

SOLUTION : Arrange the data in ascending order, we have

x	f	Cumulative frequency
150	8	8
152	6	14
156	7	21
160	5	26
161	3	29
Total	$N = \Sigma f = 29$	

Here, total number of observations N (= 29) is odd.

Thus, the median will be the value of $\left(\frac{N+1}{2}\right)$th item = $\left(\frac{29+1}{2}\right)$th item = 15th item

Now, 15th item lies in the row with 21 as cumulative frequency ($\because$ 14 < 15 < 21)

Thus, median = size of 15th item = 156

Example 2.35

Find the median of the following data:

x	100	150	80	200	250	180
f	24	26	16	20	6	30

SOLUTION : Arrange the data in ascending order, we have the following table:

x	f	Cumulative frequency (*c.f.*)
80	16	16
100	24	40
150	26	66
180	30	96
200	20	116
250	6	122
Total	$N = \Sigma f = 122$	

Here, we get N (= 122) is even and $\frac{N}{2} = 61$.

Therefore, median is the mean of values of $\left(\frac{N}{2}\right)$th and $\left(\frac{N}{2}+1\right)$th items, *i.e.*,

$$\text{Median} = \text{Mean of 61th and 62th value} = \frac{150+150}{2} = 150$$

Example 2.36

From the following data, calculate the value of median

x	20	25	28	29	33	38	42	45
f	6	20	24	28	15	4	2	1

SOLUTION : Since, the data are arranged in ascending order, we obtain the following cumulative frequency table

x	f	*c.f.*
20	6	6
25	20	26
28	24	50
29	28	78
33	15	93
38	4	97
42	2	99
45	1	100
Total	$\Sigma f = 100$	

Here, we get $N = \Sigma f = 100$, which is even and $\frac{N}{2} = 50$

Hence, median will be the mean of $\frac{N}{2}$th value and $\left(\frac{N}{2} + 1\right)$th value, *i.e.*, 50th and 51th value

Now, 50th value = 28 and 51th value = 29

Therefore, median is $\frac{28 + 29}{2} = 28.5$

2.20.3 Median for Continuous Series

In grouped data, the individual items do not have their identities. Here, we see that the value that divides the number of all items in two values lies within a class.

The value of median in grouped data can be obtained with the help of following steps.

(i) Find $\frac{N}{2}$, where N is the sum of frequencies.

(ii) Accumulate the frequencies.

(iii) Find the class whose cumulative frequency is the immediate successor of $\frac{N}{2}$. This is known as median class.

(iv) Obtain the value of median by using the formula

$$\text{Median (Md)} = L_1 + \frac{i}{f}\left(\frac{N}{2} - C\right)$$

where L_1 = lower limit of median class.

N = total frequency

C = cumulative frequency of the class preceeding the median class.

f = frequency of median class.

i = size of the class interval of the median class.

Example 2.37

Find the median of the following distribution.

Wages (in Rs.)	0—10	10—20	20—30	30—40	40—50
No. of Workers	22	38	46	35	20

SOLUTION : We have the following table:

Wages (in Rs.)	No. of workes	Cumulative frequency
0—10	22	22
10—20	38	60
20—30	46	106
30—40	35	141
40—50	20	161
total	$N = \Sigma f = 161$	

Here, $N = 161$ and $\frac{N}{2} = 80.5$

Since, the cumulative frequency 106 is the immediate successor of 80.5 (60 < 80.5 < 106). The median class is 20—30.

Thus, $L_1 = 20, f = 46, C = 60, \frac{N}{2} = 80.5$ and $i = 10$ (30—20)

$$\therefore \quad \text{Median (Md)} = L_1 + \frac{i}{f}\left(\frac{N}{2} - C\right)$$

$$= 20 + \frac{10}{46}(80.5 - 60)$$

$$= 20 + \frac{10 \times 20.5}{46} = 20 + 4.46 = 24.46$$

Example 2.38

Calculate the median of the following data:

Marks	0—10	10—20	20—30	30—40	40—50	50—60	60—70	70—80	80—90	90—100
No. of Students	4	9	15	20	16	14	10	6	4	2

SOLUTION : We construct the following table:

Marks	No. of Students	Cumulative frequency
0—10	4	4
10—20	9	13
20—30	15	28
30—40	20	48
40—50	16	64
50—60	14	78
60—70	10	88
70—80	6	94
80—90	4	98
90—100	2	100
Total	$N = \Sigma f = 100$	

Here, $N = 100$ and $\frac{N}{2} = 50$

Since, the cumulative frequency 64 is immediate successor of 50 (48 < 50 < 64). The median class is 40—50.

Thus, $L_1 = 40, f = 16, C = 48, \frac{N}{2} = 50$ and $i = 10$ (50—40)

$$\therefore \quad \text{Median (Md)} = L_1 + \frac{i}{f}\left(\frac{N}{2} - C\right)$$

$$= 40 + \frac{10}{16}(50 - 48)$$

$$= 40 + \frac{10}{16} \times 2 = 40 + 1.25 = 41.25$$

Example 2.39

From the following table, calculate the median of the cost of living index:

Cost of living index (Rs.)	140—150	150—160	160—170	170—180	180—190	190—200
No. of Weeks	5	10	20	9	6	2

SOLUTION : We construct the following table

Cost of living index (Rs.)	No. of weeks (f)	Cumulative frequency ($c.f.$)
140—150	5	5
150—160	10	15
160—170	20	35
170—180	9	44
180—190	6	50
190—200	2	52
Total	$N = \Sigma f = 52$	

Here, $N = 52$ and $\frac{N}{2} = 26$

Since, the cumulative frequency 35 is immediate successor of 26 (15 < 26 < 35)

Thus, $L_1 = 160, f = 20, C = 15, \frac{N}{2} = 26, i = 10$ (170—180)

$$\text{Median (Md)} = L_1 + \frac{i}{f}\left(\frac{N}{2} - C\right) = 160 + \frac{10}{20}(26 - 15)$$

$$= 160 + \frac{11}{2} = 160 + 5.5 = 165.5$$

Example 2.40

Find the median of the following distribution:

Class	5—10	10—15	15—20	20—25	25—30	30—35	35—40
Frequency	12	15	18	20	12	10	6

SOLUTION : We construct the following table:

Class	Frequency (f)	Cumulative frequency ($c.f.$)
5—10	12	12
10—15	15	27
15—20	18	45
20—25	20	65
25—30	12	77
30—35	10	87
35—40	6	93
Total	$N = \Sigma f = 93$	

Here, $N = 93$ and $\frac{N}{2} = \frac{93}{2} = 46.5$

Since, the cumulative frequency 65 is immediate successor of 46.5 (45 < 46.5 < 65). The median class is 20—25.

Thus, $L_1 = 20, f = 20, C = 45, \frac{N}{2} = 46.5$ and $i = 5$ (25—20)

$$\therefore \quad \text{Median (Md)} = L_1 + \frac{i}{f}\left(\frac{N}{2} - C\right)$$

$$= 20 + \frac{5}{20}(46.5 - 45)$$

$$= 20 + \frac{1.5}{4} = 20 + 0.375$$

$$= 20.375$$

Example 2.41

Calculate median for the following data:

Class	0—4	4—6	6—8	8—12	12—18	18—20
Frequency	4	6	8	12	7	2

SOLUTION : In the given distribution the class intervals are unequal. However there is no need to make them equal.

Class	Frequency (*f*)	Cumulative frequency (*c.f.*)
0—4	4	4
4—6	6	10
6—8	8	18
8—12	12	30
12—18	7	37
18—20	2	39
Total	$N = \Sigma f = 39$	

Here, $N = 39$ and $\frac{N}{2} = \frac{39}{2} = 19.5$

Since, the cumulative frequency 30 is immediate successer of 19.5 (18 < 19.5 < 30)

Thus, $L_1 = 8, f = 12, C = 18, \frac{N}{2} = 19.5, i = 4$ (12—8)

$$\text{Median (Md)} = L_1 + \frac{i}{f}\left(\frac{N}{2} - C\right)$$

$$= 8 + \frac{4}{12}(19.5 - 18)$$

$$= 8 + \frac{1}{3} \times 1.5$$

$$= 8 + 0.5 = 8.5$$

2.20.4 Median for Inclusive Series

In case of inclusive series, first we convert the given inclusive series into exclusive series and follow the same procedure as for exclusive series.

Example 2.42

Obtain the median for the following distribution

Class	20—24	25—29	30—34	35—39	40—44	45—49	50—54	55—59	60—64
f	8	10	24	46	62	20	16	10	4

Sol. First convert the given inclusive series into exclusive series as :

Class	Frequency (f)	Cumulative frequency ($c.f.$)
19.5—24.5	8	8
24.5—29.5	10	18
29.5—34.5	24	42
34.5—39.5	46	88
39.5—44.5	62	150
44.5—49.5	20	170
49.5—54.5	16	186
54.5—59.5	10	196
59.5—64.5	4	200
Total	$N = \Sigma f = 200$	

Here, $N = 200$

and $\dfrac{N}{2} = 100$

Since, the cumulative frequency 150 is immediate successor of 100 (88 < 100 < 150). The median class is (39.5—44.5).

Thus, $L_1 = 39.5, f = 62, C = 88, \dfrac{N}{2} = 100$ and $i = 5$ (44.5—39.5)

$$\text{Median (Md)} = L_1 + \frac{i}{f}\left(\frac{N}{2} - C\right)$$

$$= 39.5 + \frac{5}{62}(100 - 88)$$

$$= 39.5 + \frac{5}{62} \times 12$$

$$= 39.5 + 0.97$$

$$= 40.47$$

2.20.5 Median for Cumulative Series

In this case, first we convert the given series into continuous series and adopt the procedure for exclusive series to obtain the median.

Example 2.43

Following is the distribution of marks obtained by 50 studnets in Mathematics. Calculate the median marks:

Marks more than :	0	10	20	30	40	50
No. of students :	50	46	40	20	10	3

SOLUTION : First, we convert the given distribution into continuous distribution (series) as follows:

Marks	Frequency (f)	$c.f.$ (less than)
0—10	4 (50—60)	4
10—20	6 (46—40)	10
20—30	20 (40—20)	30
30—40	10 (20—10)	40
40—50	7 (10—3)	47
50—60	3	50
Total	$N = \Sigma f = 50$	

Here, $N = 50$ and $\frac{N}{2} = 25$

Since, the cumulative frequency 30 is immediate successor of 25 ($10 < 25 < 30$)

Thus, $L_1 = 20, f = 20, C = 10, \frac{N}{2} = 25$ and $i = 10$ (30—20)

$$\therefore \quad \text{Median (Md)} = L_1 + \frac{i}{f}\left(\frac{N}{2} - C\right)$$

$$= 20 + \frac{10}{20}(25 - 10)$$

$$= 20 + \frac{1}{2} \times 15$$

$$= 20 + 7.5 = 27.5$$

Example 2.44

Calculate the median from the following data:

Value	Less than 10	Less than 20	Less than 30	Less than 40	Less than 50	Less than 60	Less than 70	Less than 80
Frequency	04	16	40	76	96	112	120	125

SOLUTION : First, we shall convert the given cumulative frequency distribution into the continuous distribution as follows:

Value	Frequency (f)	$c.f.$ (less than)
0—10	4	4
10—20	12 (16—4)	16
20—30	24 (40—16)	40
30—40	36 (76—40)	76
40—50	20 (96—76)	96
50—60	16 (112—96)	112
60—70	8 (120—112)	120
70—80	5 (125—120)	125
Total	$N = \Sigma f = 125$	

Here, $N = 125$ and $\frac{N}{2} = \frac{125}{2} = 62.5$

Since, the cumulative frequency 76 is the immediate successor of 62.5 (40 < 62.5 < 76). The median class is 30—40.

Thus, $L_1 = 30, f = 36, C = 40, \frac{N}{2} = 62.5, i = 10\ (40—30)$

$$\therefore \quad \text{Median (Md)} = L_1 + \frac{i}{f}\left(\frac{N}{2} - C\right) = 30 + \frac{10}{36}(62.5 - 40)$$

$$= 30 + \frac{5}{18}(22.5) = 30 + 6.25 = 36.25$$

2.20.6 Median for Open End Series

In this case the end of the series are open. Here we have to convert the series into continuous series before finding the median.

Example 2.45

Calculate the median for the following statistical distribution:

Value	Less than 100	100—200	200—300	300—400	400 and above
Frequency	50	90	158	68	134

SOLUTION : First, we try to convert the given open end distribution into the continuous distribution and obtain the following table:

Value	Frequency (f)	Cumulative frequency
0—100	50	50
100—200	90	140
200—300	158	298
300—400	68	366
400—500	134	500
Total	$N = \Sigma f = 500$	

Here, $N = 500$ and $\frac{N}{2} = 250$

Since, the cumulative frequency 298 is immediate successor of 250 (140 < 250 < 298). The median class is 200—300.

Thus, $L_1 = 200, f = 158, C = 140, \frac{N}{2} = 250$ and $i = 100$ (300—200)

$$\therefore \quad \text{Median (Md)} = L_1 + \frac{i}{f}\left(\frac{N}{2} - C\right)$$

$$= 200 + \frac{100}{158}(250 - 140)$$

$$= 200 + \frac{100}{158} \times 110$$

$$= 200 + 69.62 = 269.62$$

Example 2.46

If the median of the following frequency distribution is 46, find the value of missing frequencies f_1 and f_2.

Class	10—20	20—30	30—40	40—50	50—60	60—70	70—80	Total
Frequency	12	30	f_1	65	f_2	25	18	229

SOLUTION : Here, we have the following table.

Class	Frequency (f)	Cummulative frequency ($c.f.$)
10—20	12	12
20—30	30	42
30—40	f_1	$42 + f_1$
40—50	65	$107 + f_1$
50—60	f_2	$107 + f_1 + f_2$
60—70	25	$132 + f_1 + f_2$
70—80	18	$150 + f_1 + f_2$
Total	$N = \Sigma f = 150 + f_1 + f_2$	

It is given that $N = 229 \Rightarrow \frac{N}{2} = 114.5$

$$\therefore \quad 150 + f_1 + f_2 = 229$$

$$\Rightarrow \quad f_1 + f_2 = 79 \quad \text{...(1)}$$

Now it is given that median is 46, hence, median class is 40—50.

Thus, $L_1 = 40, f = 65, C = 42 + f_1, \frac{N}{2} = 114.5$ and $i = 10$

$$\text{Median} = L_1 + \frac{i}{f}\left(\frac{N}{2} - C\right)$$

$$= 40 + \frac{10}{65}(114.5 - 42 + f_1) = 40 + \frac{(72.5 - f_1)}{13} \times 2$$

$$\Rightarrow \quad 46 = 40 + \frac{145 - 2f_1}{13}$$

$$\Rightarrow \quad 6 = \frac{145 - 2f_1}{13}$$

$$\Rightarrow \quad 78 = 145 - 2f_1$$

$$\Rightarrow \quad 2f_1 = 145 - 78 = 67$$

$$\Rightarrow \quad f_1 = 33.5 = 34 \text{ (approx.)}$$

Hence, from Eq. (1)

$$f_2 = 79 - 34 = 45$$

Therefore, the missing frequencies are $f_1 = 34$ and $f_2 = 45$.

Example 2.47

Find the missing frequencies in the following distributions if median = 50 and $n = 100$.

Marks	0–20	20–40	40–60	60–80	80–100
No. of students	14	?	27	?	15

[UPTU 2007]

SOLUTION : Here, we have the following table :

Marks	No. of students (f)	Cummulative frequency (c.f.)
0–20	14	14
20–40	f_1	$14 + f_1$
40–60	27	$41 + f_1$
60–80	f_2	$14 + f_1 + f_2$
80–100	15	$56 + f_1 + f_2$
Total	$\Sigma f = 56 + f_1 + f_2$	

We have $\quad n = 100 \Rightarrow 56 + f_1 + f_2 = 100 \Rightarrow f_1 + f_2 = 44$ ...(1)

Also median = 50. Hence, the median class = 40–60

Thus $\quad L_1 = 40, f = 27, c = 14 + f_1, \frac{n}{2} = 50, i = 20$

$$\text{Median} = L_1 + \frac{i}{f}\left(\frac{n}{2} - c\right)$$

$$50 = 40 + \frac{20}{27}(50 - 14 - f_1)$$

$$\Rightarrow \quad f_1 = 22.5 = 23 \quad \text{...(2)}$$

From (1) and (2), we get

$$f_2 = 21, f_1 = 23$$

2.21 MERITS AND DEMERITS OF MEDIAN

2.21.1 Merits

(i) It is rigidly defined.

(ii) It is easy to understand and can be calculated very easily.

(iii) It can be located by graph.

(iv) It can not be affected by exfrom values.

(v) It can be calculated for open ended series.

2.21.2 Demerits

(i) In case of even number of observations median cannot be determined exactly. It is only estimated by taking the mean two middle term.

(ii) It is not based on all observations of the given distribution.

(iii) In comparison of mean it is affected very much by fluctuations of sampling.

(iv) It cannot be subjected to algebraic treatment.

(v) To find out the median of a series, first we have to arrange the series in ascending or descending order. But it becomes very tedious in case of large number of data availabele.

PROBLEMS SET 2.2

1. Find the median of the following values of a variate:

5, 10, 3, 7, 2, 9, 6, 2, 11.

2. Let the values of a variable be:

30, 5, 21, 42, 13, 10, 27, 33, 17, 8

Find the median.

3. Find the median of the following marks obtained by 12 students:

25, 20, 23, 32, 40, 27, 30, 25, 20, 10, 15, 40.

4. Find the median from the following data:

12, 19, 40, 31, 29, 23, 40, 25, 35, 17, 22, 27, 47, 33, 39.

5. Determine the median of the data given below:

x	112	107	82	87	92	102	97
f	2	3	6	1	3	2	4

6. Find the median of the following frequency distribution:

x	3	5	7	9	11	13	15
f	3	2	8	4	12	10	4

7. Calculate the median from the following distribution:

Salary (in thousand) Rs.	5	15	25	35	45
No. of employees	22	38	46	35	20

8. Calculate the median for the following data:

Class	0–5	5–10	10–15	15–20	20–25	25–30	30–35	35–40	40–45	45–50
Frequency	1	3	6	9	15	28	20	13	3	2

9. Find the median for the following data:

Class	15–20	20–25	25–30	30–35	35–40	40–45	45–50	50–55	55–60
Frequency	4	5	11	6	5	8	9	6	4

10. Calculate the median for the following data:

Marks in Maths	0–10	10–20	20–30	30–40	40–50
No. of students	17	11	12	19	21

11. Find the median for the following frequency distribution:

Class	0–6	6–12	12–18	18–24	24–30	30–36	36–42
Frequency	5	11	25	15	18	12	6

12. Calculate the median for the following data:

Class	0–4	4–6	6–8	8–12	12–18	18–20
Frequency	4	6	8	12	7	2

13. Find the median.

Class	0–3	3–6	6–10	10–15	15–25	25–50
Frequency	10	15	30	15	10	6

14. Determine the median for the following distribution:

Wages in Rs.	No. of Workers
less than 10	15
less than 20	35
less than 30	60
less than 40	84
less than 50	96
less than 60	127
less than 70	198
less than 80	250

15. Find the median for the following data:

Marks less than	10	20	30	40	50	60
No. of students	4	10	30	40	41	50

16. Find the median for the following data:

Marks	< 5	5–10	10–15	15–20	20–25	25–30	30–35	≥ 35
No. of students	4	9	6	13	27	21	12	8

17. Obtain the median for the following distribution:

No. of days absent	No. of students
less than 5	29
5–10	15
10–15	11
15–20	21
20–25	24
25–30	17
30–35	9
35 and above	14

18. Find the missing frequency in the following distribution if $n = 161$ and median is 24.46.

Age (in years)	0–10	10–20	20–30	30–40	40–50
No. of persons	22	38	?	35	20

19. Find the missing frequencies of the following distribution if $n = 100$ and median is 30.

Marks	0–10	10–20	20–30	30–40	40–50	50–60
No. of students	10	?	25	30	?	10

20. Calculate the median for the following distribution:

Age (in years)	11–15	16–20	21–25	26–30	31–35	36–40
No. of persons	3	5	6	9	10	7

ANSWERS

1. 6 **2.** 19 **3.** 25 **4.** 29 **5.** 97 **6.** 11 **7.** 25 **8.** 27.88 **9.** 38 **10.** 30 **11.** 20 **12.** 7.12 **13.** 8.4 **14.** 59.35 **15.** 27.50 **16.** 23.33 **17.** 17.38 **18.** 46 **19.** 15, 10 **20.** 28.83 years

2.22 MODE

In a frequency distribution the value having maximum frequency is called the mode of the series.

(1) According to Zizek, mode is "the value occuring most frequently in a series of items and around which the other items are distributed **most denotely**".

(2) According to A.M. Tutl "Mode is the value which has the greatest frequency or density in its immediate neighbourhood".

2.23 COMPUTATION OF MODE

2.23.1 Ungrouped Data

(1) In case of ungrouped data, the value occuring most frequently is the mode.

Example 2.48

In the series 2, 4, 13, 17, 19, 17, 21, 41. The mode is 17.

If the number of observations is large, we convert the individual series into discrete series and obtain the mode by looking into the frequency table.

Example 2.49

Find the mode of the following series:

2, 5, 7, 5, 9, 4, 7, 3, 12, 5, 9, 7, 11, 7, 14, 17, 15, 9, 7, 11, 17, 9, 4, 9, 11, 12, 11, 9, 14, 9, 17, 9, 2, 7, 11

SOLUTION : Here, we see that the series has a large number of observations. Thus, we convert it into discrete series as follows:

x	2	3	4	5	7	9	11	12	14	15	17
y	2	1	2	3	6	8	5	2	2	1	3

The mode of the discrete series is 9.

2.24 CALCULATION OF MODE IN DISCRETE SERIES

In case of discrete series the mode can be obtained by two methods

(a) Mode by observation.

(b) Mode by grouping method

2.24.1 Mode by Observations

Mode by observation can be obtained only when the distribution under consideration satisfies the following conditions :

(i) There should be a single item having the maximum frequency.

(ii) Maximum frequency should be lying in the middle of series.

(iii) The difference between the modal frequency and the frequencies adjacent to the modal frequency is not large.

(iv) There is not a large number of fluctuation in the frequencies.

Example 2.50

Determine mode from the following distribution :

x	10	12	14	16	18	20	22
f	4	6	10	11	21	10	5

SOLUTION : In the given discrete frequency distributions the variable value 18 has the highest frequency 21. Thus, 18 is the mode of the given discrete frequency distribution.

2.24.2 Mode by Grouping Method

If mode by observations is not possible due to the failure of the distributions, in satisfying the conditions given above. We determine the mode by using grouping method. In this method, the values are first arranged in ascending order and their frequencies are noted. Then we obtain the grouping table using the following steps.

Steps to form grouping table :

Column I : It contains the original frequencies on arranging the data in ascending order. The maximum frequency is marked by putting a mark or a circle.

Column II : The frequencies are grouped in twos and put against each pair. The highest total is thus, marked.

Column III : On leaving the first frequency and combine the remaining frequencies two by two we get the column III. Then mark the highest total.

Column IV : The frequencies combined three by three to obtain this column and highest total is marked.

Column V : On leaving the first frequency and combining the remaining frequencies three by three. We obtain column V. The highest total is marked.

Column VI : Leaving the first two frequencies the remaining frequencies are grouped in threes and the highest total is marked.

After getting the grouping table, we form an analysis table to find the value or the observation which is repeated highest number of times.

Example 2.51

Determine the mode of the following table.

x	1	2	3	4	5	6	7	8	9	10	11	12
f	3	8	15	23	35	40	32	28	20	45	14	6

SOLUTION : Here we observe that distribution is not regular since, the frequencies are increasing steadily upto 40 and then starts decreasing but the frequency 45 after 20 does not seem to be consistent with the distribution. In this case we cannot say that mode is 10 even having highest frequency 45. Thus, we shall use the grouping method to obtain the mode. Using the steps mentioned above we can form the grouping table as follow :

Grouping Table

x	Frequency					
	Col. 1	Col. 2	Col. 3	Col. 4	Col. 5	Col. 6
1	3	11		26		
2	8		23		46	
3	15	38				73
4	23		58	98		
5	35	75			107	
6	40		72			100
7	32	60		80		
8	28		48		93	
9	20	65				79
10	45		59	65		
11	14	20				
12	6					

After forming grouping table, the analysis table is formed with the help of highest frequency in each column.

Analysis Table

Column Number (i)	Maximum frequency (ii)	Combination of values of x giving max frequency (iii)
1	45	10
2	75	5, 6
3	72	6, 7
4	98	4, 5, 6
5	107	5, 6, 7
6	100	6, 7, 8

From the values in column (iii) of analysis table, we find that the value 6 is repeated the maximum number of times and hence, the value of mode is 6 not 10.

2.25 MODE FOR GROUPED SERIES

(a) If maximum frequency occurs in a single class (likely at the centre), this class is known as modal class and then the mode can be obtained by the following formula.

$$\text{Mode (Mo)} = L_1 + \frac{f - f_{-1}}{2f - f_{-1} - f_1} \times i$$

where, f is the frequency of modal class.

L_1 is the lower limit of the modal class.

f_{-1} and f_1 are the frequencies of the classes preceeding and succeding the modal class respectively.

i is the width of interval.

■ **Remarks :** The above formula is applicable only for equal class interval.

Example 2.52

Determine the mode of the following distribution.

Class	0—7	7—14	14—21	21—28	28—35	35—42	42—49
Frequency	19	25	36	72	51	43	28

SOLUTION : By inspection, we find that the frequencies are regular and the highest frequency is 72. Thus the modal class is 21—28.

Here, $L_1 = 21, f = 72, f_{-1} = 36, f_1 = 51$ and $i = 7$

$$\therefore \quad \text{Mode} = L_1 + \frac{f - f_{-1}}{2f - f_{-1} - f_1} \times i$$

$$= 21 + \frac{72 - 36}{2 \times 72 - 36 - 51} \times 7$$

$$= 21 + \frac{36}{57} \times 7$$

$$= 21 + 4.42 = 25.2$$

(b) In case the maximum frequency occurs at more than one interval, we decide the modal class by using grouping method and use the same formulae given below.

Example 2.53

Compute the mode of the following distribution:

Class	10—20	20—30	30—40	40—50	50—60	60—70	70—80
Frequency	22	35	40	62	50	45	40

SOLUTION : By inspection, we find that the frequencies are regular and the highest frequency is 72. Thus, the modal class is 40—50.

Here, $L_1 = 40, f = 62, f_{-1} = 40, f_1 = 50$ and $i = 10$.

$$\therefore \quad \text{Mode} = L_1 + \frac{f - f_{-1}}{2f - f_{-1} - f_1} \times i$$

$$= 40 + \frac{62 - 40}{124 - 40 - 50} \times 10$$

$$= 40 + \frac{22}{34} \times 10$$

$$= 40 + 6.47 = 46.47.$$

Example 2.54

Determine the mode from the following data.

Class	0—2	2—4	4—6	6—8	8—10	10—12	12—14
Frequency	6	14	16	20	12	14	6

SOLUTION : Here we see the frequencies are irregular. Therefore, the modal class would be located by grouping method.

Grouping Table

Class	Frequency					
	Col. 1	Col. 2	Col. 3	Col. 4	Col. 5	Col. 6
0—2	6	20		36		
2—4	14		30		50	
4—6	16	36				48
6—8	20		32	46		
8—10	12	26			32	
10—12	14		20			
12—14	6					

Analysis table

Column Mamber (i)	Maximum Frequency (ii)	Combination of classes giving maximum number of frequency (iii)
1	20	6—8
2	36	4—6, 6—8
3	32	6—8, 8—10
4	46	6—8, 8—10, 10—12
5	50	2—4, 4—6, 6—10
6	48	4—6, 6—8, 8—10

The class 6—8 occurs maximum number of times. Hence, the modal class is 6—8

Thus, $L_1 = 6, f = 20, f_{-1} = 16, f_1 = 12$ and $i = 2$

$$\text{Mode (Mo)} = L_1 + \frac{f - f_{-1}}{2f - f_{-1} - f_1} \times i$$

$$= 6 + \frac{20 - 16}{40 - 16 - 12} \times 2$$

$$= 6 + \frac{4}{12} \times 2$$

$$= 6 + \frac{4}{6} = 6.67$$

Hence, mode is 6.67.

(c) If all the class intervals are not of equal magnitude, then firstly, we convert the series into equal class intervals and then obtain the mode.

Example 2.55

Find the mode of the following data:

Class	0—10	10—20	20—40	40—50	50—70	70—80
Frequency	5	15	40	32	28	5

SOLUTION : Here, we see that the class intervals are not equal. These classes should be converted into equal class intervals. We shall break the classes each of size 10 and the frequencies corresponding to these classes would be divided as desired. In the given distribution the class interval 20—40 and 50—70 are of length 20. Therefore, these intervals are converted as follows:

Class	0—10	10—20	20—30	30—40	40—50	50—60	60—70	70—80
Frequency	5	15	20	20	32	14	14	5

The class 40—50, having the frequency 32 is the modal class.

Thus, $L_1 = 40, f = 32, f_{-1} = 20, f_1 = 14$ and $i = 10$ (50—40)

$$\text{Mode} = L_1 + \frac{f - f_{-1}}{2f - f_{-1} - f_1} \times i$$

$$= 40 + \frac{32 - 20}{64 - 20 - 14} \times 10$$

$$= 40 + \frac{12}{30} \times 10$$

$$= 40 + 4 = 44$$

Example 2.56

Find the mode from the following data:

25, 15, 23, 40, 27, 25, 23, 25, 20

SOLUTION : First we arrange the data in ascending order:

15, 20, 23, 23, 25, 25, 25, 27, 40.

By inspection we see that item 25 occurs maximum, *i.e.*, 3 times, hence mode is 25.

Example 2.57

Calculate mode of the following distribution.

x	1	2	3	4	5	6	7	8
f	4	9	16	25	22	15	7	3

SOLUTION : Here by inspection we find that the maximum frequency is 25. Which is corresponding to the value, 4. Hence, mode is 4.

Example 2.58

Compute mode of the following series:

Age group	0—5	5—10	10—15	15—20	20—25	25—30	30—35	35—40	40—45
Frequency	20	24	32	28	20	16	34	10	8

SOLUTION : In the above distribution we find that the frequencies are irregular. Therefore, the modal class would be located by grouping method.

Grouping Table

Class	Frequency					
	Col. 1	Col. 2	Col. 3	Col. 4	Col. 5	Col. 6
0—5	20	44		76		
5—10	24		56		84	
10—15	32	60				80
15—20	28		48	64		
20—25	20	36			70	
25—30	16		50			60
30—35	34	44		52		
35—40	10		18			
40—45	8					

Analysis table

Column Number (i)	Maximum Frequency (ii)	Combination of classes giving maximum number of frequency (iii)
1	34	30—35
2	60	10—15, 15—20
3	56	5—10, 10—15
4	76	0—5, 5—10, 10—15
5	84	5—10, 10—15, 15—20
6	80	10—15, 15—20, 20—25

From the analysis table it is observed that class 10—15 occurs maximum number of times. Hence, the modal class is 10—15.

Thus, $L_1 = 10, f = 32, f_{-1} = 24, f_1 = 28$ and $i = 5$ (15—10)

$$\text{Mode (Mo)} = L_1 + \frac{f - f_1}{2f - f_{-1} - f_{-2}} \times i$$

$$= 10 + \frac{32 - 24}{64 - 24 - 28} \times 5$$

$$= 10 + \frac{8 \times 5}{12} = 10 + 3.33 = 13.33$$

Example 2.59

Calculate the mode of the following data:

Wages (Rs.)	below 100	100—200	200—300	300—400	400—500	500 and above
No. of weaks	8	12	25	15	10	5

SOLUTION : First, we shall convert the open end series (as given above) into a continuous series.

Wages (Rs.)	0—100	100—200	200—300	300—400	400—500	500—600
No. of weeks	8	12	25	15	10	5

Since, the frequencies in the above distribution are regular and by inspection we find that maximum frequency is 25. Hence, the modal class is 200—300.

Thus, $L_1 = 20, f = 25, f_{-1} = 12, f_1 = 15$ and $i = 100$ (3——200)

$$\text{Mode (Mo)} = L_1 + \frac{f - f_{-1}}{2f - f_{-1} - f_1} \times i$$

$$= 200 + \frac{25 - 12}{50 - 12 - 15} \times 100$$

$$= 200 + \frac{13}{23} \times 100 = 200 + 56.52 = 256.52$$

Example 2.60

Calculate the mode of the following distribution :

Value less than	10	20	30	40	50	60
Frequency	4	20	35	55	62	67

[UPTU 2007]

SOLUTION :

Class Interval	Frequency (f)
0–10	4
10–20	16
20–30	15
30–40	30
40–50	7
50–60	5

By Inspection we see that the frequencies are irregular. Therefore the modal class would be located by grouping method.

Grouping Table

Class	Frequency				
	Col. (1)	Col. (2)	Col. (3)	Col. (4)	Col. (5)
0–10	4	20		35	
10–20	16		31		51
20–30	15	35			
30–40	20		27	32	
40–50	7	12			
50–60	5				

Analysis Table

Column Mamber	Maximum frequency	Combination of classes giving maximum number of frequency
1	20	30–40
2	35	20–30, 30–40
3	31	10–20, 20–30
4	35	0–10, 10–20, 20–30
5	51	10–20, 20–30, 30–40

The class 20—30 occurs maximum number of times. Hence, the modal class is 20—30. Thus $L_1 = 20, f = 15, f_{-1} = 16, f_1 = 20, i = 10$

$$\text{Mode} = L_1 + \frac{f - f_{-1}}{2f - f_{-1} - f_1} \times i$$

$$= 20 + \frac{15 - 16}{30 - 16 - 20} \times 10$$

$$= 20 + \frac{10}{6} = 21.67$$

Hence, mode is 21.67.

2.26 RELATION BETWEEN MEAN, MEDIAN AND MODE

The relation between mean, median and mode is given by the formula

Mode = 3 Median – 2 Mean

2.27 MERITS AND DEMERITS OF MODE

2.27.1 Merits

(i) Mode is a qualitative measure of central tendency. It can be eaisly calculated and sometimes can be calculated even by inspection.

(ii) Mode is not affected by extreme values.

(iii) Mode can be represented graphically.

(iv) Mode can be calculated even if the frequency distribution has unequal class intervals

2.27.1 Demerits

(i) Mode is ill defined. In some cases, we come across two modes of a distribution. Such distribution one called f_1 modal. The distribution having more than two modes is called multimodal.

(ii) Since, mode is the value corresponding to highest frequency. It is based upon all the observations.

(iii) It is not capable of further mathmatical treatment.

(iv) Mode is affected to a greater extent by fluctuations of samplify in comparison of mean.

Problem Set 2.3

1. Determine mode from the following figure.

11, 25, 15, 23, 40, 27, 25, 23, 25, 20, 21, 15, 17, 25.

2. Find the mode from the following data:

10, 5, 15, 20, 25, 20, 35, 37, 25, 30, 37, 25, 27, 32, 25, 20, 28, 31, 25.

3. Compute the mode of the following:

x	2	3	4	5	6	7	8	9	10	11	12	13	14
f	3	8	10	12	16	14	10	8	17	5	4	1	3

4. Determine the mode of the following frequency distribution.

x	15	20	22	24	25	30	33	38	45
f	5	8	11	20	23	18	13	3	1

5. Calculate the mode from the following data:

Value	4	5	6	7	8	9	10	11
Frequency	1	9	3	4	8	7	8	3

6. Find the mode from the following table:

Marks	0–10	10–20	20–30	30–40	40–50	50–60	60–70	70–80
No. of students	2	18	30	45	25	20	6	3

7. Find the mode of the following:

x	0–5	5–10	10–15	15–20	20–25	25–30	30–35	35–40
f	5	7	9	18	16	15	6	3

8. Find out the mode of the following distribution:

Marks	10–25	25–40	40–55	55–70	70–85	85–100
Frequency	6	20	44	26	3	1

9. Compute the mode of the following distribution:

Class	0–10	10–20	20–30	30–40	40–50	50–60
Frequency	14	17	25	27	24	15

10. Calculate the mode from the following distribution:

Size	10–15	15–20	20–25	25–30	30–35	35–40
f	8	12	12	18	14	10

11. Estimate the mode for the following frequency distribution:

Class	1–2	2–3	3–4	4–5	5–6	6–7	7–8	8–9	9–10	10–11	11–12	12–13	13–14	14–15
Frequency	1	3	6	8	11	13	13	14	10	12	8	5	2	1

12. Calculate the mode for the following data:

Class	0–10	10–20	20–30	30–40	40–50	50–60	60–70
Frequency	3	12	5	8	12	9	2

13. Find the mode of the following:

Marks	1–5	6–10	11–15	16–20	21–25	26–30	31–35	36–40	41–45
No. of students	7	10	16	32	24	18	10	5	1

14. Find the mode of the following distribution:

Value less than	10	20	30	40	50	60
Frequency	9	20	35	55	62	67

15. For the following distribution the mode is 47.5. Find the missing frequency.

Class	0–10	10–20	20–30	30–40	40–50	50–60	60–70	70–80
Frequency	7	8	10	?	40	35	10	7

16. Find the missing frequencies in the following distribution, if $n = 100$ and mode is 52.

Expenses in Rs.	0–20	20–40	40–60	60–80	80–100
No. of families	14	?	27	?	15

17. Compute the mode of the following distribution:

Age in years less than	25	30	35	40	45	50	55	60
No. of persons	5	12	22	40	55	67	74	79

18. Find the mode of the following frequency distribution:

Class	14–15	16–17	18–20	21–24	25–29	30–34	35–39
Frequency	6	14	15	11	11	10	9

ANSWERS

1. 25 **2.** 25 **3.** 6 **4.** 25 **5.** 9 **6.** 36 **7.** 19.09 **8.** 48.53 **9.** 34
10. 28 **11.** 7 **12.** 45.71 **13.** 18.66 **14.** 32.78 **15.** 25 **16.** 21, 23
17. 38.64 **18.** 18.1

MULTIPLE CHOICE QUESTIONS

1. If the mean of a set of observations $a_1, a_2, ..., a_n$ is $\overline{A}$, then the mean of the observations $a_i + 2i; i = 1, 2, ... n$ is

(a) $\overline{A} + 2$ (b) $\overline{A} + 2n$
(c) $\overline{A} + (n + 1)$ (d) $\overline{A} + n$

2. If a varaible X can be expressed as a linear function of two variables U and V in the form $X = aU + bV$, then mean $\overline{X}$ of X is

(a) $a\overline{U} + b\overline{V}$ (b) $\overline{U} + \overline{V}$
(c) $b\overline{U} + a\overline{V}$ (d) None of these

3. The weighted AM of first n natural numbers whose weights are equal to the corresponding numbers is equal to

(a) $(2n + 1)$ (b) $\frac{1}{2}(2n + 1)$
(c) $\frac{1}{3}(2n + 1)$ (d) $\frac{2n+1}{6}$

4. Which of the following is a measure of central tendency?

(a) Mode (b) Mean deviation
(c) Standard deviation (d) Coefficient of correlation

5. The most stable measure of central tendency is

(a) The mean (b) The median
(c) The mode (d) None of these

6. The mean of the distribution, in which the values of X are 1, 2, ... n, the frequency of each being unity is

(a) $\frac{n(n+1)}{2}$ (b) $\frac{n}{2}$
(c) $\frac{n+1}{2}$ (d) None of these

7. A group of 10 items has arithmetic mean 6. If the arithmetic mean of 4 of these items is 7.5, then the mean of the remaining items is

(a) 6.5 (b) 5.5
(c) 4.5 (d) 5.0

8. The A.M. of a set of observations is $\overline{x}$. If each observation is divided by α and then increased by 10, the mean of the new series is

(a) $\frac{\overline{x}}{\alpha}$ (b) $\frac{\overline{x}+10}{\alpha}$
(c) $\frac{\overline{x}+10\alpha}{\alpha}$ (d) $\alpha\overline{x} + 10$

9. The weighted mean of first n natural numbers whose weights are equal to the squares of corresponding numbers is

(a) $\frac{n+1}{2}$ (b) $\frac{3n(n+1)}{2(2n+1)}$
(c) $\frac{(n+1)(2n+1)}{6}$ (d) $\frac{n(n+1)}{2}$

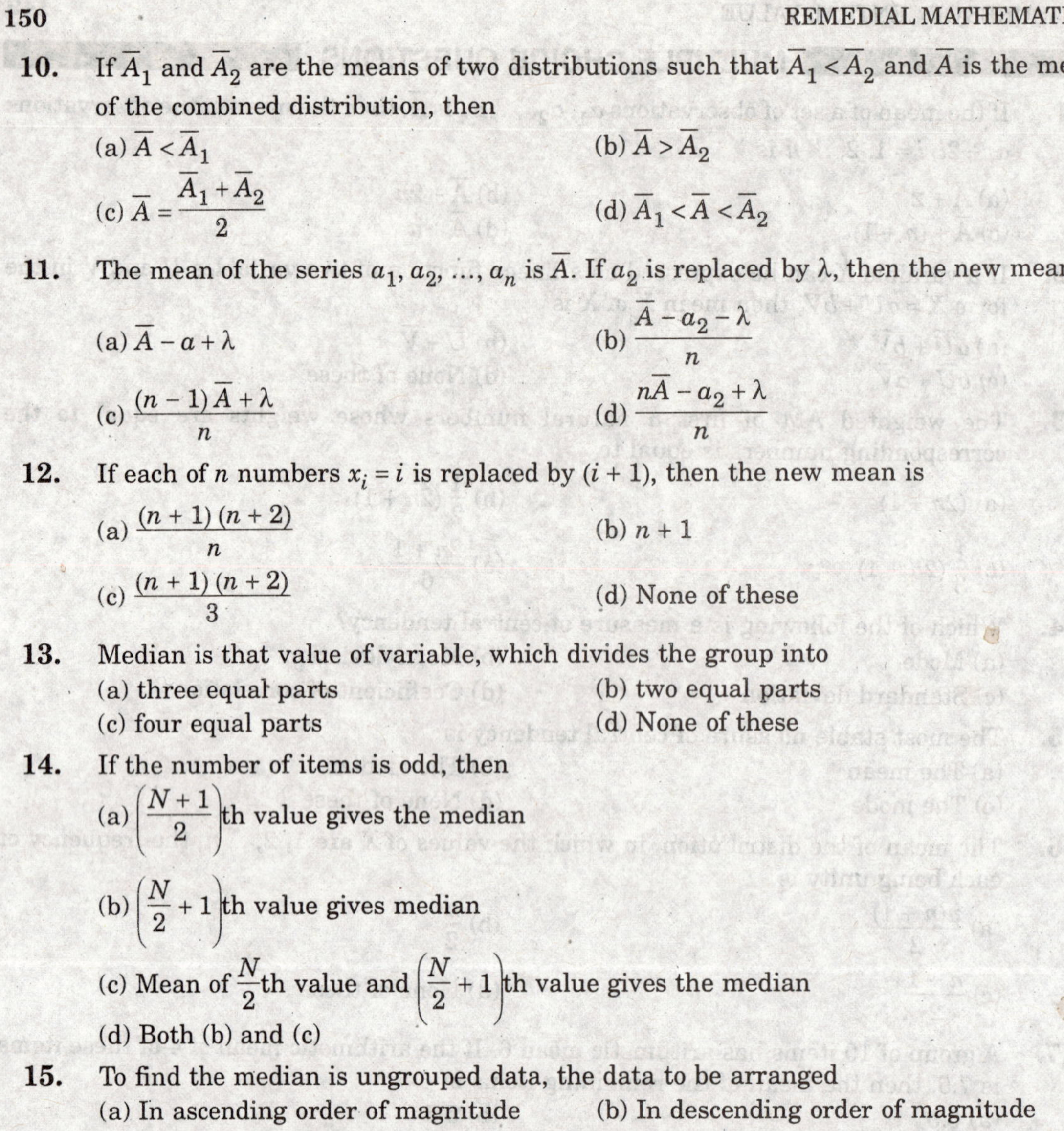

10. If $\overline{A}_1$ and $\overline{A}_2$ are the means of two distributions such that $\overline{A}_1 < \overline{A}_2$ and $\overline{A}$ is the mean of the combined distribution, then

(a) $\overline{A} < \overline{A}_1$ (b) $\overline{A} > \overline{A}_2$

(c) $\overline{A} = \dfrac{\overline{A}_1 + \overline{A}_2}{2}$ (d) $\overline{A}_1 < \overline{A} < \overline{A}_2$

11. The mean of the series $a_1, a_2, ..., a_n$ is $\overline{A}$. If a_2 is replaced by λ, then the new mean is

(a) $\overline{A} - a + \lambda$ (b) $\dfrac{\overline{A} - a_2 - \lambda}{n}$

(c) $\dfrac{(n-1)\overline{A} + \lambda}{n}$ (d) $\dfrac{n\overline{A} - a_2 + \lambda}{n}$

12. If each of n numbers $x_i = i$ is replaced by $(i + 1)$, then the new mean is

(a) $\dfrac{(n+1)(n+2)}{n}$ (b) $n + 1$

(c) $\dfrac{(n+1)(n+2)}{3}$ (d) None of these

13. Median is that value of variable, which divides the group into

(a) three equal parts (b) two equal parts

(c) four equal parts (d) None of these

14. If the number of items is odd, then

(a) $\left(\dfrac{N+1}{2}\right)$th value gives the median

(b) $\left(\dfrac{N}{2} + 1\right)$th value gives median

(c) Mean of $\dfrac{N}{2}$th value and $\left(\dfrac{N}{2} + 1\right)$th value gives the median

(d) Both (b) and (c)

15. To find the median is ungrouped data, the data to be arranged

(a) In ascending order of magnitude (b) In descending order of magnitude

(c) Either (a) or (b) (d) Both (a) and (b)

16. The median for the series 2, 10, 5, 7, 15, 17, 21, 13, 4

(a) 10 (b) 7

(c) 17 (d) 15

17. Median for the grouped data is given by the expression

(a) Median $= l_1 + \dfrac{h}{f}\left(\dfrac{N}{2} - c\right)$ (b) Median $= l_2 + \dfrac{f}{h}\left(\dfrac{N}{2} - c\right)$

(c) Median $= L_1 + \dfrac{h}{f}\left(\dfrac{N}{4} - 2c\right)$ (d) None of these

18. Median of series 18, 16, 14, 11, 13, 10, 9, 20 is

(a) 13 (b) 14

(c) 11 (d) 13.5

19. If the median of the series 22, 24, 33, 37, $x + 1$, $x + 3$, 44, 47, 51, 58 arranged in ascending order is 42. The value of x

(a) 37 (b) 41
(c) 40 (d) 43

20. Point of intersection of less than ogive and more than ogive is

(a) Median (b) Mean
(c) Quartile Deviation (d) Mode

21. Which of the average is used, if the data are qualitative

(a) mean (b) mode
(c) median (d) none

22. Which of the folloing is false?

(a) The value of median of data set gets affected by extreme values that are not representative of the data set
(b) Median is the point below which 50% of the data items fall
(c) Median can be calculated from a grouped data with open ended classes
(d) None of these

23. Most appropriate average to determine the size of oranges on a tree is

(a) Median (b) Mean
(c) Mode (d) None of these

24. The ost unstable measure of central tendency is

(a) A.M. (b) median
(c) mode (d) G.M.

25. Total wages paid to the workers into factories A and B are the same but the modal wage in factory A is larger than the modal wage in factory B. In which factory workers are well off?

(a) A (b) B
(c) Equal to both (d) None of these

26. The mode of the data set {3, 4, 4, 4, 5, 6, 6, 7, 8, 3}

(a) 2 (b) 3
(c) 4 (d) 6

27. The mode of the data set {3, 3, 3, 5, 5, 5, 7, 7, 7}

(a) 1.3 (b) 2.4
(c) 3.7 (d) None of these

28. If the mean, median and mode of a data set are equal, which of the following is correct

(a) it is negatively skewed (b) it is symmetrical
(c) it is positively skewed (d) insuficient information

29. Which of the following would be an appropriate average for determining the average size of readymade garments

(a) A.M. (b) Median
(c) Mode (d) None of these

True and False :

1. In certain cases median can be computed by inspection.
2. Median is affected by sampling fluctuations less than the mean.
3. Median × number of observations = sum of the observations.
4. Median is very useful when extreme values of the series are either not available or abnormal.
5. Histogram is used to find mode graphically.
6. For calculating mode the class intervals have to be equal.
7. The class limbs should be exclusive for determinining mode.
8. Mode of the value with minimum frequency in the data.

Fill in the blanks :

1. If the class intervals are unequal, the value of .... is not affected.
2. Sum of the absolute deviations of all the values is minimum about ....
3. .... is the value which occurs most frequently in the series.
4. If one value is zero in any obsrvation, then .... may or may not be calculated.
5. Mode is .... by fluctuations of sampling.
6. The relaton between mean, median and mode is ....

Answer

1. c	**2.** a	**3.** c	**4.** a	**5.** a	**6.** c	**7.** d	**8.** c	**9.** b	**10.** d
11. d	**12.** d	**13.** b	**14.** a	**15.** c	**16.** d	**17.** a	**18.** d	**19.** c	**20.** a
21. c	**22.** a	**23.** c	**24.** c	**25.** a	**26.** c	**27.** d	**28.** b	**29.** c	

True and False :

1. T, 2. F, 3. F, 4. T, 5. T, 6. T, 7. T, 8. F

Fill in the blanks :

1. Median, 2. Median, 3. Mode, 4. Mode, 5. Most affected,

6. Mode = 3 median - 2 mean

□ ■ □

UNIT - III

Elementary Trigonometry

3.1 INTRODUCTION

A Greek astronomer Hipparchus, born about 180 B.C in Northern Asia Minor is known as the founder of trigonometry. The word trigonometry is a combination of two Greek words "*Trigonon*", meaning triangle and "*Metron*", meaning measurement. Thus, the science, which deals with the measurement of triangles is known as trigonometry. The applications of Trigonometry are in warfare, all sorts of vibratory phenomena, sound, light, electricity, etc:

1. Angles: An angle is thought of as traced out by the rotation of a line about a fixed point O (vertex) from the intial postion OX to a terminal position OP.

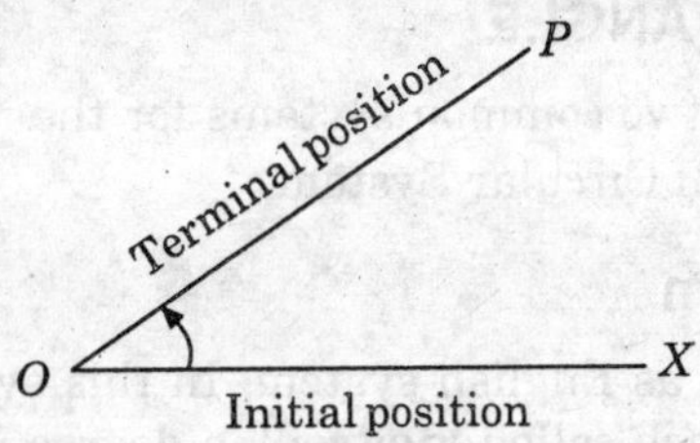

Fig. 3.1

2. Convention of signs of Angles:

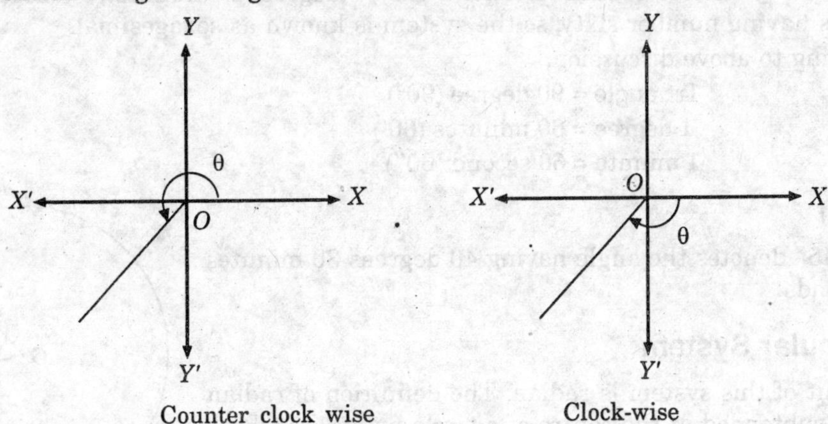

Fig. 3.2

■ **Note :** When an angle is traced out by a counter-clock wise it is positive otherwise negative.

3.2 QUADRANTS

Consider two mutually perpendicular lines XOX' and YOY' in any plane. These lines divide the plane into four equal parts known as quadrants numbered anticlockwise.

From fig. 3.3, it is clear that XOY, YOX', $X'OY'$ and $Y'OX$ are called first quadrant, second quadrent, third quadrant and fourth quadrant respectively.

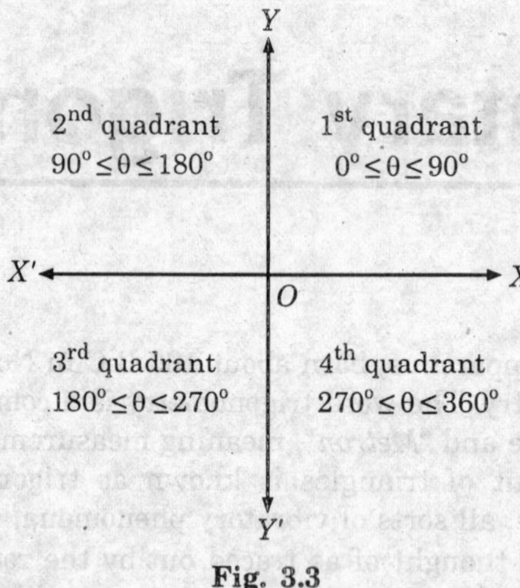

Fig. 3.3

3.3 MEASUREMENT OF ANGLE

In trigonometry there are two common systems for the measurement of angles

(1) Sexagesimal system (2) Circular System

3.3.1 Sexagesimal System

This system is also known as English system. In this system one complete revolution is divided into 360 equal parts, each called **'degree'**, a degree is divided into sixty equal parts, each called **'minute'** and a minute is again divided into sixty equal parts each called a **'second'**. The symbols for such divisions are ° (degree), ′ (minute), ″ (seconds). As the subdivisions having number sixty, so the system is known as sexagesimal.

According to above discussion,

$$1\text{rt angle} = 90 \text{ degree } (90^\circ)$$
$$1 \text{ degree} = 60 \text{ minutes } (60')$$
$$1 \text{ minute} = 60 \text{ second } (60'')$$

Example 3.1

$40^\circ 30' 45''$ denotes the angle having 40 degrees 30 minutes and 45 seconds.

3.3.2 Circular System

The unit of this system is radian. The definition of radian is an angle-subtended at the centre of a circle by an arc whose length is equal to radius of the circle.

From fig. 3.4.

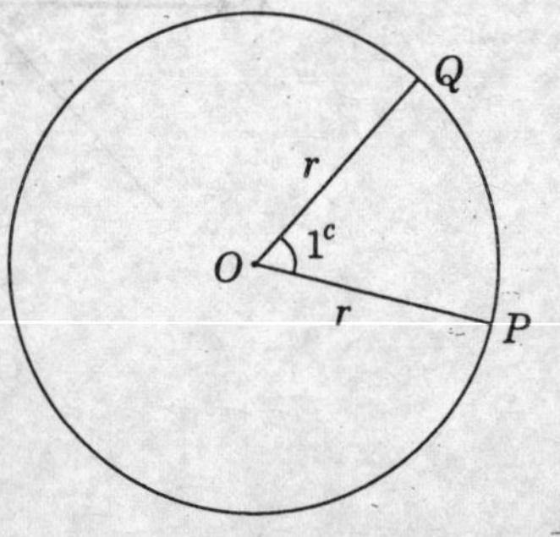

Fig. 3.4

OP = Radius of circle
$= r$
Arc $PQ = r$ [by definition]

We know that

$$\frac{\angle POQ}{\text{Total angle at } O} = \frac{\text{Arc } PQ}{\text{Circumference}}$$

$$\frac{1 \text{ radian}}{360 \text{ degree}} = \frac{r}{2\pi r}$$

or $$\frac{1 \text{ radian}}{360 \text{ degree}} = \frac{1}{2\pi}$$

$$1 \text{ radian} = \frac{360}{2\pi} \text{ degree} = \frac{180°}{\pi}$$

$$= 57.2958 \text{ degree (App.)} \qquad \left[\pi = \frac{22}{7}\right]$$

$$= 57°17'45''$$

$$180° = 3.1416 \text{ radian} = \pi^c \text{or } \pi$$

From above discussion we can measure some common angles in radian as follows:

$$90° = \frac{\pi}{2},\ 60° = \frac{\pi}{3},\ 30° = \frac{\pi}{6} \text{ etc.}$$

■ **Note :** In algebra $\pi = \frac{22}{7}$ but in trigonometry $\pi^c = 180°$.

Example 3.2

Express the following in degrees.

(i) $\frac{2\pi}{5}$ radians (ii) $\frac{\pi}{15}$ radians (iii) 0.74019 radians

SOLUTION : We know that

$$\pi \text{ radian} = 180° \Rightarrow 1 \text{ radian} = \frac{180°}{\pi}$$

(i) $\frac{2\pi}{5}$ radians $= \frac{180°}{\pi} \times \frac{2\pi}{5} = \frac{360°}{5} = 72°$

(ii) $\frac{\pi}{15}$ radians $= \frac{180°}{\pi} \times \frac{\pi}{15} = \frac{180°}{15} = 12°$

(iii) 0.74019 radian $= \frac{180°}{\pi} \times (0.74019) = 42°24'35''$ app.

Example 3.3

Express the following in radian measure.

(i) 45° (ii) 135° (iii) 2.5 right angle (iv) 40°15′36″

SOLUTION : We know that $180° = \pi$ radian $\Rightarrow 1° = \frac{\pi}{180}$ radian

(i) $45° = \frac{\pi}{180°} \times 45°$ radian $= \frac{\pi}{4}$ radian

(ii) $135° = \frac{\pi}{180} \times 135°$ radian $= \frac{3\pi}{4}$ radian

(iii) 2.5 right angle = 2.5 × 90° = 225° = $\frac{\pi}{180}$ × 225 radian

$$= \frac{5\pi}{4} \text{ radian}$$

(iv) We know that

$$60'' = 1' \Rightarrow 1'' = \frac{1'}{60} \Rightarrow 36'' = \frac{36'}{60} = 0.6'$$

$\Rightarrow$ 15′36″ = 15′ + (.6′) = 15.6′

Now, we also know that

$$60' = 1° \Rightarrow 1' = \frac{1°}{60} \Rightarrow 15.6' = \frac{15.6°}{60} = 0.26°$$

Again, 40° 15.6′ = 40° + 0.26° = 40.26°

Now 40°15′36″ = 40.26° = 40.26 × $\frac{\pi}{180}$ radian

$$= 0.2237 \text{ radian}$$

3.3.3 Length of an Arc of a Circle

Consider a circle of radius r and centre *O*. cut off an are *PQ* eaual to the radius *r*. Join *OQ*. Cut off another are *PQR* of length *l*.

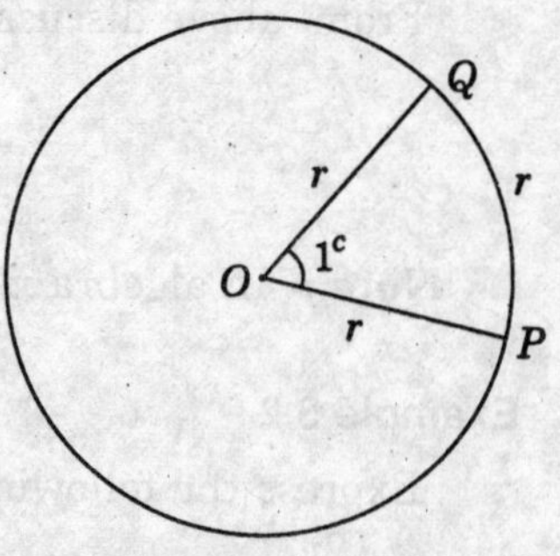

Fig. 3.5

By definition

$$\angle POQ = 1 \text{ radian}$$

Now $$\frac{\text{arc } PQR}{\text{arc } PQ} = \frac{\angle POR}{\angle POQ}$$

$$\frac{l}{r} = \frac{\theta}{1} \qquad [\text{Let } \angle POR = \theta \text{ radian}]$$

$$\Rightarrow \quad l = r\theta$$

Example 3.4

Find the length of an are of a circle of 4cm radius if the angle subtended at the centre is 45°. (π = 3.14)

SOLUTION : We know that $l = r\theta = 4 \times \frac{\pi}{4} = \pi = 3.14$cm.

Example 3.5

Find the angle subtended by an are 12 cm long at the centre of a circle whose radius is 48 cm.

SOLUTION : We know that

$$l = r\theta \Rightarrow \theta = \frac{l}{r} \text{ radians}$$

Here $\theta = \frac{12}{48}$ radians $= \frac{12}{48} \times \frac{180°}{\pi} = 14.324°$ $[\because \pi = 3.14]$

3.3.4 Area of a Sector of a Circle

Let *POQ* be a sector of a circle with *O* as centre and radius *r*. Let $\angle POQ = Q^c$ and arc $PQ = l$.

Then $\dfrac{\text{Area of the sector } POQ}{\text{Area of the circle}} = \dfrac{\angle POQ}{2\pi}$

or $\dfrac{\text{Area of the sector } POQ}{\pi r^2} = \dfrac{\theta}{2\pi}$

or Area of the sector $POQ = \dfrac{\pi r^2 \theta}{2\pi} = \dfrac{1}{2} r^2 \theta$

But $\theta = \dfrac{l}{r}$

$\therefore$ Area of the sector $POQ = \dfrac{1 r^2}{2} \cdot \dfrac{l}{r}$

$= \dfrac{1}{2} l \cdot r$

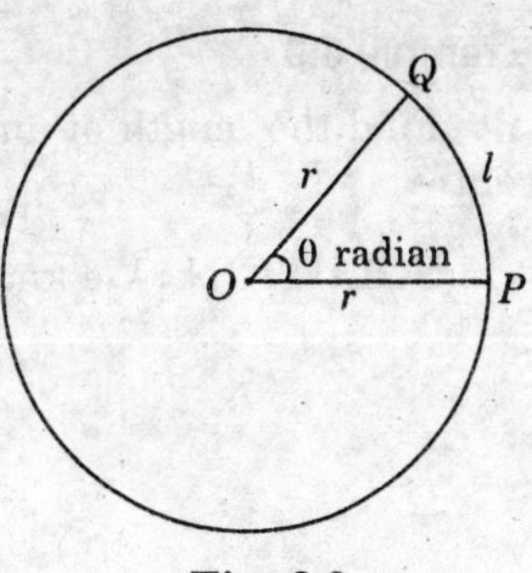

Fig. 3.6

Example 3.6

The area of a sector is 5.024 cm^2 and its angle is 36°. Find the radius ($\pi = 3.14$).

SOLUTION : We know that area of sector $= \dfrac{1}{2} r^2 \theta$

where θ is in radians

$\because$ $180° = \pi^c \Rightarrow 36° = \dfrac{\pi}{180} \times 36$ radian $= \dfrac{\pi}{5}$ radians

Now, area of sector $= \dfrac{1}{2} r^2 \left(\dfrac{\pi}{5}\right) = 5.024$

$\Rightarrow r^2 = 16$ or $r = 4$

Example 3.7

Find the area of a sector of a circle whose radius is 5m bounded by an arc of length 8m.

SOLUTION : Area of sector $= \dfrac{1}{2} r \cdot l$

$= \dfrac{1}{2} \times 5m \times 8m$

$= 20\text{m}^2$

Example 3.8

Find the number of degrees in the angle subtended at the centre of radius 10 ft. by an arc of length 20 ft. *[UPTU 2004]*

SOLUTION : We know that $l = r\theta \Rightarrow \theta = \dfrac{l}{r}$ radians

Here $l = 20$ ft, $r = 10$ ft

then $\theta = \dfrac{20}{10} = 2$ radian

In degrees $\theta = 2 \times \dfrac{180°}{\pi}$

$= \dfrac{360°}{3.14} = 114.65°$

Example 3.9

Find the length of an arc of a circle of radius 5 cms, subtending an angle of 15° at the centre. *[UPTU 2007]*

SOLUTION : We know that $l = r\theta = 5 \times \frac{\pi}{180°} \times 15° = \frac{5\pi}{12}$

$$= \frac{5}{12} \times 3.14 = 1.308 \text{ cm.}$$

PROBLEM SET 3.1

1. Express each of the following angles in radian measure.
   (i) 30° (ii) 135°
   (iii) 340° (iv) – 37° 30′
   (v) 40° 15′ 36″
2. Express each of the following measures in degrees.
   (i) $\frac{2\pi}{15}$ radian (ii) 0.25 radian
   (iii) $\frac{5\pi}{12}$ radian (iv) 0.74019 radian
   (v) – 2 radian
3. The radius of a circle is 4cm. What is the central angle in radian subtended by an arc of length 8cm?
4. Find in degrees the angle subtended at the centre of a circle of diameter 50cm by arc of length 11cm.
5. The perimeter of a certain sector of a circle is equal to the length of the semi-circle having the same radius, express the angle of the sector in degrees, minutes and seconds.
6. A horse is tied to a post by a rope. If the horse moves along a circular path always keeping the rope tight and describes 88 meters when it has traced out 72° at the centre, find the length of the rope.
7. A wheel of $\frac{8}{7}$ meter diameter is rotating at the rate of 7 revolutions per second. Find the distance travelled by a point on the rim in one second.
8. A railway train is travelling on a curve of 750 m radius at the rate of 30 km/h, through what angle has it turned in 10 seconds.
9. Find in degrees and radians the angle between the hour hand and the minute-hand of a clock at half past three.
10. The radius of circle is 30 cm. Find the length of an arc of this circle. If the length of the chord of the arc is 30cm.

ANSWERS

1. (i) 0.5236 radian (ii) 2.3562 radian (iii) $\frac{17}{9}\pi$ radian (iv) $\frac{-5}{24}$ radian

(v) 0.7027 radian **2.** (i) 24° (ii) 14°19′15″ (iii) 75° (iv) 42°24′35″

(v) −114°32′44″ **3.** 2 radian **4.** 25°12′ **5.** 65°27′16″ approximately

6. 70 m **7.** $25\frac{1}{7}$ m **8.** $\frac{1}{9}$ radian **9.** $\frac{5}{12}\pi$ radian **10.** 10π cm.

3.4 RIGHT ANGLE TRIANGLE

A triangle whose one angle out of the three angles is of 90°, *i.e.*, a right angle, is called right angle triangle. Consider an angle $\theta = \angle POQ$ as shown in Fig. 3.7. Take any point A in OQ and draw AB perpendicular to OP. A right angled triangle OBA is formed. If θ is taken as the *angle of reference*, AB the side opposite to θ is known as *the perpendicular*, OA the side opposite to the right angle is called *the hypotenuse* and OB the third side is said to be *the base* of the angle θ.

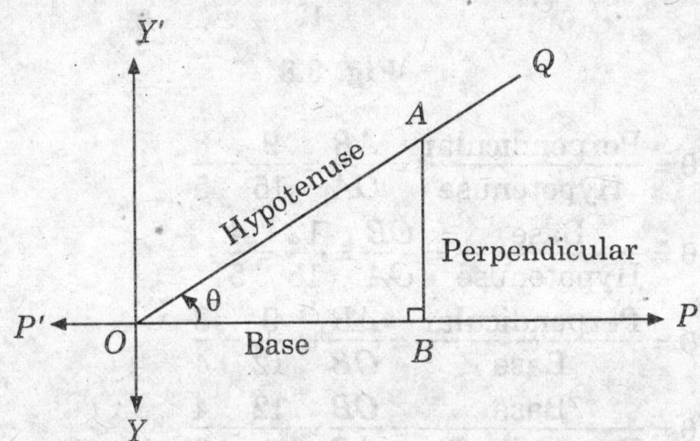

Fig. 3.7

3.5 TRIGONOMETRIC FUNCTIONS

In a right angled triangle OBA, we have base $= OB = x$, Perpendicular $= AB = y$ and hypotenuse $= OA = z$.

The ratio of any two sides in a right angle triangle are called trigonometric ratios. The three sides OB, AB and OA can be arranged two at a time in ${}^3p_2 = 6$ different ways.

These six ratios are called trigonometric ratios or trigonometric functions or t-ratios. The six trigonometric ratios in right angle triangle OBA are defined as follows:

(i) $\text{sine}\ \theta = \frac{\text{Perpendicular}}{\text{Hypotenuse}} = \frac{AB}{OA} = \frac{y}{z}$ and is denoted as $\sin\theta$

(ii) $\text{cosine}\ \theta = \frac{\text{Base}}{\text{Hypotenuse}} = \frac{OB}{OA} = \frac{x}{z}$ and is denoted as $\cos\theta$

(iii) $\text{tangent}\ \theta = \frac{\text{Perpendicular}}{\text{Base}} = \frac{AB}{OB} = \frac{y}{x}$ and is denoted as $\tan\theta$

(iv) $\text{cotangent}\ \theta = \frac{\text{Base}}{\text{Perpendicular}} = \frac{OB}{AB} = \frac{x}{y}$ and is denoted as $\cot\theta$

(v) $\text{secant } \theta = \dfrac{\text{Hypotenuse}}{\text{Base}} = \dfrac{OA}{OB} = \dfrac{z}{x}$ and is denoted as sec θ

(vi) $\text{cosecant } \theta = \dfrac{\text{Hypotenuse}}{\text{Perpendicular}} \doteq \dfrac{OA}{AB} = \dfrac{z}{y}$ and is denoted as cosec θ

■ **Note :** (1) sin θ does not mean the product of sine and θ. It means sine of angle θ.
(2) The *t*-ratios are same for same angle

Example 3.10

In Fig 3.8, *OBA* is a right angled triangle in which the sides $OB = 12$, $AB = 9$ and $OA = 15$. Find all *t*-ratios.

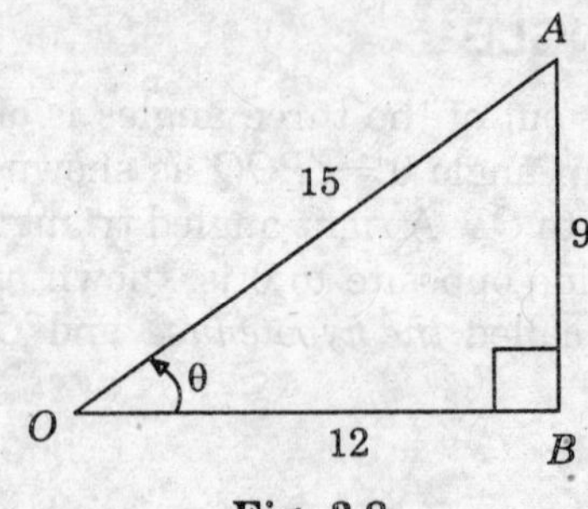

Fig. 3.8

SOLUTION : $\sin\theta = \dfrac{\text{Perpendicular}}{\text{Hypotenuse}} = \dfrac{AB}{OA} = \dfrac{9}{15} = \dfrac{3}{5}$

$$\cos\theta = \frac{\text{Base}}{\text{Hypotenuse}} = \frac{OB}{OA} = \frac{12}{15} = \frac{4}{5}$$

$$\tan\theta = \frac{\text{Perpendicular}}{\text{Base}} = \frac{AB}{OB} = \frac{9}{12} = \frac{3}{4}$$

$$\cot\theta = \frac{\text{Base}}{\text{Perpendicular}} = \frac{OB}{AB} = \frac{12}{9} = \frac{4}{3}$$

$$\sec\theta = \frac{\text{Hypotenuse}}{\text{Base}} = \frac{OA}{OB} = \frac{15}{12} = \frac{5}{4}$$

$$\text{cosec}\,\theta = \frac{\text{Hypotenuse}}{\text{Perpendicular}} = \frac{OA}{AB} = \frac{15}{9} = \frac{5}{3}$$

3.6 TRIGONOMETRIC IDENTITIES

■ **DEFINITION :** *An equation having trigonometric functions which is true for all those angles for which the functions are defined is known as trigonometric identity.*

3.6.1 Reciprocal Identities

(i) $\text{coses } \theta = \dfrac{\text{Hypotenuse}}{\text{Perpendicular}} = \dfrac{1}{\text{Perpendicular/Hypotenuse}} = \dfrac{1}{\sin\theta}$

(ii) $\sec\theta = \dfrac{\text{Hypotenuse}}{\text{Base}} = \dfrac{1}{\text{Base/Hypotenuse}} = \dfrac{1}{\cos\theta}$

(iii) $\cot\theta = \dfrac{\text{Base}}{\text{Perpendicular}} = \dfrac{1}{\text{Perpendicular/Base}} = \dfrac{1}{\tan\theta}$

In other words

(i) cosec θ is reciprocal of sin θ.

(ii) sec θ is reciprocal of cos θ.

(iii) cot θ is reciprocal of tan θ.

From the above reciprocal relations, we conclude

(i) $\sin\theta \times \text{coses}\,\theta = 1$

(ii) $\cos\theta \times \sec\theta = 1$

(iii) $\tan\theta \times \cot\theta = 1$

All these identities are true for all values of θ provided the ratio is defined.

3.6.2 Quotient Identities

(i) $\tan\theta = \dfrac{\text{Perpendicular}}{\text{Base}} = \dfrac{\dfrac{\text{Perpendicular}}{\text{Hypotenuse}}}{\dfrac{\text{Base}}{\text{Hypotenuse}}} = \dfrac{\sin\theta}{\cos\theta}$

(ii) $\cot\theta = \dfrac{\text{Base}}{\text{Perpendicular}} = \dfrac{\dfrac{\text{Base}}{\text{Hypotenuse}}}{\dfrac{\text{Perpendicular}}{\text{Hypotenuse}}} = \dfrac{\cos\theta}{\sin\theta}$

3.6.3 Square Identities

From fig. 7, we have

(i) $\sin^2\theta + \cos^2\theta = \left(\dfrac{AB}{OA}\right)^2 + \left(\dfrac{OB}{OA}\right)^2$

$= \dfrac{AB^2}{OA^2} + \dfrac{OB^2}{OA^2} = \dfrac{AB^2 + OB^2}{OA^2}$

$= \dfrac{OA^2}{OA^2}$ [using Pythagoras theorem]

$= 1$

(ii) $\sec^2\theta - \tan^2\theta = \left(\dfrac{OA}{OB}\right)^2 - \left(\dfrac{AB}{OB}\right)^2 = \dfrac{OA^2}{OB^2} - \dfrac{AB^2}{OB^2}$

$= \dfrac{OA^2 - AB^2}{OB^2} = \dfrac{OB^2}{OB^2}$ [using Pythagoras theorem]

$= 1$

(iii) $\text{cosec}^2\theta - \cot^2\theta = \left(\dfrac{OA}{AB}\right)^2 - \left(\dfrac{OB}{AB}\right)^2 = \dfrac{OA^2}{AB^2} - \dfrac{OB^2}{AB^2}$

$= \dfrac{OA^2 - OB^2}{AB^2} = \dfrac{AB^2}{AB^2}$ [using Pythagoras theorem]

$= 1$

■ **NOTE : (1)** It should be noted that $(\sin\theta)^2$ is written as $\sin^2\theta$ and is read as sine square θ.

(2) $\sin^2\theta + \cos^2\theta = 1$ gives us

(i) $\sin^2\theta = 1 - \cos^2\theta$ or $\sin\theta = \sqrt{1-\cos^2\theta}$

(ii) $\cos^2\theta = 1 - \cos^2\theta$ or $\cos\theta = \sqrt{1-\sin^2\theta}$

(3) $\sec^2\theta - \tan^2\theta = 1$ gives us

(i) $\sec^2\theta = 1 + \tan^2\theta$ or $\sec\theta = \sqrt{1+\tan^2\theta}$

(ii) $\tan^2\theta = \sec^2\theta - 1$ or $\tan\theta = \sqrt{\sec^2\theta - 1}$

(4) $\operatorname{cosec}^2\theta - \cot^2\theta = 1$ gives us

(i) $\operatorname{cosec}^2\theta = 1 + \cot^2\theta$ or $\operatorname{cosec}\theta = \sqrt{1+\cot^2\theta}$

(ii) $\cot^2\theta = \operatorname{cosec}^2\theta - 1$ or $\cot\theta = \sqrt{\operatorname{cosec}^2\theta - 1}$

Example 3.11

Prove that

(i) $(1-\cos^2\theta)(1+\tan^2\theta) = \tan^2\theta$ (ii) $\sqrt{\dfrac{1-\sin^2\theta}{1-\cos^2\theta}} = \cot\theta$

(iii) $\dfrac{\tan\theta}{1+\cot\theta} + \dfrac{\cot\theta}{1+\tan\theta} = \operatorname{cosec}\theta\sec\theta - 1$ (iv) $\dfrac{\tan\theta}{\sec\theta - 1} + \dfrac{\tan\theta}{\sec\theta + 1} = 2\operatorname{cosec}\theta$

PROOF : (i) We have $(1-\cos^2\theta)(1+\tan^2\theta) = \tan^2\theta$

$$\text{LHS} = (1-\cos^2\theta)(1+\tan^2\theta)$$

$$= \sin^2\theta\cdot\sec^2\theta \qquad [\because\ 1-\cos^2\theta = \sin^2\theta,\ 1+\tan^2\theta = \sec^2\theta]$$

$$= \sin^2\theta\cdot\left(\frac{1}{\cos\theta}\right)^2 \qquad \left[\because\ \sec\theta = \frac{1}{\cos\theta}\right]$$

$$= \frac{\sin^2\theta}{\cos^2\theta} = \left(\frac{\sin\theta}{\cos\theta}\right)^2 = \tan^2\theta$$

$$= \text{RHS}$$

(ii) We have $\sqrt{\dfrac{1-\sin^2\theta}{1-\cos^2\theta}} = \cot\theta$

$$\text{LHS} = \sqrt{\frac{1-\sin^2\theta}{1-\cos^2\theta}} = \sqrt{\frac{\cos^2\theta}{\sin^2\theta}} \qquad [\because\ \sin^2\theta + \cos^2\theta = 1]$$

$$= \sqrt{\left(\frac{\cos\theta}{\sin\theta}\right)^2} = \frac{\cos\theta}{\sin\theta} = \cot\theta$$

$$= \text{RHS}$$

(iii) We have $\dfrac{\tan\theta}{1+\cot\theta} + \dfrac{\cot\theta}{1+\tan\theta} = \operatorname{cosec}\theta\cdot\sec\theta - 1$

$$\text{LHS} = \frac{\tan\theta}{1+\cot\theta} + \frac{\cot\theta}{1+\tan\theta}$$

$$= \frac{\sin\theta/\cos\theta}{1+\cos\theta/\sin\theta} + \frac{\cos\theta/\sin\theta}{1+\sin\theta/\cos\theta}$$

$$= \frac{\sin\theta/\cos\theta}{(\sin\theta + \cos\theta)/\sin\theta} + \frac{\cos/\sin\theta}{(\cos\theta + \sin\theta)/\cos\theta}$$

$$= \frac{\sin\theta \cdot \sin\theta}{\cos\theta(\sin\theta + \cos\theta)} + \frac{\cos\theta \cdot \cos\theta}{\sin\theta(\cos\theta + \sin\theta)}$$

$$= \frac{\sin^2\theta}{\cos\theta(\sin\theta + \cos\theta)} + \frac{\cos^2\theta}{\sin\theta(\cos\theta + \sin\theta)}$$

$$= \frac{\sin^3\theta + \cos^3\theta}{\cos\theta \cdot \sin\theta(\sin\theta + \cos\theta)}$$

$$= \frac{(\sin\theta + \cos\theta)(\sin^2\theta + \cos^2\theta - \sin\theta\cos\theta)}{\cos\theta\sin\theta(\sin\theta + \cos\theta)}$$

$[\because x^3 + y^3 = (x + y)(x^2 + y^2 - xy)]$

$$= \frac{1 - \sin\theta\cos\theta}{\cos\theta\sin\theta}$$

$$= \frac{1}{\cos\theta\sin\theta} - \frac{\sin\theta\cos\theta}{\cos\theta\sin\theta} = \frac{1}{\cos\theta} \cdot \frac{1}{\sin\theta} - 1$$

$= \sec\theta \cdot \operatorname{cosec}\theta - 1 = \text{RHS}$

(iv) We have $\dfrac{\tan\theta}{\sec\theta - 1} + \dfrac{\tan\theta}{\sec\theta + 1} = 2\operatorname{cosec}\theta$

$$\text{LHS} = \frac{\tan\theta}{\sec\theta - 1} + \frac{\tan\theta}{\sec\theta + 1}$$

$$= \tan\theta\left[\frac{1}{\sec\theta - 1} + \frac{1}{\sec\theta + 1}\right]$$

$$= \tan\theta\left[\frac{\sec\theta + 1 + \sec\theta - 1}{(\sec\theta - 1)(\sec\theta + 1)}\right]$$

$$= \tan\theta\,\frac{2\sec\theta}{\sec^2\theta - 1} = \frac{2\tan\theta\sec\theta}{\tan^2\theta} \qquad [\because \sec^2\theta - \tan^2\theta = 1]$$

$$= \frac{2\sec\theta}{\tan\theta} = \frac{2/\cos\theta}{\sin\theta/\cos\theta} = \frac{2}{\sin\theta} = 2\operatorname{cosec}\theta$$

$= \text{RHS}$

Example 3.12

Prove that

(i) $\dfrac{1 - \sin\theta}{1 + \sin\theta} = (\sec\theta - \tan\theta)^2$

(ii) $\cos^4\theta - \sin^4\theta = \cos^2\theta - \sin^2\theta = 1 - 2\sin^2\theta = 2\cos^2\theta - 1$

(iii) $\cot^4\theta + \cot^2\theta = \operatorname{cosec}^4\theta - \operatorname{cosec}^2\theta$

(iv) $(\sin\theta + \operatorname{cosec}\theta)^2 + (\cos\theta + \sec\theta)^2 = \tan^2\theta + \cot^2\theta + 7$

SOLUTION : (i) We have

$$\frac{1 - \sin\theta}{1 + \sin\theta} = (\sec\theta - \tan\theta)^2$$

$$\text{LHS} = \frac{1 - \sin\theta}{1 + \sin\theta} = \frac{(1 - \sin\theta)(1 - \sin\theta)}{(1 + \sin\theta)(1 - \sin\theta)} = \frac{(1 - \sin\theta)^2}{1 - \sin^2\theta}$$

$$= \frac{(1-\sin\theta)^2}{\cos^2\theta} = \left(\frac{1-\sin\theta}{\cos\theta}\right)^2 = \left(\frac{1}{\cos\theta} - \frac{\sin\theta}{\cos\theta}\right)^2$$

$$= (\sec\theta - \tan\theta)^2 = \text{RHS}$$

(ii) We have

$$\cos^4\theta - \sin^4\theta = \cos^2\theta \cdot \sin^2\theta$$

$$\text{LHS} = \cos^4\theta - \sin^4\theta = (\cos^2\theta)^2 - (\sin^2\theta)^2$$

$$= (\cos^2\theta - \sin^2\theta)(\cos^2\theta + \sin^2\theta) \qquad [\because a^2 - b^2 = (a-b)(a+b)]$$

$$= \cos^2\theta - \sin^2\theta \qquad [\because \cos^2\theta + \sin^2\theta = 1] \qquad \text{...(1)}$$

From Eq. (1)

$$\cos^2\theta - \sin^2\theta = \cos^2\theta - (1 - \cos^2\theta)$$

$$= \cos^2\theta - 1 + \cos^2\theta$$

$$= 2\cos^2\theta - 1 \qquad \text{...(2)}$$

Again, from Eq. (1)

$$\cos^2\theta - \sin^2\theta = (1 - \sin^2\theta) - \sin^2\theta = 1 - 2\sin^2\theta$$

(iii) We have

$$\cot^4\theta + \cot^2\theta = \operatorname{cosec}^4\theta - \operatorname{cosec}^2\theta$$

$$\text{LHS} = \cot^4\theta + \cot^2\theta$$

$$= (\cot^2\theta)^2 + \cot^2\theta$$

$$= (\operatorname{cosec}^2\theta - 1)^2 + (\operatorname{cosec}^2\theta - 1) \qquad [\because 1 + \cot^2\theta = \operatorname{cosec}^2\theta]$$

$$= \operatorname{cosec}^4\theta - 2\operatorname{cosec}^2\theta + 1 + \operatorname{cosec}^2\theta - 1$$

$$= \operatorname{cosec}^4\theta - \operatorname{cosec}^2\theta$$

$$= \text{RHS}$$

(iv) We have

$$(\sin\theta + \operatorname{cosec}\theta)^2 + (\cos\theta + \sec\theta)^2 = \tan^2\theta + \cot^2\theta + 7$$

$$\text{LHS} = (\sin\theta + \operatorname{cosec}\theta)^2 + (\cos\theta + \sec\theta)^2$$

$$= (\sin^2\theta + \operatorname{cosec}^2\theta + 2\sin\theta\operatorname{cosec}\theta) + (\cos^2\theta + \sec^2\theta + 2\cos\theta\cdot\sec\theta)$$

$$= \sin^2 + \operatorname{cosec}^2\theta + 2 + \cos^2\theta + \sec^2\theta + 2$$

$$= (\sin^2\theta + \cos^2\theta) + (\operatorname{cosec}^2\theta + \sec^2\theta) + 2 + 2$$

$$= 1 + (1 + \cot^2\theta) + (1 + \tan^2\theta) + 4$$

$$= \cot^2\theta + \tan^2\theta + 7$$

$$= \text{RHS}$$

Example 3.13

Prove that

(i) $\dfrac{\cos\theta}{1+\sin\theta} = \dfrac{1-\sin\theta}{\cos\theta}$

(ii) $(1 + \cot\theta + \operatorname{cosec}\theta)(1 + \tan\theta - \sec\theta) = 2$

(iii) $\sqrt{\dfrac{1-\cos\theta}{1+\cos\theta}} = \operatorname{cosec}\theta - \cot\theta$

(iv) $\sin^2\theta\cos^2\alpha + \cos^2\theta\sin^2\alpha + \cos^2\theta\cos^2\alpha + \sin^2\theta\sin^2\alpha = 1$

SOLUTION : (i) We have

$$\frac{\cos\theta}{1+\sin\theta} = \frac{1-\sin\theta}{\cos\theta}$$

$$\text{LHS} = \frac{\cos\theta}{1+\sin\theta} = \frac{\cos\theta(1-\sin\theta)}{(1+\sin\theta)(1-\sin\theta)}$$

$$= \frac{\cos\theta(1-\sin\theta)}{1-\sin^2\theta} = \frac{\cos\theta(1-\sin\theta)}{\cos^2\theta}$$

$$= \frac{1-\sin\theta}{\cos\theta} = \text{RHS}$$

(ii) We have

$(1 + \cot\theta + \operatorname{cosec}\theta)\,(1 + \tan\theta - \sec\theta) = 2$

$$\text{LHS} = (1 + \cot\theta + \operatorname{cosec}\theta)\,(1 + \tan\theta - \sec\theta)$$

$$= \left[1 + \frac{\cos\theta}{\sin\theta} + \frac{1}{\sin\theta}\right]\left[1 + \frac{\sin\theta}{\cos\theta} - \frac{1}{\cos\theta}\right]$$

$$= \left(\frac{\sin\theta + \cos\theta + 1}{\sin\theta}\right)\left(\frac{\cos\theta + \sin\theta - 1}{\cos\theta}\right)$$

$$= \frac{[(\sin\theta + \cos\theta) + 1]\,[(\sin\theta + \cos\theta) - 1]}{\sin\theta\cdot\cos\theta}$$

$$= \frac{(\sin\theta + \cos\theta)^2 - 1}{\sin\theta\cdot\cos\theta} = \frac{\sin^2\theta + \cos^2\theta + 2\sin\theta\cos\theta - 1}{\sin\theta\cos\theta}$$

$$= \frac{1 + 2\sin\theta\cos\theta - 1}{\sin\theta\cos\theta} = \frac{2\sin\theta\cos\theta}{\sin\theta\cos\theta} = 2 = \text{RHS}$$

(iii) We have

$$\sqrt{\frac{1-\cos\theta}{1+\cos\theta}} = \operatorname{cosec}\theta - \cot\theta$$

$$\text{LHS} = \sqrt{\frac{1-\cos\theta}{1+\cos\theta}} = \sqrt{\frac{(1-\cos\theta)\,(1-\cos\theta)}{(1+\cos\theta)(1-\cos\theta)}}$$

$$= \sqrt{\frac{(1-\cos\theta)^2}{1-\cos^2\theta}} = \sqrt{\frac{(1-\cos\theta)^2}{\sin^2\theta}}$$

$$= \sqrt{\left(\frac{1-\cos\theta}{\sin\theta}\right)^2} = \frac{1-\cos\theta}{\sin\theta}$$

$$= \frac{1}{\sin\theta} - \frac{\cos\theta}{\sin\theta} = \operatorname{cosec}\theta - \cot\theta = \text{RHS}$$

(iv) We have

$\sin^2\theta\cos^2\alpha + \cos^2\theta\sin^2\alpha + \cos^2\theta\cos^2\alpha + \sin^2\theta\sin^2\alpha = 1$

$$\text{LHS} = \sin^2\theta\cos^2\alpha + \cos^2\theta\sin^2\alpha + \cos^2\theta\cos^2\alpha + \sin^2\theta\sin^2\alpha$$

$$= (\sin^2\theta\cos^2\alpha + \sin^2\theta\sin^2\alpha) + (\cos^2\theta\sin^2\alpha + \cos^2\theta\cos^2\alpha)$$

$$= \sin^2\theta\,(\cos^2\alpha + \sin^2\alpha) + \cos^2\theta(\sin^2\alpha + \cos^2\alpha)$$

$$= \sin^2\theta + \cos^2\theta$$

$$= 1$$

$$= \text{RHS}$$

Example 3.14

If $a = \tan\theta + \sin\theta$ and $b = \tan\theta - \sin\theta$

Show that $a^2 - b^2 = 4\sqrt{ab}$

SOLUTION : $\text{LHS} = a^2 - b^2 = (\tan\theta + \sin\theta)^2 - (\tan\theta - \sin\theta)^2$

$$= (\tan\theta + \sin\theta + \tan\theta - \sin\theta)(\tan\theta + \sin\theta - \tan\theta + \sin\theta)$$

$$[\because x^2 - y^2 = (x+y)(x-y)]$$

$$= 2\tan\theta \cdot 2\sin\theta$$

$$= 4\tan\theta \cdot \sin\theta \qquad \ldots(1)$$

$$\text{RHS} = 4\sqrt{ab} = 4\sqrt{(\tan\theta + \sin\theta)(\tan\theta - \sin\theta)}$$

$$= 4\sqrt{\tan^2\theta - \sin^2\theta} = 4\sqrt{\frac{\sin^2\theta}{\cos^2\theta} - \sin^2\theta}$$

$$= 4\sqrt{\frac{\sin^2(1-\cos^2\theta)}{\cos^2\theta}} = 4\sqrt{\frac{\sin^2\theta \cdot \sin^2\theta}{\cos^2\theta}} = 4\sqrt{\sin^2\theta\left(\frac{\sin\theta}{\cos\theta}\right)^2}$$

$$= 4\sqrt{\sin^2\theta\tan^2\theta} = 4\sqrt{(\sin\theta\tan\theta)^2}$$

$$= 4\sin\theta\tan\theta \qquad \ldots(2)$$

From Eqs. (1) and (2), we have $a^2 - b^2 = 4\sqrt{ab}$.

Example 3.15

If $\cos\theta - \sin\theta = \sqrt{2}\sin\theta$, prove that $\cos\theta + \sin\theta = \pm\sqrt{2}\cos\theta$

SOLUTION : We know that

$$(\cos\theta - \sin\theta)^2 + (\cos\theta + \sin\theta)^2 = 2$$

$$\Rightarrow (\sqrt{2}\sin\theta)^2 + (\cos\theta + \sin\theta)^2 = 2$$

$$\Rightarrow 2\sin^2\theta + (\cos\theta + \sin\theta)^2 = 2$$

$$\Rightarrow (\cos\theta + \sin\theta)^2 = 2 - 2\sin^2\theta$$

$$= 2(1 - \sin^2\theta)$$

$$= 2\cos^2\theta$$

$$\Rightarrow \cos\theta + \sin\theta = \pm\sqrt{2\cos^2\theta} = \pm\sqrt{2}\cos\theta$$

Example 3.16

Prove that $\tan^2\theta + \cot^2\theta + 2 = \sec^2\theta \cdot \operatorname{cosec}^2\theta$. ***[UPTU 2000]***

SOLUTION : $\text{LHS} = \tan^2\theta + \cot^2\theta + 2$

$$= \frac{\sin^2\theta}{\cos^2\theta} + \frac{\cos^2\theta}{\sin^2\theta} + 2$$

$$= \frac{\sin^4\theta + \cos^4\theta + 2\sin^2\theta\cos^2\theta}{\cos^2\theta\sin^2\theta}$$

$$= \frac{(\sin^2\theta + \cos^2\theta)^2}{\cos^2\theta \cdot \sin^2\theta}$$

$$= \frac{1}{\cos^2\theta \cdot \sin^2\theta} \qquad [\because \sin^2\theta + \cos^2\theta = 1]$$

$$= \frac{1}{\cos^2\theta} \cdot \frac{1}{\sin^2\theta}$$

$$= \sec^2\theta \cdot \text{cosec}^2\theta = \text{RHS}$$

Example 3.17

If $A = \begin{bmatrix} \cos\theta & \sin\theta \\ -\sin\theta & \cos\theta \end{bmatrix}$ Prove that $A^n = \begin{bmatrix} \cos n\theta & \sin n\theta \\ -\sin n\theta & \cos n\theta \end{bmatrix}$

for every positive integer n. ***[UPTU (C.O. 2006)]***

SOLUTION : We will prove this using induction method for n.

If $n = 1$, $A^1 = A = \begin{bmatrix} \cos\theta & \sin\theta \\ -\sin\theta & \cos\theta \end{bmatrix}$ which is true.

Let it is true for $n - 1$, *i.e.*,

$$A^{n-1} = \begin{bmatrix} \cos(n-1)\theta & \sin(n-1)\theta \\ -\sin(n-1)\theta & \cos(n-1)\theta \end{bmatrix}$$

Now
$$A^n = A^{n-1} \cdot A = \begin{bmatrix} \cos(n-1)\theta & \sin(n-1)\theta \\ -\sin(n-1)\theta & \cos(n-1)\theta \end{bmatrix} \begin{bmatrix} \cos\theta & \sin\theta \\ -\sin\theta & \cos\theta \end{bmatrix}$$

$$= \begin{bmatrix} \cos(n-1)\theta\cos\theta - \sin(n-1)\theta\sin\theta & \cos(n-1)\theta\sin\theta + \sin(n-1)\theta\cos\theta \\ -\sin(n-1)\theta\cos\theta - \cos(n-1)\theta\sin\theta & -\sin(n-1)\theta\sin\theta + \cos(n-1)\theta\cos\theta \end{bmatrix}$$

$$= \begin{bmatrix} \cos n\theta & \sin n\theta \\ -\sin n\theta & \cos n\theta \end{bmatrix}$$

Thus, it is true for every positive integer n.

Example 3.18

Prove that $\sin^6\alpha + \cos^6\alpha = 1 - 3\sin^2\alpha\cos^2\alpha$. ***[UPTU 2007]***

SOLUTION : Consider

$$(\sin^2\alpha + \cos^2\alpha)^3 = \sin^6\alpha + \cos^6\alpha + 3\sin^2\alpha\cos^2\alpha(\sin^2\alpha + \cos^2\alpha)$$

$$1 = \sin^6\alpha + \cos^6\alpha + 3\sin^2\alpha\cos^2\alpha$$

$$\Rightarrow \quad \sin^6\alpha + \cos^6\alpha = 1 - 3\sin^2\alpha\cos^2\alpha.$$

Example 3.19

If $\tan\frac{\theta}{2} = \frac{p}{q}$, find the value of $(p\sin\theta + q\cos\theta)$. ***[UPTU 2007]***

SOLUTION :
$$p\sin\theta + q\cos\theta = p\left[\frac{2\tan\frac{\theta}{2}}{1 + \tan^2\frac{\theta}{2}}\right] + q\left[\frac{1 - \tan^2\frac{\theta}{2}}{1 + \tan^2\frac{\theta}{2}}\right]$$

$$= 2p\left[\frac{\frac{p}{q}}{1+\frac{p^2}{q^2}}\right] + q\left[\frac{1-\frac{p^2}{q^2}}{1+\frac{p^2}{q^2}}\right]$$

$$= \frac{2p^2q}{p^2+q^2} + \frac{q(q^2-p^2)}{p^2+q^2}$$

$$= \frac{2p^2q+q^3-pq^2}{p^2+q^2}$$

$$= \frac{q^3+p^2q}{p^2+q^2} = \frac{q(p^2+q^2)}{p^2+q^2} = q$$

PROBLEM SET 3.2

Prove that

1. $\sqrt{\text{cosec}^2\theta - 1} = \cos\theta \cdot \text{cosec}\,\theta = \cot\theta$

2. $\dfrac{\sec^2\theta - 1}{\tan^2\theta} = 1$

3. $\dfrac{\sin^2\theta + \cos^2\theta}{\sec^2\theta - \tan^2\theta} = 1$

4. $\sin^4\theta + \sin^2\theta\cos^2\theta = \sin^2\theta$

5. $\dfrac{\sec\theta + \tan\theta}{\text{cosec}\,\theta + \cot\theta} = \dfrac{\text{cosec}\,\theta - \cot\theta}{\sec\theta - \tan\theta}$

6. $\sec\theta + \tan\theta = \dfrac{1}{\sec\theta - \tan\theta}$

7. $\sin^2\theta(\cot^2\theta - 1 + \text{cosec}^2\theta)(1 + \tan^2\theta) = 2$

8. $\dfrac{3 - 4\sin^2\theta}{\cos^2\theta} = 3 - \tan^2\theta$

9. $1 - \sin^4\theta = \cos^2\theta(1 + \sin^2\theta)$

10. $\dfrac{\sin^3\theta + \cos^3\theta}{\sin\theta + \cos\theta} + \dfrac{\sin^3\theta - \cos^3\theta}{\sin\theta - \cos\theta} = 2$

11. $\dfrac{\tan^3\theta}{1 + \tan^2\theta} + \dfrac{\cot^3\theta}{1 + \cot^2\theta} = \dfrac{1 - 2\sin^2\theta\cos^2\theta}{\sin\theta\cos\theta}$

12. $1 - \dfrac{\sin^2\theta}{1+\cot\theta} - \dfrac{\cos^2\theta}{1+\tan\theta} = \sin\theta\cos\theta$

13. $\sqrt{\dfrac{1-\sin\theta}{1+\sin\theta}} = \sec\theta - \tan\theta$

14. $\sqrt{\dfrac{1+\cos\theta}{1-\cos\theta}} = \text{cosec}\,\theta + \cot\theta$

15. $\sqrt{\dfrac{1+\sin\theta}{1-\sin\theta}} = \sec\theta + \tan\theta$

16. $\dfrac{1}{\text{cosec}\,\theta - \cot\theta} = \text{cosec}\,\theta + \cot\theta = \dfrac{1+\cos\theta}{\sin\theta}$

17. $\dfrac{\sin\theta}{1+\cos\theta} + \dfrac{1+\cos\theta}{\sin\theta} = \dfrac{2}{\sin\theta}$

18. $\dfrac{1-\sin\theta}{1+\sec\theta} - \dfrac{1+\sin\theta}{1-\sec\theta} = 2\cos\theta(\cot\theta + \text{cosec}^2\theta)$

19. If $x = a\cos\theta + b\sin\theta$and $y = a\sin\theta - b\cos\theta$
Show that $x^2 + y^2 = a^2 + b^2$.

3.7 SIGNS OF *t*-RATIOS IN DIFFERENT QUADRANTS

In the starting of this chapter, we have introduced the concept of four quadrants and six t-ratios. In coordinate system the distances measured along *OX* and *OY* axes are positive while along *OX'* and *OY'* axes are negative. The radius vector is always positive. Thus the sign of a t-ratio of an angle θ will always depend on the sign of the coordinates. The following table shows the sign of *t*-ratios in different quadrants

t-Ratios/ quadrants	Ist $x > 0$ $y > 0$	2nd $x < 0$ $y > 0$	3rd $x < 0$ $y < 0$	4th $x > 0$ $y < 0$
$\sin\theta = \dfrac{y}{z}$	+ (Positive)	+ (Positive)	– (Negative)	– (Negative)
$\cos\theta = \dfrac{x}{z}$	+ (Positive)	– (Negative)	– (Negative)	+ (Positive)
$\tan\theta = \dfrac{y}{x}$	+ (Positive)	– (Negative)	+ (Positive)	– (Negative)
$\cot\theta = \dfrac{x}{y}$	+ (Positive)	– (Negative)	+ (Positive)	– (Negative)
$\sec\theta = \dfrac{z}{x}$	+ (Positive)	– (Negative)	– (Negative)	+ (Positive)
$\text{cosec}\,\theta = \dfrac{z}{y}$	+ (Positive)	+ (Positive)	– (Negative)	– (Negative)

From the table given above it is clear that in the first quadrant all t-ratios are positive, in the second quadrant sine and its reciprocal are positive, in the third quadrant tangent and its reciprocal are positive and in the fourth quadrant cosine and its reciprocal are positive.

Thus, the fig. 3.9 indicates signs of t-ratios from quadrant first to fourth.

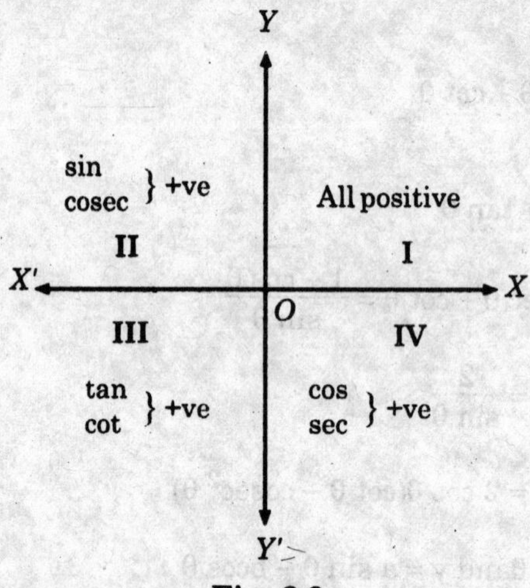

Fig. 3.9

■ **Note :** To remember the above *t*-ratios in any quadrant the word **"ADD SUGAR TO COFFEE"** are very usefull. whatever is written in a particular quadrant along with its reciprocal is positive and the rest are negative.

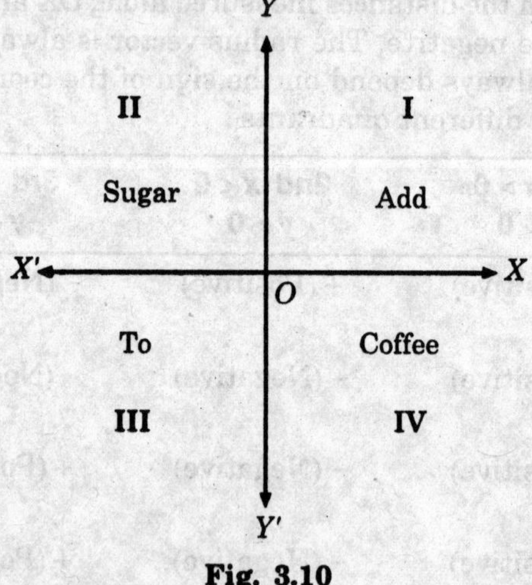

Fig. 3.10

Example 3.20

If $\sin\theta = -\frac{5}{12}$, in which quadrant does θ lie?

SOLUTION : As sin θ is negative, θ can be either in third or fourth quadrant.

Example 3.21

If $\cos\theta = \frac{5}{13}$ and θ lies in the fourth quadrant, find the value of

$$\frac{2 - 3\cot\theta}{4 - 9\sqrt{\sec^2\theta - 1}}$$

SOLUTION : As θ lies in fourth quadrant, so cot θ is negative and sec θ is positive. To find the values of cot θ and sec θ we draw a right angled triangle ABC with ∠ABC = 90° Let ∠CAB = θ.

$$\because \quad \cos\theta = \frac{\text{Base}}{\text{Hyp}} = \frac{5}{13}$$

$$\Rightarrow \quad \text{Base} = 5 \quad \text{and} \quad \text{Hyp.} = 13$$

Using pythagoras theorem,

$$(\text{Perp})^2 + (\text{Base})^2 = (\text{Hyp})^2$$

$$\Rightarrow \quad (\text{Perp})^2 = (\text{Hyp.})^2 - (\text{Base})^2 = (13)^2 - (5)^2 = 169 - 25 = 144 = (12)^2$$

$$\Rightarrow \quad \text{perp} = \pm 12$$

C, Hyp., Perp., θ, A, B, Base

Fig. 3.11

Now, in fourth quadrant

$$\cot\theta = -\frac{\text{Base}}{\text{perp}} = \frac{-5}{12}, \quad \sec\theta = \frac{\text{Hyp.}}{\text{Base}} = \frac{13}{5}$$

$$\therefore \quad \frac{2 - 3\cot\theta}{4 - 9\sqrt{\sec^2 - 1}} = \frac{2 - 3\left(-\frac{5}{12}\right)}{4 - 9\sqrt{4 - 9\left(\frac{13}{5}\right)^2 - 1}}$$

$$= \frac{2 + \frac{5}{4}}{4 - 9\sqrt{\frac{169}{25}} - 1} = \frac{\frac{13}{4}}{4 - 9 \times \frac{12}{5}}$$

$$= \frac{13}{4} \times \left(\frac{-5}{88}\right) = -\frac{65}{352}$$

3.8 *t*-RATIOS OF SOME STANDARD ANGLES

To find the t-ratios of 0°,30°,45°,60°,90° we use a table which may be formed as follows :

- **Step 1 :** write the numbers 0, 1, 2, 3, 4,
- **Step 2 :** Divide each number by 4, *i.e.*, $\frac{0}{4}, \frac{1}{4}, \frac{2}{4}, \frac{3}{4}, \frac{4}{4}$ or $0, \frac{1}{4}, \frac{1}{2}, \frac{3}{4}, 1$
- **Step 3 :** Taking square root of each number $\sqrt{0}, \sqrt{\frac{1}{4}}, \sqrt{\frac{1}{2}}, \sqrt{\frac{3}{4}}, \sqrt{1}$ or $0, \frac{1}{2}, \frac{1}{\sqrt{2}}, \frac{\sqrt{3}}{2}, 1$
- **Step 4 :** Now sin of angles 0°, 30°, 45°, 60°, 90° are $0, \frac{1}{2}, \frac{1}{\sqrt{2}}, \frac{\sqrt{3}}{2}, 1$

- **Step 5 :** write these values in the reverse order, we get cos θ.

The complete table is

Angles	0°	30°	45°	60°	90°
sin θ	0	$\frac{1}{2}$	$\frac{1}{\sqrt{2}}$	$\frac{\sqrt{3}}{2}$	1
cos θ	1	$\frac{\sqrt{3}}{2}$	$\frac{1}{\sqrt{2}}$	$\frac{1}{2}$	0
$\tan\theta = \frac{\sin\theta}{\cos\theta}$	0	$\frac{1}{\sqrt{3}}$	1	$\sqrt{3}$	∞
$\cot\theta = \frac{1}{\tan\theta}$	∞	$\sqrt{3}$	1	$\frac{1}{\sqrt{3}}$	0
$\sec\theta = \frac{1}{\cos\theta}$	1	$\frac{2}{\sqrt{3}}$	$\sqrt{2}$	2	∞
$\text{cosec}\,\theta = \frac{1}{\sin\theta}$	∞	2	$\sqrt{2}$	$\frac{2}{\sqrt{3}}$	1

Example 3.22

Prove that

$$\sin 60° \cos 30° + \cos 45° \sin 90° = \frac{3\sqrt{2}+4}{4\sqrt{2}}$$

SOLUTION : LHS = sin 60° cos 30° + cos 45° sin 90°

$$= \frac{\sqrt{3}}{2} \cdot \frac{\sqrt{3}}{2} + \frac{1}{\sqrt{2}} \cdot 1$$

$$= \frac{3}{4} + \frac{1}{\sqrt{2}} = \frac{3\sqrt{2}+4}{4\sqrt{2}} = \text{RHS}$$

Example 3.23

Prove that

$$\frac{1+\tan^2 30°}{\text{cosec } 60°} = \frac{2}{\cot 30°}$$

Solution :

$$\text{LHS} = \frac{1+\tan^2 30°}{\text{cosec } 60°} = \frac{1+\left(\frac{1}{\sqrt{3}}\right)^2}{\frac{2}{\sqrt{3}}} = \frac{1+\frac{1}{3}}{\frac{2}{\sqrt{3}}}$$

$$= \frac{4}{3} \times \frac{\sqrt{3}}{2} = \frac{2\sqrt{3}}{3} = \frac{2\sqrt{3}}{3} \times \frac{\sqrt{3}}{\sqrt{3}}$$

$$= \frac{2}{\sqrt{3}}$$

$$\text{RHS} = \frac{2}{\cot 30°} = \frac{2}{\sqrt{3}}$$

∴ LHS = RHS

Example 3.24

Prove that

$$\left(\frac{1+\cot 60^\circ}{1-\cot 60^\circ}\right)^2 = \frac{1+\cos 30^\circ}{1-\cos 30^\circ}$$

Solution : $\text{RHS} = \left(\frac{1+\cot 60^\circ}{1-\cot 60^\circ}\right)^2 = \left(\frac{1+\frac{1}{\sqrt{3}}}{1-\frac{1}{\sqrt{3}}}\right)^2 = \left(\frac{\sqrt{3}+1}{\sqrt{3}-1}\right)^2$

$$= \frac{(\sqrt{3}+1)^2}{(\sqrt{3}-1)^2} = \frac{3+1+2\sqrt{3}}{3+1-2\sqrt{3}} = \frac{4+2\sqrt{3}}{4-2\sqrt{3}} = \frac{2+\sqrt{3}}{2-\sqrt{3}}$$

$$\text{RHS} = \frac{1+\cos 30^\circ}{1-\cos 30^\circ} = \frac{1+\frac{\sqrt{3}}{2}}{1-\frac{\sqrt{3}}{2}} = \frac{2+\sqrt{3}}{2-\sqrt{3}}$$

$\therefore$ LHS = RHS

Example 3.25

Prove that $\sin(45^\circ + A)\sin(45^\circ - A) = \frac{1}{2}\cos 2A$ ***[UPTU (C.O.) 2006]***

SOLUTION : $\text{LHS} = \sin(45^\circ + A)\sin(45^\circ - A)$

$$= \sin^2 45^\circ - \sin^2 A \qquad [\because \sin(A+B)\cdot\sin(A-B) = \sin^2 A - \sin^2 B]$$

$$= \left(\frac{1}{\sqrt{2}}\right)^2 - \sin^2 A$$

$$= \frac{1}{2} - \sin^2 A = \frac{1}{2}(1 - 2\sin^2 A)$$

$$= \frac{1}{2}\cos 2A = \text{RHS}$$

PROBLEM SET 3.3

1. If $\cot\theta = \frac{4}{3}$, find the values of other *t*-ratios of θ.

2. If $\tan\theta = \frac{4}{5}$, find the value of $\frac{2\sin\theta + 3\cos\theta}{4\cos\theta + 3\sin\theta}$.

3. If θ lies in the second quadrant and $\tan\theta = \frac{-5}{12}$, find the value of $\frac{2\cos\theta}{1-\sin\theta}$.

4. If θ lies in the third quadrant and $\sin\theta = \frac{-2\sqrt{6}}{5}$, find all other *t*-ratios of θ.

5. If $\theta = \frac{\pi}{4}$, verify $\sin^2\theta + \cos^2\theta = 1$.

6. Find the value of $\tan 60° \sin^2 45° + \cos^2 60° \tan 45°$.

7. Prove that

$$4 \sin^2 45° + \tan^2 60° + \text{cosec}^2 30° = 9$$

8. Prove that

$$\sin 60° = \frac{2 \tan 30°}{1 + \tan^2 30°}$$

9. Find the value of x if

$$x \sin 30° \cos^2 45° = \frac{\cot^2 30° \sec 60° \tan 45°}{\text{cosec}^2 45° \text{ cosec } 30°}$$

10. Prove that

$$\frac{4}{3} \cot^2 30° + 3 \sin^2 60° - 2 \text{ cosec}^2 60° - \frac{3}{4} \tan^2 30° = \frac{10}{3}$$

11. Prove that

$$4\cot^2 \frac{\pi}{6} - 5 \sec^2 \frac{\pi}{4} + 6 \text{ cosec}^2 \frac{\pi}{3} - 7 \cos^2 \frac{\pi}{2} = 10$$

ANSWERS

1. $\sin \theta = \frac{3}{5}$, $\cos \theta = \frac{4}{5}$, $\tan \theta = \frac{3}{4}$, $\sec \theta = \frac{5}{4}$, $\text{cosec } \theta = \frac{5}{3}$ **2.** $\frac{23}{32}$ **3.** -3

4. $\cos \theta = -\frac{1}{5}$, $\tan \theta = 2\sqrt{6}$, $\cot \theta = \frac{1}{2\sqrt{6}}$, $\sec \theta = -5$, $\text{cosec } \theta = -\frac{5}{2\sqrt{6}}$ **6.** $\frac{2\sqrt{3}+1}{4}$

9. $x = 6$.

3.9 *t*-RATIOS OF ALLIED ANGLES

If the sum or difference of two angles are either zero or a multiple of 90° then the two angles are said to be allied angles. If θ is measured in degrees, then the angles $-\theta$, $90° \pm \theta$, $180° \pm \theta$, $360° \pm \theta$ etc. are angles allied to the angle θ. If θ is measured in radians, then the angles $-\theta$, $\frac{\pi}{2} \pm \theta$, $\pi \pm \theta$, $2\pi \pm \theta$ etc. are angles allied to the angle θ. Using t-ratios of allied angles we can find the *t*-ratios of angles of any magnitude.

(1) *t*-Ratios of $(-\theta)$ in terms of θ :

$\sin(-\theta) = -\sin \theta$ $\cos(-\theta) = \cos \theta$
$\tan(-\theta) = -\tan \theta$ $\cot(-\theta) = -\cot \theta$
$\sec(-\theta) = \sec \theta$ $\text{cosec}(-\theta) = -\text{cosec } \theta$

(2) *t*-Ratios of $(90° - \theta)$ in terms of θ :

$\sin(90° - \theta) = \cos \theta$ $\cos(90° - \theta) = \sin \theta$
$\tan(90° - \theta) = \cot \theta$ $\cot(90° - \theta) = \tan \theta$
$\sec(90° - \theta) = \text{cosec } \theta$ $\text{cosec}(90° - \theta) = \sec \theta$

(3) t-Tatios of (90° + θ) in terms of θ :

$\sin(90° + \theta) = \cos\theta$ $\cos(90° + \theta) = -\sin\theta$
$\tan(90° + \theta) = -\cot\theta$ $\cot(90° + \theta) = -\tan\theta$
$\sec(90° + \theta) = -\operatorname{cosec}\theta$ $\operatorname{cosec}(90° + \theta) = \sec\theta$

(4) t-Ratios of (180° – θ) in terms of θ :

$\sin(180° - \theta) = \sin\theta$ $\cos(180° - \theta) = -\cos\theta$
$\tan(180° - \theta) = -\tan\theta$ $\cot(180° - \theta) = -\cot\theta$
$\sec(180° - \theta) = -\sec\theta$ $\operatorname{cosec}(180° - \theta) = \operatorname{cosec}\theta$

(5) t-Ratios of (180° + θ) in terms of θ :

$\sin(180° + \theta) = -\sin\theta$ $\cos(180° + \theta) = -\cos\theta$
$\tan(180° + \theta) = \tan\theta$ $\cot(180° + \theta) = \cot\theta$
$\sec(180° + \theta) = -\sec\theta$ $\operatorname{cosec}(180° + \theta) = -\operatorname{cosec}\theta$

(6) t-Ratios of (270° – θ) in terms of θ :

$\sin(270° - \theta) = -\cos\theta$ $\cos(270° - \theta) = -\sin\theta$
$\tan(270° - \theta) = \cot\theta$ $\cot(270° - \theta) = \tan\theta$
$\sec(270° - \theta) = -\operatorname{cosec}\theta$ $\operatorname{cosec}(270° - \theta) = -\sec\theta$

(7) t-Ratios of (270° + θ) in terms of θ :

$\sin(270° + \theta) = -\cos\theta$ $\cos(270° + \theta) = \sin\theta$
$\tan(270° + \theta) = -\cot\theta$ $\cot(270° + \theta) = -\tan\theta$
$\sec(270° + \theta) = \operatorname{cosec}\theta$ $\operatorname{cosec}(270° + \theta) = -\sec\theta$

(8) The position of the revolving line after forming the angle $(n \times 360° - \theta)$ will be same as for the angle $(-\theta)$. The values of t-ratios for the angles $(n \times 360° + \theta)$ will be the same as those for θ.

■ **NOTE :** To remember the above t-ratios we use a simple procedure as follows :

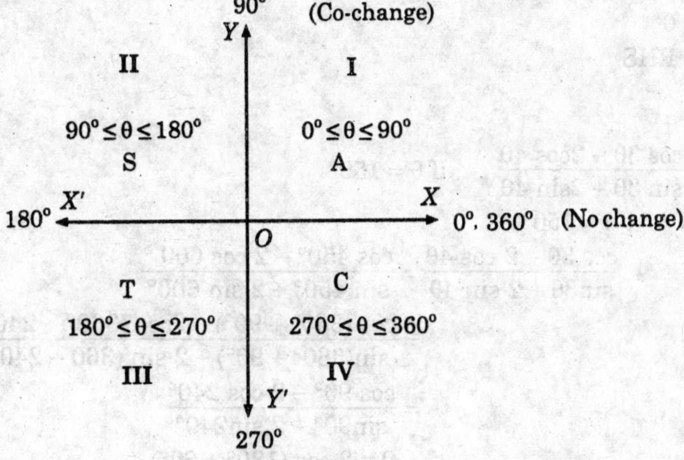

Fig. 3.12

From fig. 3.12 we have following facts :

(1) Any function of an angle expressed as 180°or 360° plus or minus an acute angle θ, *i.e.*, (along xox ′) have **no-change** in the function of θ, with the same or opposite sign according to the quadrant in which the revolving line terminates.

(2) Any function of an angle expressed as 90° or 270° plus or minus an acute angle θ (*i.e.*, along *YOY ′* have co-change in the function of θ, with the same or opposite sign according to quadrant in which the revolving line terminates. The **co-change** of sine θ is cosine θ and of cosine θ is sine θ,of tan θ is cot θ etc.

Exmple 3.26

Find the value of the following :

(i) tan 315°, (ii) sin (– 1742°), (iii) cot 270°,
(iv) tan 36° tan 54°, (v) cos 1020°, (vi) cosec 390°.

SOLUTION : (i) $\tan 315° = \tan(270° + 45°) = -\cot 45° = -1$

(ii) $\sin(-1742°) = -\sin 1742° = -\sin(360 \times 5 - 58°)$

$= -\sin(-58°) = \sin 58° = 0.84805$ (from table)

(iii) $\cot 270° = \cot(180° + 90°) = \cot 90° = 0$

(iv) $\tan 36° \tan 54° = \tan(90 - 54°) \tan 54° = \cot 54° \tan 54° = 1$

(v) $\cos 1020° = \cos(360 \times 3 - 60°) = \cos(-60) = \cos 60° = \frac{1}{2}$

(vi) $\text{cosec } 390° = \text{cosec }(360 + 30°) = \text{cosec } 30° = 2$

Example 3.27

Prove that

$$\tan 225° \cot 405° + \tan 765° \cot 675° = 0$$

SOLUTION :

$$\begin{aligned} \text{LHS} &= \tan 225° \cot 405° + \tan 765° \cot 675° \\ &= \tan(180° + 45°) \cot(360° + 45°) + \tan(360° \times 2 + 45°) \cot(360° \times 2 - 45°) \\ &= \tan 45° \cot 45° + \tan 45° \cot(-45°) \\ &= \tan 45° \cot 45° - \tan 45° \cot 45° \\ &= 0 \\ &= \text{RHS} \end{aligned}$$

Example 3.28

Evaluate $\dfrac{\cos 3\theta - 2\cos 4\theta}{\sin 3\theta + 2\sin 4\theta}$ if $\theta = 150°$

SOLUTION : At $\theta = 150°$

$$\begin{aligned} \frac{\cos 3\theta - 2\cos 4\theta}{\sin 3\theta + 2\sin 4\theta} &= \frac{\cos 450° - 2\cos 600°}{\sin 450° + 2\sin 600°} \\ &= \frac{\cos(360° + 90°) - 2\cos(360° + 240°)}{\sin(360° + 90°) + 2\sin(360 + 240°)} \\ &= \frac{\cos 90° - 2\cos 240°}{\sin 90° + 2\sin 240°} \\ &= \frac{0 - 2\cos(180° + 60°)}{1 + 2\sin(180° + 60°)} \end{aligned}$$

$$= \frac{2\cos 60^\circ}{1 - 2\sin 60^\circ} = \frac{2 \cdot \frac{1}{2}}{1 - 2 \cdot \frac{\sqrt{3}}{2}} = \frac{1}{1-\sqrt{3}}$$

$$= \frac{1}{1-\sqrt{3}} \times \frac{1+\sqrt{3}}{1+\sqrt{3}} = \frac{1+\sqrt{3}}{1-3}$$

$$= \frac{1+\sqrt{3}}{-2} = -\frac{1}{2}(1+\sqrt{3})$$

Example 3.29

Find the values of x and y, lying between 0 and 360°, which satisfy the equations

(i) $\cos 2x = 0.3572$

(ii) $\sin\left(\frac{y}{2} + 71^\circ\right) = 0.6428$

SOLUTION : (i) From table $0.3572 = \cos 69^\circ 6'$ (App.)

$\therefore \quad \cos 2x = \cos 69^\circ 6' \Rightarrow 2x = 69^\circ 6' \Rightarrow x = 34^\circ 33'$

(ii) From table $\quad 0.6428 = \sin 40^\circ$

$\therefore \quad \sin\left(\frac{y}{2} + 71^\circ\right) = \sin 40^\circ$

$\Rightarrow \quad \frac{y}{2} + 71^\circ = 40^\circ$

$\Rightarrow \quad \frac{y}{2} = -31^\circ \Rightarrow y = -62^\circ$

Example 3.30

Simplify $\cos 510^\circ \cos 330^\circ + \sin 390^\circ \cos 120^\circ$

SOLUTION : $\cos 510^\circ \cos 330^\circ + \sin 390^\circ \cos 120^\circ$

$= \cos(360^\circ + 150^\circ)\cos(360^\circ - 30^\circ) + \sin(360^\circ + 30^\circ)\cos(180 - 60^\circ)$

$= \cos 150^\circ \cos(-30^\circ) + \sin 30^\circ(-\cos 60^\circ)$

$= \cos 150^\circ \cos 30^\circ - \sin 30^\circ \cos 60^\circ$

$= \cos(180 - 30^\circ)\cos 30^\circ - \sin 30^\circ \cos 60^\circ$

$= -\cos 30^\circ \cos 30^\circ - \sin 30^\circ \cos 60^\circ$

$= -\frac{\sqrt{3}}{2} \cdot \frac{\sqrt{3}}{2} - \frac{1}{2} \cdot \frac{1}{2} = -\frac{3}{4} - \frac{1}{4} = \frac{-4}{4} = -1$

Example 3.31

In a ΔABC, prove that :

(i) $\cos(A+B) + \cos C = 0$

(ii) $\tan\frac{B+C}{2} = \cot\frac{A}{2}$

SOLUTION : (i) In ΔABC,

$A + B + C = \pi$

Now $\quad \text{LHS} = \cos(A+B) + \cos C = \cos(\pi - C) + \cos C$

$= -\cos C + \cos C = 0 = \text{RHS}$

(ii) $\text{LHS} = \tan\dfrac{B+C}{2} = \tan\dfrac{\pi - A}{2} = \tan\left(\dfrac{\pi}{2} - \dfrac{A}{2}\right)$

$= = \tan\left(90° - \dfrac{A}{2}\right)$

$= \cot\dfrac{A}{2} = \text{RHS}$

PROBLEM SET 3.4

Find the value of the following :

1. (i) cos 135°, (ii) sin 210°, (iii) tan 240°,
(iv) cot 405°, (v) cos (– 1125°), (vi) tan (– 585°).

2. (i) $\sin\dfrac{5\pi}{3}$ (ii) $\tan\dfrac{11\pi}{6}$ (iii) $\sec\dfrac{15\pi}{4}$

3. Prove that :

(i) $\sin\dfrac{8\pi}{3}\cos\dfrac{23\pi}{3} + \cos\dfrac{13\pi}{3}\sin\dfrac{35\pi}{6} = \dfrac{1}{2}$

(ii) $\sqrt{2}\sin 135°\cos 210°\tan 240°\cot 300°\sec 330° = 1$

(iii) $\sin 480°\cos 690° + \cos 780°\sin 1050° = \dfrac{1}{2}$

(iv) $\tan(-225°)\cot(-405°) - \tan(-765°)\cot(675°) = 0$

4. Prove that :

(i) $\dfrac{\cos(2\pi + \theta)\ \text{cosec}(2\pi + \theta)\tan\left(\dfrac{\pi}{2} + \theta\right)}{\sec\left(\dfrac{\pi}{2} + \theta\right)\cos\theta\cot(\pi + \theta)} = 1$

(ii) $\tan^2(90° - \theta) - \sin^2(90° - \theta) = \cos^2\theta\cot^2\theta$

(iii) $\sin^2\left(\dfrac{\pi}{18}\right) + \sin^2\dfrac{\pi}{9} + \sin^2\dfrac{7\pi}{18} + \sin^2\dfrac{4\pi}{9} = 2$

(iv) $\sec\left(\dfrac{3\pi}{2} - \theta\right)\sec\left(\theta - \dfrac{5\pi}{2}\right) + \tan\left(\dfrac{5\pi}{2} + \theta\right)\tan\left(\theta - \dfrac{3\pi}{2}\right) = -1$

5. Simplify :

(i) $\dfrac{\sin 150° - 5\cos 300° + 7\tan 225°}{\tan 135° + 3\sin 210°}$

(ii) $\dfrac{\sin(90° + \theta)\sin(180° - \theta)\tan(180° - \theta)}{\cos(90° - \theta)\cos(180° - \theta)\tan(180° + \theta)}$

6. If A, B, C are angles of a triangle, Prove that :

(i) $\cos\left(\frac{A+B}{2}\right) - \sin\frac{C}{2} = 0$

(ii) $\frac{\tan(B+C)+\tan(C+A)+\tan(A+B)}{\tan(\pi-A)+\tan(2\pi-B)+\tan(3\pi-C)} = 1$

(iii) $\frac{\text{cosec}(B+C)+\text{cosec}(C+A)+\text{cosec}(A+B)}{\sec(270°+A)+\sec(270°+B)+\sec(270°+C)} = 1$

ANSWERS

1. (i) $-\frac{1}{\sqrt{2}}$ (ii) $-\frac{1}{2}$ (iii) $\sqrt{3}$ (iv) 1 (v) $\frac{1}{\sqrt{2}}$ (vi) -1

2. (i) $\frac{-\sqrt{3}}{2}$ (ii) $-\frac{1}{\sqrt{3}}$ (iii) $\sqrt{2}$ **5.** (i) -2 (ii) 1.

3.10 *t*-RATIOS OF COMPOUND ANGLES

The algebraic sum or difference of two or more angles is known as compound angle *i.e.*, if A, B and C are three angles, then $A+B$, $A-B$, $A+B+C$, $A-B+C$ etc. are compound angles.

3.10.1 Addition Formula

Let the revolving straight line, OX rotates in anticlockwise about O from OX to OZ describe an angle $XOZ = A$and then makes an angle $ZOW = B$ with OZ. Thus

$$\angle XOW = \angle XOZ + \angle ZOW = A + B.$$

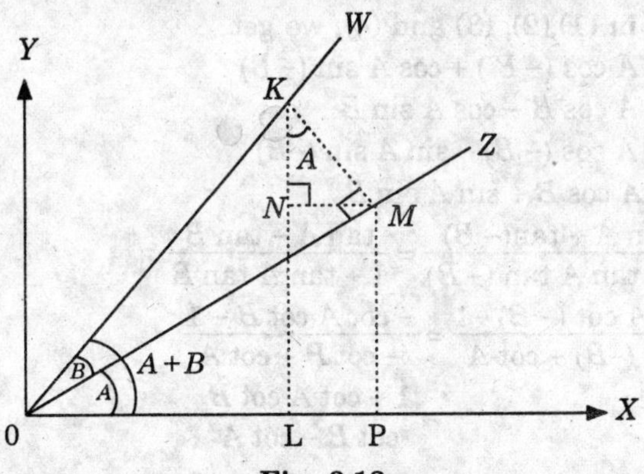

Fig. 3.13

Take a point K on OW (the final position) and draw perpendiculars KL and KM on OX and OZ respectively. From M draw perpendiculars MP and MN on OX and KL respectively.

Now, $\angle NKM = (90° - \angle KMN) = \angle NMO = \angle MOX = A$

Again in $\Delta KOL, \angle KLO = 90,°$

(1) $\sin(A+B) = \frac{KL}{OK} = \frac{KN+NL}{OK} = \frac{KN+MP}{OK}$ $[\because LN = MP]$

$$= \frac{KN}{OK} + \frac{MP}{OK} = \frac{KN}{KM} \cdot \frac{KM}{OK} + \frac{MP}{OM} \cdot \frac{OM}{OK}$$

$$= \cos A \sin B + \sin A \cos B$$

$$\therefore \quad \sin(A + B) = \sin A \cos B + \cos A \sin B$$

(2) $$\cos(A + B) = \frac{OL}{OK} = \frac{OP - LP}{OK} = \frac{OP - NM}{OK} \qquad [\because LP = NM]$$

$$= \frac{OP}{OK} - \frac{NM}{OK} = \frac{OP}{OM} \cdot \frac{OM}{OK} - \frac{NM}{KM} \cdot \frac{KM}{OK}$$

$$= \cos A \cos B - \sin A \sin B$$

$$\therefore \quad \cos(A + B) = \cos A \cos B - \sin A \sin B$$

(3) $$\tan(A + B) = \frac{\sin(A + B)}{\cos(A + B)} = \frac{\sin A \cos B + \cos A \sin B}{\cos A \cos B - \sin A \sin B}$$

$$= \frac{\dfrac{\sin A \cos B}{\cos A \cos B} + \dfrac{\cos A \sin B}{\cos A \cos B}}{\dfrac{\cos A \cos B}{\cos A \cos B} - \dfrac{\sin A \sin B}{\cos A \cos B}}$$

$$= \frac{\dfrac{\sin A}{\cos A} + \dfrac{\sin B}{\cos B}}{1 - \dfrac{\sin A}{\cos A} \cdot \dfrac{\sin B}{\cos B}}$$

$$= \frac{\tan A + \tan B}{1 - \tan A \tan B}$$

Similarly

(4) $$\cot(A + B) = \frac{\cot A \cot B - 1}{\cot B + \cot A}$$

3.10.2 Subtraction Formula

Replacing B by $-B$ in (1),(2), (3) and (4), we get

(5) $\sin(A - B) = \sin A \cos(-B) + \cos A \sin(-B)$

$= \sin A \cos B - \cos A \sin B$

(6) $\cos(A - B) = \cos A \cos(-B) - \sin A \sin(-B)$

$= \cos A \cos B + \sin A \sin B$

(7) $$\tan(A - B) = \frac{\tan A + \tan(-B)}{1 - \tan A \tan(-B)} = \frac{\tan A - \tan B}{1 + \tan A \tan B}$$

(8) $$\cot(A - B) = \frac{\cot A \cot(-B) - 1}{\cot(-B) + \cot A} = \frac{-\cot A \cot B - 1}{-\cot B + \cot A}$$

$$= \frac{1 + \cot A \cot B}{\cot B - \cot A}$$

Example 3.32

If $\sin A = \dfrac{3}{5}$ and $\cos B = \dfrac{9}{41}$, find the values of $\sin(A + B)$, $\sin(A - B)$, $\cos(A + B)$ and $\cos(A - B)$.

SOLUTION : $\because \quad \sin A = \dfrac{3}{5}$

$\Rightarrow \quad \cos A = \sqrt{1 - \sin^2 A} = \sqrt{1 - (3/5)^2}$

$$\sqrt{1-\frac{9}{25}}=\sqrt{\frac{25-9}{25}}=\sqrt{\frac{16}{25}}=\frac{4}{5}$$

and $\cos B=\frac{9}{41} \Rightarrow \sin B=\sqrt{1-\cos^2 B}=\sqrt{1-\left(\frac{9}{4}\right)^2}$

$$=\sqrt{1-\frac{81}{1681}}=\sqrt{\frac{1681-81}{1681}}=\sqrt{\frac{1600}{1681}}=\frac{40}{41}$$

Now, $\sin(A+B)=\sin A\cos B+\cos A\sin B$

$$=\frac{3}{5}\times\frac{9}{41}+\frac{4}{5}\times\frac{40}{41}=\frac{27}{205}+\frac{160}{205}=\frac{187}{205}$$

$\sin(A-B)=\sin A\cos B-\cos A\sin B$

$$=\frac{3}{5}\times\frac{9}{41}-\frac{4}{5}\times\frac{40}{41}=\frac{27}{205}-\frac{160}{205}=-\frac{133}{205}$$

$\cos(A+B)=\cos A\cos B-\sin A\sin B$

$$=\frac{4}{5}\times\frac{9}{41}-\frac{3}{5}\times\frac{40}{41}=\frac{36}{205}-\frac{120}{205}=-\frac{84}{205}$$

$\cos(A-B)=\cos A\cos B+\sin A\sin B$

$$=\frac{4}{5}\times\frac{9}{41}+\frac{3}{5}\times\frac{40}{41}=\frac{36}{205}+\frac{120}{205}=\frac{156}{205}$$

Example 3.33

Find the values of the following:

(i) sin 75° (ii) cos 15° (iii) cos 105°
(iv) tan 105° (v) cot 75° (vi) sec 75°

SOLUTION : (i) $\sin 75°=\sin(45°+30°)$

Let $A=45°$ and $B=30°$

$\because$ $\sin 75°=\sin(A+B)=\sin A\cos B+\cos A\sin B$

$$=\sin 45°\cos 30°+\cos 45°\sin 30°$$

$$=\frac{1}{\sqrt{2}}\times\frac{\sqrt{3}}{2}+\frac{1}{\sqrt{2}}\times\frac{1}{2}=\frac{\sqrt{3}}{2\sqrt{2}}+\frac{1}{2\sqrt{2}}$$

$$=\frac{\sqrt{3}+1}{2\sqrt{2}}$$

(ii) $\cos 15°=\cos(45°-30°)=\cos 45°\cos 30°+\sin 45°\sin 30°$

$$=\frac{1}{\sqrt{2}}\times\frac{\sqrt{3}}{2}+\frac{1}{\sqrt{2}}\times\frac{1}{2}=\frac{\sqrt{3}}{2\sqrt{2}}+\frac{1}{2\sqrt{2}}=\frac{\sqrt{3}+1}{2\sqrt{2}}$$

(iii) $\cos 105°=\cos(60°+45°)=\cos 60°\cos 45°-\sin 60°\sin 45°$

$$=\frac{1}{2}\times\frac{1}{\sqrt{2}}-\frac{\sqrt{3}}{2}\times\frac{1}{\sqrt{2}}=\frac{1}{2\sqrt{2}}-\frac{\sqrt{3}}{2\sqrt{2}}=\frac{1-\sqrt{3}}{2\sqrt{2}}$$

(iv) $\tan 105°=\tan(60°+45°)=\frac{\tan 60°+\tan 45°}{1-\tan 60°\tan 45°}$

$$=\frac{\sqrt{3}+1}{1-\sqrt{3}\times 1}=\frac{\sqrt{3}+1}{1-\sqrt{3}}$$

(v) $\cot 75°=\cot(30°+45°)$

$$=\frac{\cot 30°\cot 45°-1}{\cot 45°+\cot 30°}=\frac{\sqrt{3}\times 1-1}{1+\sqrt{3}}=\frac{\sqrt{3}-1}{1+\sqrt{3}}$$

(vi) $$\sec 75° = \frac{1}{\cos 75°} = \frac{1}{\cos(30° + 45°)} = \frac{1}{\cos 30° \cos 45° - \sin 30° \sin 45°}$$

$$= \frac{1}{\frac{\sqrt{3}}{2} \times \frac{1}{\sqrt{2}} - \frac{1}{2} \times \frac{1}{\sqrt{2}}} = \frac{1}{\frac{\sqrt{3}}{2\sqrt{2}} - \frac{1}{2\sqrt{2}}} = \frac{1}{\frac{\sqrt{3}-1}{2\sqrt{2}}}$$

$$= \frac{2\sqrt{2}}{\sqrt{3}-1}$$

Example 3.34

Prove that :

(i) $\cos 44° \cos 16° - \sin 44° \sin 16° = \frac{1}{2}$

(ii) $\tan 23° + \tan 22° + \tan 23° \tan 22° = 1$

(iii) $\frac{\cos 13° + \sin 13°}{\cos 13° - \sin 13°} = \tan 58°$

SOLUTION : (i) L.H.S. $= \cos 44° \cos 16° - \sin 44° \sin 16°$

$= \cos(44° + 16°)$ $\quad [\because \cos A \cos B - \sin A \sin B = \cos(A + B)]$

$= \cos 60° = \frac{1}{2} =$ RHS

(ii) We know that

$$\tan(A + B) = \frac{\tan A + \tan B}{1 - \tan A \tan B}$$

Put $A = 23°$ and $B = 22°$

$$\therefore \quad \tan(23° + 22°) = \frac{\tan 23° + \tan 22°}{1 - \tan 23° \tan 22°}$$

$$\Rightarrow \quad \tan 45° = \frac{\tan 23° + \tan 22°}{1 - \tan 23° \tan 22°}$$

$$\Rightarrow \quad 1 = \frac{\tan 23° + \tan 22°}{1 - \tan 23° \tan 22°}$$

$$\Rightarrow \quad \tan 23° + \tan 22° = 1 - \tan 23° \tan 22°$$

$$\Rightarrow \quad \tan 23° + \tan 22° + \tan 23° \tan 22° = 1$$

(iii) $$\text{RHS} = \tan 58° = \tan(45° + 13°) = \frac{\tan 45° + \tan 13°}{1 - \tan 45° \tan 13°}$$

$$= \frac{1 + \tan 13°}{1 - \tan 13°} = \frac{1 + \frac{\sin 13°}{\cos 13°}}{1 - \frac{\sin 13°}{\cos 13°}} = \frac{\cos 13° + \sin 13°}{\cos 13° - \sin 13°}$$

Example 3.35

Prove that :

(i) $\sin(A + B) \sin(A - B) = \sin^2 A - \sin^2 B$

(ii) $\cos(A + B) \cos(A - B) = \cos^2 A - \sin^2 B$

SOLUTION : (i) $\quad$ LHS $= \sin(A + B) \sin(A - B)$

$= (\sin A \cos B + \cos A \sin B)(\sin A \cos B - \cos A \sin B)$

$= (\sin A \cos B)^2 - (\cos A \sin B)^2$

$= \sin^2 A \cos^2 B - \cos^2 A \sin^2 B$

$= \sin^2 A (1 - \sin^2 B) - (1 - \sin^2 A) \sin^2 B$

$= \sin^2 A - \sin^2 A \sin^2 B - \sin^2 B + \sin^2 A \sin^2 B$

$= \sin^2 A - \sin^2 B$

$=$ RHS

(ii) LHS $= \cos (A + B) \cos (A - B)$

$= (\cos A \cos B - \sin A \sin B) (\cos A \cos B + \sin A \sin B)$

$= (\cos A \cos B)^2 - (\sin A \sin B)^2$

$= \cos^2 A \cos^2 B - \sin^2 A \sin^2 B$

$= \cos^2 A (1 - \sin^2 B) - (1 - \cos^2 A) \sin^2 B$

$= \cos^2 A - \cos^2 A \sin^2 B - \sin^2 B + \cos^2 A \sin^2 B$

$= \cos^2 A - \sin^2 B$

Example 3.36

If $A + B = 45°$, Prove that

$$(1 + \tan A)(1 + \tan B) = 2$$

SOLUTION : We know that

$$\tan (A + B) = \frac{\tan A + \tan B}{1 - \tan A \tan B}$$

$$\Rightarrow \quad \tan 45° = \frac{\tan A + \tan B}{1 - \tan A \tan B} \qquad [\because \tan 45° = 1]$$

$$1 = \frac{\tan A + \tan B}{1 - \tan A \tan B}$$

$$1 - \tan A \tan B = \tan A + \tan B$$

$$\Rightarrow \quad \tan A + \tan B + \tan A \tan B = 1$$

$$\Rightarrow \quad \tan A + \tan B + \tan A \tan B + 1 = 1 + 1$$

$$\Rightarrow \quad \tan A (1 + \tan B) + (1 + \tan B) = 2$$

$$\Rightarrow \quad (1 + \tan A)(1 + \tan B) = 2$$

Example 3.37

Prove that :

$$\tan 8A - \tan 5A - \tan 3A = \tan 8A \tan 5A \tan 3A$$

SOLUTION : We know that

$$\tan (A - B) = \frac{\tan A - \tan B}{1 + \tan A \tan B}$$

$$\Rightarrow \quad \tan (8A - 5A) = \frac{\tan 8A - \tan 5A}{1 + \tan 8A \tan 5A}$$

$$\Rightarrow \quad \tan 3A (1 + \tan 8A \tan 5A) = \tan 8A - \tan 5A$$

$$\Rightarrow \quad \tan 3A + \tan 8A \tan 5A \tan 3A = \tan 8A - \tan 5A$$

$$\Rightarrow \quad \tan 8A \tan 5A \tan 3A = \tan 8A - \tan 5A - \tan 3A$$

$$\Rightarrow \quad \tan 8A - \tan 5A - \tan 3A = \tan 8A \tan 5A \tan 3A$$

Example 3.38

Prove that :

$$\frac{\sin(A-B)+\sin A+\sin(A+B)}{\cos(A-B)+\cos A+\cos(A+B)}=\tan A$$

SOLUTION :

$$\text{LHS}=\frac{\sin(A-B)+\sin A+\sin(A+B)}{\cos(A-B)+\cos A+\cos(A+B)}$$

$$=\frac{\sin A\cos B-\cos A\sin B+\sin A+\sin A+\sin A\cos B+\cos A\sin B}{\cos A\cos B+\sin A\sin B+\cos A+\cos A\cos B-\sin A\sin B}$$

$$=\frac{2\sin A\cos B+\sin A}{2\cos A\cos B+\cos A}$$

$$=\frac{\sin A(2\cos B+1)}{\cos A(2\cos B+1)}=\frac{\sin A}{\cos A}$$

$$=\tan A=\text{RHS.}$$

Example 3.39

Prove that :

$$\frac{\tan(45^\circ+A)}{\tan(45^\circ-A)}=\left(\frac{1+\tan A}{1-\tan A}\right)^2$$

SOLUTION :

$$\text{LHS}=\frac{\tan(45^\circ+A)}{\tan(45^\circ-A)}$$

$$=\frac{\tan 45^\circ+\tan A}{1-\tan 45^\circ\tan A}\times\frac{1+\tan 45^\circ\tan A}{\tan 45^\circ-\tan A}$$

$$=\frac{1+\tan A}{1-\tan A}\times\frac{1+\tan A}{1-\tan A}=\left(\frac{1+\tan A}{1-\tan A}\right)^2$$

$$=\text{RHS}$$

Example 3.40

Find (i) tan 75° and (ii) tan 15°. ***[UPTU (C.O.) 2006]***

Solution. (i) $\tan 75^\circ=\tan(30^\circ+45^\circ)$

$$=\frac{\tan 30^\circ+\tan 45^\circ}{1-\tan 30^\circ\tan 45^\circ}$$

$$=\frac{\frac{1}{\sqrt{3}}+1}{1-\frac{1}{\sqrt{3}}\cdot 1}=\frac{1+\sqrt{3}}{\sqrt{3}-1}=\frac{\sqrt{3}+1}{\sqrt{3}-1}$$

(ii) $\tan 15^\circ=\tan(45^\circ-30^\circ)=\dfrac{\tan 45^\circ-\tan 30^\circ}{1+\tan 45^\circ\tan 30^\circ}$

$$=\frac{1-\frac{1}{\sqrt{3}}}{1+1\cdot\frac{1}{\sqrt{3}}}=\frac{\sqrt{3}-1}{\sqrt{3}+1}$$

Example 3.41

If $\tan A = \dfrac{m}{m+1}$ and $\tan B = \dfrac{1}{2m+1}$, find the value of $(A + B)$. *[UPTU 2007]*

SOLUTION : We know that

$$\tan(A+B) = \frac{\tan A + \tan B}{1 - \tan A \tan B}$$

$$= \frac{\dfrac{m}{m+1} + \dfrac{1}{2m+1}}{1 - \dfrac{m}{m+1} \cdot \dfrac{1}{2m+1}}$$

$$= \frac{2m^2 + m + m + 1}{2m^2 + 2m + m + 1 - m}$$

$$= \frac{2m^2 + 2m + 1}{2m^2 + 2m + 1} = 1$$

$\therefore \quad A + B = \tan^{-1}(1) = 45°$

PROBLEM SET 3.5

1. If $\sin A = \dfrac{3}{5}$, $\cos B = \dfrac{5}{13}$, A and B are in first quadrant. Find the values of $\sin(A+B)$, $\sin(A-B)$, $\cos(A+B)$ and $\cos(A-B)$.

2. If $\sin A = \dfrac{8}{17}$, $\tan B = \dfrac{5}{12}$, A and B are in first quadrant. Find the values of $\sin(A-B)$, $\sin(A+B)$, $\cos(A-B)$, $\cos(A+B)$, $\tan(A+B)$ and $\tan(A-B)$

3. If A and B are acute angles such that $\tan A = \dfrac{5}{6}$, $\tan B = \dfrac{1}{11}$, find $A + B$

4. If $\sin(A+B) = \dfrac{4}{5}$, $\sin(A-B) = \dfrac{5}{13}$, prove that $\tan 2A = \dfrac{63}{16}$

5. Find the value of the followings :

(i) $\cos 75°$ (ii) $\tan 75° + \cot 75°$

(iii) $\sin 135°$ (iv) $\text{cosec } 150°$

6. Find the value of

(i) $\sin 15°$ (ii) $\tan 15°$

(iii) $\cos 47° \cos 13° - \sin 47° \sin 13°$

Prove that :

7. $\cot(45° + A) = \dfrac{\cot A - 1}{\cot A + 1}$

8. $\dfrac{\cos 11° + \sin 11°}{\cos 11° - \sin 11°} = \tan 56°$

9. $\sin 51° + \cos 81° = \cos 21°$

10. $\sin(A+30°)+\cos(A+60°)=\cos A$

11. $\dfrac{\cos 23°+\sin 23°}{\cos 23°-\sin 23°}=\tan 68°$

12. $\tan(A+30°)=\dfrac{\sqrt{3}\tan A+1}{\sqrt{3}-\tan A}$

13. $\cot\left(\dfrac{\pi}{4}+A\right)\cdot\cot\left(\dfrac{\pi}{4}-A\right)=1$

14. $\sin^2(A+45°)+\sin^2(A-45°)=1$

15. $\sin 6°=\dfrac{\sqrt{3}}{2}\sin 36°-\dfrac{1}{2}\cos 36°$

16. $\dfrac{\tan(B-C)+\tan C}{1-\tan(B-C)\tan C}=\tan B$

17. $\cos A\cos B+\sin A\sin B=\dfrac{1}{2}$ if $A-B=60°$

18. $\tan 3A\tan 2A\tan A=\tan 3A-\tan 2A-\tan A$

19. Prove that $(\cot A-1)(\cot B-1)=2$ if $A+B=\pi$

20. Prove that $\sin^2\left(\dfrac{\pi}{8}+\dfrac{A}{2}\right)-\sin^2\left(\dfrac{\pi}{8}-\dfrac{A}{2}\right)=\dfrac{1}{\sqrt{2}}\sin A$

21. Prove that : $\tan 70°=\tan 20°+2\tan 50°$

22. Prove that :

(i) $\tan 8A-\tan 6A-\tan 2A=\tan 8A\tan 6A\tan 2A$

(ii) $\tan 36°+\tan 9°+\tan 36°\tan 9°=1$

(iii) $\dfrac{\tan A+\tan B}{\tan A-\tan B}=\dfrac{\sin(A+B)}{\sin(A-B)}$

23. Given that : $A=B+C$, prove that $\tan A-\tan B-\tan C=\tan A\tan B\tan C$

24. If $\tan A=\dfrac{18}{17}$ and $\tan B=\dfrac{1}{35}$, show that $A-B=45°$

25. Prove that : $\cos A+\cos(120°+A)+\cos(120°-A)=0$

26. Prove that : $\tan 15°+\cot 15°=4$

27. If $A+B+C=\pi$ and $\cos A=\cos B\cos C$. Prove that $\tan A=\tan B+\tan C$

28. Show that $\tan 75°=2+\sqrt{3}$

ANSWERS

1. $\dfrac{63}{65},\dfrac{-33}{65},\dfrac{-16}{65},\dfrac{56}{65}$ **2.** $\dfrac{21}{221},\dfrac{171}{221},\dfrac{220}{221},\dfrac{140}{221},\dfrac{171}{140},\dfrac{21}{220}$ **3.** 45°

5. (i) $\dfrac{\sqrt{3}-1}{2\sqrt{2}}$, (ii) 4, (iii) $\dfrac{1}{\sqrt{2}}$, (iv) 2. **6.** (i) $\dfrac{\sqrt{3}-1}{2\sqrt{2}}$, (ii) $\dfrac{\sqrt{3}-1}{\sqrt{3}+1}$, (iii) $\dfrac{1}{2}$

3.11 PRODUCT FORMULAE

3.11.1 Converting Products into Sum or Differnce

1. $2 \sin A \cos B = \sin (A + B) + \sin (A - B)$
2. $2 \cos A \sin B = \sin (A + B) - \sin (A - B)$
3. $2 \cos A \cos B = \cos (A + B) + \cos (A - B)$
4. $2 \sin A \sin B = \cos (A - B) - \cos (A + B)$

3.11.2 Converting Sum or Difference into Products

5. $\sin C + \sin D = 2\sin\left(\frac{C+D}{2}\right)\cos\left(\frac{C-D}{2}\right)$
6. $\sin C - \sin D = 2\sin\left(\frac{C-D}{2}\right)\cos\left(\frac{C+D}{2}\right)$
7. $\cos C + \cos D = 2\cos\left(\frac{C+D}{2}\right)\cos\left(\frac{C-D}{2}\right)$
8. $\cos C - \cos D = -2\sin\left(\frac{C+D}{2}\right)\sin\left(\frac{C-D}{2}\right)$

3.11.3 *t*-Ratios of the Angle 2*A*

9. $\sin 2A = 2 \sin A \cos A$
10. (i) $\cos 2A = \cos^2 A - \sin^2 A$
    (ii) $\cos 2A = 2 \cos^2 A - 1$
    (iii) $\cos 2A = 1 - 2 \sin^2 A$
11. $\tan 2A = \dfrac{2 \tan A}{1 - \tan^2 A}$
12. $\sin 2A = \dfrac{2 \tan A}{1 + \tan^2 A}$
13. $\cos 2A = \dfrac{1 - \tan^2 A}{1 + \tan^2 A}$

3.11.4 *t*-Ratios of the Angle 3*A*

14. $\sin 3A = 3 \sin A - 4 \sin^3 A$
15. $\cos 3A = 4 \cos^3 A - 3\cos A$
16. $\tan 3A = \dfrac{3 \tan A - \tan^3 A}{1 - 3 \tan^2 A}$

3.11.5 *t*-Ratios of the Angle $\frac{A}{2}$

17. $\sin A = 2 \sin \frac{A}{2} \cos \frac{A}{2}$
18. (i) $\cos A = \cos^2 \frac{A}{2} - \frac{\sin^2}{2}$

(ii) $\cos A = 2\cos^2\dfrac{A}{2} - 1$

(iii) $\cos A = 1 - 2\sin^2\dfrac{A}{2}$

19. $\tan A = \dfrac{2\tan\dfrac{A}{2}}{1-\tan^2\dfrac{A}{2}}$

20. $\sin A = \dfrac{2\tan A/2}{1+\tan^2\dfrac{A}{2}}$

21. $\cos A = \dfrac{1-\tan^2\dfrac{A}{2}}{1+\tan^2\dfrac{A}{2}}$

Example 3.42

Find the value of the following.

(i) 2 sin 3A cos 5A (ii) sin 75° cos 15°

SOLUTION : (i) We have $2\sin 3A\cos 5A = \sin(3A+5A) + \sin(3A-5A)$

$$= \sin 8A + \sin(-2A) = \sin 8A - \sin 2A$$

(ii) $\sin 75°\cos 15° = \dfrac{1}{2}[2\sin 75°\cos 15°]$

$$= \frac{1}{2}[\sin(75°+15°) + \sin(75°-15°)]$$

$$= \frac{1}{2}[\sin 90° + \sin 60°]$$

$$= \frac{1}{2}\left(1+\frac{\sqrt{3}}{2}\right)$$

$$= \frac{\sqrt{3}+2}{4}$$

Example 3.43

Prove that:

$$\cos 20°\cos 40°\cos 60°\cos 80° = \frac{1}{16}$$

[UPTU (C.O.) 2006]

SOLUTION :

$$\text{LHS} = \cos 20°\cos 40°\cos 60°\cos 80°$$

$$= \frac{1}{2}(2\cos 20°\cos 40°)\frac{1}{2}\cos 80° \qquad \left[\because \cos 60° = \frac{1}{2}\right]$$

$$= \frac{1}{4}\{\cos(20°+40°) + \cos(20°-40°)\}\cos 80°$$

$$= \frac{1}{4}[\cos 60° + \cos 20°]\cos 80°$$

$$= \frac{1}{4}\left(\frac{1}{2} + \cos 20°\right)\cos 80°$$

$$= \frac{1}{8}\cos 80° + \frac{1}{8}[2\cos 20° \cos 80°]$$
$$= \frac{1}{8}\cos 80° + \frac{1}{8}[\cos(20° + 80°) + \cos(20° - 80°)]$$
$$= \frac{1}{8}\cos 80° + \frac{1}{8}[\cos 100° + \cos 60°]$$
$$= \frac{1}{8}\cos 80° + \frac{1}{8}\cos 100° + \frac{1}{16}$$
$$= \frac{1}{8}\cos 80° + \frac{1}{8}\cos(180° - 80°) + \frac{1}{16}$$
$$= \frac{1}{8}\cos 80° - \frac{1}{8}(\cos 80°) + \frac{1}{16}$$
$$= \frac{1}{16}$$
$$= \text{RHS}$$

Example 3.44

Find the value of the following :-
(i) $\sin 6A - \sin 2A$
(ii) $\cos 4A + \cos 8A$

SOLUTION : (i) $\sin 6A - \sin 2A = 2\sin\left(\frac{6A - 2A}{2}\right)\cos\left(\frac{6A + 2A}{2}\right)$

$$= 2\sin 2A \cos 4A$$

(ii) $\cos 4A + \cos 8A = 2\cos\left(\frac{4A + 8A}{2}\right)\cos\left(\frac{4A - 8A}{2}\right)$

$$= 2\cos 6A \cos(-2A) = 2\cos 6A \cos 2A$$

Example 3.45

Prove that :
(i) $\sin 80° - \sin 20° = \cos 50°$
(ii) $2\cos\frac{\pi}{13}\cos\frac{9\pi}{13} + \cos\frac{3\pi}{13} + \cos\frac{5\pi}{13} = 0$ ***[UPTU (C.O.) 2006, UPTU 2007]***

SOLUTION :

(i) LHS $= \sin 80° - \sin 20°$

$$= 2\cos\left(\frac{80° + 20°}{2}\right)\sin\left(\frac{80° - 20°}{2}\right)$$

$$\left[\because \sin C - \sin D = 2\cos\frac{C+D}{2}\sin\frac{C-D}{2}\right]$$

$$= 2\cos 50° \sin 30°$$
$$= 2\cos 50° \times \frac{1}{2}$$
$$= \cos 50°$$
$$= \text{RHS}$$

(ii) $$\text{LHS} = 2\cos\frac{\pi}{13}\cos\frac{9\pi}{13} + \cos\frac{3\pi}{13} + \cos\frac{5\pi}{13}$$
$$= \cos\left(\frac{\pi}{13}+\frac{9\pi}{13}\right) + \cos\left(\frac{\pi}{13}-\frac{9\pi}{13}\right) + \cos\frac{3\pi}{13} + \cos\frac{5\pi}{13}$$
$[\because\ 2\cos A\cos B = \cos(A+B) + \cos(A-B)]$
$$= \cos\frac{10\pi}{13} + \cos\frac{8\pi}{13} + \cos\frac{3\pi}{13} + \cos\frac{5\pi}{13}$$
$$= \cos\left(\pi-\frac{3\pi}{13}\right) + \cos\left(\pi-\frac{5\pi}{13}\right) + \cos\frac{3\pi}{13} + \cos\frac{5\pi}{13}$$
$$= -\cos\frac{3\pi}{13} - \cos\frac{5\pi}{13} + \cos\frac{3\pi}{13} + \cos\frac{5\pi}{13} = 0 = \text{RHS}$$

Example 3.46

Prove that:

(i) $\sin 3A = 3\sin A - 4\sin^3 A$

(ii) $\cos 3A = 4\cos^3 A - 3\cos A$

(iii) $\tan 3A = \dfrac{3\tan A - \tan^3 A}{1 - 3\tan^2 A}$

SOLUTION :

(i) $$\text{LHS} = \sin 3A = \sin(2A + A) = \sin 2A\cos A + \cos 2A\sin A$$
$$= 2\sin A\cos A\cdot\cos A + (\cos^2 A - \sin^2 A)\sin A$$
$$= 2\sin A\cos^2 A + \cos^2 A\sin A - \sin^3 A$$
$$= 3\sin A\cos^2 A - \sin^3 A$$
$$= 3\sin A(1 - \sin^2 A) - \sin^3 A$$
$$= 3\sin A - 3\sin^3 A - \sin^3 A$$
$$= 3\sin A - 4\sin^3 A = \text{RHS}$$

(ii) $$\text{LHS} = \cos 3A$$
$$= \cos(2A + A)$$
$$= \cos 2A\cos A - \sin 2A\sin A$$
$$= (\cos^2 A - \sin^2 A)\cos A - 2\sin A\cos A\cdot\sin A$$
$$= \cos^3 A - \sin^2 A\cos A - 2\sin^2 A\cos A$$
$$= \cos^3 A - 3\sin^2 A\cos A$$
$$= \cos^3 A - 3(1 - \cos^2 A)\cos A$$
$$= 4\cos^3 A - 3\cos A = \text{RHS}$$

(iii) $$\text{LHS} = \tan 3A = \tan(2A + A) = \frac{\tan 2A + \tan A}{1 - \tan 2A\tan A}$$
$$= \frac{\dfrac{2\tan A}{1-\tan^2 A} + \tan A}{1 - \dfrac{2\tan A}{1-\tan^2 A}\cdot\tan A}$$
$$= \frac{2\tan A + \tan A - \tan^3 A}{1 - \tan^2 A - 2\tan^2 A} = \frac{3\tan A - \tan^3 A}{1 - 3\tan^2 A} = \text{RHS}$$

Example 3.47

Prove that

(i) $\dfrac{\sin 8A}{\sin A} = 8 \cos A \cos 2A \cos 4A$

(ii) $\dfrac{\sin A + \sin 3A + \sin 5A}{\cos A + \cos 3A + \cos 5A} = \tan 3A$

SOLUTION :

(i) $$\text{LHS} = \frac{\sin 8A}{\sin A} = \frac{2 \sin 4A \cos 4A}{\sin A} \qquad [\because \sin 2A = 2 \sin A \cos A]$$

$$= \frac{2\,(2 \sin 2A \cos 2A)\cos 4A}{\sin A}$$

$$= \frac{4 \sin 2A \cos 2A \cos 4A}{\sin A}$$

$$= \frac{8 \sin A \cos A \cos 2A \cos 4A}{\sin A} = 8 \cos A \cos 2A \cos 4A = \text{RHS}$$

(ii) $$\text{LHS} = \frac{\sin A + \sin 3A + \sin 5A}{(\cos A + \cos 3A) + \cos 5A}$$

$$= \frac{(\sin A + \sin 5A) + \sin 3A}{(\cos A + \cos 5A) + \cos 3A}$$

$$= \frac{2 \sin\left(\frac{A+5A}{2}\right) \cos\left(\frac{A-5A}{2}\right) + \sin 3A}{2 \cos\left(\frac{A+5A}{2}\right) \cos\left(\frac{A-5A}{2}\right) + \cos 3A}$$

$$\left[\because \sin C + \sin D = 2 \sin \frac{C+D}{2} \cos \frac{C-D}{2}\right]$$

$$= \frac{2 \sin 3A \cos 2A + \sin 3A}{2 \cos 3A \cos 2A + \cos 3A}$$

$$= \frac{\sin 3A\,(2 \cos 2A + 1)}{\cos 3A\,(2 \cos 2A + 1)}$$

$$= \frac{\sin 3A}{\cos 3A}$$

$$= \tan 3A$$

$$= \text{RHS}$$

Example 3.48

If $\sin A = \dfrac{1}{2}$, find the value of $\sin 2A$, $\cos 3A$ and $\tan 3A$

SOLUTION : Here $\sin A = \dfrac{1}{2} \Rightarrow \cos A = \sqrt{1 - \sin^2 A} = \sqrt{1 - \dfrac{1}{4}} = \sqrt{\dfrac{3}{4}} = \dfrac{\sqrt{3}}{2}$

Also, $\tan A = \dfrac{\sin A}{\cos A} = \dfrac{1}{2} \times \dfrac{2}{\sqrt{3}} = \dfrac{1}{\sqrt{3}}$

Now, (i) $\sin 2A = 2 \sin A \cos A = 2 \cdot \dfrac{1}{2} \cdot \dfrac{\sqrt{3}}{2} = \dfrac{\sqrt{3}}{2}$

(ii) $\cos 3A = 4 \cos^3 A - 3 \cos A = 4\left(\dfrac{\sqrt{3}}{2}\right)^3 - 3\left(\dfrac{\sqrt{3}}{2}\right) = 4 \cdot \dfrac{3\sqrt{3}}{8} - \dfrac{3\sqrt{3}}{2}$

$$= \frac{3\sqrt{3}}{2} - \frac{3\sqrt{3}}{2} = 0$$

(iii) $\tan 3A = \dfrac{3\tan A - \tan^3 A}{1 - 3\tan^2 A} = \dfrac{3\left(\frac{1}{\sqrt{3}}\right) - \left(\frac{1}{\sqrt{3}}\right)^3}{1 - 3\left(\frac{1}{\sqrt{3}}\right)^2}$

$$= \frac{\frac{3}{\sqrt{3}} - \frac{1}{3\sqrt{3}}}{1 - \frac{3}{3}} = \frac{\frac{9-1}{3\sqrt{3}}}{1-1} = \frac{\frac{8}{3\sqrt{3}}}{0} = \infty$$

Example 3.49

Prove that

(i) $\dfrac{\sin 2A}{1 + \cos 2A} = \tan A$, (ii) $\sin A \cdot \sin(60° - A)\sin(60° + A) = \dfrac{1}{4}\sin 3A$

SOLUTION :

(i) $\text{LHS} = \dfrac{\sin 2A}{1 + \cos 2A} = \dfrac{2\sin A\cos A}{1 + 2\cos^2 A - 1} = \dfrac{2\sin A\cos A}{2\cos^{2A}} = \dfrac{\sin A}{\cos A}$

$= \tan A$

$= \text{RHS}$

(ii) $\text{LHS} = \sin A \cdot \sin(60° - A)\sin(60° + A)$

$= \sin A(\sin^2 60° - \sin^2 A) = \sin A\left(\dfrac{3}{4} - \sin^2 A\right)$

$= \sin A\left(\dfrac{3 - 4\sin^2 A}{4}\right) = \dfrac{(3\sin A - 4\sin^3 A)}{4}$

$= \dfrac{1}{4}\sin 3A$

$= \text{RHS}$

Example 3.50

Prove that $\dfrac{\cos\theta}{1 + \sin\theta} = \tan\left(\dfrac{\pi}{4} - \dfrac{\theta}{2}\right)$ ***[UPTU 2000]***

SOLUTION : $\text{LHS} = \dfrac{\cos\theta}{1 + \sin\theta} = \dfrac{\cos^2\frac{\theta}{2} - \sin^2\frac{\theta}{2}}{\cos^2\frac{\theta}{2} + \sin^2\frac{\theta}{2} + 2\sin\frac{\theta}{2}\cos\frac{\theta}{2}}$

$$= \frac{\left(\cos\frac{\theta}{2} - \sin\frac{\theta}{2}\right)\left(\cos\frac{\theta}{2} + \sin\frac{\theta}{2}\right)}{\left(\cos\frac{\theta}{2} + \sin\frac{\theta}{2}\right)^2}$$

$$= \frac{\cos\frac{\theta}{2} - \sin\frac{\theta}{2}}{\cos\frac{\theta}{2} + \sin\frac{\theta}{2}}$$

$$= \frac{\dfrac{\cos\frac{\theta}{2}}{\cos\frac{\theta}{2}} - \dfrac{\sin\frac{\theta}{2}}{\cos\frac{\theta}{2}}}{\dfrac{\cos\frac{\theta}{2}}{\cos\frac{\theta}{2}} + \dfrac{\sin\frac{\theta}{2}}{\cos\frac{\theta}{2}}} \qquad \left[\text{Divide by } \cos\frac{\theta}{2}\right]$$

$$= \frac{1 - \tan\frac{\theta}{2}}{1 + \tan\frac{\theta}{2}} = \frac{\tan\frac{\pi}{4} - \tan\frac{\theta}{2}}{1 + \tan\frac{\pi}{4}\tan\frac{\theta}{2}}$$

$$= \tan\left(\frac{\pi}{4} - \frac{\theta}{2}\right)$$

$$= \text{RHS}$$

Example 3.51

Prove that $\dfrac{\sin A + \sin B}{\cos A + \cos B} = \tan\left(\dfrac{A+B}{2}\right)$ ***[UPTU 2000]***

SOLUTION : LHS $= \dfrac{\sin A + \sin B}{\cos A + \cos B}$

$$= \frac{2\sin\left(\frac{A+B}{2}\right)\cos\left(\frac{A-B}{2}\right)}{2\cos\left(\frac{A+B}{2}\right)\cos\left(\frac{A-B}{2}\right)}$$

$$= \frac{\sin\left(\frac{A+B}{2}\right)}{\cos\left(\frac{A+B}{2}\right)} = \tan\left(\frac{A+B}{2}\right) = \text{RHS}$$

Example 3.52

Solve that in ΔABC,

$$\tan A + \tan B + \tan C = \tan A \tan B \tan C$$ ***[UPTU 2000, 07]***

SOLUTION : Given $A + B + C = 180° \Rightarrow C = 180° - (A + B)$

$\Rightarrow \quad \tan C = \tan[180° - (A+B)] = -\tan(A+B) \qquad [\because \tan(\pi - \theta) = -\tan\theta]$

$$= -\left[\frac{\tan A + \tan B}{1 - \tan A \tan B}\right]$$

$$\tan C = \frac{-\tan A - \tan B}{1 - \tan A \tan B}$$

By cross multiplication

$$\tan C(1 - \tan A \tan B) = -\tan A - \tan B$$

$\Rightarrow \quad \tan C - \tan A \tan B \tan C = -\tan A - \tan B$

$\Rightarrow \quad \tan A + \tan B + \tan C = \tan A \tan B \tan C$

Example 3.53

Prove that $\sec\alpha + \tan\alpha = \tan\left(\frac{\pi}{4} + \frac{\alpha}{2}\right)$ *[UPTU 2001, 05]*

SOLUTION : LHS $= \sec\alpha + \tan\alpha$

$$= \frac{1}{\cos\alpha} + \frac{\sin\alpha}{\cos\alpha}$$

$$= \frac{1+\sin\alpha}{\cos\alpha}$$

$$= \frac{\cos^2\frac{\alpha}{2} + \sin^2\frac{\alpha}{2} + 2\sin\frac{\alpha}{2}\cos\frac{\alpha}{2}}{\cos^2\frac{\alpha}{2} - \sin^2\frac{\alpha}{2}}$$

$$= \frac{\left(\cos\frac{\alpha}{2} + \sin\frac{\alpha}{2}\right)^2}{\left(\cos\frac{\alpha}{2} - \sin\frac{\alpha}{2}\right)\left(\cos\frac{\alpha}{2} + \sin\frac{\alpha}{2}\right)}$$

$$= \frac{\cos\frac{\alpha}{2} + \sin\frac{\alpha}{2}}{\cos\frac{\alpha}{2} - \sin\frac{\alpha}{2}}$$

$$= \frac{\frac{\cos\frac{\alpha}{2}}{\cos\frac{\alpha}{2}} + \frac{\sin\frac{\alpha}{2}}{\cos\frac{\alpha}{2}}}{\frac{\cos\frac{\alpha}{2}}{\cos\frac{\alpha}{2}} - \frac{\sin\frac{\alpha}{2}}{\cos\frac{\alpha}{2}}}$$

$$= \frac{1+\tan\frac{\alpha}{2}}{1-\tan\frac{\alpha}{2}}$$

$$= \frac{\tan\frac{\pi}{4} + \tan\frac{\alpha}{2}}{1-\tan\frac{\pi}{4}\tan\frac{\alpha}{2}} = \tan\left(\frac{\pi}{4} + \frac{\alpha}{2}\right) = \text{RHS}$$

Example 3.54

Find the value of $\tan 22\frac{1}{2}°$ *[UPTU 2001]*

SOLUTION : We know that

$$\sin\frac{A}{2}\sqrt{\frac{1}{2}(1-\cos A)} \quad \text{and} \quad \cos\frac{A}{2} = \sqrt{\frac{1}{2}(\cos A + 1)}$$

Put $A = 45°$, we get

$$\sin\frac{45°}{2} = \sqrt{\frac{1}{2}(1-\cos 45°)} = \sqrt{\frac{1}{2}\left(1-\frac{1}{\sqrt{2}}\right)} = \sqrt{\frac{1}{2}\left(\frac{\sqrt{2}-1}{\sqrt{2}}\right)}$$

$$= \sqrt{\frac{1}{2}\left(\frac{\sqrt{2}-1}{\sqrt{2}}\right)\frac{\sqrt{2}}{\sqrt{2}}} = \sqrt{\frac{1}{2}\left(\frac{2-\sqrt{2}}{2}\right)}$$

$$\therefore \quad \sin\frac{45°}{2} = \frac{1}{2}\sqrt{2-\sqrt{2}}$$

Also, $$\cos\frac{45°}{2} = \sqrt{\frac{1}{2}(1+\cos 45°)} = \sqrt{\frac{1}{2}\left(1+\frac{1}{\sqrt{2}}\right)}$$

$$= \frac{1}{2}\sqrt{\sqrt{2}+2}$$

Now, $$\tan\frac{45°}{2} = \frac{\sin\frac{45°}{2}}{\cos\frac{45°}{2}} = \frac{\frac{1}{2}\sqrt{2-\sqrt{2}}}{\frac{1}{2}\sqrt{2+\sqrt{2}}} = \frac{\sqrt{2-\sqrt{2}}}{\sqrt{2+\sqrt{2}}}$$

$$= \frac{(\sqrt{2-\sqrt{2}})^2}{\sqrt{(2)^2-(\sqrt{2})^2}} = \frac{2-\sqrt{2}}{\sqrt{4-2}} = \frac{2-\sqrt{2}}{\sqrt{2}}$$

$$= \frac{2-\sqrt{2}}{\sqrt{2}} \times \frac{\sqrt{2}}{\sqrt{2}} = \frac{2\sqrt{2}-2}{2} = \sqrt{2}-1$$

$$\therefore \quad \tan 22\frac{1}{2}° = \sqrt{2}-1$$

Example 3.55

If $\sin 2A = \frac{4}{5}$, find the value of $\tan A$ $(0 \le A \le \frac{\pi}{4})$ ***[UPTU 2001]***

SOLUTION : Given $\sin 2A = \frac{4}{5} \Rightarrow \frac{2\tan A}{1+\tan^2 A} = \frac{4}{5}$

$$\Rightarrow \quad 10\tan A = 4 + 4\tan^2 A$$

$$\Rightarrow \quad 4\tan^2 A - 10\tan A + 4 = 0$$

or $$2\tan^2 A - 5\tan A + 2 = 0$$

$$\Rightarrow \quad \tan A = \frac{5 \pm \sqrt{25-16}}{4} = \frac{5 \pm \sqrt{9}}{4} = \frac{5 \pm 3}{4}$$

$$= \frac{5+3}{4} \text{ or } \frac{5-3}{4}$$

If $\tan A = \frac{8}{4} = 2 \Rightarrow A = \tan^{-1} 2 = 63°26'$ (App.)

If $\tan A = \frac{2}{4} = \frac{1}{2} \Rightarrow A = \tan^{-1}\frac{1}{2} = \tan^{-1}(0.5) = 26°34'$ (App.)

Given $0 \le A \le \frac{\pi}{4} \Rightarrow A = 26°34'$

Example 3.56

Prove that $\tan\left(\frac{\pi}{4}-\frac{A}{2}\right)=\sqrt{\frac{1-\sin A}{1+\sin A}}$ ***[UPTU 2002, 05]***

SOLUTION : RHS $=\sqrt{\frac{1-\sin A}{1+\sin A}}=\sqrt{\frac{\cos^2\frac{A}{2}+\sin^2\frac{A}{2}-2\sin\frac{A}{2}\cos\frac{A}{2}}{\cos^2\frac{A}{2}+\sin^2\frac{A}{2}+2\sin\frac{A}{2}\cos\frac{A}{2}}}$

$$=\sqrt{\frac{\left(\cos\frac{A}{2}-\sin\frac{A}{2}\right)^2}{\left(\cos\frac{A}{2}+\sin\frac{A}{2}\right)^2}}$$

$$=\frac{\cos A/2-\sin A/2}{\cos A/2+\sin A/2}$$

Divde both sides by $\cos\frac{A}{2}$

$$=\frac{\frac{\cos\frac{A}{2}}{\cos\frac{A}{2}}-\frac{\sin\frac{A}{2}}{\cos\frac{A}{2}}}{\frac{\cos\frac{A}{2}}{\cos\frac{A}{2}}+\frac{\sin\frac{A}{2}}{\cos\frac{A}{2}}}$$

$$=\frac{1-\tan\frac{A}{2}}{1+\tan\frac{A}{2}}=\frac{\tan\frac{\pi}{4}-\tan\frac{A}{2}}{1+\tan\frac{\pi}{4}\tan\frac{A}{2}}=\tan\left(\frac{\pi}{4}-\frac{A}{2}\right)=\text{LHS}$$

Example 3.57

Prove that $\sin\frac{5\pi}{18}-\cos\frac{4\pi}{9}=\sqrt{3}\sin\frac{\pi}{9}$. ***[UPTU 2002]***

SOLUTION : LHS $=\sin\frac{5\pi}{18}-\cos\frac{4\pi}{9}$

$$=\sin\frac{5\pi}{18}-\sin\left(\frac{\pi}{2}-\frac{4\pi}{9}\right) \qquad \left[\because\ \sin\left(\frac{\pi}{2}-\theta\right)=\cos\theta\right]$$

$$=\sin\frac{5\pi}{18}-\sin\left(\frac{9\pi-8\pi}{18}\right)$$

$$=\sin\frac{5\pi}{18}-\sin\frac{\pi}{18}$$

Now, we know that $\sin C-\sin D=2\sin\left(\frac{C-D}{2}\right)\cos\left(\frac{C+D}{2}\right)$

$$\therefore \quad \sin\frac{5\pi}{18} - \sin\frac{\pi}{18} = 2\sin\left(\frac{\frac{5\pi}{18}-\frac{\pi}{18}}{2}\right)\cos\left(\frac{\frac{5\pi}{18}+\frac{\pi}{18}}{2}\right)$$

$$= 2\sin\left(\frac{4\pi}{36}\right)\cos\left(\frac{6\pi}{36}\right)$$

$$= 2\sin\left(\frac{\pi}{9}\right)\cos\left(\frac{\pi}{6}\right)$$

$$= 2\sin\frac{\pi}{9}\cdot\frac{\sqrt{3}}{2} \qquad \left[\because \cos\frac{\pi}{6} = \frac{\sqrt{3}}{2}\right]$$

$$= \sqrt{3}\sin\frac{\pi}{9} = \text{RHS}$$

Example 3.58

Prove that $\sin 10^\circ \sin 30^\circ \sin 50^\circ \sin 70^\circ = \dfrac{1}{16}$ ***[UPTU 2002]***

SOLUTION : $\text{LHS} = \sin 10^\circ \sin 30^\circ \sin 50^\circ \sin 70^\circ$

$$= \frac{1}{2}\sin 10^\circ \sin 50^\circ \sin 70^\circ \qquad \left[\because \sin 30^\circ = \frac{1}{2}\right]$$

$$= \frac{1}{4}[2\sin 10^\circ \sin 50^\circ]\sin 70^\circ$$

$$= \frac{1}{4}[\cos(10^\circ - 50^\circ) - \cos(10^\circ + 50^\circ)]\sin 70^\circ$$

$$= \frac{1}{4}[\cos(-40^\circ) - \cos 60^\circ]\cdot\sin 70^\circ$$

$$= \frac{1}{4}\left[\cos 40^\circ - \frac{1}{2}\right]\sin 70^\circ$$

$$= \frac{1}{4}\cos 40^\circ \sin 70^\circ - \frac{1}{8}\sin 70^\circ$$

$$= \frac{1}{8}[2\cos 40^\circ \sin 70^\circ] - \frac{1}{8}\sin 70^\circ$$

$$= \frac{1}{8}[\sin(40^\circ + 70^\circ) - \sin(40^\circ - 70^\circ)] - \frac{1}{8}\sin 70^\circ$$

$$= \frac{1}{8}[\sin 110^\circ - \sin(-30^\circ)] - \frac{1}{8}\sin 70^\circ$$

$$= \frac{1}{8}\sin 110^\circ + \frac{1}{8}\sin 30^\circ - \frac{1}{8}\sin 70^\circ$$

$$= \frac{1}{8}\sin(180^\circ - 70^\circ) + \frac{1}{8}\cdot\frac{1}{2} - \frac{1}{8}\sin 70^\circ$$

$$= \frac{1}{8}\sin 70^\circ + \frac{1}{16} - \frac{1}{8}\sin 70^\circ$$

$$= \frac{1}{16}$$

$$= \text{RHS}$$

Example 3.59

Evaluate $\cos \frac{\pi}{5}$. *[UPTU 2002]*

SOLUTION : Let $\theta = \frac{\pi}{5} = 36^\circ$

$2\theta = 72^\circ$ and $3\theta = 108^\circ$

Now, $2\theta = 180^\circ - 3\theta$

$\sin 2\theta = \sin (180^\circ - 3\theta) = \sin 3\theta$

$2 \sin \theta \cos \theta = 3 \sin \theta - 4 \sin^3 \theta$

As $\sin 36^\circ \neq 0$, dividing both sides by $\sin \theta$, we get

$2 \cos \theta = 3 - 4 \sin^2 \theta$

$\Rightarrow \quad 4 \sin^2 \theta + 2 \cos \theta - 3 = 0$

$\Rightarrow \quad 4 (1 - \cos^2 \theta) + 2 \cos \theta - 3 = 0$

$\Rightarrow \quad 4 \cos^2 \theta - 2 \cos \theta - 1 = 0$

$\Rightarrow \quad \cos \theta = \frac{2 \pm \sqrt{4 + 16}}{8} = \frac{2 \pm \sqrt{20}}{8} = \frac{2 \pm 2\sqrt{5}}{8}$

$= \frac{1 \pm \sqrt{5}}{4}$

$\therefore \quad \cos 36^\circ = \frac{1 \pm \sqrt{5}}{4}$

Example 3.60

Prove that $\frac{\tan \theta + \sec \theta - 1}{\tan \theta - \sec \theta + 1} = \frac{1 + \sin \theta}{\cos \theta}$ *[UPTU 2003]*

SOLUTION : LHS $= \frac{\tan \theta + \sec \theta - 1}{\tan \theta - \sec \theta + 1}$

$= \frac{(\tan \theta + \sec \theta) - (\sec^2 \theta - \tan^2 \theta)}{\tan \theta - \sec \theta + 1}$

$= \frac{(\tan \theta + \sec \theta)(1 - \sec \theta + \tan \theta)}{\tan \theta - \sec \theta + 1}$

$= \tan \theta + \sec \theta$

$= \frac{\sin \theta}{\cos \theta} + \frac{1}{\cos \theta} = \frac{1 + \sin \theta}{\cos \theta}$

= RHS

Example 3.61

If $A + B + C = \frac{\pi}{2}$, prove that

$\cos^2 A + \cos^2 B + \cos^2 C = 2 + 2 \sin A \sin B \sin C$ *[UPTU 2003, 05]*

SOLUTION : LHS $= \frac{1}{2} [(1 + \cos 2A) + (1 + \cos 2B) + (1 + \cos 2C)]$

$$= \frac{3}{2} + \frac{1}{2} [\cos 2A + \cos 2B + \cos 2C]$$

$$= \frac{3}{2} + \frac{1}{2} [2 \cos (A + B) \cdot \cos (A - B) + (1 - 2 \sin^2 C)]$$

$$= \frac{3}{2} + \cos (A + B) \cos (A - B) + \frac{1}{2} - \sin^2 C$$

$$= 2 + \cos\left(\frac{\pi}{2} - C\right) \cos (A - B) - \sin^2 C$$

$$= 2 + \sin C \cos (A - B) - \sin^2 C$$

$$= 2 + \sin C [\cos (A - B) - \sin C]$$

$$= 2 + \sin C \left[\cos (A - B) - \sin\left(\frac{\pi}{2} - (A + B)\right)\right]$$

$$= 2 + \sin C [\cos (A - B) - \cos (A + B)]$$

$$= 2 + \sin C[2 \sin A \sin B]$$

$$= 2 + 2 \sin A \sin B \sin C$$

$$= \text{RHS}$$

Example 3.62

Prove that $\tan 13A - \tan 9A - \tan 4A = \tan 13A \tan 9A \tan 4A$ *[UPTU 2004]*

SOLUTION : We know that

$$13A = 9A + 4A$$

$$\Rightarrow \quad \tan 13A = \tan (9A + 4A)$$

$$= \frac{\tan 9A + \tan 4A}{1 - \tan 9A \tan 4A}$$

$$\Rightarrow \quad \tan 13A (1 - \tan 9A \tan 4A) = \tan 9A + \tan 4A$$

$$\Rightarrow \quad \tan 13A - \tan 13A \tan 9A \tan 4A = \tan 9A + \tan 4A$$

$$\Rightarrow \quad \tan 13A - \tan 9A - \tan 4A = \tan 13A \tan 9A \tan 4A$$

PROBLEM SET 3.6

Prove that :

1. $\sin \theta \sin 2\theta + \sin 3\theta \sin 6\theta - \sin 4\theta \sin 5\theta = 0$

2. $\sin 20° \sin 40° \sin 60° \sin 80° = \frac{3}{16}$ **[UPTU 2003, 05]**

3. $\frac{\sin \theta + \sin 3\theta}{\cos \theta + \cos 3\theta} = \tan 2\theta$

4. $\frac{2 \sin (A - C) \cos C - \sin (A - 2C)}{2 \sin (B - C) \cos C - \sin (B - 2C)} = \frac{\sin A}{\sin B}$

5. $\cos 140° + \cos 20° + \cos 100° = 0$

6. $\cos 10° + \cos 20° + \cos 40° + \cos 50° = 2\sqrt{3}\cos 15° \cos 5°$

7. $\dfrac{1-\cos 2A}{1+\cos 2A} = \tan^2 A$

8. $\dfrac{1+\sin 2A+\cos 2A}{1+\sin 2A-\cos 2A} = \cot A$

9. $\dfrac{\cos A}{1+\sin A} = \tan\left(\dfrac{\pi}{4}-\dfrac{A}{2}\right)$

10. $\cos 6A = 32\cos^6 A - 48\cos^4 A + 18\cos^2 A - 1$

11. $\sin A\cdot\sin(60°-A)\sin(60°+A) = \dfrac{1}{4}\sin 3A$

12. $\cos 4A = 1 - 8\cos^2 A + 8\cos^4 A$

13. $\cos 20° \cos 40° \cos 80° = \dfrac{1}{8}$ **[UPTU 2004]**

14. If $\cos A + \cos B = \dfrac{1}{3}$ and $\sin A + \sin B = \dfrac{1}{4}$, Prove that

$$\tan\left(\frac{A+B}{2}\right) = \frac{3}{4}$$

Prove that :

15. $\tan\left(\dfrac{\pi}{4}+\dfrac{A}{2}\right)+\tan\left(\dfrac{\pi}{4}-\dfrac{A}{2}\right) = 2\sec A$

16. $4\sin A\sin(60°-A)\sin(60°+A) = \sin 3A$

17. $\tan^2 A - \sin^2 A = \tan^2 A \sin^2 A$

18. $\tan 20° \tan 40° \tan 80° = \sqrt{3}$

19. $\dfrac{\sin 6A\cos 3A - \sin 8A\cos A}{\sin 3A\sin 4A - \cos 2A\cos A} = \tan 2A$

20. Prove that $\dfrac{\sin(A-B)}{\sin A\sin B}+\dfrac{\sin(B-C)}{\sin B\sin C}+\dfrac{\sin(C-A)}{\sin C\sin A} = 0$

21. Prove that $\dfrac{\cos 29° + \sin 29°}{\cos 29° - \sin 29°} = \tan 74°$

22. If A, B, C are angles of a triangle, show that

(i) $\sin(B+C)+\sin(C+A)+\sin(A+B) = \sin A + \sin B + \sin C$

(ii) $\cos\dfrac{A}{2}+\cos\dfrac{B}{2}+\cos\dfrac{C}{2} = 4\cos\dfrac{\pi-A}{4}\cos\dfrac{\pi-B}{4}\cdot\cos\dfrac{\pi-C}{4}$

(iii) $\sin\dfrac{A}{2}+\sin\dfrac{B}{2}+\sin\dfrac{C}{2} = 1+4\sin\dfrac{\pi-A}{4}\sin\dfrac{\pi-B}{4}\cdot\sin\dfrac{\pi-C}{4}$

(iv) $\dfrac{1-\cos A+\cos B+\cos C}{1-\cos C+\cos A+\cos B} = \tan\dfrac{A}{2}\cdot\cot\dfrac{C}{2}$

(v) $\dfrac{\cot A+\cot B}{\tan A+\tan B}+\dfrac{\cot B+\cot C}{\tan B+\tan C}+\dfrac{\cot C+\cot A}{\tan C+\tan A} = 1$

(vi) $\sin^2 \frac{A}{2} + \sin^2 \frac{B}{2} - \sin^2 \frac{C}{2} = 1 - 2 \cos \frac{A}{2} \sin \frac{B}{2} \sin \frac{C}{2}$

(vii) $\frac{\sin 2A + \sin 2B + \sin 2C}{\sin A + \sin B + \sin C} = 8 \sin \frac{A}{2} \sin \frac{B}{2} \cdot \sin \frac{C}{2}$

(viii) $\sin A + \sin B - \sin C = 4 \sin \frac{A}{2} \sin \frac{B}{2} \cos \frac{C}{2}$

(ix) $\frac{\cos A}{\sin B \sin C} + \frac{\cos B}{\sin A \sin C} + \frac{\cos C}{\sin A \sin B} = 2$

(x) $\cos 2A + \cos 2B - \cos 2C = 1 - 4 \sin A \sin B \sin C$

23. If $A + B + C = \frac{\pi}{2}$, prove that

(i) $\sin^2 A + \sin^2 B + \sin^2 C = 1 - 2 \sin A \sin B \sin C$

(ii) $\cos^2 A + \cos^2 B + \cos^2 C = 2 + 2 \sin A \sin B \sin C$

24. If $\tan \theta = \frac{\tan \alpha + \tan \beta}{1 + \tan \alpha \tan \beta}$, prove that

$$\sin 2\theta = \frac{\sin 2\alpha + \sin 2\beta}{1 + \sin 2\alpha \sin 2\beta}$$

25. Prove that $\sin 12^\circ \sin 48^\circ \sin 54^\circ = \frac{1}{8}$

26. Prove that $\cos 12^\circ \cos 24^\circ \cos 48^\circ \cos 96^\circ = -\frac{1}{16}$

27. If $2 \cos \theta = x + \frac{1}{x}$, prove that $2 \cos 3\theta = x^3 + \frac{1}{x^3}$

28. Find the value of $\cos 22\frac{1}{2}^\circ$, $\sin 22\frac{1}{2}^\circ$ and $\cot 22\frac{1}{2}^\circ$.

29. Prove that $\tan\left(\frac{\pi}{6} + \frac{\theta}{2}\right) \cdot \tan\left(\frac{\pi}{6} - \frac{\theta}{2}\right) = \frac{2 \cos \theta - 1}{2 \cos \theta + 1}$

30. Prove that

(i) $\frac{1 + \cos \theta + \sin \theta}{1 - \cos \theta + \sin \theta} = \cot \frac{\theta}{2}$

(ii) $\tan^2\left(\frac{\pi}{4} + \frac{\theta}{2}\right) = \frac{1 + \sin \theta}{1 - \sin \theta}$

(iii) $\cos^2 \frac{\pi}{16} + \cos^2 \frac{3\pi}{16} + \cos^2 \frac{5\pi}{16} + \cos^2 \frac{7\pi}{16} = 2$

(iv) $\sin^2 \frac{\pi}{8} + \sin^2 \frac{3\pi}{8} + \sin^2 \frac{5\pi}{8} + \sin^2 \frac{7\pi}{8} = 2$

(v) $\sin 3A + \sin 2A - \sin A = 4 \sin A \cos \frac{A}{2} \cos \frac{3A}{2}$

MULTIPLE CHOICE QUESTIONS

1. The radian measure of 45° is

(a) $\frac{\pi}{4}$ (b) $\frac{\pi}{3}$

(c) $\frac{\pi}{6}$ (d) $\frac{\pi}{2}$

2. The radian measure of 12°26′30″ is

(a) 2.3 (b) 3.1

(c) 0.023 (d) 0.069

3. Which of following is true

(a) 1 radian $= \frac{180^\circ}{\pi}$ (b) 1 radian $= \frac{90^\circ}{\pi}$

(c) 1 radian $= \frac{145^\circ}{\pi}$ (d) None of these

4. Which of the following is true?

(a) Angle $= \frac{\text{radius}}{\text{arc}}$ (b) angle $= \frac{\text{arc}}{\text{radius}}$

(c) arc $= \frac{\text{angle}}{\text{radius}}$ (d) None of these

5. The area of a sector of a circle with radius 6 m bounded by an arc of length 12 m is

(a) 20 m^2 (b) 12 m^2

(c) 10 m^2 (d) 36 m^2

6. If the rotation is anticlockwise, the angle measured is

(a) negative (b) positive

(c) not affected by rotation (d) None of these

7. In a right angled triangle

(a) $\sin\theta = \frac{\text{Base}}{\text{Hypoteneous}}$ (b) $\sin\theta = \frac{\text{Perpendicular}}{\text{Hypotenuse}}$

(c) $\cos\theta = \frac{\text{Base}}{\text{Hypotenuse}}$ (d) Both (b) and (c)

8. In a right angled triangel OBA, $\angle AOB = \theta$ and OB (base) = 15, BA (perpendicular) = 10 and OA (Hypotenuse) = 18. The value of $\tan\theta$ is

(a) $\frac{2}{3}$ (b) $\frac{3}{5}$

(c) $\frac{5}{9}$ (d) $\frac{5}{6}$

9. In a right angled triangle which of the following is true?

(a) $\cot\theta = \frac{\cos\theta}{\sin\theta}$ (b) $\tan\theta = \frac{\sin\theta}{\cos\theta}$

(c) Both (a) and (b) (d) None of these

10. Which of the following is incorrect?

(a) $\sin^2\theta + \cos^2\theta = 1$ (b) $\sec^2\theta - \tan^2\theta = 1$

(c) $\operatorname{cosec}^2\theta - \cot^2\theta = 1$ (d) $\sin^2\theta - \cos^2\theta = 0$

11. If $\tan\theta = -\frac{4}{3}$ then $\sin\theta$ is

(a) $-\frac{4}{5}$ but not $\frac{4}{5}$ (b) $-\frac{4}{5}$ or $\frac{4}{5}$

(c) $\frac{4}{5}$ but not $-\frac{4}{5}$ (d) None of these

12. Which of the following is correct

(a) $\sin 1^\circ > \sin 1$ (b) $\sin 1^\circ < \sin 1$

(c) $\sin 1^\circ = \sin 1$ (d) $\sin 1^\circ = \frac{\pi}{180}\sin 1$

13. Maximum and minimum value of $a\cos 2\theta + b\sin 2\theta$ are

(a) $\sqrt{a^2 - b^2}$ and $-\sqrt{a^2 + b^2}$ (b) $a + b$ and $a - b$

(c) $\sqrt{a^2 + b^2}$ and $-\sqrt{a^2 - b^2}$ (d) $a^2 + b^2$ and $-(a^2 + b^2)$

14. If the angle α is in the third quadrant and $\tan\alpha = 2$, then $\sin\alpha$ is equal to

(a) $\frac{2}{\sqrt{5}}$ (b) $-\frac{2}{\sqrt{5}}$

(c) $\frac{\sqrt{5}}{2}$ (d) None of these

15. $|\sin x + \cos x|$ is

(a) ≤ 2 (b) $\leq \frac{1}{\sqrt{2}}$

(c) $\leq \sqrt{2}$ (d) $\geq \sqrt{2}$

16. Which of the following is true

(a) $(1 - \cos^2\theta)(1 + \tan^2\theta) = \tan^2\theta$ (b) $\frac{1 - \sin^2\theta}{1 - \cos^2\theta} = \cot\theta$

(c) $\frac{1 + \cot\theta}{\tan\theta} + \frac{\tan\theta}{1 + \cot\theta} = \operatorname{cosec}\theta$ (d) None of these

17. If $\sec\theta = -\frac{2}{3}$, in which quadrant θ, lies?

(a) First (b) Second or third

(c) First or fourth (d) None of these

18. If $\sin\theta = \frac{3}{5}$, what is the value of $\frac{2\cos\theta}{1 - \sin\theta}$?

(a) 2 (b) 16

(c) 12 (d) 4

19. The value of $\sin 135^\circ$ is

(a) $\frac{1}{\sqrt{2}}$ (b) $\frac{1}{2}$

(c) $\frac{\sqrt{3}}{2}$ (d) $-\frac{1}{\sqrt{2}}$

20. The value of $\tan \frac{15\pi}{4}$ is

(a) 1
(b) -1
(c) $\frac{1}{\sqrt{3}}$
(d) $\sqrt{3}$

21. The expression for $\tan 3\theta$ is

(a) $\frac{\tan\theta - 3\tan^3\theta}{3 - \tan^2\theta}$
(b) $\frac{3\tan\theta - \tan^3\theta}{3 - 2\tan^2\theta}$
(c) $\frac{3\tan\theta - \tan^3\theta}{1 - 3\tan^2\theta}$
(d) None of these

22. In a triangle ABC, $\sin A - \cos B = \cos C$, then angle B is

(a) $\frac{\pi}{2}$
(b) $\frac{\pi}{3}$
(c) $\frac{\pi}{4}$
(d) $\frac{\pi}{6}$

23. If θ lies in the first quadrant, which of the following is not true

(a) $\frac{\theta}{2} < \tan\frac{\theta}{2}$
(b) $\sin\frac{\theta}{2} > \frac{\theta}{2}$
(c) $\theta \sin\frac{\theta}{2} < 2\sin\frac{\theta}{2}$
(d) $\theta\cos^2\frac{\theta}{2} < \sin\theta$

24. Minimum value of $5\cos\theta + 3\cos\left(\theta + \frac{\pi}{3}\right) + 3$ is

(a) 5
(b) 10
(c) -4
(d) -7

25. The value of $\sqrt{3}\cot 20° - 4\cos 20°$ is

(a) 0
(b) 1
(c) -1
(d) 2

26. If $\frac{\sin(x+y)}{\sin(x-y)} = \frac{a+b}{a-b}$, then $\frac{\tan x}{\tan y}$ is equal to

(a) $\frac{b}{a}$
(b) $\frac{a}{b}$
(c) ab
(d) $-\frac{a}{b}$

27. The value of $\sqrt{3}\operatorname{cosec} 20° - \sec 20°$ is equal to

(a) 2
(b) 1
(c) 4
(d) -4

28. If $\cot A = \frac{3}{4}$, then $32\sin\left(\frac{A}{2}\right)\sin\frac{5A}{2}$ is equal to

(a) 11
(b) 8
(c) 7
(d) 4

29. If $\sin 2\theta = \cos 3\theta$ and θ is an acute angle, then $\sin \theta$ equals

(a) $\frac{1+\sqrt{2}}{4}$ (b) $\frac{\sqrt{5}-1}{4}$

(c) $\frac{\sqrt{5}+1}{4}$ (d) $\frac{\sqrt{3}-1}{4}$

30. The value of $\log \tan 1° + \log \tan 2° + ... + \log \tan 89°$ is

(a) 0 (b) −1

(c) ∞ (d) 1

31. The value of $\cos \frac{2\pi}{15} \cos \frac{4\pi}{15} \cos \frac{8\pi}{15} \cos \frac{14\pi}{15}$ is

(a) $\frac{1}{2}$ (b) $\frac{1}{4}$

(c) $\frac{1}{8}$ (d) $\frac{1}{16}$

32. The value of $\sin 3\theta$ is

(a) $3 \sin \theta - 4 \sin^3 \theta$ (b) $4 \sin^3 \theta - 3 \sin \theta$

(c) $3 \sin \theta + 4 \sin^3 \theta$ (d) None of these

33. The value of $\sin 10° + \sin 20° + \sin 30° + ... + \sin 36°$ is

(a) 1 (b) −1

(c) 0 (d) 2

34. The value of $e^{\log_{10} \tan 1° + \log_{10} \tan 2° + \log_{10} \tan 3° + ... + \log_{10} \tan 89°}$ is

(a) 0 (b) 1

(c) e (d) $\frac{1}{e}$

35. The value of $\cot 36° \cot 72°$ is

(a) $\frac{1}{5}$ (b) $\frac{1}{\sqrt{5}}$

(c) 1 (d) 2

36. The value of $\sin 12° \sin 48° \sin 54°$ is

(a) $\frac{1}{4}$ (b) $\frac{1}{8}$

(c) $\frac{1}{16}$ (d) None of these

37. The value of $\sin \frac{\pi}{7} + \sin \frac{2\pi}{7} + \sin \frac{3\pi}{7}$ is

(a) $\cot \frac{\pi}{14}$ (b) $\frac{1}{2} \cot \frac{\pi}{14}$

(c) $\tan \frac{\pi}{14}$ (d) $\frac{1}{2} \tan \frac{\pi}{14}$

38. If $\sin A = \frac{336}{625}$, where $450° < A < 540°$, then $\sin \frac{A}{4}$ is equal to

(a) $\frac{3}{5}$ (b) $-\frac{3}{5}$

(c) $\frac{4}{5}$ (d) $-\frac{4}{5}$

39. $\tan 9^\circ - \tan 27^\circ - \tan 63^\circ + \tan 81^\circ$ is equal to

(a) 0 (b) 4

(c) –1 (d) 1

40. If $\cos\theta = \frac{3}{4}$, then the value of $\sin\frac{\theta}{2}\sin\frac{5\theta}{2}$ is

(a) $\frac{1}{32}$ (b) $\frac{11}{8}$

(c) $\frac{11}{32}$ (d) $\frac{11}{16}$

41. The minimum value of $9\tan^2\theta + 4\cot^2\theta$ is

(a) 13 (b) 9

(c) 6 (d) 12

42. The expression $\text{cosec}^2\theta\cot^2\theta - \sec^2\theta\tan^2\theta - (\cot^2\theta - \tan^2\theta)(\sec^2\theta\,\text{cosec}^2\theta - 1)$ is equal to

(a) 1 (b) –1

(c) 0 (d) none of these

43. If $\cos(A - B) = \frac{3}{5}$ and $\tan A\tan B = 2$, then

(a) $\cos A\cos B = \frac{1}{5}$ (b) $\sin A\sin B = -\frac{2}{5}$

(c) $\cos(A + B) = -\frac{1}{5}$ (d) none of these

44. The value of $\frac{(3 + \cot 76^\circ\cot 16^\circ)}{(\cot 76^\circ + \cot 16^\circ)}$ is equal to

(a) $\cot 44^\circ$ (b) $\tan 44^\circ$

(c) $\tan 2^\circ$ (d) $\cot 40^\circ$

45. If $\sin A = \sin B$ and $\cos A = \cos B$, then

(a) $\sin\left(\frac{A+B}{2}\right) = 0$ (b) $\cos\left(\frac{A+B}{2}\right) = 0$

(c) $\sin\left(\frac{A-B}{2}\right) = 0$ (d) $\cos\left(\frac{A-B}{2}\right) = 0$

46. The value of $\tan 5A\tan 3A\tan 2A$ is equal to

(a) $\tan 5A - \tan 3A - \tan 2A$ (b) $\frac{\sin 5A - \sin 3A - \sin 2A}{\cos 5A - \cos 3A - \cos 2A}$

(c) 0 (d) none of these

47. The value of $\sin 12^\circ\sin 24^\circ\sin 48^\circ\sin 84^\circ$ is equal to

(a) $\cos 20^\circ\cos 40^\circ\cos 60^\circ\cos 80^\circ$ (b) $\sin 20^\circ\sin 40^\circ\sin 60^\circ\sin 80^\circ$

(c) $\frac{1}{16}$ (d) None of these

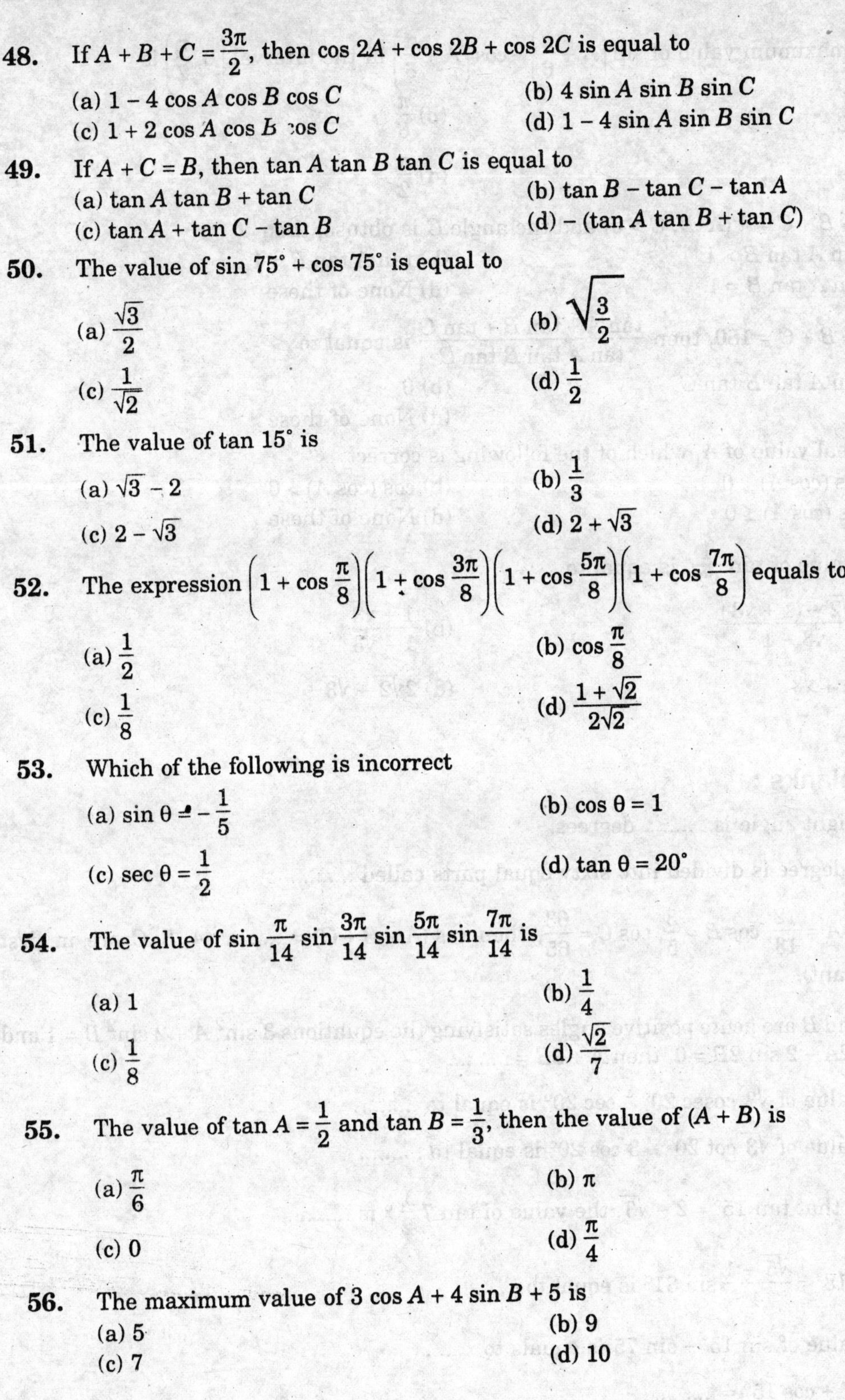

48. If $A + B + C = \frac{3\pi}{2}$, then $\cos 2A + \cos 2B + \cos 2C$ is equal to

(a) $1 - 4 \cos A \cos B \cos C$ (b) $4 \sin A \sin B \sin C$

(c) $1 + 2 \cos A \cos B \cos C$ (d) $1 - 4 \sin A \sin B \sin C$

49. If $A + C = B$, then $\tan A \tan B \tan C$ is equal to

(a) $\tan A \tan B + \tan C$ (b) $\tan B - \tan C - \tan A$

(c) $\tan A + \tan C - \tan B$ (d) $-(\tan A \tan B + \tan C)$

50. The value of $\sin 75^\circ + \cos 75^\circ$ is equal to

(a) $\frac{\sqrt{3}}{2}$ (b) $\sqrt{\frac{3}{2}}$

(c) $\frac{1}{\sqrt{2}}$ (d) $\frac{1}{2}$

51. The value of $\tan 15^\circ$ is

(a) $\sqrt{3} - 2$ (b) $\frac{1}{3}$

(c) $2 - \sqrt{3}$ (d) $2 + \sqrt{3}$

52. The expression $\left(1 + \cos \frac{\pi}{8}\right)\left(1 + \cos \frac{3\pi}{8}\right)\left(1 + \cos \frac{5\pi}{8}\right)\left(1 + \cos \frac{7\pi}{8}\right)$ equals to

(a) $\frac{1}{2}$ (b) $\cos \frac{\pi}{8}$

(c) $\frac{1}{8}$ (d) $\frac{1 + \sqrt{2}}{2\sqrt{2}}$

53. Which of the following is incorrect

(a) $\sin \theta = -\frac{1}{5}$ (b) $\cos \theta = 1$

(c) $\sec \theta = \frac{1}{2}$ (d) $\tan \theta = 20^\circ$

54. The value of $\sin \frac{\pi}{14} \sin \frac{3\pi}{14} \sin \frac{5\pi}{14} \sin \frac{7\pi}{14}$ is

(a) 1 (b) $\frac{1}{4}$

(c) $\frac{1}{8}$ (d) $\frac{\sqrt{2}}{7}$

55. The value of $\tan A = \frac{1}{2}$ and $\tan B = \frac{1}{3}$, then the value of $(A + B)$ is

(a) $\frac{\pi}{6}$ (b) π

(c) 0 (d) $\frac{\pi}{4}$

56. The maximum value of $3 \cos A + 4 \sin B + 5$ is

(a) 5 (b) 9

(c) 7 (d) 10

57. The maximum value of $\sin\left(A+\frac{\pi}{6}\right)+\cos\left(A+\frac{\pi}{6}\right)$ in the interval $\left(0, \frac{\pi}{2}\right)$ is

(a) $\frac{\pi}{12}$ (b) $\frac{\pi}{6}$

(c) $\frac{\pi}{3}$ (d) $\frac{\pi}{2}$

58. If $A+B+C=\pi$ $(A, B, C>0)$ and the angle C is obtuse, then

(a) $\tan A \tan B > 1$ (b) $\tan A \tan B < 1$

(c) $\tan A \tan B = 1$ (d) None of these

59. If $A+B+C=180$, then $\frac{\tan A+\tan B+\tan C}{\tan A \tan B \tan C}$ is equal to

(a) $\tan A \tan B \tan C$ (b) 0

(c) 1 (d) None of these

60. For real value of A, which of the following is correct

(a) $\cos(\cos A)<0$ (b) $\cos(\cos A)>0$

(c) $\cos(\cos A)\leq 0$ (d) None of these

61. The value of $\tan \frac{1}{\sqrt{2}}^{\circ}$ is equal to

(a) $\frac{2\sqrt{2}-(1+\sqrt{3})}{\sqrt{3}-1}$ (b) $\frac{1+\sqrt{3}}{1-\sqrt{3}}$

(c) $\frac{1}{\sqrt{3}}+\sqrt{3}$ (d) $2\sqrt{2}+\sqrt{3}$

Fill in the blanks :

1. One right angle is .... degrees.
2. Each degree is divided into sixty equal parts called ....
3. If $\cos A=\frac{12}{13}$, $\cos B=\frac{3}{5}$, $\cos C=\frac{63}{65}$, then $\cos(A+B+C)=$ .... $(A, B, C$ are in first quadrant).
4. If A and B are acute positive angles satisfying the equations $3\sin^2 A+2\sin^2 B=1$ and $3\sin 2A-2\sin 2B=0$, then $A+2B=$ ....
5. The value of $\sqrt{3}$ cosec $20^\circ-\sec 20^\circ$ is equal to ....
6. The value of $\sqrt{3}\cot 20^\circ-3\cos 20^\circ$ is equal to ....
7. Given that $\tan 15^\circ=2-\sqrt{3}$, the value of $\tan 7\frac{1}{2}^\circ$ is ....
8. If $\sin 18^\circ=\frac{\sqrt{5}-1}{4}$ $\sin 81^\circ$ is equal to ....
9. The value of $\sin 15^\circ+\sin 75^\circ$is equals to ....
10. $\sin 15^\circ+\cos 15^\circ=$ ....

11. The value of $\dfrac{\sin 70° + \cos 40°}{\cos 70° + \sin 40°}$ = ...

12. If $\tan A = \dfrac{\cos 9° + \sin 9°}{\cos 9° - \sin 9°}$, then A is equal to ....

13. If $A + B + C = 720°$, then $\tan A + \tan B + \tan C$ is equal to ....

14. $\sin 12° \sin 48° \sin 54°$ is equal to ....

15. The value of $\tan 20° + \tan 40° + \tan 60° +$ .... $+ \tan 180°$ is ....

16. If $\cos A = \dfrac{3}{4}$, then $32 \sin \dfrac{A}{2} \cos \dfrac{5}{2} A$ is equal to ....

17. If $\tan A = \dfrac{1 - \cos B}{\sin B}$, then $\tan 2A$ is equal to ....

18. The value of $4 \sin A \cos^3 A - 4 \cos A \sin^3 A$ is equal to ....

True and False :

1. There exists a value of A between 0 and 2π, which satisfy the equation $\sin^4 A - 2 \sin^2 A - 1 = 0$

2. The expression $\text{cosec}^2 A \cot^2 A - \sec^2 A \tan^2 A - (\cot^2 A - \tan^2 A)(\sec^2 A \,\text{cosec}^2 A - 1)$ is independent of A.

3. $\sin 2x \cdot (\tan x + \cot x) = 2$

4. $\sin A + \cos A = 2$

5. $2 \sin A - 3 \cos A = 4$

6. If $\tan A = \dfrac{1 - \cos B}{\sin B}$, then $\tan 2A = \tan B$

7. $\dfrac{\tan 3x}{\tan x}$ never lies between $\dfrac{1}{3}$ and 3.

8. If $\sin A = \sin B$, then $\sin 2A = \sin 2B$ always.

9. The value of $\cos \dfrac{\pi}{7} \cos \dfrac{4\pi}{7} \cos \dfrac{5\pi}{7}$ is $\dfrac{1}{2}$.

10. $x = 0$ is only the value for which $2^x + 2^{-x} = 2 \cos \left(\dfrac{x}{2}\right)$.

11. If $\sin x + \sin y = \sqrt{3}(\cos y - \cos x)$, then $\sin 3y + \sin 3x = 0$.

12. $2^{\sin^2 x} = \sin x$ does not hold good for any real x.

13. $4^{\sin^2 x} + 4^{\cos^2 x} \geq 4$ for all real values of x.

14. sin 2 is greater than cos 3.

Answers

Objective type questions :

1. a	**2.** d	**3.** a	**4.** b	**5.** d	**6.** b	**7.** d	**8.** a	**9.** c	**10.** d
11. b	**12.** b	**13.** c	**14.** b	**15.** c	**16.** a	**17.** b	**18.** d	**19.** a	**20.** b
21. c	**22.** a	**23.** b	**24.** c	**25.** b	**26.** b	**27.** c	**28.** a	**29.** b	**30.** a
31. d	**32.** a	**33.** c	**34.** b	**35.** b	**36.** b	**37.** b	**38.** c	**39.** b	**40.** c
41. d	**42.** c	**43.** a	**44.** a	**45.** c	**46.** a	**47.** c	**48.** d	**49.** b	**50.** b
51. c	**52.** c	**53.** c	**54.** c	**55.** d	**56.** d	**57.** a	**58.** b	**59.** c	**60.** b
61. a									

Fill in the blanks :

1. 90, **2.** minutes, **3.** 0, **4.** $\frac{\pi}{2}$, **5.** 4, **6.** 1, **7.** $\sqrt{6} + \sqrt{2} - 2 - \sqrt{3}$

8. $\frac{\sqrt{3} + 5\sqrt{5 - \sqrt{5}}}{4}$, **9.** $\frac{\sqrt{3}}{2}$, **10.** $\frac{\sqrt{3}}{2}$, **11.** $\sqrt{3}$ **12.** 54°, **13.** $\tan A \tan B \tan C$,

14. $\frac{1}{8}$, **15.** 0, **16.** $-\sqrt{7}$, **17.** $\tan B$, **18.** $\sin 4A$

True and False :

1. F, **2.** T, **3.** T, **4.** F, **5.** F, **6.** T, **7.** T, **8.** F, **9.** F, **10.** T

11. T, **12.** T, **13.** T, **14.** T

UNIT - IV

Analytical Plain Geometry

4.1 INTRODUCTION

Analytical geometry was invented by the French mathematician and philosopher Rene Descartes in 1637, when he published his book *La Geometric* in which he used algebra in the study of geometry. Prior to this, mathematician confined themselves to Euclidean geometry. In fact co-ordinate geometry is that branch of geometry in which two numbers, called co-ordinates are used to indicate the position of a point in a plane and the study of geometric figures is made with the help of algebraic methods.

4.2 CARTESIAN CO-ORDINATE SYSTEM

The position of a point in a plane is fixed by selecting the axes of reference, which is formed by combining two mutually perpendicular lines $X'OX$ and $Y'OY$ through a point O in the plane. The point O is called the origin. The horizontal axis $X'OX$ is called the x-axis and the vertical axis $Y'OY$ is called the as y-axis. The two lines taken together are called the co-ordinate axes or the axes of coordinate.

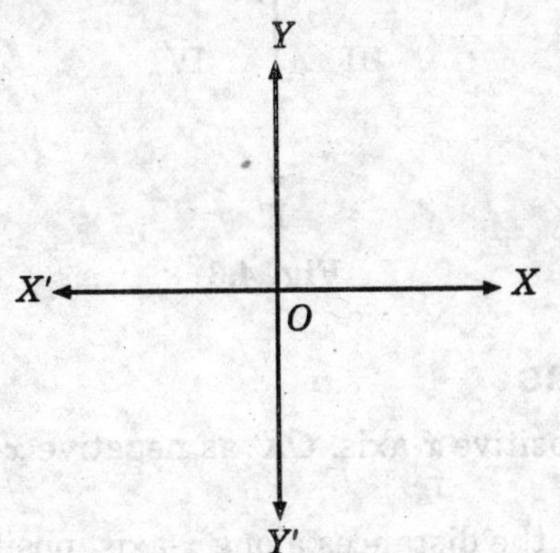

Fig. 4.1

4.2.1 Co-ordinate of a Point

The position of a point P of the plane is determined with reference to the rectangular axes $X'OX$ and $Y'OY$ by means of an ordered pair of numbers (x, y) called the co-ordinate of the point P.

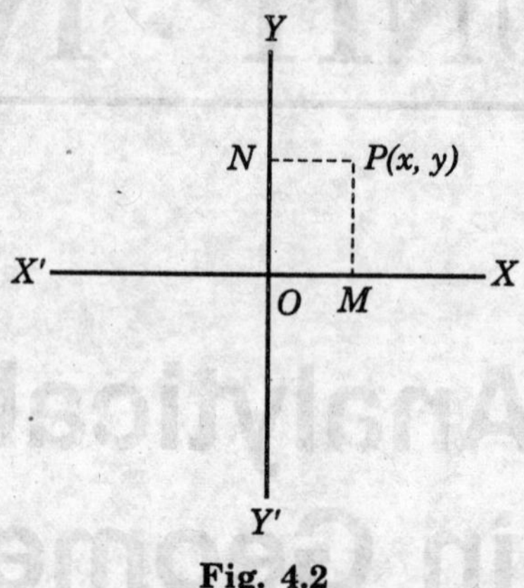

Fig. 4.2

Here, x represents the length of line segment OM, *i.e.*, the distance of point P from y-axis. This distance x is called the x-coordinate or **abscissa** of point P. In a similar manner the length of line segment ON is y, which is the distance of point P from x-axis. This distnace y is called y-coordinate or **ordinate** of point P in XY-plane.

4.2.2 Quadrants

We observe that the two axes $X'OX$ and $Y'OY$ separates the Euclidean plane into four regions called the quadrants. By custom the regions $XOY, X'OY, X'OY'$ and XOY' are numbered as I, II, III and IV quadrants respectively in the counter-clockwise direction as shown in fig. 4.3.

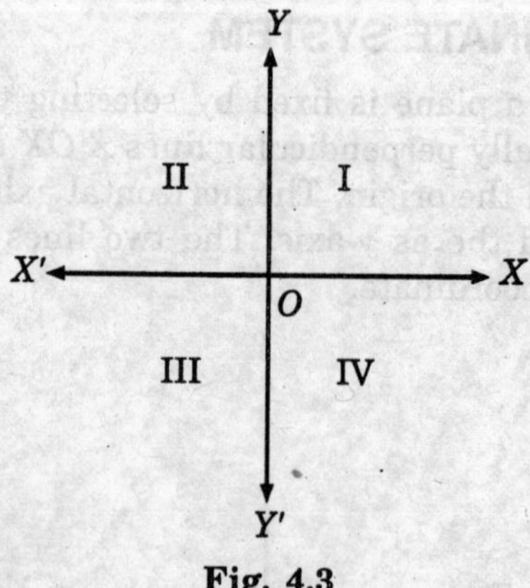

Fig. 4.3

4.2.3 Convention of Signs

The ray OX is taken as positive x-axis, OX' as negative x-axis, OY as positive y-axis and OY' as negative y-axis :

(i) For a point in plane for the distances along x-axis, positive values are measured to the right of the origin and negative values to its left.

(ii) Similarly, for distances along y-axis positive values are measured upward and negative values downward.

In view of the above sign convention the four quadrants are characterised as shown in Fig. 4.4.

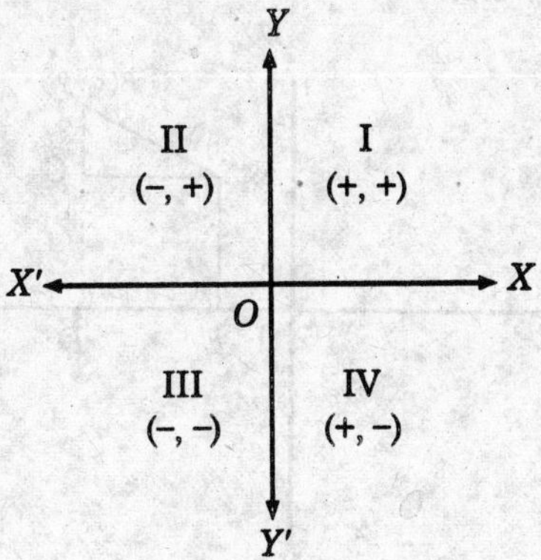

Fig. 4.4

I quadrant : $x > 0,\ y > 0$
II qudrant : $x < 0,\ y > 0$
III qudrant : $x < 0, y < 0$
IV quadrant : $x > 0, y < 0$

The co-ordinates of origin are taken to be (0, 0). The co-ordinates of any point on x-axis and y-axis are of the form $(x, 0)$ and $(0, y)$ respectively.

Example 4.1

In which quadrant do the points (2, 3), (3, – 1) and (– 1 , 4) lie? Plot them on a graph paper.

SOLUTION : In case of the point (2, 3) both abscissa and ordinate are positive so it lies in the first quadrant.

The point (3, – 1) lies in the fourth (IV) quadrant because abscissa is positive and ordinate is negative.

Similarly, the point (– 1, 4) lies in the second (II) quadrants since, abscissa is negative and ordinate is positive.

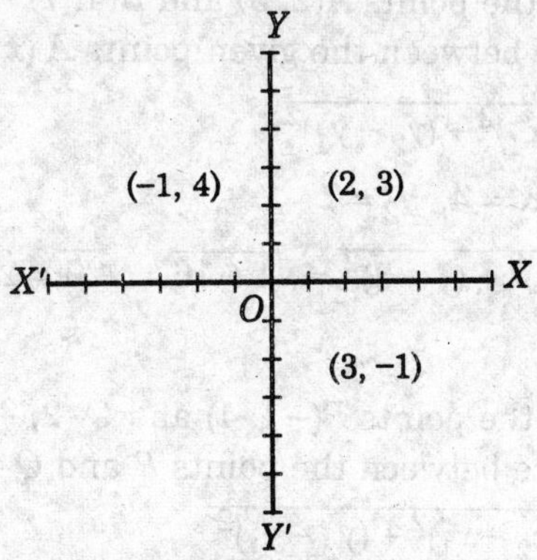

Fig. 4.5

4.3 DISTANCE BETWEEN TWO POINTS

Let (x_1, y_1) and (x_2, y_2) be the co-ordinates of the points A and B respectively in XY plane with origin at O as shown in Fig. 4.6.

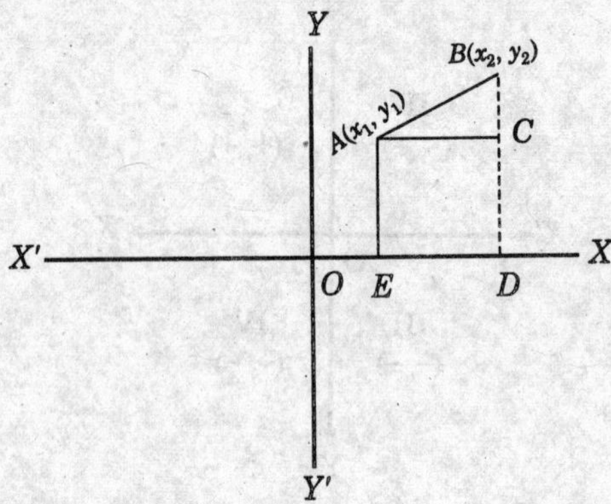

Fig. 4.6

Let AE and BD are perpendiculars drawn from point A and B on the X-axis respectively. Also AC is a perpendicular from A on the line BD.

Then,

$$OE = x_1,\ OD = x_2,\ AE = y_1 \text{ and } BD = y_2$$

$$\therefore \quad AC = ED = OD - OE = x_2 - x_1 \text{ and } BC = BD - CD = y_2 - y_1$$

Sine ABC is a right angled triangle. Thus, with the help of Pythagorous theorem, we get

$$AB^2 = AC^2 + BC^2 = (x_2 - x_1)^2 + (y_2 - y_1)^2$$

$$\Rightarrow \quad AB = \sqrt{(x_2 - x_1)^2 + (y_2 - y_1)^2}$$

In particular, the distance of any point $A(x, y)$ from the origin O (0, 0) can be given as.

$$OA = \sqrt{(x - 0)^2 + (y - 0)^2} = \sqrt{x^2 + y^2}$$

Example 4.2

Find the distance between the points $A(2, 3)$ and $B(4, 7)$.

SOLUTON : The distance between the given points $A\ (x_1, y_1)$ and $B\ (x_2, y_2)$.

$$= \sqrt{(x_2 - x_1)^2 + (y_2 - y_1)^2}$$

Here, $x_1 = 2,\ y_1 = 3$ and $x_2 = 4,\ y_2 = 7$

$$\therefore \quad AB = \sqrt{(4 - 2)^2 + (7 - 3)^2} = \sqrt{4 + 16} = \sqrt{20} = 2\sqrt{5}.$$

Example 4.3

Find the distance between the points $P(-1, 4)$ and $Q\ (2, -3)$.

SOLUTION : The distance between the points P and Q

$$PQ = \sqrt{(x_2 - x_1)^2 + (y_2 - y_1)^2}$$

Here, $x_1 = -1,\ y_1 = 4$ and $x_2 = 2,\ y_2 = -3$

$$\therefore \quad PQ = \sqrt{2 - (-1)^2 + (-3 - 4)^2} = \sqrt{9 + 49} = \sqrt{58}$$

Example 4.4

Compute the length of the line segment joining points P_1 (0, – 1) and P_2 (1, 1).

SOLUTON : As we know that the length of line segment joining points P_1 (0, – 1) and P_2 (1, 1) is equal to the distance between the points.

Thus, length of line segment $P_1P_2 = \sqrt{(1-0)^2 + (1-(-1)^2}$

$$= \sqrt{1+4} = \sqrt{5}$$

Example 4.5

Find the value of a when the distance between the points (3, a) and (4, 1) is $\sqrt{10}$.

[UPTU 2002, 03]

SOLUTION : Using distance formula, we can write

$$(\sqrt{10}) = \sqrt{(4-3)^2 + (1-a)^2} = \sqrt{1+(1-a)^2}$$

On squaring both sides, we get

$$10 = 1 + (1-a)^2$$

$$\Rightarrow \quad (1-a)^2 = 9$$

$$\Rightarrow \quad (1-a) = \pm 3 \Rightarrow a = -2, 4.$$

4.4 SOME IMPORTANT DEFINITIONS

4.4.1 Equidistant Points

A point P is said to be equidistant from points Q and R if the distnace between the points P and Q is equal to the distance between the points P and R.

Example 4.6

Show that the point P(5, 2) is equidistant from Q(8 – 2) and R(2, – 2).

SOLUTION : Distance between the points P(5, 2) and Q(8, – 2) is given by

$$PQ = \sqrt{(8-5)^2 + (-2-2)^2}$$

$$= \sqrt{9+16} = 5$$

Distance between the points P (5, 2) and R (2, – 2) is given by

$$PR = \sqrt{(2-5)^2 + (-2-2)^2}$$

$$= \sqrt{9+16} = 5$$

Since, the distance between points P and Q is equal to the distance between P and R. The point P is equidistant from points Q and R.

4.4.2 Isosceles Triangle

A triangle having two of its sides equal is called an isosceles triangle, *i.e.,* if we have a triangle ABC, then ΔABC will be isosceles, if either $AB = BC$ or $BC = CA$ or $AB = AC$.

4.4.3 Equilateral Triangle

As name suggests, an equilateral triangle is of that kind whose all sides are equal. For a triangle ABC to be equilateral, all its sides AB, BC and CA must be equal, *i.e.,* $AB = BC = CA$.

4.4.4 Particular Cases of a Quadrilateral

A quadrilateral, *i.e.*, a structure in *XY* plane with four sides is said to form

(i) a **rectangle** if its opposite sides are of equal length and the diagonals are also equal in length.

(ii) a **square** if all its sides are of equal length and both the diagonals are also equal.

(iii) a **parallelogram** if its opposite sides are equal in length but both the diagonal are not equal.

(iv) a rhombus if all its sides are equal in length but the diagonals are not of equal length.

4.5 SECTION FORMULAE

4.5.1 Formula for Internal Division

(i) The co-ordinates of the point *R* dividing the line segment joining two points *P* and *Q* internally in the ratio $m_1 : m_2$ are given by

$$\left(\frac{m_1x_2 + m_2x_1}{m_1 + m_2}, \frac{m_1y_2 + m_2y_1}{m_1 + m_2}\right)$$

4.5.2 Formula for External Division

(ii) The co-ordinates of the point *R* dividing the line segment joining two points *P* and *Q* externally in the ratio $m_1 : m_2$ are given by

$$\left(\frac{m_1x_2 - m_2x_1}{m_1 - m_2}, \frac{m_1y_2 - m_2y_1}{m_1 - m_2}\right)$$

4.5.3 Co-ordinates of Middle Point of a Line Segment

The co-ordinates of the middle point *R* of the line segment joining the points $P(x_1, y_1)$ and $Q(x_2, y_2)$ are

$$\left(\frac{x_1 + x_2}{2}, \frac{y_1 + y_2}{2}\right)$$

4.6 CENTROID OF A TRIANGLE

The co-ordinates of the centroid *G* of a trianlge *ABC* whose vertices are $A(x_1, y_1)$, $B(x_2, y_2)$ and $C(x_3, y_3)$ are given as

$$G = \left(\frac{x_1 + x_2 + x_3}{3}, \frac{y_1 + y_2 + y_3}{3}\right),$$

4.7 IN-CENTRE OF A TRIANGLE

The co-ordinates of the in-centre of a triangle whose vertices are (x_1, y_1), (x_2, y_2), (x_3, y_3)

are $\left(\frac{ax_1 + bx_2 + cx_3}{a + b + c}, \frac{ay_1 + by_2 + cy_3}{a + b + c}\right)$, where a, b and c are lengths of the sides *BC*, *CA*, *AB* respectively.

Example 4.7

Show that the following set of points are the vertices of an isosceles triangle

$$(-2, 3), (3, 8), (4, 1)$$

SOLUTION : Let $P(-2, 3)$, $Q(3, 8)$ and $R(4, 1)$ are the given points. Then

$$PQ = \sqrt{(3-(-2))^2 + (8-3)^2} = \sqrt{25+25} = \sqrt{50} = 5\sqrt{2}$$

and $$QR = \sqrt{(4-3)^2 + (1-8)^2} = \sqrt{1+49} = \sqrt{50} = 5\sqrt{2}$$

Thus, we find length $PQ = QR$. Which shows that the points P, Q and R are the vertices of an isosceles triangle.

Example 4.8

Show that the points (a, a), $(-a, -a)$ and $(-\sqrt{3}a, \sqrt{3}a)$ are the vertices of an equilateral triangle.

SOLUTION : Let $A(a, a)$, $B(-a, -a)$ and $C(-\sqrt{3}a, \sqrt{3}a)$ are the given points. Then

$$AB = \sqrt{(-a-a)^2 + (-a-a)^2}$$

$$= \sqrt{4a^2 + 4a^2} = \sqrt{8a^2} = 2\sqrt{2}a$$

$$BC = \sqrt{(-\sqrt{3}a + a)^2 + (\sqrt{3}a + a)^2} = \sqrt{a^2(1-\sqrt{3})^2 + a^2(\sqrt{3}+1)^2}$$

$$= a[\sqrt{(1-\sqrt{3})^2 + (1+\sqrt{3})^2}] = a[\sqrt{1+3-2\sqrt{3}+1+3+2\sqrt{3}}]$$

$$= a[\sqrt{4+4}] = a\sqrt{8} = 2\sqrt{2}a$$

$$AC = \sqrt{(-\sqrt{3}a - a)^2 + (\sqrt{3}a - a)^2} = \sqrt{a^2(\sqrt{3}+1)^2 + a^2(\sqrt{3}-1)^2}$$

$$= a[\sqrt{(\sqrt{3}+1)^2 + (\sqrt{3}-1)^2}] = a[\sqrt{3+1+2\sqrt{3}+3+1-2\sqrt{3}}]$$

$$= a[\sqrt{4+4}] = a\sqrt{8} = 2\sqrt{2}a$$

It is observed that $AB = BC = AC$, Thus, the triangle ABC formed by the given points is an equilateral triangle.

Example 4.9

Show that the points $(0, -1)$, $(-2, 3)$, $(6, 7)$ and $(8, 3)$ are the vertices of a rectangle.

SOLUTION : Let $P(0, -1)$, $Q(-2, 3)$, $R(6, 7)$ and $S(8, 3)$ are the given points.

Then $$PQ = \sqrt{(-2-0)^2 + (3-(-1))^2} = \sqrt{4+16} = 2\sqrt{5}$$

$$QR = \sqrt{(6-(-2))^2 + (7-3)^2} = \sqrt{64+15} = \sqrt{80} = 4\sqrt{5}$$

$$RS = \sqrt{(8-6)^2 + (3-7)^2} = \sqrt{4+16} = \sqrt{20} = 2\sqrt{5}$$

$$SP = \sqrt{(0-8)^2 + (-1-3)^2} = \sqrt{64+16} = \sqrt{80} = 4\sqrt{5}$$

Thus, we find that

$$PQ = RS = 2\sqrt{5} \quad \text{and} \quad QR = SP = 4\sqrt{5}$$

Also the length of diagonals

$$PR = \sqrt{(6-0)^2 + (7-(-1))^2} = \sqrt{36+64} = 10$$

$$QS = \sqrt{(8-(-2))^2 + (3-3)^2} = \sqrt{100+0} = 10$$

Which shows that diagonal PR and QS are also of equal length.

Hence, $PQRS$ is a rectangle.

Example 4.10

Show that the points (0, – 1), (2, 1), (0, 3) and (– 2, 1) are the vertices of a square.

SOLUTION : Let $P(0, -1)$, $Q(2, 1)$, $R(0, 3)$ and $S(-2, 1)$ be the given points

Then $PQ = \sqrt{(2-0)^2 + (1-(-1))^2} = \sqrt{4+4} = \sqrt{8} = 2\sqrt{2}$

$QR = \sqrt{(0-2)^2 + (3-1)^2} = \sqrt{4+4} = \sqrt{8} = 2\sqrt{2}$

$RS = \sqrt{(-2-0)^2 + (1-3)^2} = \sqrt{4+4} = \sqrt{8} = 2\sqrt{2}$

$SP = \sqrt{(0-(-2)^2 + (-1-1)^2} = \sqrt{4+4} = \sqrt{8} = 2\sqrt{2}$

Here we observe that

$PQ = QR = RS = SP = 2\sqrt{2}$

Also the length of diagonals

$PR = \sqrt{0 + (3-(-1))^2} = \sqrt{16} = 4$

$QS = \sqrt{(-2-2)^2 + (1-1)^2} = \sqrt{16} = 4$

Thus, $PR = QS = 4$

Hence, $PQRS$ is a square.

Example 4.11

Show that the points (– 1, 0), (0, 3) (1, 3) and (0, 0) are the vertices of a parallelogram.

SOLUTION : Let $P(-1, 0)$, $Q(0, 3)$, $R(1, 3)$ and $S(0, 0)$ are the given points

Then $PQ = \sqrt{(0-(-1))^2 + (3-0)^2} = \sqrt{1+9} = \sqrt{10}$

$QR = \sqrt{(1-0)^2 + (3-3)^2} = \sqrt{1} = 1$

$RS = \sqrt{(0-1)^2 + (0-3)^2} = \sqrt{10}$

$SP = \sqrt{(-1-0)^2 + (0-0)^2} = \sqrt{1} = 1$

Here we find that

$PQ = RS = \sqrt{10}$ and $QR = SP = 1$

Also the length of diagonals

$PR = \sqrt{(1-(-1))^2 + (3-0)^2}$

$= \sqrt{4+9} = \sqrt{13}$

$QS = \sqrt{(0-0)^2 + (0-3)^2} = \sqrt{9} = 3$

Thus, $PR \neq QS$

Hence, the given points P, Q, R, S are the vertices of a parallelogram.

Example 4.12

Show that the points (0, 5), (– 2, – 2), (5, 0) and (7, 7) form a rhombus.

SOLUTION : Let $P(0, 5)$, $Q(-2, -2)$, $R(5, 0)$ and $S(7, 7)$ be the given points.

Then $PQ = \sqrt{(-2-0)^2 + (-2-5)^2} = \sqrt{4+49} = \sqrt{53}$

$QR = \sqrt{(5-(-2))^2 + (0-(-2))^2} = \sqrt{49+4} = \sqrt{53}$

$RS = \sqrt{(7-5)^2 + (7-0)^2} = \sqrt{4+49} = \sqrt{53}$

$SP = \sqrt{(0-7)^2 + (5-7)^2} = \sqrt{49+4} = \sqrt{53}$

Thus, $PQ = QR = RS = SP = \sqrt{53}$, *i.e.*, all the sides are equal.

Also the diagonals $PR = \sqrt{(5-0)^2 + (0-5)^2} = \sqrt{25+25} = \sqrt{50} = 5\sqrt{2}$

and $QS = \sqrt{(7-(-2))^2 + (7-(-2))^2} = \sqrt{81+81} = \sqrt{162}$

So $PR \neq QS$

Therefore, *PQRS* is a rhombus.

Example 4.13

Find the co-ordinates of a point *R* dividing internally the line segment joining two given points $P(2, 3)$ and $Q(5, 9)$ in the ratio 1 : 2.

SOLUTION : As we know that the co-ordinate of the point *R* dividing *PQ* in the ratio $m_1 : m_2$ (internally)

$$= \left(\frac{m_1x_2 + m_2x_1}{m_1 + m_2}, \frac{m_1y_2 + m_2y_1}{m_1 + m_2}\right)$$

Thus, the co-ordinates of the point *R* dividng $P(0, 3)$ and $Q(5, 9)$ in the ratio 1 : 2 (internally)

$$= \left(\frac{1\times 5 + 2\times 2}{1+2}, \frac{1\times 9 + 2\times 3}{1+2}\right) = \left(\frac{5+4}{3}, \frac{9+6}{3}\right) = (3, 5)$$

Example 4.14

Obtain the co-ordinates of a point *R* dividng the segment joining two points $P(2, 4)$ and $Q(8, 9)$ in the ratio 2 : 3 externally.

SOLUTION : Since, the co-ordinates of a point *R* dividng the segment joining two points $P(x_1, y_1)$ and $Q(x_2, y_2)$ in the ratio $m_1 : m_2$ (externally)

$$= \left(\frac{m_1x_2 - m_2x_1}{m_1 - m_2}, \frac{m_1y_2 - m_2y_1}{m_1 - m_2}\right)$$

Thus, the co-ordinates of the point *R* dividng the points $P(2, 4)$ and $Q(8, 9)$ in the ratio 2 : 3 (externally)

$$= \left(\frac{2\times 8 - 3\times 2}{2-3}, \frac{2\times 3 - 3\times 4}{2-3}\right)$$

$$= \left(\frac{10}{-1}, \frac{6}{-1}\right) = (-10, -6)$$

Example 4.15

Find the co-ordinates of the mid point *R* of the line segment *PQ*, where co-ordinates of *P* and *Q* are $P(3, 5)$ and $Q(9, 7)$ respectively.

SOLUTION : As we know that the co-ordinates of the mid point of the line segment. joining $P(x_1, y_1)$ and $Q(x_2, y_2)$

$$= \left(\frac{x_1 + x_2}{2}, \frac{y_1 + y_2}{2}\right)$$

Thus, the co-ordinates of the mid point of the line segment joining $P(3, 5)$ and $Q(9, 7)$

$$= \left(\frac{3+9}{2}, \frac{5+7}{2}\right) = (6, 6)$$

Example 4.16

A quadrilateral has the vertices at the points (– 4, 2), (2, 6), (8, 5) and (9, – 7). Show that the mid points of the sides of the quadrilateral are the vertices of a parallelogram.

SOLUTION : Let $P(-4, 2)$, $Q(2, 6)$, $R(8, 5)$ and $S(9, -7)$ be the given points. Then

$$\text{Mid point of the side } PQ = \left(\frac{-4+2}{2}, \frac{2+6}{2}\right) = (-1, 4) = A \text{ (say)}$$

$$\text{Mid point of the side } QR = \left(\frac{2+8}{2}, \frac{6+5}{2}\right) = \left(5, \frac{11}{2}\right) = B \text{ (say)}$$

$$\text{Mid point of the side } RS = \left(\frac{8+9}{2}, \frac{5-7}{2}\right) = \left(\frac{17}{2}, -1\right) = C \text{ (say)}$$

$$\text{Mid point of the side } SP = \left(\frac{9-4}{2}, \frac{-7+2}{2}\right) = \left(\frac{5}{2}, -\frac{5}{2}\right) = D \text{ (say)}$$

Now we shall show that these points A, B, C and D are vertices of a parallelogram.

$$AB = \sqrt{(5-(-1))^2 + \left(\frac{11}{2} - 4\right)^2} = \sqrt{36 + \frac{9}{4}}$$

$$= \sqrt{\frac{144+9}{4}} = \sqrt{\frac{153}{4}} = \sqrt{\frac{153}{2}}$$

$$BC = \sqrt{\left(\frac{17}{2} - 5\right)^2 + \left(-1 - \frac{11}{2}\right)^2} = \sqrt{\left(\frac{7}{2}\right)^2 + \left(\frac{-13}{2}\right)^2}$$

$$= \sqrt{\frac{49}{4} + \frac{169}{4}} = \sqrt{\frac{218}{4}} = \sqrt{\frac{218}{2}}$$

$$CD = \sqrt{\left(\frac{5}{2} - \frac{17}{2}\right)^2 + \left(\frac{-5}{2} - (-1)\right)^2} = \sqrt{(-6)^2 + \left(\frac{-3}{2}\right)^2}$$

$$= \sqrt{36 + \frac{9}{4}} = \sqrt{\frac{153}{4}} = \sqrt{\frac{153}{2}}$$

$$DA = \sqrt{\left(-1 - \frac{5}{2}\right)^2 + \left(4 - \left(\frac{-5}{2}\right)\right)^2}$$

$$= \sqrt{\left(\frac{-7}{2}\right)^2 + \left(\frac{13}{2}\right)^2}$$

$$= \sqrt{\frac{218}{4}} = \sqrt{\frac{218}{2}}$$

Thus, $AB = CD$ and $BC = DA$, *i.e.*, opposite sides are equal.

Now, diagonals $AC = \sqrt{\left(\frac{17}{2} - (-1)\right)^2 + (-1-4)^2} = \sqrt{\left(\frac{19}{2}\right)^2 + 25}$

$$= \sqrt{\frac{361}{4} + 25} = \sqrt{\frac{461}{4}} = \sqrt{\frac{461}{2}}$$

and $BD = \sqrt{\left(\frac{5}{2} - 5\right)^2 + \left(\frac{-5}{2} \frac{-11}{2}\right)^2} = \sqrt{\left(\frac{-5}{2}\right)^2 + (-8)^2}$

$$= \sqrt{\frac{24}{4} + 64} = \sqrt{\frac{281}{2}}$$

Since, $AC \neq BD$

Therefore, $ABCD$ is a parallelogram.

Example 4.17

Find the centroid of the triangle ABC whose vertices are (3, – 5), (– 7, 4) and (10, – 2) respectively.

SOLUTION : Let G denote the centroid of ΔABC. Then, the co-ordinates of centroid of the given triangle

$$= \left(\frac{x_1 + x_2 + x_3}{3}, \frac{y_1 + y_2 + y_3}{3}\right)$$

Here, $x_1 = 3,\ x_2 = -7,\ x_3 = 10$ and $y_1 = -5,\ y_2 = 4,\ y_3 = -2$

$$\text{Thus, } G = \left(\frac{3 + (-7) + 10}{3}, \frac{-5 + 4 - 2}{3}\right)$$

$$= \left(\frac{6}{3}, -\frac{3}{3}\right) = (2, -1)$$

Example 4.18

Find the co-ordinates of the circum-centre of the triangle whose vertices are (8, 6), (8, –2) and (2, –2).

SOLUTION : Let $P(8, 6)$, $Q(8, -2)$ and $R(2, -2)$ be the vertices of the given triangle and let $A(x, y)$ be the circum-centre of the triangle. Then

$$AP^2 = AQ^2 = AR^2$$

Now, $AP^2 = AQ^2 \Rightarrow (x-8)^2 + (y-6)^2 = (x-8)^2 + (y+2)^2$

$$\Rightarrow 16y = 32 \Rightarrow y = 2$$

Also, $AQ^2 = AR^2 \Rightarrow (x-8)^2 + (y+2)^2 = (x-2)^2 + (y+2)^2$

$$\Rightarrow 12x = 60 \Rightarrow x = 5$$

So, the co-ordinates of the circum-centre A one (5, 2).

PROBLEM SET 4.1

1. In which quadrant do the following points lie.
(i) $(-4, -6)$ (ii) $(3, -5)$
(iii) $(-1, +5)$ (iv) $(1, 2)$

2. Find the distance between following pair of points
(i) $(-6, 7)$ and $(-1, -5)$
(ii) $(a+b, b+c)$ and $(a-b, c-b)$
(iii) $(a \sin \alpha, -b \cos \alpha)$ and $(-a \cos \alpha, b \sin \alpha)$
(v) $(a, 0)$ and $(0, b)$

3. Find the value of 'a' when the distance between the points $(3, a)$ and $(4, 1)$ is $\sqrt{10}$.

4. Find the values of x and y if the distances of the point (x, y) from $(-3, 0)$ as well as from $(3, 0)$ are 4.

5. The length of a line segment is of 10 units and the co-ordinates of one end point are $(2, -3)$. If the abscissa of the other end is 10, find the ordinate of the other end.

6. Prove that $(2, -2)$, $(-2, 1)$ and $(5, -2)$ are the vertices of a right angled triangle.

7. Two vertices of an isosceles triangle are $(2, 0)$ and $(2, 5)$. Find the third vertex if the length of the equal sides is 3.

8. Prove that the points $(2a, 4a)$, $(2a, 6a)$ and $(2a + \sqrt{3}a, 5a)$ are the vertices of an equilateral triangle.

9. Which point of x-axis is equidistant from $(5, 9)$ and $(-4, 6)$?

10. Show that the points $A(2, -2)$, $B(14, 10)$, $C(11, 13)$ and $D(-1, 1)$ form a rectangle.

11. Show that the points $A(0, 1)$, $B(1, 4)$, $C(4, 3)$ and $D(3, 0)$ are the vertices of a square.

12. Show that the following points are the vertices of a parallelogram.
(i) $(1, -2)$, $(3, 6)$, $(5, 10)$ and $(3, 2)$
(ii) $(-1, 0)$, $(0, 3)$, $(1, 3)$ and $(0, 0)$.

13. Show that the quadrilateral with vertices $(2, -1)$, $(3, -4)$, $(-2, 3)$ and $(-3, -2)$ is a rhombus.

14. Find the co-ordinates of the points which divides the line joining of the point $(8, 9)$ and $(-7, 4)$
(i) internally in the ratio 2 : 3 and
(ii) externally in the ratio 4 : 3.

15. Find the co-ordinates of circum-centre and in-centre of the triangle whose vertices are $(-2, -3)$, $(-1, 0)$, $(7, -6)$.

16. If P and Q are two points whose co-ordinates are $(at^2, 2at)$ and $\left(\frac{a}{t^2}, \frac{2a}{t}\right)$ respectively and S is the point $(a, 0)$. Show that $\frac{1}{SP} + \frac{1}{SQ}$ is independent of t.

ANSWERS

1. (i) III qudrant (ii) IV quadrant (iii) II qudrant (iv) I qudrant

2. (i) 13 (ii) $2\sqrt{2}b$ (iii) $(\sqrt{a^2+b^2})$ (sin α + cos α) (iv) $(\sqrt{a^2+b^2}$

3. 4, – 2 **4.** $x = 0, y = \pm\sqrt{7}$ **5.** 3, – 9 **7.** $\left(2-\frac{\sqrt{11}}{2}, \frac{5}{2}\right), \left(2+\frac{\sqrt{11}}{2}, \frac{5}{2}\right)$

9. $(3, a)$ **14.** (i) (2, 7) (ii) (– 52, – 11) **15.** (3, – 3)

4.8 APPLICATIONS OF DETERMINANTS TO CO-ORDINATE GEOMETRY

4.8.1 Area of A Triangle

We know that the area of triangle whose vertices are (a_1, b_1), (a_2, b_2) and (a_3, b_3) is given by

$$A = \frac{1}{2}[a_1(b_2 - b_3) + a_2(b_3 - b_1) + a_3(b_1 - b_2)] \quad ...(1)$$

Also, $$\begin{vmatrix} a_1 & b_1 & 1 \\ a_2 & b_2 & 1 \\ a_3 & b_3 & 1 \end{vmatrix} = a_1(b_2 - b_3) + a_2(b_3 - b_1) + a_3(b_1 - b_2) \quad ...(2)$$

From (1) and (2), we get

$$A = \frac{1}{2}\begin{vmatrix} a_1 & b_1 & 1 \\ a_2 & b_2 & 1 \\ a_3 & b_3 & 1 \end{vmatrix}$$

4.8.2 Condition of Collinearity of Three Points

Let $A(a_1, b_1)$, $B(a_2, b_2)$ and $C(a_3, b_3)$ be any three points. Then A, B, C are collinear if and only if area of triangle ABC is equal to zero *i.e.*,

Area of $\Delta ABC = 0$

Since area of $\Delta ABC = \frac{1}{2}\begin{vmatrix} a_1 & b_1 & 1 \\ a_2 & b_2 & 1 \\ a_3 & b_3 & 1 \end{vmatrix}$

Hence, three points A, B and C are collinear if $\frac{1}{2}\begin{vmatrix} a_1 & b_1 & 1 \\ a_2 & b_2 & 1 \\ a_3 & b_3 & 1 \end{vmatrix} = 0$

Example 4.19

If the points (a, b), (a', b') and $(a - a', b - b')$ are collinear, prove that

$$ab' = a'b.$$

(UPTU 2005)

SOLUTION : Let A be the area of given triangle, then

$$A = \begin{vmatrix} a & b & 1 \\ a' & b' & 1 \\ a-a' & b-b' & 1 \end{vmatrix}$$

$$= \begin{vmatrix} a & b & 1 \\ a-a' & b-b' & 0 \\ a-a' & b-b' & 1 \end{vmatrix} \quad R_2 \to R_1 - R_2$$

$$= \begin{vmatrix} a & b & 1 \\ a-a' & b-b' & 0 \\ 0 & 0 & 1 \end{vmatrix} \quad R_3 \to R_3 - R_2$$

$$= 1 \begin{vmatrix} a & b \\ a-a' & b-b' \end{vmatrix}$$

$$= a(b-b') - b(a-a')$$
$$= ab - ab' - ab + a'b = a'b - ab'$$

But, we have $A = 0 \Rightarrow a'b - ab' = 0 \Rightarrow a'b = ab'$

Example 4.20

Prove that the points $(a, b+c)$, $(b, c+a)$, $(c, a+b)$ are collinear.

[UPTU 2000, UPTU 2007]

SOLUTION : We have

$$A = \begin{vmatrix} a & b+c & 1 \\ b & c+a & 1 \\ c & a+b & 1 \end{vmatrix} = \begin{vmatrix} a & a+b+c & 1 \\ b & a+b+c & 1 \\ c & a+b+c & 1 \end{vmatrix} \quad C_2 \to C_1 + C_2$$

$$= (a+b+c) \begin{vmatrix} a & 1 & 1 \\ b & 1 & 1 \\ c & 1 & 1 \end{vmatrix} = (a+b+c) \times 0 = 0$$

Hence, the given points are collinear.

Example 4.21

For what value of k, are the points $(1, 5)$, $(k, 1)$ and $(11, 7)$ collinear ? ***[UPTU 2002]***

SOLUTION : The given points are collinear if and only if

$$\begin{vmatrix} 1 & 5 & 1 \\ k & 1 & 1 \\ 11 & 7 & 1 \end{vmatrix} = 0$$

$$\Rightarrow \quad \begin{vmatrix} 1 & 5 & 1 \\ k-1 & -4 & 0 \\ 10 & 2 & 0 \end{vmatrix} = 0 \quad \begin{matrix} R_2 \to R_2 - R_1 \\ R_3 \to R_3 - R_1 \end{matrix}$$

$$\Rightarrow \quad 1 \begin{vmatrix} k-2 & -4 \\ 10 & 2 \end{vmatrix} = 0 \quad \text{[By expanding along } C_3\text{]}$$

$$\Rightarrow \quad 2k - 4 + 40 = 0$$

$$\Rightarrow \quad 2k + 36 = 0$$

$$\Rightarrow \quad k = -18$$

Example 4.22

Find the area of triangle with vertices at the point (3, 8), (–4, 2) and (5, –1).

SOLUTION : Let A be the area of the triangle, then

$$A = \frac{1}{2}\begin{vmatrix} 3 & 8 & 1 \\ -4 & 2 & 1 \\ 5 & -1 & 1 \end{vmatrix}$$

$$= \frac{1}{2}\begin{vmatrix} 3 & 8 & 1 \\ -7 & -6 & 0 \\ 2 & -9 & 0 \end{vmatrix} \quad \begin{matrix} R_2 \to R_2 - R_1 \\ R_3 \to R_3 - R_1 \end{matrix}$$

$$= \frac{1}{2}\begin{vmatrix} -7 & -6 \\ 2 & -9 \end{vmatrix} \qquad \text{[By expanding along } C_3\text{]}$$

$$= \frac{1}{2}[63 + 12]$$

$$= \frac{1}{2}(75) = 37\frac{1}{2} \text{ sq. units.}$$

Example 4.23

Find the area of a triangle whose vertices are (0, 5), (2, 3) and (4, 5).

SOLUTION : Let the co-ordinates of three points $A(x_1, y_1)$, $B(x_2, y_2)$ and $C(x_3, y_3)$ as (0, 5), (2, 3) and (4, 5).

Then the area of triangle ABC

$$= \frac{1}{2}[x_1(y_2 - y_3) + x_2(y_3 - y_1) + x_3(y_1 - y_2)]$$

$$= \frac{1}{2}[0(3 - 5) + 2\,(5 - 5) + 4\,(5 - 3)]$$

$$= \frac{1}{2}[0 + 0 + 8] = 4 \text{ square units.}$$

Example 4.24

Show that the points (2, 0), (11, 6) and (– 4, – 4) are collinear.

SOLUTION : Area of the triangle formed by the points $A(2, 0)$, $B(11, 6)$ and $C(-4, -4)$ is

$$= \frac{1}{2}\begin{vmatrix} 2 & 0 & 1 \\ 11 & 6 & 1 \\ -4 & -4 & 1 \end{vmatrix}$$

$$= \frac{1}{2}[2(6 + 4) - 0\,(11 - (-4)) + 1\,(-44 - (-24))]$$

$$= \frac{1}{2}[20 - 0 + (-20)]$$

$$= \frac{1}{2}[20 - 20] = 0$$

Since, the area of triangle ABC is zero, the points A, B and C are collinear.

Example 4.25

Find the value of k, so that the points $(k, 1)$, $(5, 5)$ and $(10, 7)$ are collinear.

SOLUTION : If the points $(k, 1)$, $(5, 5)$ and $(10, 7)$ are collinear, then area of the triangle formed by these points is zero.

$$\Rightarrow \quad \frac{1}{2}\begin{vmatrix} k & 1 & 1 \\ 5 & 5 & 1 \\ 10 & 7 & 1 \end{vmatrix} = 0$$

$$\Rightarrow \quad k(5-7) - 1(5-10) + 1(35-50) = 0$$

$$\Rightarrow \quad -2k + 5 - 15 = 0$$

$$\Rightarrow \quad -2k = 10 \Rightarrow k = -5$$

Example 4.26

Find the condition that the point (x, y) may lie on the line joining $(3, 4)$ and $(-5, -6)$.

SOLUTION : Let the given points are $P(x, y)$, $Q(3, 4)$ and $R(-5, -6)$. It is given that the point $P(x, y)$ lie on the line joining $Q(3, 4)$ and $R(-5, -6)$, therefore, P, Q and R are collinear.

$$\Rightarrow \quad \frac{1}{2}\begin{vmatrix} x & y & 1 \\ 3 & 4 & 1 \\ -5 & -6 & 1 \end{vmatrix} = 0$$

$$\Rightarrow \quad x(4+6) - y(3+5) + 1(-18+20) = 0$$

$$\Rightarrow \quad 10x - 8y + 2 = 0$$

$$\Rightarrow \quad 5x - 4y + 1 = 0$$

Hence, point (x, y) lies on the line joining $(3, 4)$ and $(-5, -6)$ if $5x - 4y + 1 = 0$

Example 4.27

If the co-ordinates of two points A and B are $(3, 4)$ and $(5, -2)$ respectively. Find the co-ordinate of any point P if $PA = PB$ and Area of $\Delta PAB = 10$.

SOLUTION : Let (x, y) be the co-ordinates of point P. Then

$$PA = PB$$

$$\Rightarrow \quad PA^2 = PB^2$$

$$\Rightarrow \quad (x-3)^2 + (y-4)^2 = (x-5)^2 + (y+2)^2$$

$$\Rightarrow \quad x^2 - 6x + 9 + y^2 - 8y + 16 = x^2 - 10x + 25 + y^2 + 4y + 4$$

$$\Rightarrow \quad x - 3y - 1 = 0 \quad \text{...(1)}$$

Now, Area of $\Delta PAB = 10$

$$\Rightarrow \quad \frac{1}{2}\begin{vmatrix} x & y & 1 \\ 3 & 4 & 1 \\ 5 & -2 & 1 \end{vmatrix} = \pm 10$$

$$\Rightarrow \quad x(4+2) - y(3-5) + 1(-6-20) = \pm 20$$

$$\Rightarrow \quad 6x + 2y - 26 = \pm 20$$

$$\Rightarrow \quad 3x + y - 23 = 0$$

or $3x + y - 3 = 0$ ...(2)

On solving $x - 3y - 1 = 0$ and $3x + y - 23 = 0$,

We get $x = 7, y = 2$.

Similarly, on solving $x - 3y - 1 = 0$ and $3x + y - 3 = 0$, we get $x = 1, y = 0$. Thus, the co-ordinates of point P are (7, 2) or (1, 0).

4.9 AREA OF A QUADRILATERAL

The area of a quadrilateral can be obtained by spliting up the quadrilateral into two triangles and adding up their areas.

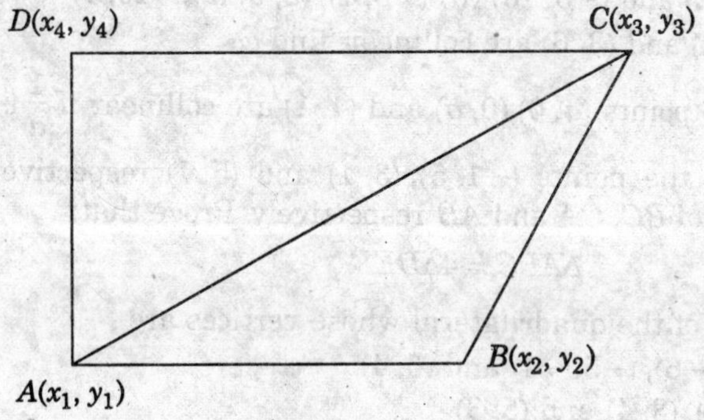

Fig. 4.7

Thus, area of the quadrilateral whose vertices (in order) are $A(x_1, y_1)$, $B(x_2, y_2)$, $C(x_3, y_3)$ and $D(x_4, y_4)$ would be calculated as follows:

$$\text{quad. } ABCD = \Delta ABC + \Delta ACD$$

$$= \frac{1}{2}[(x_1y_2 - x_2y_1) + (x_2y_3 - x_3y_2) + (x_3y_1 - x_1y_3)]$$

$$+ \frac{1}{2}[(x_1y_3 - x_3y_1) + (x_3y_4 - x_4y_3) + (x_4y_1 - x_1y_4)]$$

$$= \frac{1}{2}[(x_1y_2 - x_2y_1) + (x_2y_3 - x_3y_2) + (x_3y_4 - x_4y_3) + (x_4y_1 - x_1y_4)]$$

Example 4.28

Find the area of the quadrilateral whose vertices are (3, 4), (0, 5), (2, – 1) and (3, – 2).

SOLUTION : Let $ABCD$ be the quadrilateral with vertices $A(3, 4)$, $B(0, 5)$, $C(2, -1)$ and $D(3, -2)$.

The area of quadrilateral $ABCD$.

$$= \frac{1}{2}[3 \times 5 - 0 \times 4 + 0 \times (-1) - 2 \times 5 + 2 \times (-2) - 3 \times (-1) + 3 \times 4 - 3 \times (-2)]$$

$$= \frac{1}{2}[15 - 0 - 0 - 10 - 4 + 3 + 12 + 6]$$

$$= \frac{1}{2} \times 22 = 11$$

PROBLEM SET 4.2

1. Find the area of a triangle whose vertices are:
(i) (4, 2), (4, 5) and (– 2, 2) (ii) $(a, 0)$, $(0, b)$ and (x, y)
(iii) $(at_1^2, 2at_1)$, $(at_2^2, 2at_2)$ and $(at_3^2, 2at_3)$
(iv) $(a, c + a)$, (a, c) and $(-a, c - a)$

2. Show that the following points are collinear.
(i) (1, 4), (3, – 2) and (– 3, 16) (ii) (– 5, 1), (5, 5) and (10, 7)

3. If $(7, a)$, (– 5, 2) and (3, 6) are collinear, find a.

4. Prove that the points $(a, 0)$,$(0, b)$ and (1, 1) are collinear if $\frac{1}{a} + \frac{1}{b} = 1$

5. If A, B, C are the points (– 1, 5), (3, 1) and (5, 7) respectively. D, E and F are the middle points of BC, CA and AB respectively. Prove that:

$$\Delta ABC = 4\Delta DEF$$

6. Find the area of the quadrilateral whose vertices are
(i) (– 3, 2), (7, – 6), (– 5, – 4) and (5, 4)
(ii) (1, 2), (6, 2), (3, 4) and (5, 3).

7. The area of a triangle is 5. Two of its vertices are (2, 1) and (3, – 2). The third vertex lies on $y = x + 3$. Find the third vertex.

8. The four vertex of a quadrilateral are (1, 2), (– 5, 6), (7, – 4) and $(k, -2)$ taken in order. If the area of the quadrilateral is zero, find the value of K.

9. If $a \neq b \neq c$, show that (a, a^2), (b, b^2) and (c, c^2) can never be collinears.

10. Find the area of the triangle with vertices
(i) (5, 4), (–2, 4) and (2, –6)
(ii) (–2, 4), (2, –6) and (5, 4)
(iii) (2, 7), (1, 1) and (10, 8)
(iv) (5, 5), (–5, 1) and (10, 7)

11. Using determinants. Prove that the following points are collinear.
(i) (1, –1), (2, 1) and (4, 5)
(ii) (3, –2), (8, 8) and (5, 2)

12. If the points (x_1, y_1), (x_2, y_2) and $(x_1 + x_2, y_1 + y_2)$ are collinear, prove that $x_1y_2 = x_2y_1$.

13. If the points (2, –3), $(k, -1)$ and (0, 4) are collinear, find the value of k.

14. If (a_1, b_1), (a_2, b_2) and (a_3, b_3) are vertices of an equilateral triangle of side x, prove that

$$\begin{vmatrix} a_1 & b_1 & 2 \\ a_2 & b_2 & 2 \\ a_3 & b_3 & 2 \end{vmatrix}^2 = 3x^4.$$

15. If the points $(x, 0)$, $(0, y)$ and $(1, 1)$ are collinear, prove that $x + y = xy$.

16. Find the area of the triangle with vertices $(-3, 5)$, $(3, -6)$ and $(7, 2)$.

17. Find the value of 'a' if the area of triangle is 35 sq. cms with vertices $(a, 4)$, $(2, -6)$ and $(5, 4)$.

ANSWERS

1. (i) 9 (ii) $\frac{1}{2}(bx + ay - ab)$ (iii) $a^2(t_1 - t_2)(t_2 - t_3)(t_3 - t_1)$ (iv) a^2

3. 8 **6.** (i) 60 sq. units (ii) $\frac{11}{2}$ sq. units **7.** $\left(\frac{7}{2}, \frac{13}{2}\right)$ or $\left(\frac{-3}{2}, \frac{3}{2}\right)$. **8.** $k = 3$.

10. (i) 35 sq. units; (ii) 35 sq. units; (iii) $\frac{47}{2}$ sq. units; (iv) 0 sq. units

13. 10/7 **16.** 8.46 sq. units **17.** $a = -2, 12$.

4.10 LOCUS AND ITS EQUATION

4.10.1 Locus

Locus of a point is the path which it describes when moving under a certain given condition or conditions.

Example 4.29

Let $A(-1, 3)$ and $B(3, 3)$ are two fixed points. Then the locus of the moving point P such that $PA = PB$ is the perpendicular bisector MN of the line segment AB. Points $P_1, P_2, P_3, \ldots$ are different positions of the moving point P.

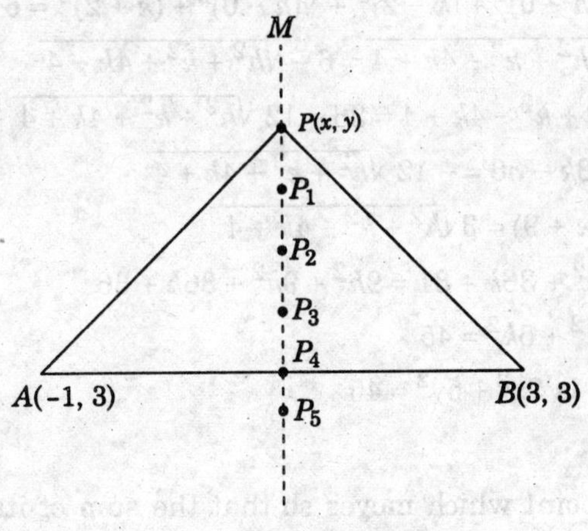

Fig. 4.8

4.10.2 Equation of the Locus

The equation of the locus of a point $P(x, y)$ is a relation between x and y, which is satisfied by the co-ordinates of all points of the locus.

4.10.3 Method to Find the Equation of the Locus

The equation of locus of a point can be obtained by following steps :

- **Step I :** Take the co-ordinates of the point whose locus is to be found as (h, k).
- **Step II :** Use the given condition to find relation between h and k.
- **Step III :** Eliminate the variable, if any (*i.e.*, simplify the relation).
- **Step IV :** Replace h by x and k by y in the result obtained in step III.

The equation thus, obtained is the required equation of the locus.

Example 4.30

Find the equation to the locus of a point equidistant from the points $A(1, 3)$ and $B(-2, 1)$

SOLUTION : Let (h, k) be any point on the locus. Then

$$PA = PB \qquad \text{(given)}$$

$$\Rightarrow \quad PA^2 = PB^2$$

$$\Rightarrow \quad (h-1)^2 + (k-3)^2 = (h+2)^2 + (k-1)^2$$

$$\Rightarrow \quad 6h + 4k = 5$$

Hence, locus of (h, k) is $6x + 4y = 5$.

Example 4.31

Find the locus of a point such that the sum of its distances from the points (0, 2) and (0, – 2) is 6.

SOLUTION : Let $P(h, k)$ be any point on the locus and $A(0, 2)$ and $B(0, -2)$ be the given points.

Given $PA + PB = 6$

$$\Rightarrow \quad \sqrt{(h-0)^2 + (k-2)^2} + \sqrt{(h-0)^2 + (k+2)^2} = 6$$

$$\Rightarrow \quad \sqrt{(h^2 + k^2 - 4k + 4} = 6 - \sqrt{h^2 + k^2 + 4k + 4}$$

$$\Rightarrow \quad h^2 + k^2 - 4k + 4 = 36 - 12\sqrt{h^2 + k^2 + 4k + 4} + (h^2 + k^2 + 4k + 4)$$

$$\Rightarrow \quad -8k - 36 = -12\sqrt{h^2 + k^2 + 4k + 4}$$

$$\Rightarrow \quad (2k + 9) = 3\sqrt{h^2 + k^2 + 4k + 4}$$

$$\Rightarrow \quad 4k^2 + 36k + 81 = 9h^2 + 9k^2 + 36k + 36$$

$$\Rightarrow \quad 9h^2 + 5k^2 = 45$$

Hence, locus of (h, k) is $9x^2 + 5y^2 = 45$.

Example 4.32

Find the locus of a point which moves so that the sum of its distances from (3, 0) and (– 3, 0) is less than 9. ***[UPTU 2002, 03]***

SOLUTION : Let $A(3, 0)$ and $B(-3, 0)$ be the given points and let $P(h, k)$ be any point on the locus.

Then $\quad PA + PB < 9$

$\Rightarrow \quad \sqrt{(h-3)^2+(k-0)^2}+\sqrt{(h+3)^2+(k-0)^2}<9$

$\Rightarrow \quad \sqrt{(h-3)^2+k^2}+\sqrt{(h+3)^2+k^2}<9$

$\Rightarrow \quad \sqrt{(h-3)^2+k^2}<9-\sqrt{(h+3)^2+k^2}$

$\Rightarrow \quad (h-3)^2+k^2<81+(h+3)^2+k^2-18\sqrt{(h+3)^2+k^2}$

$\Rightarrow \quad h^2-6h+9+k^2<81+h^2+6h+9+k^2-18\sqrt{(h+3)^2+k^2}$

$\Rightarrow \quad -12h-81<-18\sqrt{(h+3)^2+k^2}$

$\Rightarrow \quad 4h+27>6\sqrt{(h+3)^2+k^2}$

$\Rightarrow \quad (4h+27)^2>36\,[(h^2+6h+9+k^2]$

$\Rightarrow \quad 16h^2+216h+729>36h^2+36k^2+216h+324$

$\Rightarrow \quad -20h^2-36k^2>-405$

$\Rightarrow \quad 20k^2+36k^2<405$

Hence, locus of (h, k) is $20x^2+36y^2<405$

Example 4.33

A point moves so that its distance from (3, 8) is twice the distance from (– 3, 0). Find the equation of the locus. ***[UPTU 2004]***

SOLUTION : Let $P(h, k)$ be any point on the locus and $A(3, 0)$ and $B(-3, 0)$ are the given points.

Then, we have $\quad PA = 2PB \Rightarrow PA^2 = 4PB^2$

$\Rightarrow \quad (h-3)^2+(k-0)^2=4[(h+3)^2+(k-0)^2]$

$\Rightarrow \quad h^2-6h+9+k^2=4\,[h^2+6h+9+k^2]$

$\Rightarrow \quad 3h^2+3k^2+30h+27=0$

$\Rightarrow \quad h^2+k^2+10h+9=0$

Hence, the locus of (h, k) is $x^2+y^2+10x+9=0$

Example 4.34

If $A(1, 1)$ and $B(-2, 3)$ are two fixed points, find the locus of a point P so that the area of ΔPAB = 9 square units.

SOLUTION : Let $P(h, k)$ be any point on the locus.

Then $\quad \dfrac{1}{2}\begin{vmatrix} h & k & 1 \\ 1 & 1 & 1 \\ -2 & 3 & 1 \end{vmatrix} = \pm 9$

$\Rightarrow \quad \dfrac{1}{2}\,[h\,(1-3)-k\,(1+2)+1\,(3+2)]=\pm 9$

$\Rightarrow \quad -2h-3k+5=\pm 18$

$\Rightarrow \quad -2h-3k+5+18=0 \text{ or } -2h-3k+5-18=0$

$\Rightarrow \quad 2h+3k-23=0 \text{ or } 2h+3k+13=0$

Hence, the locus of (h, k) are given by the equations.

$$2x+3y-23=0 \text{ or } 2x+3y+13=0$$

PROBLEM SET 4.3

1. Find the locus of a point which moves so that its distance from the x-axis is c^2.
2. Find the locus of a point which is equidistant from (1, 0) and (– 1, 0).
3. Find the locus of a point which moves so that its distance from (3, 0) is always 2.
4. A rod of length l slides between the two perpendicular lines. Find the locus of the point on the rod which divides it in the ratio 1 : 2.
5. Find the locus of a point such that the sum of its distances from (0, 2) and (0, – 2) is 6.
6. Find the locus of a point such that the line segment having end points (2, 0) and (– 2, 0) subtend a right angle at the point.
7. $A(2, 0)$ and $B(4, 0)$ are two given points. A point P moves so that $PA^2 + PB^2 = 10$. Find the locus of P.
8. Find the locus of a point which is equidistant from (1, 3) and x-axis.

ANSWERS

1. $y = c^2$ **2.** $x = 0$ **3.** $x^2 + y^2 - 6x = 0$ **4.** $x^2 + \frac{y^2}{4} = \frac{l^2}{9}$

5. $\frac{x^2}{5} + \frac{y^2}{9} = 1$ **6.** $x^2 + y^2 = 4$ **7.** $x^2 + y^2 - 6x + 5 = 0$ **8.** $x^2 - 2x - 6y + 10 = 0$

4.11 THE STRAIGHT LINES

A striaght line is a curve such that every point on the line segment joining two points on it lies on it.

■ **NOTE :** Every first degree equation in x and y represents a straight line.

4.11.1 Slope of a Line

The slope or gradient of a line is equal to the trigonometrical tangent of the angle that a line makes with the positive direction of x-axis in anticlockwise direction.

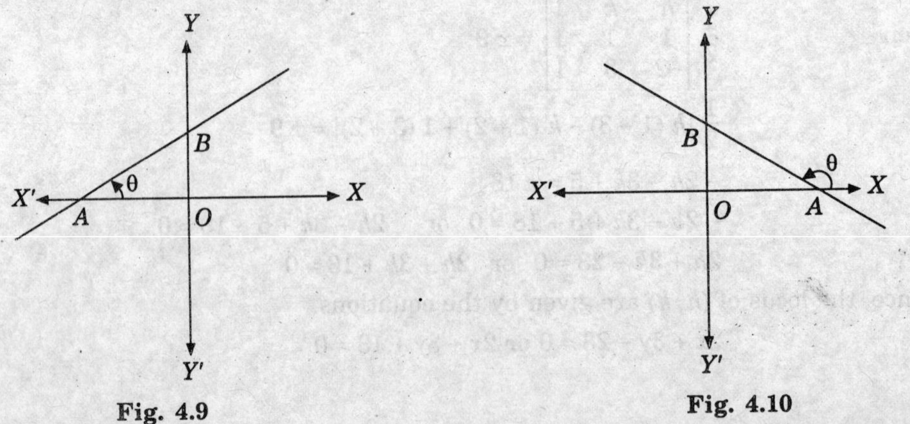

Fig. 4.9 Fig. 4.10

If θ be the angle that a line AB makes with x-axis (in anticlockwise direction), then the slope the line AB is equal to $\tan\theta$ and it is denoted by m, *i.e.*, $m = \tan\theta$

■ **NOTE :** The angle of inclination of a line with the positive direction of x-axis in anticlockwise direction always lies between 0° and 180°, *i.e.*, $0° < \theta < 180°$.

Example 4.35

Find the slope of a line whose inclination to the positive direction of x-axis in anticlockwise sense is :

(i) 45° (ii) 60° (iii) 120°.

SOLUTION : (i) Slope $= \tan 45° = 1$

(ii) Slope $= \tan 60° = \sqrt{3}$

(iii) Slope $= \tan 120° = \tan(90° + 30°) = -\cot 30° = -\frac{1}{\sqrt{3}}$.

Example 4.36

What can be said regarding a line if its slope is

(i) positive, (ii) zero, (iii) negative?

SOLUTION : Let θ be the angle of inclination of the given line with the positive direction of x-axis in anticlockwise direction.

Then its slope is given by $m = \tan\theta$.

(i) Slope is + ive $\Rightarrow$ $m = \tan\theta < 0$

$\Rightarrow$ θ lies between 0° and 90°

$\Rightarrow$ θ is an acute angle.

(ii) Slope is zero $\Rightarrow$ $m = \tan\theta = 0$

$\Rightarrow$ $\theta = 0$

$\Rightarrow$ the line is either x-axis or parallel to x-axis.

(iii) Slope is negative $\Rightarrow m = \tan\theta < 0$

$\Rightarrow$ θ lies between 90° and 180°

$\Rightarrow$ θ is an obtuse angle.

4.11.2 Slope of a Line Passing Through Two Points

Let $P(x_1, y_1)$ and $Q(x_2, y_2)$ be two points on a line making angle θ with the positive direction of x-axis.

Then, the slope of the line is

$$= \frac{y_2 - y_1}{x_2 - x_1} = \frac{\text{Difference of ordinates}}{\text{Difference of abscissae}} = m \text{ (say)}$$

Example 4.37

Find the slope of line passing through points (3, 2) and (– 1, 5).

SOLUTION : Let $A(3, 2)$ and $B(-1, 5)$ be the given points. Then the slope of line passing through A and B

$$m = \frac{y_2 - y_1}{x_2 - x_1} = \frac{5 - 2}{-1 - 3} = \frac{3}{-4} = -\frac{3}{4}$$

Example 4.38

Find the slope of a line passing through $(at_1^2, 2at_1)$ and $(at_2^2, 2at_2)$.

SOLUTION : Let $A(at_1^2, 2at_1)$ and $B(at_2^2, 2at_2)$ be the given points. Then the slope of line passing through A and B is

$$m = \frac{2at_2 - 2at_1}{at_2^2 - at_1^2} = \frac{2a(t_2 - t_1)}{a(t_2^2 - t_1^2)} = \frac{2(t_2 - t_1)}{(t_2 + t_1)(t_2 - t_1)} = \frac{2}{(t_2 + t_1)}$$

Example 4.39

Show that the line through (0, 0) and (2, 3) is parallel to the line through (2, – 2) and (6, 4).

SOLUTION : Let $A(0, 0)$, $B(2, 3)$, $C(2, -2)$ and $D(6, 4)$ are the given points then,

$$\text{Slope of } AB = \frac{3-0}{2-0} = \frac{3}{2}$$

$$\text{and slope of } CD = \frac{4-(-2)}{6-2} = \frac{3}{2}$$

$\because$ Slope of AB = Slope of CD.

$\therefore$ AB and CD are parallel.

4.11.3 Angle between Two Lines

Let θ be the angle between two lines having slopes m_1 and m_2,

$$\text{Then} \quad \tan\theta = \left|\frac{m_1 - m_2}{1 + m_1 m_2}\right|$$

$$\text{or} \quad \theta = \tan^{-1}\left|\frac{m_1 - m_2}{1 + m_1 m_2}\right|$$

Example 4.40

Find the angle between the lines AB and AC where the co-ordinates of points A, B and C are (– 2, 1), (2, 3) and (3, – 4) respectively.

SOLUTION : Let m_1 and m_2 are the slopes of line AB and AC respectively. Then

$$m_1 = \frac{3-1}{2-(-2)} = \frac{2}{4} = \frac{1}{2} \quad \text{and} \quad m_2 = \frac{-4-1}{3-(-2)} = \frac{-5}{5} = -1$$

If θ be the angle between AB and AC, then

$$\theta = \tan^{-1}\left|\frac{\frac{1}{2}+1}{1+\frac{1}{2}\times -1}\right|$$

$$= \tan^{-1}\left|\frac{\frac{3}{2}}{\frac{1}{2}}\right| = \tan^{-1} 3$$

4.11.4 Condition of Parallelism of Two Lines

If two lines having slopes m_1 and m_2 are parallel, then the angle θ between them is of 0°.

$$\therefore \quad \tan\theta = \tan 0° = 0$$

$$\Rightarrow \quad \frac{m_1 - m_2}{1 + m_1 m_2} = 0$$

$$\Rightarrow \quad m_1 - m_2 = 0$$

$$\Rightarrow \quad m_1 = m_2$$

Thus, if two lines are parallel, their slopes are equal.

Example 4.41

Let $A(-1, 4)$, $B\left(5, \frac{11}{2}\right)$, $C\left(\frac{11}{2}, -1\right)$ and $D\left(\frac{5}{2}, -\frac{5}{2}\right)$ are four points. Show that the line AB and CD are parallel to each other.

SOLUTION : Let m_1 and m_2 are the slope of line AB and CD respectively.

$$\text{Then,} \quad m_1 = \frac{\frac{11}{2} - 4}{5 - (-1)} = \frac{3/2}{6} = \frac{1}{4}$$

$$\text{and} \quad m_2 = \frac{-\frac{5}{2} + 1}{\frac{5}{2} - \frac{17}{2}} = \frac{-3/2}{-6} = \frac{1}{4}$$

We observe that $m_1 = m_2 = \frac{1}{4}$, *i.e.*, the slope of line AB and CD are equal and hence, two line AB and CD are parallel.

4.11.5 Condition of Perpendicularity of Two Lines

If two lines having slopes m_1 and m_2 are perpendicular, then the angle between them is of 90°.

$$\therefore \quad \tan\theta = \tan 90° = \frac{m_1 - m_2}{1 + m_1 m_2}$$

$$\Rightarrow \quad \frac{1}{\cot 90°} = \frac{m_1 - m_2}{1 + m_1 m_2}$$

$$\Rightarrow \quad \cot 90° = \frac{1 + m_1 m_2}{m_1 - m_2}$$

$$\Rightarrow \quad 0 = \frac{1 + m_1 m_2}{m_1 - m_2}$$

$$\Rightarrow \quad 1 + m_1 m_2 = 0$$

$$\Rightarrow \quad m_1 m_2 = -1$$

Thus, if two lines are perpendicular, then the product of their slopes is – 1.

Example 4.42

Let $A(4, 4), B(3, 5)$ and $C(-1, -1)$ are three points. Show that line AB and AC are perpendicular lines.

SOLUTION : Let m_1 be the slope of line AB and m_2 be the slope of line AC.

Then, $$m_1 = \frac{5-4}{3-4} = \frac{1}{-1} = -1$$

and $$m_2 = \frac{-1-4}{-1-4} = \frac{-5}{-5} = 1$$

It is clear that $m_1 m_2 = -1 \times 1 = -1$

Thus, the lines AB and AC are perpendicular to each other.

Example 4.43

Determine x so that the line passing through (3, 4) and $(x, 5)$ makes 135° angle with the positive direction of x-axis.

SOLUTION : The slope of the line passing through (3, 4) and $(x, 5)$

$$= \frac{5-4}{x-3} = \frac{1}{x-3}$$

But since the line makes 135° angle with the positive direction of x-axis therefore,

Slope of line = tan 135° = – 1

Hence, $$\frac{1}{x-3} = -1$$

$$\Rightarrow \quad x - 3 = -1 \Rightarrow x = 2$$

Example 4.44

Without using Pythagorous theorem, show that the points $A(0, 4)$, $B(1, 2)$ and $C(3, 3)$ are the vertices of a right angled triangle.

SOLUTION : In ΔABC, we find:

The slope of $BA = m_1 = \frac{4-2}{0-1} = -2$

The slope of $BC = m_2 = \frac{3-2}{3-1} = \frac{1}{2}$

$$\therefore \quad m_1 m_2 = -2 \times \frac{1}{2} = -1$$

This shows that BA and BC are perpendicular. Hence, the given point A, B and C are vertices of a right angled triangle.

PROBLEM SET 4.4

1. Find the slope of a line whose inclination is

(i) 15° (ii) 30°

(iii) 45° (iv) 60°

(v) 135°

2. Find the slope and inclination of the line through each pair of the following points.

(i) (1, 2) and (5, 6) (ii) (0, 0) and $(-\sqrt{3}, 3)$

(iii) (10, 4) and (2, – 2) (iv) (– 1, – 8) and (5, 7)

3. Find y if the slope of the line joining (– 8, 11) and $(2, y)$ is $-\frac{4}{3}$.

4. Find x so that the line through $(x, -9)$ and (2, 7) is parallel to the line through (– 1, 4) and (0, 6).

5. Find the slope of a line perpendicular to the line whose slope is

(i) $\frac{1}{3}$ (ii) 5

(iii) $2\frac{1}{3}$ (iv) 0

(v) infinite slope

6. Find the angle between the lines whose slope are

(i) 2 and – 1

(ii) 2 and $\frac{3}{4}$.

7. Using slopes determine which of the following sets of three points are collinear?

(i) (5, – 2), (7, 6), (0, – 2)

(ii) (– 2, 3), (8, – 5), (6, 4)

8. Show that the line joining $\left(4, -\frac{4}{3}\right)$ and (1, – 3) is perpendicular to the line joining (– 5, 7) and (0, – 2).

ANSWERS

1. (i) $\frac{\sqrt{3}-1}{\sqrt{3}+1}$ (ii) $\frac{1}{\sqrt{3}}$ (iii) 1 (iv) $\sqrt{3}$ (v) – 1

2. (i) 1, 45° (ii) $-\sqrt{3}$, 120° (iii) $\frac{1}{2}$, 26° (iv) $\frac{5}{2}$, $68^\circ 12'$

3. $-\frac{7}{3}$ **4.** 3 **5.** (i) – 3 (ii) $\frac{-1}{5}$ (iii) $\frac{3}{7}$ (iv) Infinite slope (v) 0

6. (i) $71^\circ 34'$ (ii) $26^\circ 34'$; **7.** Neither (i) nor (ii).

4.12 INTERCEPTS OF A LINE ON THE AXES

If a straight line cuts x-axis at A and y-axis at B, then OA and OB are known to be the intercepts of the line on x-axis and y-axis respectively.

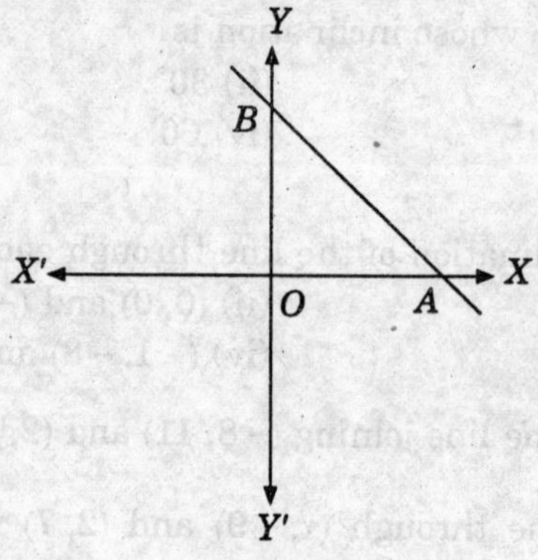

Fig. 4.11

4.12.1 Equation of Line Parallel to x-Axis

The line parallel to x-axis is the locus of a point at a distance b from x-axis. Hence, equation of line parallel to x-axis is $y = b$ shown in fig 4.12.

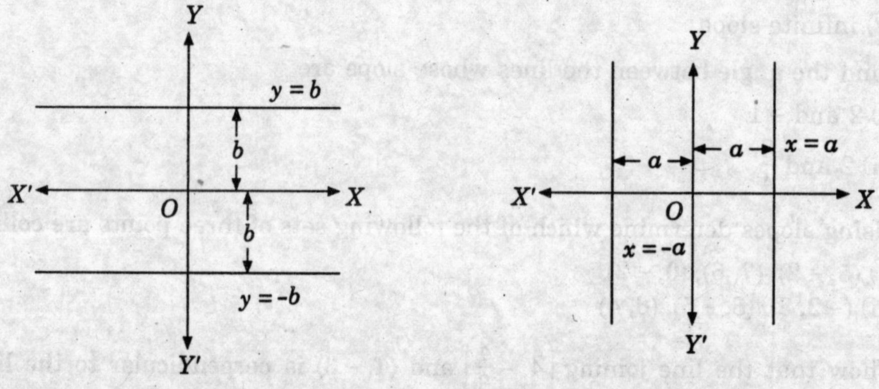

Fig. 4.12 **Fig. 4.13**

4.12.2 Equation of Line Parallel to y-Axis

The line parallel to y-axis is the locus of a point at a distance a from y-axis. Hence, equation of line parallel to y-axis is $x = a$ shown in fig 4.13.

Example 4.45

Find the equation of a line parallel to x-axis and passing through $(3, -5)$.

SOLUTION : We know that equation of a line parallel to x-axis is $y = b$. Since, it passes through the point $(3, -5)$. So $-5 = b$ and hence, the equation of line is $y = -5$.

Example 4.46

Obtain the equation of line parallel to y-axis passing through (1, 2).

SOLUTION : We know that equation of a line parallel to y-axis is $x = a$. Since, it passes through (1, 2). So $1 = a$ and hence, $x = 1$ is the required equation of line parallel to y-axis.

4.13 THE SLOPE INTERCEPT FORM OF A LINE

The equation of a line having slope m and making intercept c on y-axis is given by $y = mx + c$.

Let l be the straight line making an angle θ with OX and the intercept OL on y-axis is c.

Thus, $m = \tan\theta$. Let $B(x, y)$ be any arbitrary point on line l. Draw perpendicular BN to the x-axis.

From L draw LM, a perpendicular to BN.

Then it is clear that $\angle MLB = \theta$, $LM = x$ and $BM = BN - MN = y - c$

Thus, from triangle BLM, we have

$$\tan\theta = \frac{BM}{LM} = \frac{y-c}{x} = m$$

$$\Rightarrow \quad y - c = mx$$

$$\Rightarrow \quad y = mx + c$$

which is required equation.

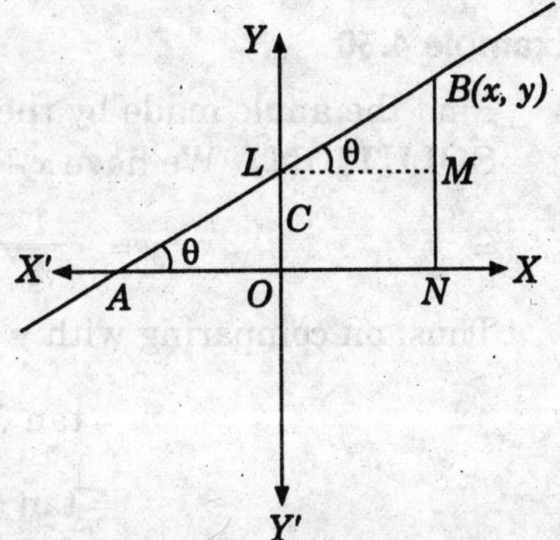

Fig. 4.14

■ **NOTE :** (i) If the equation of a straight line is put in the form $y = mx + c$, then m (the coefficient of x) is the slope and c (the constant term) is the intercept of the straight line.

(ii) The equation of a line with slope m, passing through origin is $y = mx$.

Example 4.47

Find the equation of a line with slope -1 and cutting off an intercept of 4 units on negative direction of y-axis.

SOLUTION : Here, it is given that $m = -1$ and $c = -4$.

So the equation of the line is

$$y = mx + c$$

$$\text{or} \quad y = -x - 4$$

$$\text{or} \quad x + y + 4 = 0$$

Example 4.48

Find the equation of a straight line which cuts off an intercept of 5 units on negative side of y-axis and makes an angle of 135° with the positive direction of x-axis.

SOLUTION : Here, it is given that

$$m = \tan 135° = \tan(90° + 45°) = -\cot 45° = -1$$

$$\text{and} \quad c = -5$$

Thus, the equation of the line is

$$y = -x - 5 \text{ or } x + y + 5 = 0$$

Example 4.49

Find the slope and the y-intercept of the line whose equation is $4x + 6y = 11$.

SOLUTION : The given equation is $5x + 6y = 11$

$$\Rightarrow \qquad y = \left(\frac{-5}{6}\right)x + \left(\frac{11}{6}\right)$$

Thus, slope $= -\frac{5}{6}$ and y-intercept $= \frac{11}{6}$

Example 4.50

Find the angle made by the line $x + \sqrt{3}y - 6 = 0$ with the positive direction of x-axis.

SOLUTION : We have $x + \sqrt{3}y - 6 = 0$

$$\Rightarrow \qquad y = -\frac{1}{\sqrt{3}}x + \frac{6}{\sqrt{3}}$$

Thus, on comparing with $y = mx + c$, we get

$$m = \tan\theta = \frac{-1}{\sqrt{3}} = -\tan 30°$$

$$= \tan(180° - 30°)$$

$$= \tan 150°$$

Hence, $\theta = 150°$, *i.e.*, the given line makes an angle of 150° with positive direction of x-axis.

Example 4.51

Show that the lines $27x - 18y + 25 = 0$ and $2x + 3y + 7 = 0$ are perpendicular to each other.

SOLUTION : The equation of lines are

$$27x - 18y + 25 = 0$$

and $$2x + 3y + 7 = 0$$

or $$y = \frac{3x}{2} - \frac{25}{27} \text{ and } y = \frac{-2}{3}x - \frac{7}{3}$$

It is observed that the slope of first line is $\frac{3}{2}(= m_1)$ and the slope of second line is $-\frac{2}{3}(= m_2)$. Then $m_1m_2 = \frac{3}{2} \times -\frac{2}{3} = -1$

Therefore, the given lines are perpendicular to each other.

4.14 THE POINT SLOPE FORM OF A LINE

The equation of a straight line passing through a given point (x_1, y_1) and having slope m is given by

$$y - y_1 = m(x - x_1)$$

Let $L(x_1, y_1)$ be the point through which the line passes and $B(x, y)$ be any point on the line.

Thus, the slope of the line is $\frac{y - y_1}{x - x_1}$

But slope of the line is m (given). Therefore,

$$m = \frac{y - y_1}{x - x_1}$$

$$\Rightarrow \quad y - y_1 = m\,(x - x_1)$$

which is the required equation.

Example 4.52

Find the equation of a line passing through (3, – 2) and inclined at an angle of 120° with positive direction of x-axis.

SOLUTION : Here, the slope of the line $m = \tan 120°$

$$= \tan (90° + 30°) = - \cot 30° = - \sqrt{3}$$

Thus, the equation of line is

$$y - (-2) = (-\sqrt{3})\,(x - 3)$$

or $\quad y + 2 = -\sqrt{3}x + 3\sqrt{3}$

or $\quad \sqrt{3}x + y = 3\sqrt{3} - 2$

Example 4.53

Determine the equation of a line parallel to x-axis and passing through the point (– 4, 3).

SOLUTION : The slope of the line $= m = 0$

Thus, equation of line will be

$$y - 3 = 0\,(x - (-4)) \Rightarrow y = 3$$

4.15 THE TWO POINT FORM OF A LINE

The equation of the straight line passing through two given points (x_1, y_1) and (x_2, y_2) is

$$y - y_1 = \left(\frac{y_2 - y_1}{x_2 - x_1}\right)(x - x_1)$$

If m be the slope of the line then

$$m = \frac{y_2 - y_1}{x_2 - x_1}$$

Then the equation of line is

$$y - y_1 = m\,(x - x_1)$$

or $\quad y - y_1 = \left(\frac{y_2 - y_1}{x_2 - x_1}\right)(x - x_1)$

which is the required equation :

Example 4.54

Determine the equation of the line joining the points (– 1, 3) and (4, – 2).

SOLUTION : The given points are $(x_1, y_1) = (-1, 3)$ and $(x_2, y_2) = (4, -2)$.

Thus, the equation of the line passing through two points is

$$y - 3 = \left(\frac{-2-3}{4-(-1)}\right)(x - (-1))$$

$$\Rightarrow \quad y - 3 = -\frac{5}{5}(x + 1)$$

$$\Rightarrow \quad y - 3 = -x - 1$$

$$\Rightarrow \quad x + y - 2 = 0$$

Example 4.55

Determine the equation of the altitude AD of a triangle with vertices $A(7, -1)$, $B(-2, 8)$ and $C(1, 2)$.

SOLUTION : Slope of line BC

$$m_1 = \frac{(2-8)}{1-(-2)} = \frac{-6}{3} = -2$$

$\because$ The altitude AD is perpendicular to the line BC.

$\therefore$ The slope of line $AD = m_2 = \frac{-1}{m_1} = -\left(\frac{1}{-2}\right) = \frac{1}{2}$

Thus, equation of the altitude AD is

$$y - (-1) = \left(\frac{1}{2}\right)(x - 7) \Rightarrow y + 1 = \frac{1}{2}(x - 7)$$

$$\Rightarrow \quad 2y + 2 = x - 7 \Rightarrow x - 2y - 9 = 0$$

Example 4.56

Find the slope and the equation of the straight line joining the points $(2, -5)$ and $(4, 1)$. *[UPTU 2001]*

SOLUTION : The given points are $(x_1, y_1) = (2, -5)$ and $(x_2, y_2) = (4, 1)$

Thus, the equation of the straight line through the given points is

$$y - (-5) = \frac{1-(-5)}{4-2}(x - 2)$$

$$\Rightarrow \quad y + 5 = \frac{6}{2}(x - 2) \Rightarrow y + 5 = 3x - 6$$

$$\Rightarrow \quad 3x - y - 11 = 0 \text{ or } y = 3x - 11$$

Hence slope of the line is 3.

Example 4.57

Find the equation of perpendicular bisector of the line joining the points $A(2, 3)$ and $B(6, -5)$. *[UPTU 2003]*

SOLUTION : The slope of the line $AB = m_1 = \frac{-5-3}{6-2} = -2$

Let the slope of the straight line is m_2.

Then $\quad m_2 = \frac{-1}{m_1} = \frac{-1}{(-2)} = \frac{1}{2}$

Also the mid point of AB is

$$= \left(\frac{2+6}{2}, \frac{3-5}{2}\right) = (4, -1)$$

Thus, the equation of the line with slope $\frac{1}{2}$, passing through (4, – 1) is

$$y - (-1) = \frac{1}{2}(x - 4)$$

$$\Rightarrow \quad y + 1 = \frac{1}{2}x - 2 \Rightarrow 2y + 2 = x - 4$$

$$\Rightarrow \quad x - 2y - 6 = 0$$

Example 4.58

In what ratio is the line joining the points (2, 3) and (4, 1) divides the segment joining the point (1, 2) and (4, 3)?

SOLUTION : The equation of the line joining the points (2, 3) and (4, 1) is

$$y - 3 = \left(\frac{1-3}{4-2}\right)(x - 2) \Rightarrow y - 3 = -(x - 2)$$

$$\Rightarrow \quad x + y - 5 = 0 \qquad \text{...(i)}$$

Now suppose the line joining (2, 3) and (4, 1) divides the segment joining (1, 2) and (4, 3) at point L in ratio ($\lambda : 1$). Then the co-ordinates of L are

$$L\left(\frac{4\lambda+1}{\lambda+1}, \frac{3\lambda+2}{\lambda+1}\right)$$

But it is clear that the point L lies on the line (i)

$$\therefore \quad \frac{4\lambda+1}{\lambda+1} + \frac{3\lambda+2}{\lambda+1} - 5 = 0$$

$$\Rightarrow \quad 4\lambda + 1 + 3\lambda + 2 - 5\lambda - 5 = 0 \Rightarrow 2\lambda - 2 = 0 \Rightarrow \lambda = 1$$

Hence, the required ratio is 1 : 1.

Example 4.59

Find the equations of the medians of the triangle ABC whose vertices are $A(2, 5)$, $B(-4, 9)$ and $C(-2, -1)$.

SOLUTION : Let the mid points of BC, CA and AB are D, E and F respectively. Then the co-ordinates of these points are (– 3, 4), (0, 2) and (– 1, 7) respectively.

The median AD is the straight line passes through the points $A(2, 5)$ and $D(-3, 4)$.

Thus, equation of median AD is

$$y - 5 = \frac{4-5}{-3-2}(x - 2)$$

$$\Rightarrow \quad y - 5 = \frac{-1}{-5}(x - 2)$$

$$\Rightarrow \quad y - 5 = \frac{1}{5}(x - 2)$$

$$\Rightarrow \quad 5y - 25 = x - 2$$

$$\Rightarrow \quad x - 5y + 23 = 0$$

The median BE is the straight line passes through the points B (– 4, 9) and E (0, 2). Thus, equation of the median BE is

$$y - 9 = \left(\frac{2-9}{0+4}\right)(x+4)$$

$$\Rightarrow \quad 4y - 36 = -7x - 28 \Rightarrow 7x + 4y - 8 = 0$$

In a similar manner the equation of the median CF is

$$(y+1) = \frac{7+1}{-1+2}(x+2)$$

$$\Rightarrow \quad (y+1) = 8(x+2)$$

$$\Rightarrow \quad 8x - y + 15 = 0$$

PROBLEM SET 4.5

1. Find the equation of a striaght line passing through the point (8, 3) and having slope – 8.

2. Find the equation of a line which makes an angle of $\tan^{-1}\left(\frac{1}{2}\right)$ and passing through (2, 3).

3. Find the equation of a straight line passing through the following pair of points.
(i) $(0, -a)$ and $(b, 0)$ (ii) (2, 2) and (3, 7)
(iii) $\left(ab_1, \frac{a}{b_1}\right)$ and $\left(ab_2, \frac{a}{b_2}\right)$ (iv) (a, b) and $(a + b, a - b)$.

4. Find the equation of line which is parallel to x-axis and passes through the point (11, 11).

5. Find the equations of the medians of a triangle, the co-ordinates of whose vertices are (– 1, 6), (– 3, – 9) and (5, – 8).

6. Prove that the line $y - x + 2 = 0$ divides the line joining of points (3, – 1) and (8, 9) in the ratio 2 : 3.

7. Prove that the points (– 2, 3), (1, 2) and (7, 0) lie on the same straight line. Also find the equation of line.

8. The vertices of a quadrilateral are A(– 2, 6), B (1, 2), C(10, 4) and D(7, 8). Find the equations of its diagonals.

ANSWERS

1. $8x + y - 61 = 0$ **2.** $x - 2y + 4 = 0$

3. (i) $ax - by = ab$ (ii) $5x - y - 8 = 0$ (iii) $b_1b_2y + x = a(b_1 + b_2)$

(iv) $(a - 2b)x - by + b^2 + 2ab - a^2 = 0$

4. $y = 11$ **5.** $29x + 4y + 5 = 0$, $8x - 5y - 21 = 0$, $13x + 14y + 47 = 0$

7. $x + 3y - 7$ **8.** $x + 6y - 3y = 0$; $x - y + 1 = 0$

4.16 THE INTERCEPT FORM OF A LINE

The equation of the straight line which makes intercept 'a' and 'b' on the x-axis and y-axis respectively is

$$\frac{x}{a} + \frac{y}{b} = 1$$

Let the straight line AB meets the x-axis and y-axis at A and B respectively such that $OA = a$ and $OB = b$. The line AB may be considered as the locus of a point $P(x, y)$.

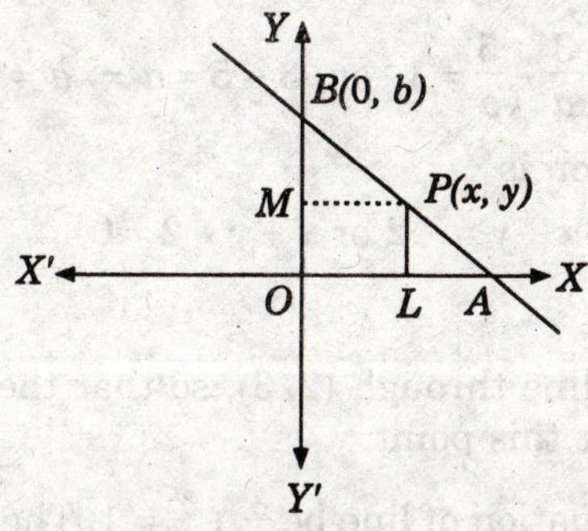

Fig. 4.15

Draw the perpendicular PL to x-axis. Then from similar triangles ALP and AOB; we get

$$\frac{LA}{OA} = \frac{LP}{OB}$$

or $$\frac{a - x}{a} = \frac{y}{b} \Rightarrow 1 - \frac{x}{a} = \frac{y}{b} \Rightarrow \frac{x}{a} + \frac{y}{b} = 1$$

which is the required equation.

Example 4.60

Determine the equation of a line which makes an intercept on the positive direction of x-axis and an intercept 8 on the negative direction of y-axis.

SOLUTION : It is given that $a = 7$ and $b = -8$. Thus, the equation of the straight line is

$$\frac{x}{a} + \frac{y}{b} = 1 \text{ or } \frac{x}{7} + \frac{y}{(-8)} = 1 \text{ or } 8x - 7y - 1 = 0$$

Example 4.61

Find the intercepts cut off by the line $2x - 7y = 14$ on the axes.

SOLUTION : The equation of the given line is

$$2x - 7y = 14 \Rightarrow \frac{2x}{14} - \frac{7y}{14} = 1$$

$$\Rightarrow \frac{x}{7} - \frac{y}{2} = 1 \Rightarrow \frac{x}{7} + \frac{y}{(-2)} = 1$$

Therefore, the given line makes the intercept of 7 units on positive direction of x-axis and 2 units on negative direction of y-axis.

Example 4.62

Find the equation of the straight line which makes equal intercepts on both the axes and passes through (3, – 5). ***[UPTU 2001]***

SOLUTION : It is given that the line makes equal intercept on both the axes, let it be a. Then the equation of the line is

$$\frac{x}{a}+\frac{y}{a}=1$$

But is also given that this line passes through the point (3, – 5), therefore,

$$\frac{3}{a}-\frac{5}{a}=1 \quad \text{or} \quad 3-5=a \Rightarrow a=-2$$

Thus, the required equation is

$$x+y=-2 \text{ or } x+y+2=0$$

Example 4.63

Find the equation of the line through (2, 3), so that the segment of the line intercepted between the axes is bisected at this point.

SOLUTION : Let the equation of line be $\frac{x}{a}+\frac{y}{b}=1$, The above equation of line intersects x-axis at $A(a, 0)$ and y-axis at $B(0, b)$. Also the co-ordinates of mid point of AB are $\left(\frac{a}{2}, \frac{b}{2}\right)$.

But it is given that the point (2, 3) bisects AB. Therefore,

$$\frac{a}{2}=2 \text{ and } \frac{b}{2}=3$$

$$\Rightarrow \quad a=4 \text{ and } b=6$$

Hence, the equation of the required line is

$$\frac{x}{4}+\frac{y}{6}=1 \quad \text{or} \quad 3x+2y=12$$

Example 4.64

A straight line passes through the point (– 2, 6) and the portion of the line intecepted between the axes is divided at this point in the ratio 3 : 2. Find the equation of the line. ***[UPTU 2004]***

SOLUTION : Let the equation of line be $\frac{x}{a}+\frac{y}{b}=1$ which intersects the axes at $A(a, 0)$ and $B(0, b)$. It is given that the point (– 2, 6) divides the segment AB in the ratio 3 : 2.

Thus, $$\frac{3\times 0+2\times a}{3+2}=-2 \quad \text{and} \quad \frac{3\times b+2\times 0}{3+2}=6$$

$$\Rightarrow \quad 2a=-10 \quad \text{and} \quad 36=30$$

$$\Rightarrow \quad a=-5 \quad \text{and} \quad b=10$$

Hence, the required equation of line is

$$\frac{x}{-5}+\frac{y}{10}=1$$

or $$2x-y+10=0$$

4.17 THE NORMAL (PERPENDICULAR) FORM OF A LINE

The equation of a straight line in terms of the length of the perpendicular 'p' from the origin to the line and the angle 'α' which the perpendicular makes with x-axis is given by

$$x \cos \alpha + y \sin \alpha = p$$

Let we take a line AB in such a manner that the length of the perpendicular OQ from O to the line is p and α be the angle made by this perpendicular with x-axis.

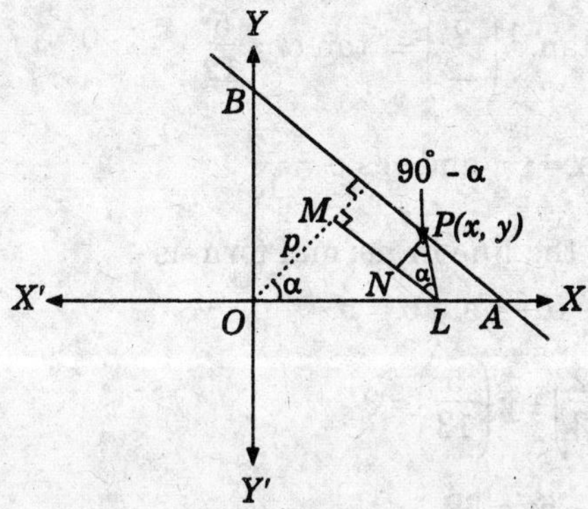

Fig. 4.16

Also let $P(x, y)$ be any point on the line. Draw perpendicular PL to x-axis and PN to the line segment LM. Then $OL = x$ and $LP = y$

$\therefore$ In ΔOLM we have

$$\cos \alpha = \frac{OM}{OL} \Rightarrow OM = OL \cos \alpha \;\Rightarrow\; OM = x \cos \alpha$$

Again in ΔPNL,

$$\sin \alpha = \frac{PN}{PL} \;\Rightarrow\; PN = PL \sin \alpha = y \sin \alpha$$

$$\Rightarrow \qquad MQ = PN = y \sin \alpha$$

Since, $\qquad OQ = OM + MQ \;\Rightarrow\; p = x \cos \alpha + y \sin \alpha$

Hence, the equation of line is $x \cos \alpha + y \sin \alpha = p$.

Example 4.65

Find the equation of the line which is at a distance of 3 from the origin and the perpendicular from the origin to the line makes an angle 30° with the positive direction of the x-axis.

SOLUTION : It is given that $p = 3$ and $\alpha = 30°$. Thus, equation of line in normal form is

$$x \cos 30° + y \sin 30° = 3$$

or $\qquad x\left(\frac{\sqrt{3}}{2}\right) + y\left(\frac{1}{2}\right) = 3$

or $\qquad \sqrt{3}x + y = 6$

Example 4.66

Find the equation of the straight line at a distance of 3 units from the origin, such that the perpendicular from the origin to the line makes an angle $\tan^{-1}\left(\frac{5}{12}\right)$ with positive direction of x-axis.

SOLUTION : It is given that $p = 3$ and $\alpha = \tan^{-1}\left(\frac{5}{12}\right)$

$$\because \quad \alpha = \tan^{-1}\left(\frac{5}{12}\right) \Rightarrow \tan\alpha = \frac{5}{12}$$

$$\therefore \quad \sin\alpha = \frac{5}{13} \text{ and } \cos\alpha = \frac{12}{13}$$

Hence, the equation of the line in normal form is

$$x\cos\alpha + y\sin\alpha = p$$

$$\text{or} \quad x\left(\frac{12}{13}\right) + y\left(\frac{5}{13}\right) = 3$$

$$\Rightarrow \quad 12x + 5y = 39$$

PROBLEM SET 4.6

1. Find the equation of the straight line
   (i) cutting off intercepts 3 and 2 from the axes.
   (ii) cutting of intercepts 2, – 7 from the axes.
2. Find the equation of a straight line which passes through the point (5, 6) and has intercepts equal in magnitude but opposite in sign.
3. The area of the triangle formed by the co-ordinate axes and a line is 6 square units. The length of hypotenuse is 5 units. Find the equation of the line.
4. Find the equation of a straight line which makes an angle of 30° with the x-axis and cuts off an intercept of 2 with y-axis.
5. Find the equation of a straight line which passes through the point (– 3, 8) and cuts off positive intercepts on the co-ordinate axes whose sum is 7.
6. Find the equation of a line for which
   (i) $p = 5$, $\alpha = 60°$
   (ii) $p = 8$ and $\alpha = 300°$
7. The slope of a straight line through $A(3, 2)$ is $\frac{3}{4}$. Find the co-ordinates of the points on the line that are 5 units away from it.

8. Find the equation of line at a distance 3 units from the origin such that the perpendicular from the origin to the line makes an angle $\tan^{-1}\left(\dfrac{3}{4}\right)$ with the positive direction of x-axis.

9. Find the equation of the straight line on which the length of the perpendicular from the origin is 2 and the perpendicular makes an angle α with x-axis such that $\sin \alpha = \dfrac{1}{3}$.

10. The length of perpendicular from origin to the line is 9 and the line makes an angle 135°. Find the equation of line.

ANSWERS

1. (i) $2x + 3y = 6$ (ii) $72 - 2y = 14$ **2.** $x - y = -1$

3. $3x + 4y = 12$ or $3x + 4y = -12$ or $4x + 3y = 12$ or $-4x - 3y = 12$

4. $\sqrt{3}x - y + 2 = 0$ **5.** $4x + 3y = 12$ **6.** (i) $x + \sqrt{3}y = 10$, (ii) $x - \sqrt{3}y = 16$

7. (7, 5) and (– 1, – 1) **8.** $4x + 3y = 15$ **9.** $2\sqrt{2}x + y = 6$ **10.** $x - y + 9\sqrt{2} = 0$

4.18 GENERAL FORM OF THE STRAIGHT LINE

The general equation of a straight line is $ax + by + c = 0$ provided both a and b are not zero simultaneously, *i.e.*, a linear equation in x and y always represents a straight line.

4.19 TRANSFORMATION OF GENERAL FORM

4.19.1 Slope-Intercept Form

Let $ax + by + c = 0$ be the given equation of straight line.

$$\Rightarrow \quad by = -ax - c \Rightarrow y = -\frac{a}{b}x - \frac{c}{b}$$

which is the equation of straight line in slope-intercept form with slope $-\dfrac{a}{b}$ and intecept on y-axis $-\dfrac{c}{b}$.

4.19.2 Intercepts From

We have $ax + by + c = 0$

$$\Rightarrow \quad ax + by = -c \Rightarrow \frac{ax}{-c} + \frac{by}{-c} = 1 \Rightarrow \frac{x}{-\frac{c}{a}} + \frac{y}{-\frac{c}{b}} = 1$$

which is the equation of line in intercept form with intercepts $-\dfrac{c}{a}$ and $-\dfrac{c}{b}$ on x-axis and y-axis respectively.

4.19.3 Normal Form

We have

$$ax + by + c = 0 \qquad \text{...(i)}$$

Let $$x \cos \alpha + y \sin \alpha - p = 0 \qquad \text{...(ii)}$$

be the normal form of the given equation. Since, equations (i) and (ii) represent the same straight line. Therefore,

$$\frac{a}{\cos \alpha} = \frac{b}{\sin \alpha} = \frac{c}{-p}$$

$$\Rightarrow \quad \cos \alpha = -\frac{ap}{c} \text{ and } \sin \alpha = -\frac{bp}{c}$$

$$\Rightarrow \quad \cos^2 \alpha + \sin^2 \alpha = \frac{a^2p^2}{c^2} + \frac{b^2p^2}{c^2}$$

$$\Rightarrow \quad \frac{p^2}{c^2}(a^2 + b^2) = 1$$

$$\Rightarrow \quad p^2 = \frac{c^2}{a^2 + b^2}$$

$$\Rightarrow \quad p = \pm\sqrt{\frac{c^2}{(a^2 + b^2)}} = \pm\frac{c}{\sqrt{a^2 + b^2}}$$

Since, p is the length of perpendicular. So it is always positive.

$$\therefore \quad p = \frac{c}{\sqrt{a^2 + b^2}}$$

Thus, we have

$$\cos \alpha = -\frac{a}{\sqrt{a^2 + b^2}}$$

and $$\sin \alpha = -\frac{b}{\sqrt{a^2 + b^2}}$$

Hence, the Eq. (ii) reduces to the equation.

$$-\frac{a}{\sqrt{a^2 + b^2}}x - \frac{b}{\sqrt{a^2 + b^2}}y - \frac{c}{\sqrt{a^2 + b^2}} = 0$$

$$\Rightarrow \quad -\frac{a}{\sqrt{a^2 + b^2}}x - \frac{b}{\sqrt{a^2 + b^2}}y = \frac{c}{\sqrt{a^2 + b^2}}$$

which is the required normal form of the equation $ax + by + c = 0$

Example 4.67

Transform the equation of the line $\sqrt{3} \cdot x + y - 8 = 0$ to :

(i) Slope intercept form and find its slope and y-intercept.

(ii) Intercept form and find intercept on both the axes.

(iii) Normal form and find the inclination of the perpendicular of segment from the origin on the line with the axis and its length.

SOLUTION : (i) The given equation of straight line is

$$\sqrt{3}x + y - 8 = 0$$

$$\Rightarrow \quad y = -\sqrt{3}x + 8,$$

which is the slope intercept form with slope $-\sqrt{3}$ and y-intercept 8.

(ii) The given equation is

$$\sqrt{3}x + y - 8 = 0 \Rightarrow \frac{\sqrt{3}x}{8} + \frac{y}{8} = 1$$

$$\Rightarrow \quad \frac{x}{8/\sqrt{3}} + \frac{y}{8} = 1$$

This is the intercept form of the given line; the x-intercept is $\frac{8}{\sqrt{3}}$ and the y-intercept is 8.

(iii) We have $\quad \sqrt{3}x + y - 8 = 0$

$$\Rightarrow \quad \sqrt{3}x + y = 8$$

$$\Rightarrow \quad \frac{\sqrt{3}x}{\sqrt{(\sqrt{3})^2 + 1^2}} + \frac{y}{\sqrt{(\sqrt{3})^2 + 1^2}} = \frac{8}{\sqrt{(\sqrt{3})^2 + 1^2}}$$

$$\Rightarrow \quad \frac{\sqrt{3}}{2}x + \frac{y}{2} = \frac{8}{2}$$

$$\Rightarrow \quad \frac{\sqrt{3}}{2}x + \frac{y}{2} = 4$$

which is the normal form of the given line.

On comparison with standard form we have :

$$\cos\alpha = \frac{\sqrt{3}}{2}, \sin\alpha = \frac{1}{2} \text{ and } p = 4$$

Thus, $\alpha = \frac{\pi}{6}$ and $p = 4$.

Example 4.68

Reduce $4x + 3y - 9 = 0$ to the normal form and find the distance (perpendicular distance p) from origin. *[UPTU 2003]*

SOLUTION : The given equation is

$$4x + 3y - 9 = 0$$

$$\Rightarrow \quad 4x + 3y = 9$$

$$\Rightarrow \quad \frac{4}{\sqrt{4^2 + 3^2}}x + \frac{3}{\sqrt{4^2 + 3^2}}y = \frac{9}{\sqrt{4^2 + 32}}$$

$$\Rightarrow \quad \frac{4}{5}x + \frac{3}{5}y = \frac{9}{5}$$

which is the normal form of the given equation.

On comparison with standard form

$$\cos\alpha = \frac{4}{5}, \sin\alpha = \frac{3}{5} \text{ and } p = \frac{9}{5} \text{ (the distance from origin).}$$

Example 4.69

Find the equation of a line with slope 2 and length of perpendicular from origin equal to $\sqrt{5}$.

SOLUTION : The equation of straight line with slope 2 is

$$y = 2x + c \Rightarrow -2x + y = c$$

$$\Rightarrow \quad \frac{-2}{\sqrt{4+1}}x + \frac{1}{\sqrt{4+1}}y = \frac{c}{\sqrt{4+1}}$$

$$\Rightarrow \quad \frac{-2}{\sqrt{5}}x + \frac{1}{\sqrt{5}}y = \frac{c}{\sqrt{5}}$$

which is of the form $x \cos \alpha + y \sin \alpha = p$

Hence, $\quad p = \frac{c}{\sqrt{5}}$

But it is given that

$$p = \sqrt{5} \Rightarrow \frac{c}{\sqrt{5}} = \sqrt{5} \Rightarrow c = 5$$

Therefore, the equation of the straight line is

$$y = 2x + 5$$

Example 4.70

Reduce the line $3x - 4y + 4 = 0$ and $4x - 3y + 12 = 0$ to the normal form and hence, determine which line is nearer to the origin.

SOLUTION : We are given

$$3x - 4y + 4 = 0 \Rightarrow -3x + 4y = 4$$

$$\Rightarrow \quad \frac{-3}{\sqrt{(-3)^2 + 4^2}}x + \frac{4}{\sqrt{(-3)^2 + 4^2}}y = \frac{4}{\sqrt{(-3)^2 + 4^2}}$$

$$\Rightarrow \quad -\frac{3}{5}x + \frac{4}{5}y = \frac{4}{5}$$

This is the normal form of $3x - 4y + 4 = 0$ and the length of the perpendicular from the origin to it is given by $p_1 = \frac{4}{5}$

Now, $\quad 4x - 3y + 12 = 0$

$$\Rightarrow \quad -4x + 3y = 12$$

$$\Rightarrow \quad -\frac{4}{\sqrt{(-4)^2 + 3^2}}x + \frac{3}{\sqrt{(-4)^2 + 3^2}}y = \frac{12}{\sqrt{(-4)^2 + 3^2}}$$

$$\Rightarrow \quad -\frac{4}{5}x + \frac{3}{5}y = \frac{12}{5}$$

This is the normal form of $4x - 3y + 12 = 0$ and the length of the perpendicular from the origin to it is given by $p_2 = \frac{12}{5}$

It can be observed that $p_1 < p_2$, thus, the line $3x - 4y + 4 = 0$ is nearer to the origin.

PROBLEM SET 4.7

1. Reduce the equation $x + \sqrt{3}y + 7 = 0$ into slope intercept form and find the slope and y-intercept.
2. Reduce the equation $\frac{x}{a} + \frac{y}{b} = 1$ into the slope intercept form and find its slope and y-intercept
3. Reduce the equation $3x - 2y + 6 = 0$ to the intercept form and find x and y intercepts.
4. Reduce the following equations to the normal form and find p and α. (p and α are same as in standard form):
   (i) $x + \sqrt{3}y - 4 = 0$ (ii) $x + y + \sqrt{2} = 0$
   (iii) $x - y + 2\sqrt{2} = 0$ (iv) $y - 2 = 0$
5. The perpendicular distance of a line from the origin is 5 units and its slope is – 1. Find the equation of the line.
6. Reduce the lines $3x - 4y + 4 = 0$ and $2x + 4y - 5 = 0$ to the normal form and hence, find which line is nearer to the origin.
7. Show that the origin is equidistant from the lines $4x + 3y + 10 = 0$, $5x - 12y + 26 = 0$ and $7x + 24y = 50$.
8. Find the length of perpendicular from the origin to the line $x + y - \sqrt{2} = 0$

ANSWERS

1. Slope $= -\frac{1}{\sqrt{3}}$, y-intercept $= -\frac{7}{3}$ **2.** Slope $= -\frac{b}{a}$, y-intercept $= b$

3. x-intercept = 2, y-intercept = 3

4. (i) $p = 2, \alpha = \frac{\pi}{3}$ (ii) $p = 1, \alpha = 225°$ (iii) $p = 2, \alpha = 135°$ (iv) $p = 2, \alpha = \frac{\pi}{2}$.

5. $x + y - 5\sqrt{2} = 0$ **6.** $3x - 4y + 4 = 0$ **8.** $p = 1$

4.20 POINT OF INTERSECTION OF TWO LINES

Let we are given two straight lines

$$a_1x + b_1y + c_1 = 0 \quad ...(i)$$

and $$a_2x + b_2y + c_2 = 0 \quad ...(ii)$$

Suppose the point of intersection of these two lines is $p(x_1, y_1)$

Then $a_1x_1 + by_1 + c_1 = 0$

and $a_2x_1 + b_2y_1 + c_2 = 0$

Solving these two equations, we get

$$x_1 = \frac{b_1c_2 - b_2c_1}{a_1b_2 - a_2b_1}, \quad y_1 = \frac{c_1a_2 - c_2a_1}{a_1b_2 - a_2b_1}$$

Hence, the co-ordinates of the point of intersection of (i) and (ii) are

$$\left(\frac{b_1c_2 - b_2c_1}{a_1b_2 - a_2b_1}, \frac{c_1a_2 - c_2a_1}{a_1b_2 - a_2b_1}\right)$$

4.20.1 Concurrency of Three Lines

Three straight lines will be concurrent; if they passes through a common point.

4.20.2 Coincident Lines

Two lines $a_1x + b_1y + c_1 = 0$ and $a_2x + b_2y + c_2 = 0$ are coincident, if

$$\frac{a_1}{a_2} = \frac{b_1}{b_2} = \frac{c_1}{c_2}$$

4.20.3 Parallel Lines

Two lines $a_1x + b_1y + c_1 = 0$ and $a_2x + b_2y + c_2 = 0$ will be parallel if

$$\frac{a_1}{a_2} = \frac{b_1}{b_2} \neq \frac{c_1}{c_2}$$

4.20.4 Perpendicular Lines

Straight lines $a_1x + b_1y + c_1 = 0$ and $a_2x + b_2y + c_2 = 0$ are said to be perpendicular lines if

$$a_1a_2 + b_1b_2 = 0$$

4.20.5 Intersecting Lines

Two straight lines will be interesting if they are neither coincident nor parallel.

Example 4.71

Find the co-ordinates of the point of intersection of the lines $2x - y + 3 = 0$ and $x + 2y - 4 = 0$

SOLUTION : The given equations are

$$2x - y + 3 = 0 \quad \text{...(i)}$$

and $$x + 2y - 4 = 0 \quad \text{...(ii)}$$

On solving eqs. (i) and (ii), we get

$$\frac{x}{4-6} = \frac{y}{3+8} = \frac{1}{4+1} \Rightarrow \frac{x}{-2} = \frac{y}{11} = \frac{1}{5} \Rightarrow x = -\frac{2}{5}, y = \frac{11}{5}$$

Hence, the point of intersection can be written as $\left(-\frac{2}{5}, \frac{11}{5}\right)$.

Example 4.72

Find the area of a triangle formed by the lines $y = 2x$, $y = x$ and $y = 3x + 4$.

[UP TU 2000]

SOLUTION : The given equations are

$$y = 2x \quad \text{...(i)}$$

$$y = x \quad \text{...(ii)}$$

and $y = 3x + 4$ ...(iii)

Let the given Eqs. (i), (ii) and (iii) represent the sides AB, BC and CA respectively of a triangle ABC.

Solving (i) and (ii), we get

$$\frac{x}{0} = \frac{y}{0} = \frac{1}{-2+1}$$

$\Rightarrow x = 0$ and $y = 0$, Thus, AB and BC intersect at point, $B(0, 0)$.

Solving (ii) and (iii), we get

$$\frac{x}{-4} = \frac{y}{-4} = \frac{1}{-1+3}$$

$$\Rightarrow \qquad x = -2 \text{ and } y = -2$$

So, BC and CA intersect at the point $C(-2, 2)$.

Solving Eq. (i) and (iii), we get

$$\frac{x}{-4} = \frac{y}{-8} = \frac{1}{-2+3}$$

$$\Rightarrow \qquad x = -4 \text{ and } y = -8$$

Therefore, $A(-4, -8)$ is the point of intersection of line AB and CA.

Thus, $A(-4, -8)$, $B(0, 0)$ and $C(-2, 2)$ are the vertices of the triangle ABC.

$$\therefore \quad \text{Area of } \Delta ABC = \frac{1}{2}\begin{vmatrix} -4 & -8 & 1 \\ -2 & -2 & 1 \\ 0 & 0 & 1 \end{vmatrix}$$

$$= \frac{1}{2}[8 - 16] = -4$$

$= 4$ (neglecting– ve sign) or (area can not be –ve)

Hence, the required area is 4 square unit.

Example 4.73

Show that the lines $x - y - 6 = 0$, $4x - 3y - 20 = 0$ and $6x + 5y + 8 = 0$ are concurrent. Also find the point of intersection.

SOLUTION : The given lines are:

$$x - y - 6 = 0 \qquad \text{...(i)}$$

$$4x - 3y - 20 = 0 \qquad \text{...(ii)}$$

$$6x + 5y + 8 = 0 \qquad \text{...(iii)}$$

On solving Eqs. (i) and (ii), we get

$$\frac{x}{20-18} = \frac{y}{-24+20} = \frac{1}{-3+4}$$

$$\Rightarrow \qquad x = 2 \text{ and } y = -4$$

Thus, the point of intersection of first two lines is $(2, -4)$. Substituting $x = 2$ and $y = -4$ in Eq. (iii), we get

$$6 \times 2 + 5 \times -4 + 8 = 0$$

So, $(2, -4)$ lies on the Eq. (iii).

Hence, the given lines are concurrent and their common point of intersection is $(2, -4)$.

Example 4.74

Find the value of λ, if the lines $3x - 4y - 13 = 0$, $8x - 11y - 33 = 0$ and $2x - 3y + \lambda = 0$ are concurrent.

SOLUTION : The given equations are

$$3x - 4y - 13 = 0 \quad \text{...(i)}$$
$$8x - 11y - 33 = 0 \quad \text{...(ii)}$$
$$2x - 3y + \lambda = 0 \quad \text{...(iii)}$$

Solving Eqs. (i) and (ii), we get

$$\frac{x}{-4 \times -33 - (-13) \times (-11)} = \frac{-y}{3(-33) - 8(-13)} = \frac{1}{3(-11) - 8(-4)}$$

$$\Rightarrow \quad \frac{x}{132 - 143} = \frac{-y}{-99 + 104} = \frac{1}{-33 + 32}$$

$$\Rightarrow \quad \frac{x}{-11} = \frac{-y}{5} = \frac{1}{-1}$$

$$\Rightarrow \quad x = 11$$

$$\Rightarrow \quad y = 5$$

Thus, (11, 5) is the point of intersection of line (i) and (ii).

The given lines will be concurrent if they passes through the common point.

Therefore, (11, 5) lies on line (iii),

$$\Rightarrow \quad 2 \times 11 - 3 \times 5 + \lambda = 0$$

$$\Rightarrow \quad 22 - 15 + \lambda = 0$$

$$\Rightarrow \quad \lambda = -7$$

Example 4.75

Classify whether following lines are coincident, parallel or intersecting:

(i) $4x - 5y + 2 = 0$ and $12x - 15y + 6 = 0$

(ii) $x - 2y = 7$ and $4y - 2x = 13$

(iii) $3x - 4y = 8$ and $3x + 4y = 11$

SOLUTION : (i) For the given lines

$$4x - 5y + 2 = 0 \quad \text{and} \quad 12x - 15y + 6 = 0$$

$$a_1 = 4, b_1 = -5, c_1 = 2 \text{ and } a_2 = 12, b_2 = -15, c_2 = 6$$

$$\therefore \quad \frac{a_1}{a_2} = \frac{4}{12} = \frac{1}{3}, \frac{b_1}{b_2} = \frac{-5}{-15} = \frac{1}{3} \text{ and } \frac{c_1}{c_2} = \frac{2}{6} = \frac{1}{3}$$

$$\because \quad \frac{a_1}{a_2} = \frac{b_1}{b_2} = \frac{c_1}{c_2} = \frac{1}{3}$$

The lines are coincident.

(ii) For the given lines $x - 2y = 7$ and $4y - 2x = 13$

$a_1 = 1, b_1 = -2, c_1 = -7$ and $a_2 = -2, b_2 = 4, c_2 = -13$

Thus, $\frac{a_1}{a_2} = \frac{1}{-2} = -\frac{1}{2}, \frac{b_1}{b_2} = \frac{-2}{4} = -\frac{1}{2}$ and $\frac{c_1}{c_2} = \frac{-7}{-13} = \frac{7}{13}$

$\because \quad \frac{a_1}{a_2} = \frac{b_1}{b_2} = -\frac{1}{2} \neq \frac{7}{13} = \frac{c_1}{c_2}$

$\Rightarrow \quad \frac{a_1}{a_2} = \frac{b_1}{b_2} \neq \frac{c_1}{c_2}$

Hence, the given lines are parallel to each other.

(iii) For the given lines $3x - 4y = 8$ and $3x + 4y = 11$

$$a_1 = 3, b_1 = -4, c_1 = -8 \text{ and } a_2 = 3, b_2 = 4, c_2 = -11$$

Thus, $\frac{a_1}{a_2} = \frac{3}{3} = 1, \frac{b_1}{b_2} = \frac{-4}{4} = -1$ and $\frac{c_1}{c_2} = \frac{-8}{-11} = \frac{8}{11}$

Here it is observed that

$$\frac{a_1}{a_2} = 1 \neq -1 = \frac{b_1}{b_2} \neq \frac{c_1}{c_2} = \frac{8}{11}$$

i.e., $$\frac{a_1}{a_2} \neq \frac{b_1}{b_2} \neq \frac{c_1}{c_2}.$$

Hence, the lines are neither coincident nor parallel but intersecting.

Example 4.76

A straight line drawn through the point $A(2, 1)$ makes an angle $\frac{\pi}{4}$ with + ive x-axis and intersects another line $x + 2y + 1 = 0$ at point B. Find the length of AB.

SOLUTION : The equation of a line passing through $A(2, 1)$ and making angle $\frac{\pi}{4}$ with +ve x-axis is

$$y - y_1 = m(x - x_1)$$

where $m = \tan\frac{\pi}{4} = 1$ and $(x_1, y_1) = (2, 1)$

$\therefore \quad y - 1 = 1(x - 2)$

$\Rightarrow \quad x - y - 1 = 0$

Now the point of intersection of line $x - y - 1 = 0$ and $x + 2y + 1 = 0$ can be obtained by solving these equations,

$$\frac{x}{-1+2} = \frac{y}{-1-1} = \frac{1}{2+1}$$

$\Rightarrow \quad \frac{x}{1} = \frac{y}{-2} = \frac{1}{3}$

$\Rightarrow \quad x = \frac{1}{3}$ and $y = -\frac{2}{3}$

Thus, the point of intersection of these line is $B\left(\frac{1}{3}, -\frac{2}{3}\right)$

Therefore, the required length AB = Distance between points A and B

$$= \sqrt{\left(\frac{1}{3}-2\right)^2+\left(\frac{-2}{3}-1\right)^2} = \sqrt{\left(\frac{-5}{3}\right)^2+\left(\frac{-5}{3}\right)^2}$$

$$= \sqrt{\frac{50}{9}} = \frac{5\sqrt{2}}{3}$$

PROBLEM SET 4.8

1. Find the point of intersection of the following lines

(i) $2x - y + 3 = 0$ and $x + y - 5 = 0$

(ii) $y = m_1 x + \frac{a}{m_1}$ and $y = m_2 x + \frac{a}{m_2}$

2. Determine the co-ordinates of the vertices of a triangle, the equatioin of whose sides are $x + y - 4 = 0$, $2x - y + 3 = 0$ and $x - 3y + 2 = 0$

3. Calculate the area of the triangle formed by the lines $y = 0$, $x = 2$ and $x + 2y = 3$

4. Prove that the lines $y = \sqrt{3}x + 1$, $y = 4$ and $y = -\sqrt{3}x + 2$ form an equilateral triangle.

5. Prove that the following lines are concurrent.

(i) $3x - 5y - 11 = 0$, $5x + 3y - 7 = 0$ and $x + 2y = 0$

(ii) $\frac{x}{a} + \frac{y}{b} = 1$, $\frac{x}{b} + \frac{y}{a} = 1$ and $y = x$

6. For what value of λ, the line $5x - 9y + \lambda = 0$, $2x - 5y + 3 = 0$ and $x - 2y + 1 = 0$ are concurrent?

7. Classify whether the following pair of lines are coincident, parallel, perpendicular or intersecting:

(i) $2x + y - 1 = 0$ and $3x + 2y + 5 = 0$

(ii) $x - y = 0$ and $3x - 3y + 5 = 0$

(iii) $3x + 2y - 4 = 0$ and $6x + 4y - 8 = 0$

8. Find the equation of the line joining the point (3, 5) to the point of intersection of the lines $4x + y - 1 = 0$ and $7x - 3y - 35 = 0$

ANSEWRS

1. (i) $\left(\frac{2}{3}, \frac{13}{3}\right)$ (ii) $\left(\frac{a}{m_1 m_2}, a\left(\frac{1}{m_1}+\frac{1}{m_2}\right)\right)$ **2.** $\left(\frac{1}{3}, \frac{11}{3}\right)$, $\left(\frac{-7}{5}, \frac{1}{5}\right)$, $\left(\frac{5}{2}, \frac{3}{2}\right)$

3. 0, **6.** $\lambda = 4$ **7.** (i)intersecting, (ii) parallel, (iii) coincident

8. $12x - y - 31 = 0$

4.21 DISTANCE OF A POINT FROM A LINE

Distance of a point $P(x_1, y_1)$ from the line $ax + by + c = 0$ is equal to the length of perpendicular drawn from the point $P(x_1, y_1)$ to the given line. Let it be d, then

$$d = \left| \frac{ax_1 + by_1 + c}{\sqrt{a^2 + b^2}} \right|$$

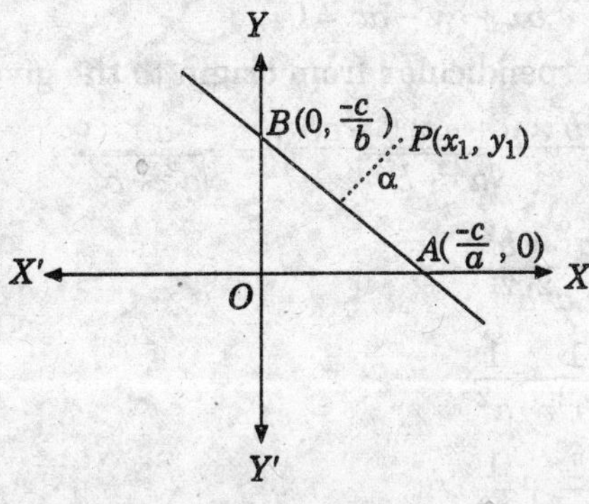

Fig. 4.17

■ **NOTE :** In particular, the length of perpendicular from the origin to the line $ax + by + c = 0$ is $\frac{|c|}{\sqrt{a^2 + b^2}}$.

4.21.1 Rule to find the Length of Perpendicular from a Point (x_1, y_1) to the given Straight Line $ax + by + c = 0$

Following are the steps to find out the perpendicular distance from point (x_1, y_1) to the given line $ax + by + c = 0$

- **Step I:** Write the equation of line in the form $ax + by + c = 0$
- **Step II :** Replace x and y by x_1 and y_1 respectively in the expression $ax + by + c$.
- **Step III :** Divide the result obtained from Step II, *i.e.*, $ax_1 + by_1 + c$. by the square root of the sum of the squares of the coefficients of x and y, *i.e.*, $\sqrt{a^2 + b^2}$.
- **Step IV :** Take the modulus of the value obtained from Step III.

Example 4.77

Find the distance of point $P(2, -3)$ from the line $12x - 7y + 11 = 0$.

SOLUTION : The required distance $= \left| \frac{12 \times 2 + (-7) \times (-3) + 11}{\sqrt{144 + 49}} \right|$

$$= \frac{24 + 21 + 11}{\sqrt{193}} = \frac{56}{\sqrt{193}}$$

Example 4.78

If p is the length of perpendicular from origin to the line $\frac{x}{a}+\frac{y}{b}=1$, then show that

$$\frac{1}{p^2}=\frac{1}{a^2}+\frac{1}{b^2}$$

SOLUTION : The given line can be written as:

$$bx+ay-ab=0 \qquad \ldots(i)$$

We have p = length of perpendicular from origin to the given line (i)

$$\Rightarrow \quad p=\frac{|b\times 0+a\times 0-ab|}{\sqrt{a^2+b^2}}=\frac{ab}{\sqrt{a^2+b^2}}$$

$$\Rightarrow \quad \frac{1}{p^2}=\frac{a^2+b^2}{a^2b^2}$$

$$\Rightarrow \quad \frac{1}{p^2}=\frac{1}{b^2}+\frac{1}{a^2}$$

$$\Rightarrow \quad \frac{1}{p^2}=\frac{1}{a^2}+\frac{1}{b^2}$$

Example 4.79

What are the points on x-axis whose perpendicular distance from the line $4x+3y=12$ is 4?

SOLUTION : Let the point be $(a, 0)$. Then the length of perpendicular from $P(a, 0)$ upon $4x+3y-12=0$ is 4

$$\Rightarrow \quad \left|\frac{4a+3\times 0-12}{\sqrt{16+9}}\right|=4$$

$$\Rightarrow \quad \left|\frac{4a-12}{5}\right|=4 \Rightarrow \frac{|a-3|}{5}=1$$

$$\Rightarrow \quad |a-3|=5 \Rightarrow a-3=5 \text{ or } 3-a=5$$

$$\Rightarrow \quad a=8 \text{ or } -2$$

Hence, the points are $(-2, 0)$ and $(8, 0)$.

Example 4.80

Find the co-ordinates of a point on $x+y+3=0$ whose distance from $x+2y+2=0$ is $\sqrt{5}$.

SOLUTION : Let the point be (a, b). It is given that the point (a, b) lies on $x+y+3=0$. So

$$a+b+3=0 \qquad \ldots(i)$$

Also, $\sqrt{5}$ = the length of perpendicular from (a, b) to the line $x+2y+2=0$

$$\Rightarrow \quad \sqrt{5}=\frac{a+2b+2}{\sqrt{1+4}}$$

$$\Rightarrow \quad \sqrt{5}=\frac{a+2b+2}{\sqrt{5}}$$

$$\Rightarrow \quad 5 = a + 2b + 2$$

$$\Rightarrow \quad a + 2b = 3 \qquad \text{...(ii)}$$

On solving equation (i) and (ii), we get

$$a = -9 \text{ and } b = 6$$

Thus, (– 9, 6) is the required point.

4.22 DISTANCE BETWEEN TWO PARALLEL LINES

If two lines are parallel, they have the same distance between them throughout.

4.22.1 Rule to find the Distance between Two Lines

Let we have two parallel lines, $ax + by + c_1 = 0$ and $ax + by + c_2 = 0$.

The distance between these lines can be obtained as follows:

- **Step I :** Take a point on any one of the line, preferably by putting $x = 0$ or $y = 0$.
- **Step II :** Find the length of perpendicular drawn from the point (from Step I) to the other line.
- **Step III :** The length obtained in step II is the required distance between two parallel lines.

Example 4.81

Find the perpendicular distance between two lines

$$3x + 4y - 5 = 0 \text{ and } 6x + 8y - 45 = 0$$

SOLUTION : Putting $x = 0$ in the first equation, we get

$$4y = 5 \Rightarrow y = \frac{5}{4}$$

Hence, $\left(0, \frac{5}{4}\right)$ is a point on the line $3x + 4y - 5 = 0$.

Now, the perpendicular distance of the point $\left(0, \frac{5}{4}\right)$ from the line $6x + 8y - 45 = 0$ is

$$= \frac{\left|6 \times 0 + 8 \times \frac{5}{4} - 45\right|}{\sqrt{36 + 64}} = \frac{|10 - 45|}{10}$$

$$= \frac{|-35|}{10} = \frac{35}{10} = \frac{7}{2}$$

Therefore, the distance between the given lines is $\frac{7}{2}$.

Example 4.82

Find the distance between the parallel lines $2x - 3y + 9 = 0$ and $4x - 6y + 1 = 0$.

[UPTU 2004]

SOLUTION : Put $x = 0$ in first equation, we get

$$-3y = -9 \Rightarrow y = 3$$

Hence, (0, 3) is a point on the line $2x - 3y + 9 = 0$

Now, the perpendicular distance of the point (0, 3) from the line $4x - 6y + 1 = 0$ is

$$= \frac{|4 \times 0 - 6 \times 3 + 1|}{\sqrt{16+36}}$$

$$= \frac{|-17|}{\sqrt{52}} = \frac{17}{\sqrt{52}}$$

Therefore, the distance between the given lines is $\frac{17}{\sqrt{52}}$.

Example 4.83

Find the distance between the parallel lines $3x + 4y = 12$ and $3x + 4y = 3$.

SOLUTION : Put $x = 0$ in the first equation, we get

$$4y = 12 \Rightarrow y = 3$$

Thus, (0, 3) is a point on the first line $3x + 4y - 12 = 0$.

Now the perpendicular distance of the point (0, 3) from the second line $3x + 4y = 3$ is

$$= \frac{|3 \times 0 + 4 \times 3 - 3|}{\sqrt{9+16}} = \frac{|9|}{\sqrt{25}} = \frac{9}{5}$$

Hence, the distance between the given lines is $\frac{9}{5}$.

Example 4.84

Two sides of a square lie on the lines $x + y = 1$ and $x + y + 2 = 0$. What is its area?

SOLUTION : It is clear that the length of the side of the square is equal to the distance between the parallel lines

$$x + y - 1 = 0 \quad \ldots(i)$$

$$\text{and} \quad x + y + 2 = 0 \quad \ldots(ii)$$

Putting $x = 0$ in (i), we get $y = 1$. So (0, 1) is a point on line (i).

Thus, distance between the parallel lines

= length of perpendicular from (0, 1) to the line (ii)

$$= \frac{|0+1+2|}{\sqrt{1^2+1^2}} = \frac{|3|}{\sqrt{2}} = \frac{3}{\sqrt{2}}$$

So, the length of the side of the square is $\frac{3}{\sqrt{2}}$ and hence, its area $= \left(\frac{3}{\sqrt{2}}\right)^2 = \frac{9}{2}$ sq. unit.

Example 4.85

Find the locus of a point equidistant from the lines $x + y + 4 = 0$ and $7x + y + 20 = 0$.

SOLUTION : Let $P(h, k)$ be the point, equidistant from the given lines, then

$$\frac{h+k+4}{\sqrt{1^2+1^2}} = \frac{7h+k+20}{\sqrt{7^2+1^2}}$$

$$\Rightarrow \quad \frac{h+k+4}{\sqrt{2}} = \frac{7h+k+20}{\sqrt{50}}$$

$$\Rightarrow \quad \frac{h+k+4}{\sqrt{2}} = \frac{7h+k+20}{5\sqrt{2}}$$

$$\Rightarrow \quad 5h + 5k + 20 = 7h + k + 20$$

$$\Rightarrow \quad 2h - 4k = 0$$

Hence, the locus of point $P(h, k)$ is $2x - 4y = 0$

4.23 POSITIONS OF POINTS RELATIVE TO A LINE

The points $P(x_1, y_1)$ and $Q(x_2, y_2)$ lie:

(i) on the same side of the line $ax + by + c = 0$ if $ax_1 + by_1 + c$ and $ax_2 + by_2 + c$ have the same sign.

(ii) on the opposite side of the line $ax + by + c = 0$, if $ax_1 + by_1 + c$ and $ax_2 + by_2 + c$ have opposite signs.

Example 4.86

Are the points (3, 4) and (2, – 6) on the same or opposite sides of the line $3x - 4y = 8$?

SOLUTION : Let $l = 3x - 4y - 8$. Then the value of l at (3, 4) is

$$3 \times 3 - 4 \times 4 - 8 = l_1 \quad \text{(say)}$$

$$\Rightarrow \quad l_1 = -15 (-\text{ve})$$

Similarly, value of l at (2, – 6) is

$$3 \times 2 - 4 \times 6 - 8 = l_2 \quad \text{(say)}$$

$$\Rightarrow \quad l_2 = 22 (+\text{ve})$$

Since, l_1 and l_2 are of opposite signs therefore, the two points are on opposite sides of the given line.

Example 4.87

Show that the points (2, – 3) and (3, – 5) lie on the same side of the line $2x - 3y - 7 = 0$.

SOLUTION : Let $l = 2x - 3y - 7$. Then the value of l at (2, – 3) is

$$2 \times 2 - 3 \times -3 - 7 = l_1 \text{ (say)}$$

$$\Rightarrow \quad l_1 = 4 + 9 - 7 = 6 \quad (+\text{ve})$$

Similarly, value of l at (3, – 5) is

$$2 \times 3 - 3 \times (-5) - 7 = l_2 \text{ (say)}$$

$$\Rightarrow \quad l_2 = 6 + 15 - 7 = 14 (+\text{ve})$$

Since, l_1 and l_2 are of same sign, therefore, the two points lie on the same side of the given line.

PROBLEM SET 4.9

1. Find the distance of a point (2, 3) from the straight line $3x + 4y - 2 = 0$.
2. Find the distance of a point of intersection of the lines $2x + 3y = 21$ and $3x - 4y + 11 = 0$ from the line $8x + 6y + 5 = 0$
3. Find the distance between the parallel lines.
   (i) $x + 2y + 7 = 0$ and $x + 2y + 9 = 0$
   (ii) $2x - y - 1 = 0$ and $2x - y + 7 = 0$
4. Find the distance between two parallel lines $y = mx + c$ and $y = mx + d$.
5. Show that the lines $2x + 3y = 19$ and $2x + 3y + 7 = 0$ are equidistant from the line $2x + 3y = 6$.
6. Are the points (3, – 4) and (2, 6) lie on the same or opposite sides of the line $3x - 4y = 8$.
7. Which one of the points (1, 1), (– 1, 2) and (2, 3) lies on that side of the line $4x + 3y - 5 = 0$ on which the origin lies?
8. What are the points on x-axis whose perpendicular distance from straight line $\frac{x}{a} + \frac{y}{b} = 1$ is a?

ANSWERS

1. $\frac{16}{5}$ **2.** $\frac{59}{10}$ **4.** $\frac{|c-d|}{\sqrt{1+m^2}}$ **6.** opposite sides **7.** (– 1, 2)

8. $\left(\frac{a}{b}(b \pm \sqrt{a^2+b^2}), 0\right)$

4.24 EQUATION OF STRIAGHT LINE PASSING THROUGH A GIVEN POINT AND PARALLEL/ PERPENDICULAR TO THE GIVEN LINE

(i) The equation of a line parallel to a given line $ax + by + c = 0$ is $ax + by + \lambda = 0$, where λ is constant.

(ii) The equation of a line perpendicular to a given line $ax + by + c = 0$ is $bx - ay + \lambda = 0$, where λ is a constant.

Example 4.88

Find the equation of a line passing through the pont (– 1, 1) and parallel to the line $2x + 5y = 7$.

SOLUTION : The equation of a line parallel to the line $2x + 5y = 7$ is given as

$$2x + 5y + \lambda = 0,$$

where λ is a constant.

Since, the line $2x + 5y + \lambda = 0$ passes through $(-1, 1)$.

Thus, $$-2 + 5 + \lambda = 0 \Rightarrow \lambda = -3$$

Hence, the equation of required line is

$$2x + 5y - 3 = 0 \text{ or } 2x + 5y = 3$$

Example 4.89

Find the equation of a line passing through the point (1, 2) and perpendicular to the straight line $4x - 3y = 8$.

SOLUTION : The equation of line perpendicular to the line $4x - 3y = 8$ will be $-3x - 4y + \lambda = 0$.

It is given that the line obtained, passes through the point (1, 2). Therefore,

$$-3 \times 1 - 4 \times 2 + \lambda = 0$$

$$\Rightarrow \lambda = 11$$

Hence, the equation of required line is

$$3x + 4y = 11$$

Example 4.90

Find the equations of the two straight lines through the point $(2, -1)$ making an angle of 45° with the line $6x + 5y - 1 = 0$.

SOLUTION : The equation of given line is

$$6x + 5y - 1 = 0$$

$$\Rightarrow y = -\frac{6}{5}x + \frac{1}{5} \quad \text{...(i)}$$

$\therefore$ Slope of this line is $\frac{-6}{5}$.

Now the equation of line passing through $(2, -1)$ is

$$y + 1 = m(x - 2)$$

or $$y = m(x - 2) - 1 \quad \text{...(ii)}$$

It is given that line (ii) makes an angle of 45° with the line (i). Thus,

$$\tan 45^\circ = \left| \frac{m - \left(\frac{-6}{5}\right)}{1 + m\left(\frac{-6}{5}\right)} \right|$$

$$\Rightarrow 1 = \pm \frac{5m + 6}{5 - 6m}$$

$$\Rightarrow 5 - 6m = 5m + 6 \text{ and } 5 - 6m = -5m - 6$$

$$\Rightarrow -11m = 1 \text{ and } m = 11$$

$$\Rightarrow m = -\frac{1}{11} \text{ and } m = 11$$

On substituting these values of m in eq. (ii), we get

$$y = \frac{-1}{11}(x-2) - 1$$

and $$y = 11(x-2) - 1$$

$$\Rightarrow \quad 11y = -x + 2 - 1$$

and $$y = 11x - 22 - 1$$

$$\Rightarrow \quad x + 11y = 1$$

and $$11x - y = 23$$

which are the required equations of lines.

PROBLEM SET 4.10

1. Find the equation of a line passing through (2, 3) and parallel to $2x + 3y + 7 = 0$.
2. Find the equation of a line passing through (7, 9) and perpendicular to the line $3x + 5y + 3 = 0$.
3. Find the equation of a line passing through the point (1, 3) and perpendicular to $x + y = 3$.
4. Find the equation of a line passing through (1, 1) and parallel to $5x + 9y = 15$.
5. Find the equation of a straight line passing through origin and making an angle of 45° with the straight line $\sqrt{3}x + y = 11$.
6. Find the equation of straight line pasing through (2, 1) and making an angle of 45° with $6x + 5y - 8 = 0$.

ANSWERS

1. $2x + 3y - 18 = 0$ **2.** $5x - 3y - 8 = 0$ **3.** $x - y + 2 = 0$

4. $5x + 2y - 14 = 0$ **5.** $y = (\sqrt{3} \pm 2)x$ **6.** $11x - y - 23 = 0$

MULTIPLE CHOICE QUESTIONS

1. The point (–1, –3) lies in which quadrant?
(a) I (b) II
(c) III (d) IV

2. Which of the point lies in IV quadrant?
(a) (1, 2) (b) (2, –7)
(c) (–3, 4) (d) (–1, –5)

3. The distance between the points $A(2, 1)$ and $B(3, 5)$ is
(a) $\sqrt{14}$ (b) 4
(c) $\sqrt{17}$ (d) 3

4. The length of the line segment joining $S_1(2, 3)$ and $S_2(3, 5)$
(a) $\sqrt{10}$ (b) $\sqrt{5}$
(c) 3 is (d) 1

5. Which of the following set of points forms the vertices of an isosceles triangle
(a) (–1, 3), (5, 8), (4, 1) (b) (–1, 1), (5, 8), (1, 4)
(c) (–2, 3), (3, 8), (4, 1) (d) (0, 0), (2, 3), (7, 8)

6. The coordinates of any point on AB that divides the line joining the pionts $A(x_1, y_1)$ and (x_2, y_2) in the ratio $\lambda : 1$.
(a) $\left(\frac{\lambda x_1 + x_2}{\lambda + 1}\right), \left(\frac{\lambda y_2 + y_1}{\lambda + 1}\right)$ (b) $\left(\frac{\lambda x_2 + x_1}{\lambda + 1}, \frac{\lambda y_2 + y_1}{\lambda + 1}\right)$
(c) $\left(\frac{\lambda x_2 + x_1}{\lambda - 1}, \frac{\lambda y_2 + y_1}{\lambda - 1}\right)$ (d) None of these

7. The coordinates of mid point of the line segment joining points $A(1, 4)$ and $B(3, 6)$.
(a) (1, 2) (b) (2, 3)
(c) (2, 5) (d) (3, 3)

8. The coordinates of the centroid of a triangle with vertices $A(x_1, y_1)$, $B(x_2, y_2)$ and $C(x_3, y_3)$ are
(a) $\left(\frac{x_1 + x_2 + x_3}{3}, \frac{y_1 + y_2 + y_3}{3}\right)$ (b) $\left(\frac{x_1 - x_2 + x_3}{3}, \frac{y_1 + y_2 - y_3}{2}\right)$
(c) $\left(\frac{x_1 + x_2 + x_3}{2}, \frac{y_1 + y_2 + y_3}{2}\right)$ (d) None of these

9. The centroid of the triangle ABC with vertices $A(1, 1)$, $B(2, 2)$ and $C(3, 3)$ is
(a) (1, 2) (b) (2, 1)
(c) (2, 2) (d) (3, 3)

10. Which of the following points are vertices of a parallelogram?
(a) (1, –2), (3, 6), (5, 10) and (3, 2) (b) (–1, 0), (0, 3), (1, 3) and (0, 0)
(c) Both (a) and (b) (d) None of these

11. The points $(-a, -b)$, $(0, 0)$, (a, b) and (a^2, ab) are
(a) Colinear (b) Vertices of a parallelogram
(c) Vertices of a rectangle (d) None of these

12. The points $\left(0, \frac{8}{3}\right)$, (1, 3) and (82, 30) are vertices of

(a) a right angled triangle (b) an isosceles triangle
(c) an obtuse angled triangle (d) none of these

13. Length of the median from B on AC where $A(-1, 3)$, $B(1, -1)$ and $C(5, 1)$ is

(a) $\sqrt{18}$ (b) $\sqrt{10}$
(c) $2\sqrt{3}$ (d) 4

14. If (3, –4) and (–6, 5) are the extermities of the diagonal of a parallelogram and (–2, 1) is its third vertex, then its fourth vertex is

(a) (–1, 0) (b) (0, –1)
(c) (–1 , 1) (d) None of these

15. If the coordinates of the middle point of the sides of a triangle are (4, 2), (3, 3) and (2, 2), then the coordinates of its centroid are

(a) $\left(3, \frac{7}{3}\right)$ (b) (3, 3)
(c) (4, 3) (d) None of these

16. The incentre of the triangle whose vertices are (–36, 7), (20,7) and (0, –8) is

(a) (0, –1) (b) (–1, 0)
(c) $\left(\frac{1}{2}, 1\right)$ (d) None of these

17. The triangle with vertices (2, 4), (2, 6) and $(2 + \sqrt{3}, 5)$ is

(a) right angled (b) right angled and isosceles
(c) equilateral (d) none of these

18. Area of the triangle with vertices (x_1, y_1), (x_2, y_2) and (x_3, y_3) is given by

(a) $\begin{vmatrix} x_1 & y_1 & 1 \\ x_2 & y_2 & 1 \\ x_3 & y_3 & 1 \end{vmatrix}$ (b) $\frac{1}{2}\begin{vmatrix} x_1 & x_2 & 1 \\ y_1 & y_2 & 1 \\ x_3 & y_3 & 1 \end{vmatrix}$

(c) $\frac{1}{2}\begin{vmatrix} x_1 & y_1 & 1 \\ x_2 & y_2 & 1 \\ x_3 & y_3 & 1 \end{vmatrix}$ (d) None of these

19. The value of K for which the points (1, 5), (K, 1) and (11, 7) are

(a) 1 (b) 9
(c) –11 (d) –18

20. The points (x, y) may lie on the line joining (3, 4) and (–5, –6) if

(a) $x + y + 1 = 0$ (b) $2x - 3y + 1 = 0$
(c) $5x - 4y + 1 = 0$ (d) None of these

21. Area of the triangle with vertices (4, 2), (4, 5) and (–2, 2) is

(a) 9 (b) 18
(c) 7 (d) 5

22. The points (a, a^2), (b, b^2) and (c, c^2) can never be collinear if

(a) $a = b = c$ (b) $a \neq b = c$

(c) $a = b \neq c$ (d) $a \neq b \neq c$

23. Area of the triangle with vertices $(a, b + c)$, $(b, c + a)$, $(c, a + b)$ is

(a) 0 (b) $a + b + c$

(c) $ab + bc + ca$ (d) 1

24. If $P(1, 0)$, $Q(-1, 0)$ and $R(2, 0)$ are three given points, then the locus of point S satisfying the equation $SQ^2 + SR^2 = 2SP^2$ is

(a) straight line parallel to x-axis (b) circle through origin

(c) circle with centre at origin (d) straight line parallel to y-axis

25. The locus of the midpoint of the portion intercepted between the axes by the line $x \cos \alpha + y \sin \alpha = p$, where p is a constant is

(a) $x^2 + y^2 = 4p^2$ (b) $\frac{1}{x^2} + \frac{1}{y^2} = \frac{4}{p^2}$

(c) $x^2 + y^2 = \frac{4}{p^2}$ (d) $\frac{1}{x^2} + \frac{1}{y^2} = \frac{2}{p^2}$

26. The locus of the point of intersection of lines $x \cos \alpha + y \sin \alpha = a$ and $x \sin \alpha - y \cos \alpha = b$ is (α is a variable)

(a) $2(x^2 + y^2) = a^2 + b^2$ (b) $x^2 - y^2 = a^2 - b^2$

(c) $x^2 + y^2 = a^2 + b^2$ (d) None of these

27. If A and B are two fixed points, then the locus of a point which moves in such a way that the triangle APB is right angled is

(a) circle (b) ellipse

(c) parabola (d) None of these

28. If a variable line passes through the point of intersection of the lines $x + 2y - 1 = 0$ and $2x - y - 1 = 0$ and meets the coordinate axes in A and B, then the locus of mid point of AB is

(a) $x + 3y = 0$ (b) $x + 3y = 10$

(c) $x + 3y = 10xy$ (d) none of these

29. The nearest point on the line from the origin is

(a) $(-4, 5)$ (b) $(3, -4)$

(c) $(3, 4)$ (d) $(3, 5)$

30. The slope of a straight line whose inclination is 225

(a) $\sqrt{3}$ (b) 1

(c) $\frac{1}{\sqrt{3}}$ (d) None of these

31. The slope of the line joining points (1, 2) and (3, 5) is

(a) 3 (b) $\frac{1}{2}$

(c) $\frac{3}{2}$ (d) $\frac{3}{5}$

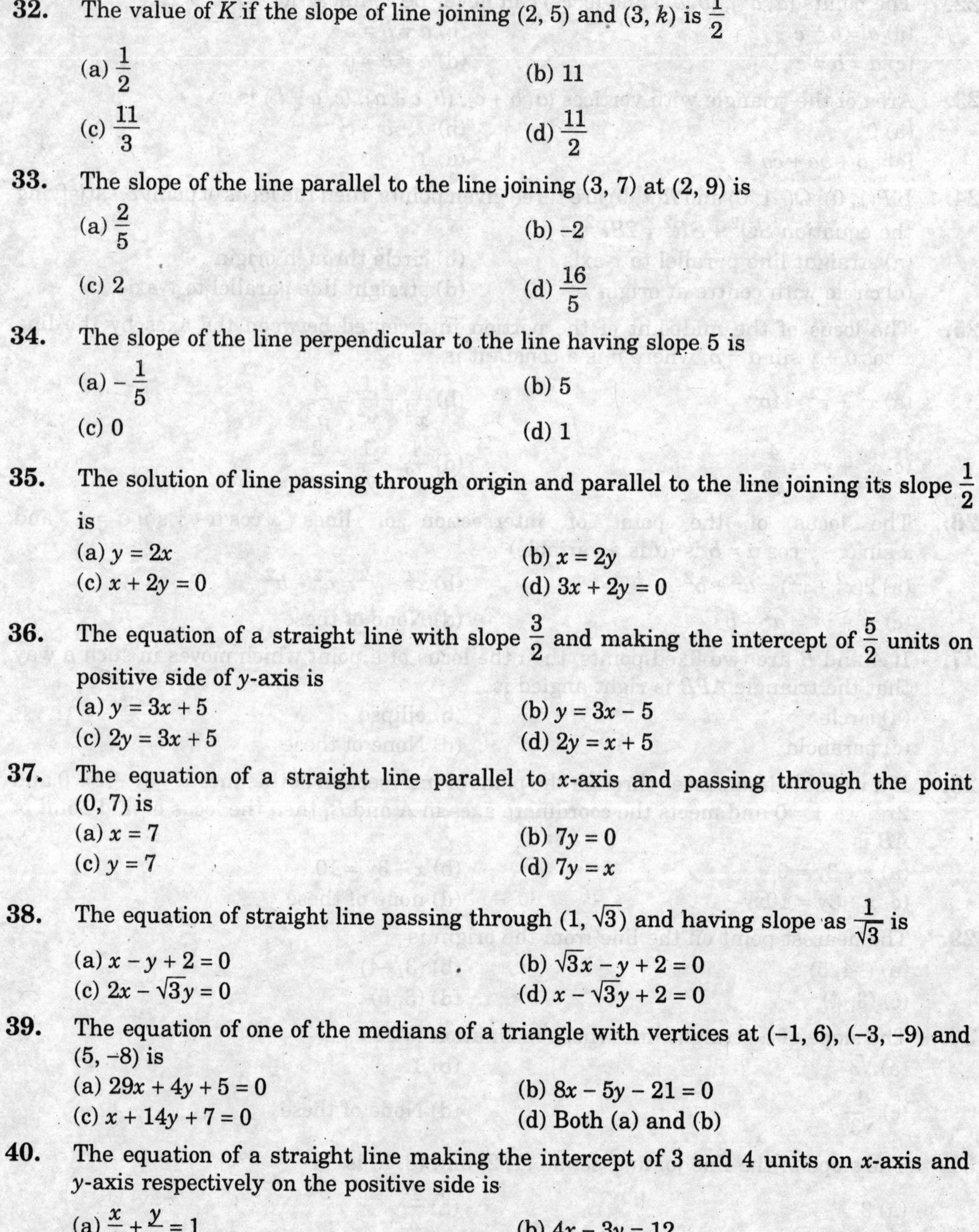

32. The value of K if the slope of line joining (2, 5) and (3, k) is $\frac{1}{2}$

(a) $\frac{1}{2}$ (b) 11

(c) $\frac{11}{3}$ (d) $\frac{11}{2}$

33. The slope of the line parallel to the line joining (3, 7) at (2, 9) is

(a) $\frac{2}{5}$ (b) -2

(c) 2 (d) $\frac{16}{5}$

34. The slope of the line perpendicular to the line having slope 5 is

(a) $-\frac{1}{5}$ (b) 5

(c) 0 (d) 1

35. The solution of line passing through origin and parallel to the line joining its slope $\frac{1}{2}$ is

(a) $y = 2x$ (b) $x = 2y$

(c) $x + 2y = 0$ (d) $3x + 2y = 0$

36. The equation of a straight line with slope $\frac{3}{2}$ and making the intercept of $\frac{5}{2}$ units on positive side of y-axis is

(a) $y = 3x + 5$ (b) $y = 3x - 5$

(c) $2y = 3x + 5$ (d) $2y = x + 5$

37. The equation of a straight line parallel to x-axis and passing through the point (0, 7) is

(a) $x = 7$ (b) $7y = 0$

(c) $y = 7$ (d) $7y = x$

38. The equation of straight line passing through $(1, \sqrt{3})$ and having slope as $\frac{1}{\sqrt{3}}$ is

(a) $x - y + 2 = 0$ (b) $\sqrt{3}x - y + 2 = 0$

(c) $2x - \sqrt{3}y = 0$ (d) $x - \sqrt{3}y + 2 = 0$

39. The equation of one of the medians of a triangle with vertices at (–1, 6), (–3, –9) and (5, –8) is

(a) $29x + 4y + 5 = 0$ (b) $8x - 5y - 21 = 0$

(c) $x + 14y + 7 = 0$ (d) Both (a) and (b)

40. The equation of a straight line making the intercept of 3 and 4 units on x-axis and y-axis respectively on the positive side is

(a) $\frac{x}{3} + \frac{y}{4} = 1$ (b) $4x - 3y = 12$

(c) $x - y = 12$ (d) None of these

41. The intercepts cut off by the straight line $2x + 5y = 10$ on x-axis and y-axis respectively are

(a) 5, 1 (b) 1, 3
(c) 5, 2 (d) 2, 3

42. The equation of a straight line making equal intercept on both the axis and passing through the point (1, 1) is

(a) $x + y = 1$ (b) $x + y = 2$
(c) $x - y = 2$ (d) $y - x = 2$

43. The equation of a straight line which is at a distance of 8 from the origin and perpendicular from the origin makes an angle 45° with the positive direction of x-axis

(a) $x + y = 8\sqrt{2}$ (b) $x - y = 8\sqrt{2}$
(c) $2x + 3y = 8$ (d) None of these

44. The slope of a straight line through (3, 2) is $\frac{3}{4}$. The coordinates of the point on the line that are 5 unit away from it are

(a) (7, 5) and (1, 1) (b) (4, 4) and (2, 2)
(c) (7, 5) and (–1, 1) (d) None of these

45. The straight line $x + y = 0$, $3x + y = 4$, $x + 3y - 4 = 0$ from a triangle which is

(a) isosceles (b) equilateral
(c) right angled (d) none of these

46. The point of intersection of the lines $\frac{x}{a} + \frac{y}{b} = 1$ and $\frac{x}{b} + \frac{y}{a} = 1$ lies on the line

(a) $x - y = 0$ (b) $(x - y)(a + b) = 2ab$
(c) $(lx + ny)(a + b) = (l - m)ab$ (d) None of these

47. The distance of the point (3, 5) from the line $2x + 3y - 14 = 0$ measured parallel to the line $x - 2y = 1$ is

(a) $\frac{7}{\sqrt{5}}$ (b) $\frac{7}{\sqrt{13}}$
(c) $\sqrt{5}$ (d) $\sqrt{13}$

48. Points (1, 2) and (–2, 1) are

(a) on the same side of the line $4x + 2y = 1$
(b) on the line $4x + 2y = 1$
(c) on opposite side of $4x + 2y = 1$
(d) none of these

49. Points on the line $x + y = 4$ that lie at a unit distance from the line $4x + 3y - 10 = 0$ are

(a) (3, 1) and (–7, 11) (b) (–3, 7) and (2, 2)
(c) (–3, 7) and (–7, 11) (d) none of these

50. The equation of a line passing through the intersection of lines $x - 3y + 1 = 0$ and $2x + 5y - 9 = 0$ and at distance $\sqrt{5}$ from the origin is

(a) $2x - y = 5$ (b) $x + 2y = 5$
(c) $2x + y = 5$ (d) $x + 2y = 1$

51. The line $x + y = 4$ divides the line joining $(-1, 1)$ and $(5, 7)$ in the ratio $\lambda : 1$, then the value of λ is

(a) 2 (b) $\frac{1}{2}$

(c) 3 (d) none of these

52. The distance between two parallel lines $y = 2x + 4$ and $6x = 3y + 5$ is

(a) $\frac{17}{\sqrt{3}}$ (b) 1

(c) $\frac{3}{\sqrt{5}}$ (d) $\frac{17\sqrt{5}}{15}$

53. The equation of the straight line, perpendicular to $y = x$ and passes through $(3, 2)$ is given by

(a) $x - y = 5$ (b) $x + y = 5$

(c) $x + y = 1$ (d) $x - y = 1$

54. The equation of the straight line passing through the point $(1, -2)$ and cuts off equal intercepts from the axes will be

(a) $x + y = 1$ (b) $x - y = 1$

(c) $x + y + 1 = 0$ (d) $x - y - 2 = 0$

55. The equation of the straight line passing through $(1, 2)$ and perpendicular to $x + y + 1 = 0$ is

(a) $y - x + 1 = 0$ (b) $y - x - 1 = 0$

(c) $y - x + 2 = 0$ (d) $y - x - 2 = 0$

Fill in the Blanks:

56. The point $(-2, -3)$ lies in .... quadrant.

57. Distance between $(2, 0)$ and $(0, 3)$ is ....

58. Length of the line segment joining $(-3, 7)$ and $(-1, -3)$ is ....

59. The coordinates of circumcentre of the triangle with vertices $(-2, -3)$, $(-1, 0)$, $(7, -6)$ are ....

60. The points $(1, 4)$, $(3, -2)$ and $(-3, 16)$ are ....

61. The orthocentre of the triangle formed by the lines $x + y = 1$, $2x + 3y = 6$ and $4x - y + 9 = 0$ lies in quadrant number ....

62. The area enclosed by the lines $|x| + |y| = 1$ is ....

63. If the points $(-2, -5)$, $(2, -2)$, $(8, a)$ are collinear, then the value of a is ....

64. If the vertices of a triangle have integral coordinates, then the triangle cannot be ....

65. The equations of the sides of a square whose centre is at origin and vertex at $(1, 2)$ are ....

66. A line is such that its segment between the lines $5x - y + 4 = 0$ and $3x + 4y - 4 = 0$ is bisected at $(1, 5)$. Its equation is ....

67. The points $(1, 3)$ and $(5, 1)$ are two opposite vertices of a rectangle and other two vertices lie on the line $y = 2x + c$. Then the value of c is ....

True and False :

68. The point (–2, –3) lies in V quadrant.

69. The distance between (0, 0) and (3, 4) is 5.

70. The length of line segment with vertices (1, 2) and (3, 4) is 2.

71. Point $P(2, -2)$, $Q(14, 10)$, $R(11, 13)$ and $S(-1, 1)$ form a rectangle.

72. The coordinates of the vertices of an equilateral triangle are integers.

73. If $\begin{vmatrix} x_1 & y_1 & 1 \\ x_2 & y_2 & 1 \\ x_3 & y_3 & 1 \end{vmatrix} = \begin{vmatrix} a_1 & b_1 & 1 \\ a_2 & b_2 & 1 \\ a_3 & b_3 & 1 \end{vmatrix}$, then the two triangles with vertices (x_1, y_1), (x_2, y_2), (x_3, y_3) and (a_1, b_1), (a_2, b_2), (a_3, b_3) must be congruent.

74. The point $P(-3, 2)$ lies inside the triangle whose sides are given by the equations $x + y = 4$, $4x - y = 31$, $3x - 7y + 8 = 0$.

75. The family of lines $ax + by + c = 0$ where $3a + 2b + 4c = 0$ passes through the point (–3, –2).

76. Only two straight lines can be drawn through the origin at equal distance from the points $P(2, 2)$ and $Q(4, 0)$.

77. The lines $ax + by + c = 0$, $bx + cy + a = 0$ and $cx + ay + b = 0$ are concurrent if $a^3 + b^3 + c^3 = 3abc$.

78. The points (3, 4) and (2, –6) are on same side.

79. The distance between the straight lines $2x - 3y + 9 = 0$ and $4x - 6y + 1 = 0$ is $\frac{17}{\sqrt{52}}$.

Answers

Objective Type Questions :

1. c	**2.** b	**3.** c	**4.** b	**5.** c	**6.** b	**7.** c	**8.** a	**9.** c	**10.** c
11. a	**12.** d	**13.** b	**14.** a	**15.** a	**16.** b	**17.** b	**18.** c	**19.** d	**20.** c
21. a	**22.** d	**23.** a	**24.** d	**25.** b	**26..** c	**27.** a	**28.** c	**29.** b	**30.** b
31. c	**32.** d	**33.** b	**34.** a	**35.** b	**36.** c	**37.** c	**38.** d	**39.** d	**40.** a
41. c	**42.** b	**43.** a	**44.** c	**45.** a	**46.** a	**47.** c	**48.** c	**49.** a	**50.** c
51. b	**52.** d	**53.** b	**54.** c	**55.** b					

Fill in the blanks :

56. III, **57.** $\sqrt{13}$, **58.** $2\sqrt{26}$, **59.** (3, –3), **60.** collinear, **61.** II

62. 2 square units, **63.** $\frac{5}{2}$, **64.** equilateral,

65. $3x + y = 5 = 0$, $x - 3y - 5 = 0$, $x - 3y + 5 = 0$, $3x + y - 5 = 0$

66. $107x - 3y - 92 = 0$, **67.** –4

True and False :

68. F, **69.** T, **70.** F, **71.** T, **72.** F, **73.** F, **74.** F, **75.** F, **76.** T

77. T, **78.** F, **79.** T.

□ ■ □

UNIT - V

Calculus

5.1 FUNCTIONS

A function is a special type of relation between two variables x and y such that for every value of the variable x there exists a definite value of another variable y. If y is a function of x then we write $y = f(x)$. here the variable x is independent variable while the variable y is a dependent variable which depends on the variable x. To find the value of $f(x)$ at $x = a$ say $f(a)$, replace x by a in $f(x)$.

Example 5.1

Let $f(x) = 3x^2 + 4x + 5$. Find the value of $f(0)$, $f(-1)$ and $f(2/3)$

SOLUTION : Given

$$f(x) = 3x^2 + 4x + 5 \qquad ...(1)$$

(i) $$f(0) = 3(0)^2 + 4(0) + 5 = 5 \qquad \text{[put } x = 0 \text{ in (1)]}$$

(ii) $$f(-1) = 3(-1)^2 + 4(-1) + 5 = 3 - 4 + 5 = 4 \qquad \text{[put } x = -1 \text{ in (1)]}$$

(iii) $$f(2/3) = 3\left(\frac{2}{3}\right)^2 + 4\left(\frac{2}{3}\right) + 5 \qquad \left[\text{put } x = \frac{2}{3} \text{ in (1)}\right]$$

$$= \frac{4}{3} + \frac{8}{3} + 5 = \frac{12}{3} + 5$$

$$= 4 + 5 = 9$$

Example 5.2

If $f(x) = \log \frac{1+x}{1-x}$, show that $f\left(\frac{2x}{1+x^2}\right) = 2f(x)$

[UPTU (C.O.) 2006]

SOLUTION : Given $f(x) = \log \frac{1+x}{1-x}$

$$\therefore \quad f\left(\frac{2x}{1+x^2}\right) = \log\left[\frac{1+\dfrac{2x}{1+x^2}}{1-\dfrac{2x}{1+x^2}}\right] = \log\left[\frac{1+x^2+2x}{1+x^2-2x}\right]$$

$$= \log\left[\frac{(1+x)^2}{(1-x)^2}\right] = \log\left(\frac{1+x}{1-x}\right)^2 = 2\log\left(\frac{1+x}{1-x}\right) = 2f(x)$$

5.2 TYPES OF FUNCTIONS

5.2.1 Explicit and Implicit Functions

A function is said to be explicit function if it is expressed directly in terms of the independent variable alone. For example, $y = 2x^3 + 3x + 4$

A function is said to be implicit function if it is not expressed directly in terms of the independent variable alone. For example,

$$x^2 + 3xy + y^2x + 3y^2 = 0,\; 3y - 2x - 5 = 0,\; x^2 + y^2 = 25$$

5.2.2 Single Valued and Multi-valued Functions

If for each value of x there exists one and only one value of y then $y = f(x)$ is said to be single-valued function of x. For example, $f(x) = 3 + 2x$

If for each value of x there exists more than one value of y, then $y = f(x)$ is said to be multi-valued function of x. For example, $f(x) = x^2 + 3x + 4$

5.2.3 Even and Odd Functions

A function $f(x)$ is said to be an even function if $f(-x) = f(x)$ for all x. For example, $\cos x$, x^2 etc.

A function $f(x)$ is said to be an odd function if $f(-x) = f(x)$ for all x. For example, $\sin x$, x^3 etc.

5.2.4 Algebraic Functions

If a function contains finite number of terms involving powers and roots of the independent variable x and the four fundemental operations such as addition, subtraction, multiplication and division then the function is said to be an algebraic function.

For example,

$$3x^2 + 2x - 5,\; \frac{3x-2}{5x^2+4x+2},\; \frac{\sqrt{2x}-3}{4x^2+3} \text{ etc. are all algebraic functions.}$$

There are three types of algebraic functions :

(i) Polynomials or rational integral functions : A function of the from

$$a_0x^n + a_1x^{n-1} + a_2x^{n-2} + \ldots + a_{n-1}x + a_n$$

where n is a positive integer and $a_0, a_1, a_2, \ldots, a_{n-1}, a_n$ are all real numbers is said to be a polynomial or rational integral function of x. For example, $3x^4 - 4x^3 + 2x^2 + 3$ etc.

(ii) Rational functions : An algebraic function of the from $\dfrac{f(x)}{g(x)}$ where $f(x)$ and $g(x)$ are polynomials in x is said to be rational.

For example, $\dfrac{2x^2 + 4x + 3}{x^2 + 1}$, etc.

(iii) Irrational functions : A function which is not rational is said to be an irrational function.

For example, $\sqrt{2x + 4}$, $\dfrac{x^3 + 3x + 2}{\sqrt{x^2 + 2x + 4}}$ etc.

5.2.5 Transcendental Functions

A function which is not an algebraic function is said to be a transcendental function. Such as

(i) Trigonometric functions
(ii) Inverse trigonometric functions
(iii) Exponential functions
(iv) Logarithmic functions

5.2.6 Modulus Functions

The function $f(x) = |x| = \begin{cases} x \text{ if } x \geq 0 \\ -x \text{ if } x < 0 \end{cases}$ is called modulus function.

5.2.7 The Greatest Integer Function

The function $f(x) = [x]$ = greatest integer less than or equal to x is known as the greatest integer function or step function. For example, [4.3] = 4, [4] = 4, [4.95] = 4,[–1.3] = –2 etc.

5.2.8 Constant Functions

A function $f(x)$ is constant if for all values of x, $f(x) = k$ where k is a constant. For example, $f(x) = 7$.

5.3 LIMIT OF A FUNCTION

Consider a function $f(x) = \dfrac{x^2 - 9}{x - 3}$, $x \neq 3$. This function is defined for all values of x except $x = 3$. At $x = 3$ we are unable to find the value of $f(x)$ by putting $x = 3$ in $f(x)$ because at $x = 3$, $f(x) = \dfrac{0}{0}$ which is an indeterminate form. But if $x \neq 3$, $f(x) = \dfrac{(x-3)(x+3)}{x-3} = x + 3$

The following table shows the values of $f(x)$ at the points which are closer to 3 from lower (left) and higher (right) sides.

x	2.5	2.6	2.7	2.8	2.9	3	3.1	3.2	3.3	3.4	3.5
$f(x)$	5.5	5.6	5.7	5.8	5.9	$\frac{0}{0}$	6.1	6.2	6.3	6.4	6.5

Left side ⟷ Right side

From this table we can see that as x approaches to 3 from lower (left) side then the function $f(x)$ approaches the limit 6.

In other words, if x tends to 3 from left hand side then the function $f(x)$ tends to the limit 6. and denoted by $x = 3-$. Thus $\lim_{x \to 3-} f(x) = 6$ or left hand limit of $f(x)$ at $x = 3$ is 6

Similarly, if x approaches to 3 from higher (right) side then the function $f(x)$ tends to the limit 6 and denoted by $x \to 3+$. Thus, $\lim_{x \to 3+} f(x) = 6$ or right hand limit of $f(x)$ at $x = 3$ is 6.

■ **DEFINITION :** *Let $f(x)$ be a function of x defined for all values of x in the neighbourhood of a fixed number a. A real number 'l' is said to be the limit of f(x) as x tends to a if for given positive real number epsilon (ε), however small, there exists a number δ (delta) such that $|f(x) - l| < \varepsilon$ for all values of x satisfying $|x - a| < \delta$. Symbolically $\lim_{x \to a} f(x) = l$.*

■ **NOTE :** It is very clear that $f(x)$ is exact value of $f(x)$ at $x = a$ but $\lim f(x)$ is the limiting value of $f(x)$ at $x = a$. In general $\lim_{x \to a} f(x) = f(a)$.

5.3.1 Left Hand and Right Hand Limits

(1) When x approaches to a through values greater than a than we can say that x tends to a from the right side and denoted by $x \to a+$. Thus, $\lim_{x \to a+} f(x)$ is the right hand limit of $f(x)$ at $x = a$.

To find right hand limit of $f(x)$ at $x = a$, we substitete $x = a + h$ and take $h \to 0$. Thus

$$\lim_{x \to a+} f(x) = \lim_{h \to 0} f(a + h)$$

(2) When x approaches to a through values smaller than a then we can say thet x tends to a from the left hand side and denoted by $x \to a-$. Thus, $\lim_{x \to a-} f(x)$ is the left hand limit of $f(x)$ at $x = a$

To find left hand limit of $f(x)$ at $x = a$ we substitute $x = a - h$ and take $h \to 0$. Thus,

$$\lim_{x \to a-} f(x) = \lim_{h \to 0} f(a - h)$$

5.3.2 Existence of Limit

The limit of a function $f(x)$ at $x = a$ exists iff $\lim_{x \to a-} f(x) = \lim_{x \to a+} f(x)$

i.e., if the left hand and right hand limits of a function $f(x)$ exists and are equal the limit of $f(x)$, then limit of $f(x)$ exists.

5.3.3 Working Rule for Finding the Left and Right Hand Limits

(1) To find the left hand limit as $x \to a$

(i) Put $x = a - h$ in $f(x)$

(ii) Take the limit of $f(a - h)$ as $h \to 0$

(2) To find the right hand limit as $x \to a$

(i) Put $x = a + h$ in $f(x)$

(ii) Take the limit of $f(a + h)$ as $h \to 0$

Example 5.3

If $f(x) = \dfrac{x^2 - 4}{x - 2}$, find $\lim\limits_{x \to 2} f(x)$

SOLUTION : Here $f(x) = \dfrac{x^2 - 4}{x - 2}$,

$$\text{Left Hand Limit (LHL)} = \lim_{x \to 2-} \left(\frac{x^2 - 4}{x - 2}\right) = \lim_{h \to 0} \frac{(2 - h)^2 - 4}{(2 - h) - 2}$$

$$= \lim_{h \to 0} \frac{4 - 4h + h^2 - 4}{2 - h - 2} = \lim_{h \to 0} \frac{h^2 - 4h}{-h}$$

$$= \lim_{h \to 0} (-h + 4) = 4$$

$$\text{Right Hand Limit (RHL)} = \lim_{x \to 2+} \left(\frac{x^2 - 4}{x - 2}\right) = \lim_{h \to 0} \frac{(2 + h)^2 - 4}{(2 + h) - 2}$$

$$= \lim_{h \to 0} \frac{4 + 4h + h^2 - 4}{2 + h - 2} = \lim_{h \to 0} \frac{4h - h^2}{h}$$

$$= \lim_{h \to 0} (4 + h) = 4$$

Since, LHL= RHL = 4, Thus, $\lim\limits_{x \to 2} f(x)$ exists and is equal to 4.

Example 5.4

$$\text{If } f(x) = \begin{cases} x & \text{if } 0 \le x < \frac{1}{2} \\ 0 & \text{if } x = \frac{1}{2} \\ x - 1 & \text{if } \frac{1}{2} < x \le 1 \end{cases}$$

Find $\lim\limits_{x \to \frac{1}{2}} f(x)$, if it exists.

SOLUTION : $\text{LHL} = \lim\limits_{x \to \frac{1}{2}-} f(x) = \lim\limits_{h \to 0} f\left(\frac{1}{2} - h\right)$

$$= \lim_{h \to 0} \left(\frac{1}{2} - h\right) = \frac{1}{2} \qquad \left[\because \frac{1}{2} - h < \frac{1}{2}\right]$$

Again, $\text{RHL} = \lim\limits_{x \to \frac{1}{2}+} f(x) = \lim\limits_{h \to 0} f\left(\frac{1}{2} + h\right)$

$$= \lim_{h \to 0} \left(\frac{1}{2} + h - 1\right) = \frac{1}{2} - 1 = -\frac{1}{2} \qquad \left[\because \frac{1}{2} + h > \frac{1}{2}\right]$$

Hence, LHL ≠ RHL. Thus, $\lim\limits_{x \to \frac{1}{2}} f(x)$ does not exists.

5.4 PROPERTIES OF LIMITS

The following results (without proof) are of fundamental importance and are often used for evaluating the limits of a function. Let f and g are two functions such that

$$\lim_{x \to a} f(x) = l \quad \text{and} \quad \lim_{x \to a} g(x) = m \text{ , then}$$

(1) $\lim\limits_{x \to a} [f(x) \pm g(x)] = \lim\limits_{x \to a} f(x) \pm \lim\limits_{x \to a} g(x) = l \pm m$

(2) $\lim\limits_{x \to a} [f(x) \cdot g(x)] = \lim\limits_{x \to a} f(x) \cdot \lim\limits_{x \to a} g(x) = l.m$

(3) $\lim\limits_{x \to a} \left(\dfrac{f(x)}{g(x)}\right) = \dfrac{\lim\limits_{x \to a} f(x)}{\lim\limits_{x \to a} g(x)} = \dfrac{l}{m}$, provided $\lim\limits_{x \to a} g(x) \neq 0$

(4) $\lim\limits_{x \to a} k\, f(x) = k \cdot \lim\limits_{x \to a} f(x) = k.l$, where k is a constant

(5) $\lim\limits_{x \to a} \log f(x) = \log \lim\limits_{x \to a} f(x) = \log l$.

(6) $\lim\limits_{x \to a} k = k$, where k is constant.

5.5 SOME IMPORTANT SERIES

5.5.1 Binomial Theorem for Positive Integral Index :

$$(x + a)^n = x^n + {}^nc_1\, x^{n-1.}\, a + {}^nc_2\, x^{n-2.}\, a^2 + \dots + {}^nc_n\, a^n$$

for all x and $n \in N$.

5.5.2 Binomial Theorem for any Index other than a Positive Integer

$$(1 + x)^n = 1 + nx + \frac{n(n-1)}{2\,!} x^2 + \frac{n(n-1)(n-2)}{3!} x^3 + \dots$$

for $|x| < 1$.

5.5.3 Exponential Series

(i) $a^x = 1 + x \log_e a + \dfrac{x^2}{2!} (\log_e a)^2 + \dots$

(ii) $e^x = 1 + x + \dfrac{x^2}{2!} + \dfrac{x^3}{3!} + \dots$

5.5.4 Logarithmic Series

(i) $\log (1 + x) = x - \dfrac{x^2}{2} + \dfrac{x^3}{3} - \dfrac{x^4}{4} + \dots, \quad -1 < x < 1$

(ii) $\log (1 - x) = -x - \dfrac{x^2}{2} - \dfrac{x^3}{3} - \dfrac{x^4}{4} - \dots, \quad -1 < x < 1$

5.5.5 Trigonometric Series

(i) $\sin x = x - \frac{x^3}{3!} + \frac{x^5}{5!} - \frac{x^7}{7!} + \dots$

(ii) $\cos x = 1 - \frac{x^2}{2!} + \frac{x^4}{4!} - \frac{x^6}{6!} + \dots$

(iii) $\tan x = x + \frac{x^3}{3} + \frac{2}{15}x^5 + \dots$

5.6 INDETERMINATE FORMS

The following forms are known as indeterminate form such as

$$\frac{0}{0}, \frac{\infty}{\infty}, 0 \times \infty, \infty - \infty, 1^\infty, 0^\infty, \infty^0 \text{ etc.}$$

5.7 SOME IMPORTANT LIMITS

(1) $\lim\limits_{x \to a} \frac{x^n - a^n}{x - a} = na^{n-1} \ (a > 0)$

(2) $\lim\limits_{x \to a} \frac{x^m - a^m}{x^n - a^n} = \frac{m}{n} a^{m-n}$, m and n are rational numbers.

(3) $\lim\limits_{x \to \infty} \left(1 + \frac{1}{x}\right)^x = e$

(4) $\lim\limits_{x \to 0} (1 + x)^{1/x} = e$

(5) $\lim\limits_{x \to 0} \frac{e^x - 1}{x} = 1$

(6) $\lim\limits_{x \to 0} \sin x = 0$ and $\lim\limits_{x \to 0} \cos x = 1$

(7) $\lim\limits_{x \to 0} \frac{\sin x}{x} = 1$

(8) $\lim\limits_{x \to 0} \frac{\tan x}{x} = 1$

(9) $\lim\limits_{x \to 0} \sec x = 1$

(10) $\lim\limits_{x \to 0} \frac{\sin^{-1} x}{x} = 1$

(11) $\lim\limits_{x \to 0} \frac{\tan^{-1} x}{x} = 1$

5.8 EVALUATION OF LIMITS

There are two types of limits such as algebraic and non-algebraic limits. If a problems of limits does not contain trigonometric, inverse trigonometric, exponential and logarithmic functions then it is a problem of algebraic limits. Otherwise it is a problem of non-algebraic limits.

Example 5.5

$\lim_{x \to 1} \frac{x^2 - 1}{x + 1}$, $\lim_{x \to 0} (3x^2 + 4x + 5)$ are algebraic limits.

$\lim_{x \to 0} (\sin x + 2e^x)$, $\lim_{x \to 1} (\sin^{-1} x + 4x^2 + 3 \cos x)$ are non-algebraic limits.

There are various methods of evaluating the limits of a function such as :

(1) Substitution method

(2) Factorisation method

(3) Rationalisation method

(4) $x \to \infty$ type

5.8.1 Substitution Method

If we get a finite number by direct substitution of a point in the given function, then that finite number is the limits of the given function.

Example 5.6

Evaluate : (i) $\lim_{x \to 3} (x^3 + 4x + 5)$ (ii) $\lim_{x \to 0} \frac{\sin x}{x^2 + 2}$

(iii) $\lim_{x \to 1} (x^2 + 4)(3x - 2)$ (iv) $\lim_{x \to 1} \frac{2x^2 + x - 3}{x^2 + 9}$

SOLUTION : (i) $\lim_{x \to 3} (x^3 + 4x + 5) = (3)^3 + 4(3) + 5 = 27 + 12 + 5 = 44$

(ii) $\lim_{x \to 0} \frac{\sin x}{x^2 + 2} = \frac{\sin 0}{(0)^2 + 2} = \frac{0}{2} = 0$

(iii) $\lim_{x \to 1} (x^2 + 4)(3x - 2) = [(1)^2 + 4][3(1) - 2] = (1 + 4)(3 - 2) = 5$

(iv) $\lim_{x \to 1} \frac{2x^2 + x - 3}{x^2 + 9} = \frac{2(1)^2 + 1 - 3}{(1)^2 + 9} = \frac{2 + 1 - 3}{1 + 9} = \frac{0}{10} = 0$

5.8.2 Factorisation Method

Example 5.7

(i) $\lim_{x \to 2} \frac{x^2 - 5x + 6}{x^2 - 4}$ (ii) $\lim_{x \to 1} \frac{x - 1}{2x^2 - 7x + 5}$

(iii) $\lim_{x \to 1} \frac{(2x - 3)(\sqrt{x} - 1)}{2x^2 + x - 3}$ (iv) $\lim_{x \to 3} \frac{x^2 + 2x - 15}{x^2 - 9}$

SOLUTION :

(i) $\lim_{x \to 2} \frac{x^2 - 5x + 6}{x^2 - 4}$

At $x = 2$, $\frac{x^2 - 5x + 6}{x^2 - 4} = \frac{4 - 10 + 6}{4 - 4} = \frac{0}{0}$ (indeterminate form)

So, its limit is evaluated by proper factorisation.

$$\therefore \quad \lim_{x \to 2} \frac{x^2 - 5x + 6}{x^2 - 4} = \lim_{x \to 2} \frac{(x-2)(x-3)}{(x-2)(x+2)}$$

$$= \lim_{x \to 2} \frac{x-3}{x+2} = \frac{2-3}{2+2} = -\frac{1}{4}$$

(ii) $$\lim_{x \to 1} \frac{x-1}{2x^2 - 7x + 5}$$

At $x = 1, \dfrac{x-1}{2x^2 - 7x + 5} = \dfrac{0}{0}$ (indeterminate form)

So, its limit is evaluated by proper factorisation.

$$\therefore \quad \lim_{x \to 1} \frac{x-1}{2x^2 - 7x + 5} = \lim_{x \to 1} \frac{x-1}{(x-1)(2x-5)}$$

$$= \lim_{x \to 1} \frac{1}{2x-5} = \frac{1}{2(1)-5} = \frac{1}{2-5} = -\frac{1}{3}$$

(iii) $\lim\limits_{x \to 1} \dfrac{(2x-3)(\sqrt{x}-1)}{2x^2 + x - 3} = \dfrac{0}{0}$ (indeterminate from)

$$\therefore \quad \lim_{x \to 1} \frac{(2x-3)(\sqrt{x}-1)}{(2x+3)(x-1)} = \lim_{x \to 1} \frac{(2x-3)(\sqrt{x}-1)}{(2x+3)(\sqrt{x}+1)(\sqrt{x}-1)}$$

$$= \lim_{x \to 1} \frac{(2x-3)}{(2x+3)(\sqrt{x}+1)} = \frac{2-3}{(2+3)(1+1)} = -\frac{1}{10}$$

(iv) $\lim\limits_{x \to 3} \dfrac{x^2 + 2x - 15}{x^2 - 9} = \dfrac{0}{0}$ (indeterminate form)

$$\therefore \quad \lim_{x \to 3} \frac{(x-3)(x+5)}{(x-3)(x+3)} = \lim_{x \to 3} \frac{x+5}{x+3} = \frac{3+5}{3+3} = \frac{8}{6} = \frac{4}{3}$$

5.8.3 Rationalisation Method

This is used when either the numerator or denominator or both having functions of square roots.

Example 5.8

Evaluate : (i) $\lim\limits_{h \to 0} \left(\dfrac{\sqrt{x+h} - \sqrt{x}}{h} \right)$ (ii) $\lim\limits_{x \to 0} \dfrac{\sqrt{1+x} - \sqrt{1+x^2}}{x}$

(iii) $\lim\limits_{x \to 0} \dfrac{x}{\sqrt{a+x} - \sqrt{a-x}}$ (iv) $\lim\limits_{x \to 0} \dfrac{\sqrt{1+x} - 1}{x}$ ***[UPTU 2004]***

SOLUTION :

(i) $$\lim_{h \to 0} \left(\frac{\sqrt{x+h} - \sqrt{x}}{h} \right) = \lim_{h \to 0} \left(\frac{\sqrt{x+h} - \sqrt{x}}{h} \times \frac{\sqrt{x+h} + \sqrt{x}}{\sqrt{x+h} + \sqrt{x}} \right)$$

$$= \lim_{h \to 0} \left(\frac{x+h-x}{h(\sqrt{x+h} + \sqrt{x})} \right)$$

$$= \lim_{h \to 0} \frac{h}{h(\sqrt{x+h} + \sqrt{x})}$$

$$= \lim_{h \to 0} \frac{1}{\sqrt{x+h} + \sqrt{x}}$$

$$= \frac{1}{\sqrt{x} + \sqrt{x}} = \frac{1}{2\sqrt{x}}$$

(ii) $$\lim_{x \to 0} \left(\frac{\sqrt{1+x} - \sqrt{1+x^2}}{x} \right) = \lim_{x \to 0} \left(\frac{\sqrt{1+x} - \sqrt{1+x^2}}{x} \times \frac{\sqrt{1+x} + \sqrt{1+x^2}}{\sqrt{1+x} + \sqrt{1+x^2}} \right)$$

$$= \lim_{x \to 0} \left(\frac{1+x-1-x^2}{x(\sqrt{1+x} + \sqrt{1+x^2})} \right)$$

$$= \lim_{x \to 0} \frac{x - x^2}{x(\sqrt{1+x} + \sqrt{1+x^2})} = \lim_{x \to 0} \frac{1-x}{\sqrt{1+x} + \sqrt{1+x^2}}$$

$$= \frac{1}{\sqrt{1} + \sqrt{1}} = \frac{1}{1+1} = \frac{1}{2}$$

(iii) $$\lim_{x \to 0} \left(\frac{x}{\sqrt{a+x} - \sqrt{a-x}} \right) = \lim_{x \to 0} \left(\frac{x}{\sqrt{a+x} - \sqrt{a-x}} \times \frac{\sqrt{a+x} + \sqrt{a-x}}{\sqrt{a+x} + \sqrt{a-x}} \right)$$

$$= \lim_{x \to 0} \left(\frac{x(\sqrt{a+x} + \sqrt{a-x})}{a+x-a+x} \right)$$

$$= \lim_{x \to 0} \left(\frac{x(\sqrt{a+x} + \sqrt{a-x})}{2x} \right)$$

$$= \frac{1}{2} \lim_{x \to 0} (\sqrt{a+x} + \sqrt{a-x}) = \frac{1}{2} (\sqrt{a} + \sqrt{a}) = \frac{2\sqrt{a}}{2} = \sqrt{a}$$

(iv) $$\lim_{x \to 0} \frac{\sqrt{1+x} - 1}{x} = \lim_{x \to 0} \frac{\sqrt{1+x} - 1}{x} \times \frac{\sqrt{1+x} + 1}{\sqrt{1+x} + 1}$$

$$= \lim_{x \to 0} \frac{1+x-1}{x(\sqrt{1+x} + 1)}$$

$$= \lim_{x \to 0} \frac{x}{x(\sqrt{1+x} + 1)} = \lim_{x \to 0} \frac{1}{\sqrt{1+x} + 1}$$

$$= \frac{1}{\sqrt{1} + 1} = \frac{1}{1+1} = \frac{1}{2}$$

5.8.4 Evaluation of Algebraic Limits when $x \to \infty$

Method 1 **:** First divide the numerator and denominator by the highest power of x

Example 5.9

Evaluate : (i) $\lim_{x \to \infty} \frac{(x+1)(2x+3)}{(x+2)(3x+4)}$ (ii) $\lim_{x \to \infty} (\sqrt{x^2+x+1} - \sqrt{x^2+1})$

SOLUTION :

(i) $$\lim_{x \to \infty} \frac{(x+1)(2x+3)}{(x+2)(3x+4)} = \lim_{x \to \infty} \frac{2x^2+5x+3}{3x^2+10x+8}$$

Divide numerator and denominator by x^2

$$= \lim_{x \to \infty} \frac{\frac{2x^2 + 5x + 3}{x^2}}{\frac{3x^2 + 10x + 8}{x^2}} = \lim_{x \to \infty} \frac{2 + \frac{5}{x} + \frac{3}{x^2}}{3 + \frac{10}{x} + \frac{8}{x^2}}$$

$$= \frac{2 + 0 + 0}{3 + 0 + 0} \qquad \left[\because \frac{1}{x} = \frac{1}{x^2} = 0 \text{ at } x = \infty\right]$$

$$= \frac{2}{3}$$

(ii) $\lim_{x \to \infty} \left(\sqrt{x^2 + x + 1} - \sqrt{x^2 + 1}\right)$

$$= \lim_{x \to \infty} \left(\sqrt{x^2 + x + 1} - \sqrt{x^2 + 1}\right) \times \left(\frac{\sqrt{x^2 + x + 1} + \sqrt{x^2 + 1}}{\sqrt{x^2 + x + 1} + \sqrt{x^2 + 1}}\right)$$

$$= \lim_{x \to \infty} \left(\frac{x^2 + x + 1 - x^2 - 1}{\sqrt{x^2 + x + 1} + \sqrt{x^2 + 1}}\right)$$

$$= \lim_{x \to \infty} \left(\frac{x}{\sqrt{x^2 + x + 1} + \sqrt{x^2 + 1}}\right)$$

Divide numerator and denominator by x

$$= \lim_{x \to \infty} \left(\frac{\frac{x}{x}}{\sqrt{\frac{x^2}{x^2} + \frac{x}{x^2} + \frac{1}{x^2}} + \sqrt{\frac{x^2}{x^2} + \frac{1}{x^2}}}\right)$$

$$= \lim_{x \to \infty} \left(\frac{1}{\sqrt{1 + \frac{1}{x} + \frac{1}{x^2}} + \sqrt{1 + \frac{1}{x^2}}}\right)$$

$$= \frac{1}{\sqrt{1} + \sqrt{1}} = \frac{1}{1 + 1} = \frac{1}{2}$$

Example 5.10

Evaluate : (i) $\lim_{n \to \infty} \left(\frac{1}{3} + \frac{1}{3^2} + \frac{1}{3^3} + \ldots + \frac{1}{3^n}\right)$ (ii) $\lim_{n \to \infty} \frac{\Sigma n^2}{n^3}$

SOLUTION : (i) We have $\frac{1}{3} + \frac{1}{3^2} + \frac{1}{3^3} + \ldots + \frac{1}{3^n}$

This is a G.P. up to n terns

$$\therefore \quad \frac{1}{3} + \frac{1}{3^2} + \frac{1}{3^3} + \ldots + \frac{1}{3^n} = \frac{1}{2}\left(1 - \frac{1}{3^n}\right)$$

Now, $\lim_{n \to \infty} \left(\frac{1}{3} + \frac{1}{3^2} + \frac{1}{3^3} + \ldots + \frac{1}{3^n}\right) = \lim_{n \to \infty} \frac{1}{2}\left(1 - \frac{1}{3^n}\right)$

$$= \frac{1}{2}\left(\lim_{n \to \infty} 1 - \lim_{n \to \infty} \left(\frac{1}{3}\right)^n\right) \qquad \because \frac{1}{3} < 1 \quad \therefore \lim_{n \to \infty} \left(\frac{1}{3}\right)^n = 0$$

So, $$\lim_{n \to \infty}\left(\frac{1}{3} + \frac{1}{3^2} + \frac{1}{3^3} + \dots + \frac{1}{3^n}\right) = \frac{1}{2}\cdot[1 - 0] = \frac{1}{2}$$

(ii) $$\lim_{n \to \infty} \frac{\Sigma n^2}{n^3} = \lim_{n \to \infty} \frac{1^2 + 2^2 + 3^2 + \dots + n^2}{n^3}$$

$$= \lim_{n \to \infty} \frac{\frac{1}{6} n\,(n+1)\,(2n+1)}{n^3}$$

$$= \lim_{n \to \infty} \frac{(n+1)\,(2n+1)}{6n^2} = \lim_{n \to \infty} \frac{2n^2 + 3n + 1}{6n^2}$$

$$= \frac{1}{6} \lim_{n \to \infty}\left(2 + \frac{3}{n} + \frac{1}{n^2}\right)$$

$$= \frac{1}{6}(2 + 0 + 0) \qquad \left(\because \frac{1}{n} = \frac{1}{n^2} = 0 \text{ at } n = \infty\right)$$

$$= \frac{1}{3}$$

5.8.5 Some Standard Results

Example 5.11

Evaluate : (i) $\lim_{x \to a} \frac{x\sqrt{x} - a\sqrt{a}}{x - a}$ (ii) $\lim_{x \to a} \frac{(x+2)^{5/3} - (a+2)^{5/3}}{x - a}$

(iii) $\lim_{x \to 1} \frac{(x + x^2 + x^3 + \dots + x^n) - n}{x - 1}$ (iv) $\lim_{x \to a} \frac{x^n - a^n}{x^m - a^m}$

SOLUTION :

(i) $$\lim_{x \to a} \frac{x\sqrt{x} - a\sqrt{a}}{x - a} = \lim_{x \to a} \frac{x^{3/2} - a^{3/2}}{x - a}$$

$$= \frac{3}{2} a^{3/2 - 1} \qquad \left(\because \lim_{x \to a} \frac{x^n - a^n}{x - a}\right) = na^{n-1}$$

$$= \frac{3}{2} a^{1/2}$$

(ii) $$\lim_{x \to a} \frac{(x+2)^{5/3} - (a+2)^{5/3}}{x - a}$$

$$= \lim_{x \to a} \frac{(x+2)^{5/3} - (a+2)^{5/3}}{(x+2) - (a+2)}$$

$$= \frac{5}{3}(a+2)^{5/3 - 1} \qquad \left(\because \lim_{x \to a} \frac{x^n - a^n}{x - a} = na^{n-1}\right)$$

$$= \frac{5}{3}(a+2)^{2/3}$$

(iii) $\lim\limits_{x \to 1} \dfrac{(x + x^2 + x^3 + \ldots + x^n) - n}{x - 1}$

$$= \lim_{x \to 1} \frac{(x + x^2 + x^3 + \ldots + x^n) - (1 + 1 + \ldots + 1 \; n \text{ terms})}{x - 1}$$

$$= \lim_{x \to 1} \frac{(x - 1) + (x^2 - 1) + (x^3 - 1) + \ldots + (x^n - 1)}{x - 1}$$

$$= \lim_{x \to 1} \left(\frac{x-1}{x-1} + \frac{x^2-1}{x-1} + \frac{x^3-1}{x-1} + \ldots + \frac{x^n-1}{x-1} \right)$$

$$= \lim_{x \to 1} \frac{x-1}{x-1} + \lim_{x \to 1} \frac{x^2-1}{x-1} + \lim_{x \to 1} \frac{x^3-1}{x-1} + \ldots + \lim_{x \to 1} \frac{x^n-1}{x-1}$$

$$= 1 + 2\,(1)^{2-1} + 3\,(1)^{3-1} + \ldots + n\,(1)^{n-1} \qquad \left(\because \lim_{x \to a} \frac{x^n - a^n}{x - a} = na^{n-1} \right)$$

$$= 1 + 2 + 3 + \ldots + n$$

$$= \frac{n\,(n+1)}{2}$$

(iv) $\lim\limits_{x \to a} \dfrac{x^n - a^n}{x^m - a^m}$

$$= \lim_{x \to a} \frac{\dfrac{x^n - a^n}{x - a}}{\dfrac{x^m - a^m}{x - a}} = \frac{\lim\limits_{x \to a} \dfrac{x^n - a^n}{x - a}}{\lim\limits_{x \to a} \dfrac{x^m - a^m}{x - a}}$$

$$= \frac{na^{n-1}}{ma^{m-1}} = \frac{n}{m}\, a^{n-1-m+1} = \frac{n}{m}\, a^{n-m}$$

5.8.6 Trigonometric Limits

Example 5.12

Evaluate : (i) $\lim\limits_{x \to 0} \dfrac{\sin 2x}{3x}$ (ii) $\lim\limits_{x \to 0} \dfrac{\sin 3x}{\sin 5x}$

(iii) $\lim\limits_{x \to 0} \dfrac{x^3}{\sin x^2}$ (iv) $\lim\limits_{x \to 0} \dfrac{\sin 2x\,(1 - \cos 2x)}{x^3}$

(v) $\lim\limits_{x \to 0} \dfrac{\sin 3x \cos 2x}{\sin 2x}$

SOLUTION :

(i) $$\lim_{x \to 0} \frac{\sin 2x}{3x} = \lim_{x \to 0} \frac{2}{3} \frac{\sin 2x}{2x} = \frac{2}{3} \lim_{x \to 0} \frac{\sin 2x}{2x} = \frac{2}{3} \times 1 = \frac{2}{3}$$

(ii) $$\lim_{x \to 0} \frac{\sin 3x}{\sin 5x} = \lim_{x \to 0} \frac{\dfrac{\sin 3x}{x}}{\dfrac{\sin 5x}{x}} = \frac{\lim\limits_{x \to 0} \dfrac{3 \sin 3x}{3x}}{\lim\limits_{x \to 0} \dfrac{5 \sin 5x}{5x}}$$

$$= \frac{3}{5}\frac{\lim\limits_{x\to 0}\dfrac{\sin 3x}{3x}}{\lim\limits_{x\to 0}\dfrac{\sin 5x}{5x}} = \frac{3}{5}\cdot\frac{1}{1} = \frac{3}{5}$$

(iii) $$\lim_{x\to 0}\frac{x^3}{\sin x^2} = \lim_{x\to 0}\left(\frac{x^2}{\sin x^2}\cdot x\right) = \lim_{x\to 0}\frac{x^2}{\sin x^2}\cdot\lim_{x\to 0} x$$

$$= 1\times 0 = 0$$

(iv) $$\lim_{x\to 0}\frac{\sin 2x\,(1-\cos 2x)}{x^3}$$

$$= \lim_{x\to 0}\frac{2\sin x\cos x\,(1-1+2\sin^2 x)}{x^3}$$

$$= \lim_{x\to 0}\frac{4\sin^3 x\cos x}{x^3} = 4\lim_{x\to 0}\left(\frac{\sin x}{x}\right)^3\cos x$$

$$= 4\times 1\times 1 = 4$$

(v) $$\lim_{x\to 0}\frac{\sin 3x\cos 2x}{\sin 2x} = \lim_{x\to 0}\left(\frac{\dfrac{3\sin 3x}{3x}}{\dfrac{2\sin 2x}{2x}}\cdot\cos 2x\right)$$

$$= \frac{3\times 1}{2\times 1}\times 1$$

$$= \frac{3}{2}$$

Example 5.13

Evaluate : (i) $\lim\limits_{x\to 0}\dfrac{\tan x-\sin x}{x^3}$ (ii) $\lim\limits_{x\to 0}\dfrac{1-\cos x}{x^2}$

(iii) $\lim\limits_{x\to 0}\dfrac{\tan x-\sin x}{\sin^3 x}$ (iv) $\lim\limits_{x\to 0}\dfrac{(x+y)\sec(x+y)-y\sec y}{x}$

SOLUTION :

(i) $$\lim_{x\to 0}\frac{\tan x-\sin x}{x^3} = \lim_{x\to 0}\frac{\dfrac{\sin x}{\cos x}-\sin x}{x^3} = \lim_{x\to 0}\frac{\sin x-\sin x\cos x}{x^3\cos x}$$

$$= \lim_{x\to 0}\frac{\sin x\,(1-\cos x)}{x^3\cos x} = \lim_{x\to 0}\frac{\sin x\,(1-1+2\sin^2 x/2)}{x^3\cos x}$$

$$= \lim_{x\to 0} 2\cdot\frac{\sin x}{x}\cdot\frac{\sin^2\dfrac{x}{2}}{x^2}\cdot\frac{1}{\cos x}$$

$$= \lim_{x\to 0}\frac{2}{4}\cdot\frac{\sin x}{x}\cdot\left(\frac{\sin\dfrac{x}{2}}{\dfrac{x}{2}}\right)^2\cdot\frac{1}{\cos x}$$

$$= \frac{1}{2} \times 1 \times 1^2 \times \frac{1}{1} = \frac{1}{2}$$

(ii) $\lim\limits_{x \to 0} \dfrac{1 - \cos x}{x^2} = \lim\limits_{x \to 0} \dfrac{1 - 1 + 2\sin^2 \frac{x}{2}}{x^2} = \lim\limits_{x \to 0} \dfrac{2\sin^2 \frac{x}{2}}{x^2}$

$$= \lim_{x \to 0} \frac{2}{4} \left(\frac{\sin \frac{x}{2}}{\frac{x}{2}} \right)^2 = \frac{2}{4} \times 1 = \frac{1}{2}$$

(iii) $\lim\limits_{x \to 0} \dfrac{\tan x - \sin x}{\sin^3 x} = \lim\limits_{x \to 0} \dfrac{\sin x - \sin x \cos x}{\sin^3 x \cos x} = \lim\limits_{x \to 0} \dfrac{\sin x\,(1 - \cos x)}{\sin^3 x \cos x}$

$$= \lim_{x \to 0} \frac{1 - \cos x}{\sin^2 x \cos x} = \lim_{x \to 0} \frac{2\sin^2 \frac{x}{2}}{\sin^2 x} . \frac{1}{\cos x}$$

$$= \lim_{x \to 0} 2\, \frac{\sin^2 \frac{x}{2}}{x^2} \times \frac{x^2}{\sin^2 x} . \frac{1}{\cos x}$$

$$= \lim_{x \to 0} 2\, \frac{\sin^2 \frac{x}{2}}{\frac{4x^2}{4}} \times \frac{x^2}{\sin^2 x} . \frac{1}{\cos x}$$

$$= \lim_{x \to 0} \frac{2}{4} \left(\frac{\sin \frac{x}{2}}{\frac{x}{2}} \right)^2 \left(\frac{x}{\sin x} \right)^2 \frac{1}{\cos x}$$

$$= \frac{2}{4} \times 1^2 \times 1^2 \times \frac{1}{1} = \frac{1}{2}$$

(iv) $\lim\limits_{x \to 0} \dfrac{(x + y)\sec(x + y) - y \sec y}{x}$

$$= \lim_{x \to 0} \frac{x \sec(x + y) + y\,[\sec(x + y) - \sec y]}{x}$$

$$= \lim_{x \to 0} \frac{x \sec(x + y)}{x} + \lim_{x \to 0} y \left(\frac{\sec(x + y) - \sec y}{x} \right)$$

$$= \lim_{x \to 0} \sec(x + y) + \lim_{x \to 0} y \left(\frac{\cos y - \cos(x + y)}{x \cos y \cos(x + y)} \right)$$

$$= \lim_{x \to 0} \sec(x + y) + \lim_{x \to 0} y \left(\frac{2\sin\left(y + \frac{x}{2}\right) \sin\left(\frac{x}{2}\right)}{x} \frac{1}{\cos y \cos(x + y)} \right)$$

$$= \lim_{x \to 0} \sec(x + y) + \lim_{x \to 0} 2y \sin\left(y + \frac{x}{2}\right) \cdot \frac{\sin \frac{x}{2}}{\frac{2x}{2}} \cdot \frac{1}{\cos y \cos(x + y)}$$

$$= \sec y + y \sin y \times 1 \times \frac{1}{\cos^2 y}$$

$$= \sec y + \frac{y \sin y}{\cos^2 y} = \sec y + y \tan y \sec y$$

$$= \sec y\,(1 + y \tan y)$$

Example 5.14

Evaluate :

(i) $\lim\limits_{x \to \frac{\pi}{2}} \dfrac{1 + \cos 2x}{(\pi - 2x)^2}$ (ii) $\lim\limits_{x \to \frac{\pi}{2}} (\sec x - \tan x)$

(iii) $\lim\limits_{x \to \frac{\pi}{2}} (\pi/2 - x) \tan x$ (iv) $\lim\limits_{x \to \frac{\pi}{4}} \dfrac{\sec^2 x - 2}{\tan x - 1}$

SOLUTION : (i) $\lim\limits_{x \to \frac{\pi}{2}} \dfrac{1 + \cos 2x}{(\pi - 2x)^2} = \lim\limits_{h \to 0} \dfrac{1 + \cos 2\left(\frac{\pi}{2} + h\right)}{\left(\pi - 2\left(\frac{\pi}{2} + h\right)\right)^2}$

$$= \lim_{h \to 0} \frac{1 + \cos(\pi + 2h)}{4h^2}$$

$$= \lim_{h \to 0} \frac{1 - \cos 2h}{4h^2} = \lim_{h \to 0} \frac{1 - 1 + 2\sin^2 h}{4h^2}$$

$$= \frac{1}{2} \lim_{h \to 0} \left(\frac{\sin h}{h}\right)^2 = \frac{1}{2}$$

(ii) $\lim\limits_{x \to \frac{\pi}{2}} (\sec x - \tan x) = \lim\limits_{x \to \frac{\pi}{2}} \left(\dfrac{1}{\cos x} - \dfrac{\sin x}{\cos x}\right)$

$$= \lim_{x \to \frac{\pi}{2}} \left(\frac{1 - \sin x}{\cos x}\right) = \lim_{h \to 0} \left(\frac{1 - \sin\left(\frac{\pi}{2} + h\right)}{\cos\left(\frac{\pi}{2} + h\right)}\right)$$

$$= \lim_{h \to 0} \left(\frac{1 - \cos h}{-\sin h}\right) = \lim_{h \to 0} \left(\frac{1 - 1 + 2\sin^2 \frac{h}{2}}{-2 \sin \frac{h}{2} \cos \frac{h}{2}}\right)$$

$$= -\lim_{h \to 0} \frac{\sin \frac{h}{2}}{\cos \frac{h}{2}} = 0$$

(iii) $$\lim_{x \to \frac{\pi}{2}} \left(\frac{\pi}{2} - x\right) \tan x = \lim_{h \to 0} \left(\frac{\pi}{2} - \left(\frac{\pi}{2} + h\right)\right) \tan\left(\frac{\pi}{2} + h\right)$$

$$= \lim_{h \to 0} [-h(-\cot h)]$$

$$= \lim_{h \to 0} (h \cot h) = \lim_{h \to 0} \frac{h}{\tan h} = 1$$

(iv) $$\lim_{x \to \frac{\pi}{4}} \frac{\sec^2 x - 2}{\tan x - 1}$$

$$= \lim_{x \to \frac{\pi}{4}} \frac{1 + \tan^2 x - 2}{\tan x - 1} = \lim_{x \to \frac{\pi}{4}} \frac{\tan^2 x - 1}{\tan x - 1}$$

$$= \lim_{x \to \frac{\pi}{4}} \frac{(\tan x - 1)(\tan x + 1)}{(\tan x - 1)} = \lim_{x \to \frac{\pi}{4}} (\tan x + 1)$$

$$= \tan \frac{\pi}{4} + 1 = 1 + 1 = 2$$

Example 5.15

Evaluate : (i) $\lim_{x \to 0} \frac{\sqrt{1+x} - \sqrt{1-x}}{\sin^{-1} x}$ (ii) $\lim_{x \to 1} \frac{\cos^{-1} x}{\sqrt{1-x^2}}$

SOLUTION :

(i) $$\lim_{x \to 0} \frac{\sqrt{1+x} - \sqrt{1-x}}{\sin^{-1} x}$$

$$= \lim_{x \to 0} \frac{\sqrt{1+x} - \sqrt{1-x}}{\sin^{-1} x} \times \frac{\sqrt{1+x} + \sqrt{1-x}}{\sqrt{1+x} + \sqrt{1-x}}$$

$$= \lim_{x \to 0} \frac{(1+x) - (1-x)}{\sin^{-1} x (\sqrt{1+x} + \sqrt{1-x})}$$

$$= \lim_{x \to 0} \frac{2x}{\sin^{-1} x (\sqrt{1+x} + \sqrt{1-x})}$$

$$= 2 \lim_{x \to 0} \frac{x}{\sin^{-1} x} \cdot \frac{1}{\sqrt{1+x} + \sqrt{1-x}}$$

$$= 2 \times 1 \times \frac{1}{\sqrt{1} + \sqrt{1}} = \frac{2}{1+1} = 1$$

(ii) $$\lim_{x \to 1} \frac{\cos^{-1} x}{\sqrt{1-x^2}}$$

Let $x = \cos\theta$ then $x = 1 \Rightarrow \theta = 0$

$$\therefore \quad \lim_{x \to 1} \frac{\cos^{-1} x}{\sqrt{1-x^2}} = \lim_{\theta \to 0} \frac{\theta}{\sqrt{1 - \cos^2\theta}} = \lim_{\theta \to 0} \frac{\theta}{\sin\theta} = 1$$

Example 5.16

Evaluate $\lim\limits_{x \to a} \frac{\cos x - \cos a}{x - a}$ *[UPTU 2003]*

SOLUTION : $\lim\limits_{x \to a} \frac{\cos x - \cos a}{x - a} \quad \left(\frac{0}{0}\right)$

By L'hospital rule

$$\lim_{x \to a} \frac{-\sin x - 0}{1 - 0} = \lim_{x \to a} (-\sin x) = -\sin a$$

Example 5.17

Evaluate $\lim\limits_{x \to a} \frac{a^x - 1}{x}$ *[UPTU 2005]*

SOLUTION : $\lim\limits_{x \to a} \frac{a^x - 1}{x} \quad \left(\frac{0}{0}\right)$

By L'hospital rule

$$\lim_{x \to 0} \frac{a^x \log_e a}{1} = \lim_{x \to 0} a^x \log_e a = \log_e a$$

Example 5.18

Evaluate $\lim\limits_{x \to 0} \frac{\cos x - 1 + \frac{x^2}{2}}{x^4}$ *[UPTU (C.O.) 2006]*

SOLUTION : We have

$$\lim_{x \to 0} \frac{\cos x - 1 + \frac{x^2}{2}}{x^4} \qquad \left(\frac{0}{0}\right)$$

By L'Hospital Rule

$$= \lim_{x \to 0} \frac{-\sin x - 0 + \frac{2x}{2}}{4x^3} = \lim_{x \to 0} \frac{-\sin x + x}{4x^3} \qquad \left(\frac{0}{0}\right)$$

Again using the same rule

$$= \lim_{x \to 0} \frac{-\cos x + 1}{12x^2} \qquad \left(\frac{0}{0}\right)$$

Again using the same rule

$$= \lim_{x \to 0} \frac{\sin x}{24x} \qquad \left(\frac{0}{0}\right)$$

$$= \lim_{x \to 0} \frac{\cos x}{24}$$

$$= \frac{1}{24}$$

PROBLEM SET 5.1

1. Prove that $\lim\limits_{x \to 0} \frac{|x|}{x}$, $x \neq 0$ does not exist.

2. Show that $\lim\limits_{x \to 0} \sin \frac{1}{x}$ does not exist.

3. Evaluate the following :

(i) $\lim\limits_{x \to 1} \frac{x^4 - 3x^3 + 2}{x^3 - 5x^2 + 3x + 1}$ (ii) $\lim\limits_{x \to \frac{1}{2}} \frac{8x^3 - 1}{16x^4 - 1}$

(iii) $\lim\limits_{x \to \sqrt{3}} \frac{x^4 - 9}{x^2 + 4\sqrt{3}x - 15}$

4. Evaluate the following :

(i) $\lim\limits_{x \to 5} \frac{x - 5}{\sqrt{6x - 5} - \sqrt{4x + 5}}$ (ii) $\lim\limits_{x \to 0} \frac{\sqrt{1 + x + x^2} - \sqrt{x + 1}}{2x^2}$

(iii) $\lim\limits_{x \to 2} \frac{x^2 - 4}{\sqrt{3x - 2} - \sqrt{x + 2}}$

5. Evaluate the following :

(i) $\lim\limits_{x \to 0} \frac{(1 - x)^n - 1}{x}$ (ii) $\lim\limits_{x \to \infty} (\sqrt{x + 1} - \sqrt{x})$

(iii) $\lim\limits_{n \to \infty} \left(\frac{1}{n^2} + \frac{2}{n^2} + \frac{3}{n^2} + \ldots + \frac{n-1}{n^2}\right)$ (iv) $\lim\limits_{n \to \infty} \frac{1^3 + 2^3 + \ldots + n^3}{n^4}$

6. Evaluate the following :

(i) $\lim\limits_{x \to 0} \frac{1 - \cos mx}{1 - \cos nx}$ (ii) $\lim\limits_{x \to 0} \frac{\sin 2x + \sin 6x}{\sin 5x - \sin 3x}$

(iii) $\lim\limits_{x \to 0} \frac{\sec 4x - \sec 2x}{\sec 3x - \sec x}$ (iv) $\lim\limits_{x \to 0} \frac{\tan x + 4 \tan 2x - 3 \tan 3x}{x^2 \tan x}$

7. Evaluate the following :

(i) $\lim\limits_{x \to 0} \frac{\sin (2 + x) - \sin (2 - x)}{x}$ (ii) $\lim\limits_{x \to 0} \frac{1 - \cos 2x + \tan^2 x}{x \sin x}$

(iii) $\lim\limits_{x \to 0} \frac{x \tan x}{1 - \cos 2x}$ (iv) $\lim\limits_{x \to 0} \frac{\text{cosce}\, x - \cot x}{x}$

8. Evaluate the following :

(i) $\lim_{x \to \frac{\pi}{4}} \dfrac{\sin x - \cos x}{x - \frac{\pi}{4}}$ (ii) $\lim_{x \to \frac{\pi}{2}} \dfrac{\cos^2 x}{1 - \sin x}$

(iii) $\lim_{x \to \frac{\pi}{4}} \dfrac{1 - \tan x}{1 - \sqrt{2} \sin x}$ (iv) $\lim_{x \to \frac{\pi}{4}} \dfrac{\text{cosce}^2 x - 2}{\cot x - 1}$

ANSWERS

5. (i) $\frac{5}{4}$ (ii) $\frac{3}{4}$ (iii) 2 **6.** (i) 5 (ii) $\frac{1}{4}$ (iii) 8

5.9 CONTINUITY

5.9.1 Continuous Functions

A function $f(x)$ is said to be continuous at a point if the graph of the function is unbroken, *i.e.*, there is no gap in the graph of the function. If there is any gap or break in the graph at a point, then the function is not continuous or it is discontinuous at that point.

Example 5.19

Consider $f(x) = 3x$. The graph of $f(x) = 3x$ is continuous at every point because there is no gap in the graph of $f(x) = 3x$.

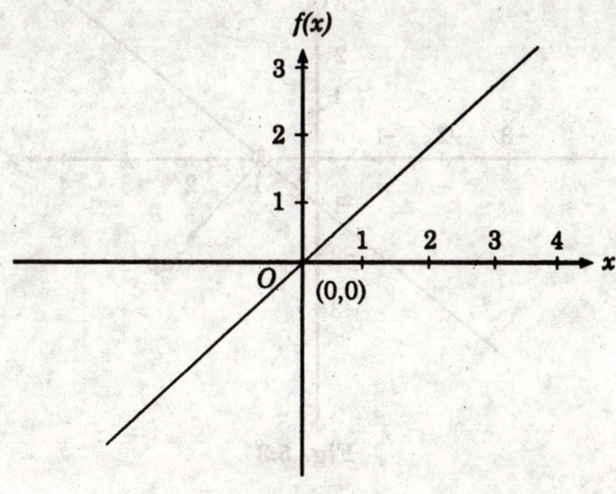

Fig. 5.1

Example 5.20

If $f(x) = \begin{cases} 3 \text{ for } x > 0 \\ 1 \text{ for } \leq 0 \end{cases}$, then the function is discontinuous at $x = 0$, as there is a jump in the graph at $x = 0$.

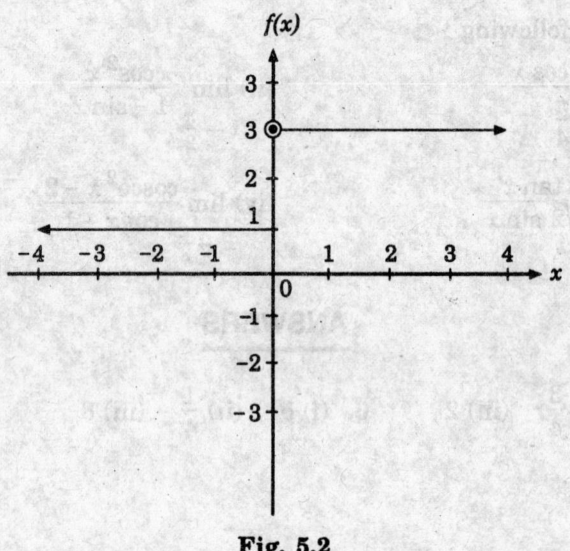

Fig. 5.2

Example 5.21

The graph of the function $f(x) = \dfrac{x^2 - 2^{x+1}}{x-1}$ is discontinuous at $x = 1$ because at $x = 1$, $f(x)$ does not exist. Here $f(x)$ is continuous at $x \neq 1$.

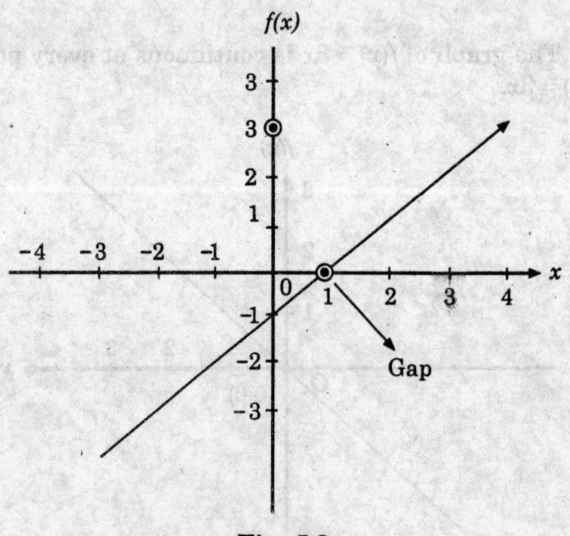

Fig. 5.3

5.10 DEFINITIONS

5.10.1 Continuity at a Point

A function $f(x)$ is said to be continuous at $x = a$ if

(i) $f(a)$ exist, *i.e.*, $f(x)$ is defined at $x = a$

(ii) $\lim\limits_{x \to a+} f(x) = \lim\limits_{x \to a-} f(x) = f(a)$

5.10.2 Discontinuity at a Point

A function $f(x)$ is discontinuous at $x = a$ if

(i) $f(a)$ does not exists, *i.e.*, $f(x)$ is not defined at $x = a$.

(ii) $\lim_{x \to a+} f(x)$ or $\lim_{x \to a-} f(x)$ or both does not exists at $x = a$.

(iii) Both $\lim_{x \to a+} f(x)$ and $\lim_{x \to a-} f(x)$ exists at $x = a$ but are not equal.

(iv) Both $\lim_{x \to a+} f(x)$ and $\lim_{x \to a-} f(x)$ exists at $x = a$ and are equal but not equal to $f(a)$.

Example 5.22

Show that the function at $x = 0$ is discontinuous.

$$f(x) = \begin{cases} \dfrac{|x|}{x}; & x \neq 0 \\ 1; & x = 0 \end{cases}$$

SOLUTION : Left hand limit at $x = 0 = \lim_{x \to 0-} f(x) = \lim_{h \to 0} f(0 - h)$

$$= \lim_{h \to 0} f(-h) = \lim_{h \to 0} \frac{|-h|}{-h} = \lim_{h \to 0} \frac{h}{-h}$$

$$= \lim_{h \to 0} (-1) = -1$$

Right Hand Limit at $x = 0 = \lim_{x \to 0+} f(x) = \lim_{h \to 0} f(0 + h)$

$$= \lim_{h \to 0} f(h) = \lim_{h \to 0} \frac{|h|}{h} = \lim_{h \to 0} \frac{h}{h} = \lim_{h \to 0} 1 = 1$$

Here, LHL $\neq$ RHL $\Rightarrow$ $f(x)$ is not continuous at $x = 0$.

Example 5.23

Discuss the continuity of the function $f(x)$ at $x = \frac{1}{2}$, where

$$f(x) = \begin{cases} \dfrac{1}{2} - x; & 0 \leq x < \dfrac{1}{2} \\ 1; & x = \dfrac{1}{2} \\ \dfrac{3}{2} - x; & \dfrac{1}{2} < x \leq 1 \end{cases}$$

SOLUTION : LHL at $x = \frac{1}{2} = \lim_{x \to \frac{1}{2}-} f(x)$

$$= \lim_{x \to \frac{1}{2}} \left(\frac{1}{2} - x\right) = \frac{1}{2} - \frac{1}{2} = 0$$

$$\text{RHL at } x = \frac{1}{2} = \lim_{x \to \frac{1}{2}+} f(x) = \lim_{x \to \frac{1}{2}} \left(\frac{3}{2} - x\right)$$

$$= \frac{3}{2} - \frac{1}{2} = 1$$

Here, LHL $\neq$ RHL, so function is discontinuous at $x = \frac{1}{2}$.

Example 5.24

If $f(x) = \begin{cases} \dfrac{x^2 - 1}{x - 1}; & \text{for } x \neq 1 \\ 2; & \text{for } x = 1 \end{cases}$. Find whether $f(x)$ is continuous at $x = 1$

SOLUTION : LHL at $x = 1 = \lim_{x \to 1-} f(x) = \lim_{h \to 0} f(1 - h)$

$$= \lim_{h \to 0} \frac{(1-h)^2 - 1}{1 - h - 1} = \lim_{h \to 0} \frac{1 + h^2 - 2h - 1}{-h}$$

$$= \lim_{h \to 0} \frac{h^2 - 2h}{-h} = \lim_{h \to 0} (-h + 2) = 2$$

RHL at $x = 1 = \lim_{x \to 1+} f(x) = \lim_{h \to 0} f(1 + h)$

$$= \lim_{h \to 0} \frac{(1+h)^2 - 1}{1 + h - 1} = \lim_{h \to 0} \frac{1 + h^2 + 2h - 1}{h}$$

$$= \lim_{h \to 0} (h + 2) = 2$$

Here, $\lim_{x \to 1-} f(x) = \lim_{x \to 1+} f(x) = f(1)$

So, $f(x)$ is continuous at $x = 1$.

PROBLEM SET 5.2

1. Let $f(x) = \begin{cases} x^2 & \text{if } x \leq 1 \\ x + 5 & \text{if } x > 1 \end{cases}$

   Show that $f(x)$ is discontinuous at $x = 1$.

2. Let $f(x) = \begin{cases} \cos x & \text{if } x \geq 0 \\ -\cos x & \text{if } x < 0 \end{cases}$

   Show that $f(x)$ is discontinuous at $x = 0$.

3. Let $f(x) = \begin{cases} \dfrac{\sin 3x}{x} & \text{if } x \neq 0 \\ 1 & \text{if } x = 0 \end{cases}$

   Find whether $f(x)$ is continuous at $x = 0$.

4. If $f(x) = \begin{cases} x \cos \dfrac{1}{x} \ ; & x \neq 0 \\ 0 \ ; & x = 0 \end{cases}$

Examine its continuity at $x = 0$.

5. Show that the function

$$f(x) = \begin{cases} x \ ; & 0 \leq x < \dfrac{1}{2} \\ 1 \ ; & x = \dfrac{1}{2} \\ 1 - x \ ; & \dfrac{1}{2} < x \leq 1 \end{cases}$$

is discontinuous at $x = \dfrac{1}{2}$

6. Let $f(x) = \begin{cases} \dfrac{1 - \cos x}{x^2} \ ; & x \neq 0 \\ 1 \ ; & x = 0 \end{cases}$

Show that $f(x)$ is discontinuous at $x = 0$.

7. Evaluate the continuity of $f(x)$ at $x = 0$, where

$$f(x) = \begin{cases} 2x - 1 \ ; & x < 0 \\ 2x + 1 \ ; & x \geq 0 \end{cases}$$

8. Show that $f(x) = 2x - |x|$ is continuous at $x = 0$.

9. Examine the continuity of $f(x)$ at $x = 0$, where $f(x) = |x| + |x - 1|$

10. Find 'a' such that $f(x) = \begin{cases} ax^2 - 2 \ ; & x < 1 \\ -1 \ ; & x \geq 1 \end{cases}$

is continuous at $x = 1$.

ANSWERS

3. Discontinuous **4.** Continuous **7.** Discontinuous **9.** Continuous

10. $a = 3$

5.11 DIFFERENTIATION

5.11.1 Differentiability at a Point

Let $y = f(x)$ be a continuous function of a variable x and δx, δy are small increments in x and y respectively. The incremental ratio $\dfrac{\delta y}{\delta x}$ is called average rate of change of $y = f(x)$ with respect to x.

Let $\quad y = f(x)$

$\Rightarrow \quad y + \delta y = f(x + \delta x)$

$\Rightarrow \quad y + \delta y - y = f(x + \delta x) - f(x)$

$\Rightarrow \quad \delta y = f(x + \delta x) - f(x)$

$$\Rightarrow \quad \frac{\delta y}{\delta x} = \frac{f(x+\delta x) - f(x)}{\delta x}; \quad \delta x \neq 0$$

$$\Rightarrow \quad \lim_{\delta x \to 0} \frac{\delta y}{\delta x} = \lim_{\delta x \to 0} \frac{f(x+\delta x) - f(x)}{\delta x}$$

If $\lim\limits_{\delta x \to 0} \frac{\delta y}{\delta x}$ exists, then it is called derivative of $f(x)$ with respect to x and denoted by

$$\frac{dy}{dx} \text{ or } Df(x) \text{ or } f'(x)$$

$$\text{or } Dy \text{ or } y' \text{ or } \frac{d}{dx} f(x)$$

Hence,
$$\frac{dy}{dx} = \lim_{\delta x \to 0} \frac{\delta y}{\delta x} = \lim_{\delta x \to 0} \frac{f(x+\delta x) - f(x)}{\delta x}$$

The derivative of $f(x)$ is also called the differential conefficient of $f(x)$. But we shall be using the words differentiation or derivative only.

5.11.2 Geometrical Meaning of Derivative at a Point

Take a curve $y = f(x)$. Let $f(x)$ is differentiable at $x = a$. Let $P(a, f(a))$ be a point on the curve and let $Q(x, f(x))$ be a point in the neighbourhood of the curve.

Slope of chord $PQ = \frac{f(x) - f(a)}{x - a}$

Taking limit as

$Q \to P$ *i.e.*, $x \to a$, we get

$$\lim_{Q \to P} (\text{ Slope of the chord } PQ) = \lim_{x \to a} \frac{f(x) - f(a)}{x - a}$$

As $Q \to P$, chord PQ becomes tangent at P. Therefore,

Slope of the tangent at

$$P = \lim_{x \to a} \frac{f(x) - f(a)}{x - a} = \left[\frac{d}{dx} f(x)\right]_{x=a} = f'(a)$$

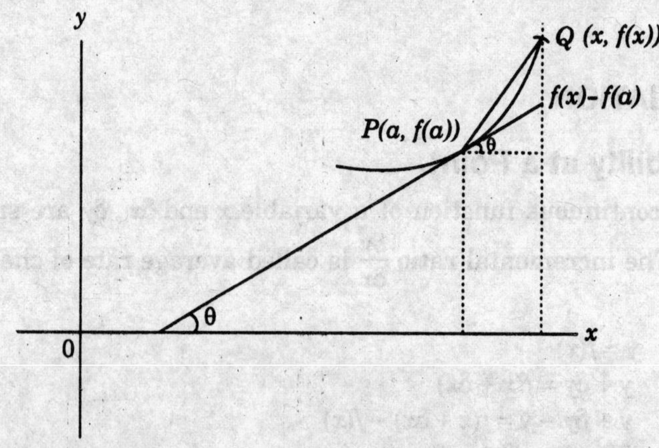

Fig. 5.4

Thus, the derivative of a function at a point $x = a$ is the slope of the tangent to the curve $y = f(x)$ at the point $(a, f(a))$. Also, if a function is not differentiable at $x = a$ only if the point $(a, f(a))$ is a corner point of the curve $y = f(x)$ *i.e.*, the curve suddenly changes its direction at point $(a, f(a))$.

5.11.3 Differentiation by First Principles

The process of finding the differential co-efficient of a function $f(x)$ with respect to the independent variable x is called differentiation.

To find the derivative of $f(x)$ from first principle, we will use the following steps :

- **Step 1 :** Let $y = f(x)$
- **Step 2 :** $y + \delta y = f(x + \delta x)$
- **Step 3 :** $\delta y = f(x + \delta x) - f(x)$
- **Step 4 :** $\dfrac{\delta y}{\delta x} = \dfrac{f(x + \delta x) - f(x)}{\delta x}$
- **Step 5 :** Find $\lim\limits_{\delta x \to 0} \dfrac{\delta y}{\delta x} = \lim\limits_{\delta x \to 0} \dfrac{f(x + \delta x) - f(x)}{\delta x}$, which is required $\dfrac{dy}{dx}$.

Example 5.25

Find the derivative of x^n with respect to x by first principle where $n \in R$.

SOLUTION : Let $y = x^n$

$$\Rightarrow \quad y + \delta y = (x + \delta x)^n$$

$$\Rightarrow \quad \delta y = (x + \delta x)^n - x^n$$

$$\Rightarrow \quad \frac{\delta y}{\delta x} = \frac{(x + \delta x)^n - x^n}{\delta x}$$

$$\Rightarrow \quad \lim_{\delta x \to 0} \frac{\delta y}{\delta x} = \lim_{\delta x \to 0} \frac{(x + \delta x)^n - x^n}{\delta x}$$

or

$$\frac{dy}{dx} = \lim_{\delta x \to 0} \left(\frac{x^n + {}^nc_1\, x^{n-1}\, \delta x + {}^nc_2\, x^{n-2}\, (\delta x)^2 + \ldots + {}^nc_n\, (\delta x)^n - x^n}{\delta x} \right)$$

$$\frac{dy}{dx} = \lim_{\delta x \to 0} \frac{x^n + n\, x^{n-1}\, \delta x + \frac{n(n-1)}{2}\, x^{n-2}\, (\delta x)^2 + \ldots + (\delta x)^n - x^n}{\delta x}$$

$$= \lim_{\delta x \to 0} \frac{n\, x^{n-1}\, \delta x + \frac{n(n-1)}{2}\, x^{n-2}\, (\delta x)^2 + \ldots + (\delta x)^n}{\delta x}$$

$$= \lim_{\delta x \to 0} \left(n\, x^{n-1} + \frac{n(n-1)}{2}\, x^{n-2}\, \delta x + \ldots + (\delta x)^{n-1} \right)$$

$$\frac{dy}{dx} = n\, x^{n-1}$$

Example 5.26

Find the derivative of the following functions w. r. to x.

(i) x^9 (ii) $\frac{1}{x^2}$ (iii) $\frac{1}{x^{3/2}}$ (iv) $x^{5/3}$

SOLUTION :

(i) Let $y = x^9$

$$\therefore \quad \frac{dy}{dx} = \frac{d}{dx}(x^9) = 9\,x^{9-1} = 9\,x^8 \qquad \left(\because \frac{d}{dx}(x^n) = n\,x^{n-1}\right)$$

(ii) Let $y = \frac{1}{x^2} = x^{-2}$

$$\therefore \quad \frac{dy}{dx} = \frac{d}{dx}(x^{-2}) = -2x^{-2-1} = -2x^{-3} = \frac{-2}{x^3}$$

(iii) Let $y = \frac{1}{x^{3/2}} = x^{-3/2}$

$$\therefore \quad \frac{dy}{dx} = \frac{d}{dx}(x^{-3/2}) = \frac{-3}{2}\,x^{-3/2-1} = \frac{-3}{2}\,x^{-5/2} = \frac{-3}{2\,x^{5/2}}$$

(iv) Let $y = x^{5/3}$

$$\Rightarrow \quad \frac{dy}{dx} = \frac{d}{dx}(x^{5/3}) = \frac{5}{3}\,x^{5/3-1} = \frac{5}{3}\,x^{2/3}$$

Example 5.27

Find the derivative of e^x from first principle :

SOLUTION : Let $y = e^x$

$$\Rightarrow \quad y + \delta y = e^{x+\delta x}$$

$$\Rightarrow \quad \delta y = e^{x+\delta x} - e^x$$

$$\Rightarrow \quad \frac{\delta y}{\delta x} = \frac{e^{x+\delta x} - e^x}{\delta x}$$

or

$$\frac{dy}{dx} = \lim_{\delta x \to 0} \frac{\delta y}{\delta x} = \lim_{\delta x \to 0} \frac{e^{x+\delta x} - e^x}{\delta x}$$

$$= \lim_{\delta x \to 0} \frac{e^x . e^{\delta x} - e^x}{\delta x} = \lim_{\delta x \to 0} e^x \left(\frac{e^{\delta x} - 1}{\delta x}\right)$$

$$= \lim_{\delta x \to 0} e^x \left(\frac{1 + \delta x + \frac{(\delta x)^2}{2} + \frac{(\delta x)^3}{3} + \ldots - 1}{\delta x}\right)$$

$$= \lim_{\delta x \to 0} e^x \left(1 + \frac{\delta x}{2} + \frac{(\delta x)^2}{3} + \ldots\right)$$

$$= e^x (1 + 0 + 0 + \ldots) = e^x$$

$$\therefore \quad \frac{d(e^x)}{dx} = e^x$$

Example 5.28

Find the derivative of a^x from first principal, where $a > 0, a \neq 1$.

SOLUTION : Let $y = a^x \Rightarrow y + \delta y = a^{x+\delta x}$

$$\Rightarrow \quad \delta y = a^{x+\delta x} - a^x$$

$$\frac{\delta y}{\delta x} = \frac{a^{x+\delta x} - a^x}{\delta x}$$

or $$\frac{dy}{dx} = \lim_{\delta x \to 0} \frac{\delta y}{\delta x} = \lim_{\delta x \to 0} \frac{a^{x+\delta x} - a^x}{\delta x}$$

$$= \lim_{\delta x \to 0} \frac{a^x \cdot a^{\delta x} - a^x}{\delta x} = \lim_{\delta x \to 0} a^x \left(\frac{a^{\delta x} - 1}{\delta x}\right)$$

$$= a^x \lim_{\delta x \to 0} a^x \left(\frac{a^{\delta x} - 1}{\delta x}\right)$$

$$= a^x \cdot \log_e a \qquad \left(\because \lim_{h \to 0} \frac{a^h - 1}{h} = \log_e a\right)$$

$$\therefore \quad \frac{d(a^x)}{dx} = a^x \cdot \log_e a$$

Example 5.29

Find the derivative of sin x from first principle :

SOLUTION : Let $y = \sin x$

$$y + \delta y = \sin(x + \delta x)$$

$$\delta y = \sin(x + \delta x) - \sin x$$

$$\frac{\delta y}{\delta x} = \frac{\sin(x + \delta x) - \sin x}{\delta x}$$

or $$\frac{dy}{dx} = \lim_{\delta x \to 0} \frac{\delta y}{\delta x} = \lim_{\delta x \to 0} \frac{\sin(x + \delta x) - \sin x}{\delta x}$$

$$\frac{dy}{dx} = \lim_{\delta x \to 0} \frac{2\sin\left(\frac{x + \delta x - x}{2}\right) \cos\left(\frac{x + \delta x + x}{2}\right)}{\delta x}$$

$$\left(\because \sin C - \sin D = 2\sin\frac{C-D}{2}\cos\frac{C+D}{2}\right)$$

$$= \lim_{\delta x \to 0} \frac{2\sin\frac{\delta x}{2} \cos\left(x + \frac{\delta x}{2}\right)}{\delta x}$$

$$= \lim_{\delta x \to 0} \frac{2\sin\frac{\delta x}{2}}{\frac{2\delta x}{2}} \cdot \cos\left(x + \frac{\delta x}{2}\right)$$

$$= \lim_{\delta x \to 0} \frac{\sin \frac{\delta x}{2}}{\frac{\delta x}{2}} \cdot \lim_{\delta x \to 0} \cos\left(x + \frac{\delta x}{2}\right)$$

$$= 1 \cdot \cos x = \cos x$$

$$\therefore \quad \frac{d}{dx}(\sin x) = \cos x$$

5.11.4 Differential Co-efficients of Some Standard Functions

(1) $\frac{d}{dx}(x^n) = nx^{n-1}$

(2) $\frac{d}{dx}(e^x) = e^x$

(3) $\frac{d}{dx}(a^x) = a^x \log_e a$

(4) $\frac{d}{dx}(\log_e x) = \frac{1}{x}$

(5) $\frac{d}{dx}(\log_a x) = \frac{1}{x \log_e a}$

(6) $\frac{d}{dx}(\sin x) = \cos x$

(7) $\frac{d}{dx}(\cos x) = -\sin x$

(8) $\frac{d}{dx}(\tan x) = \sec^2 x$

(9) $\frac{d}{dx}(\cot x) = -\operatorname{cosec}^2 x$

(10) $\frac{d}{dx}(\sec x) = \sec x \tan x$

(11) $\frac{d}{dx}(\operatorname{cosec} x) = -\operatorname{cosec} x \cot x$

(12) $\frac{d}{dx}(\sin^{-1} x) = \frac{1}{\sqrt{1-x^2}}$

(13) $\frac{d}{dx}(\cos^{-1} x) = -\frac{1}{\sqrt{1-x^2}}$

(14) $\frac{d}{dx}(\tan^{-1} x) = \frac{1}{1+x^2}$

(15) $\frac{d}{dx}(\cot^{-1} x) = -\frac{1}{1+x^2}$

(16) $\frac{d}{dx}(\sec^{-1} x) = \frac{1}{|x|\sqrt{x^2-1}}$

(17) $\frac{d}{dx}(\text{cosecc}^{-1} x) = \frac{-1}{|x|\sqrt{x^2-1}}$

(18) $\frac{d}{dx}(ax+b)^n = n(ax+b)^{n-1} a$

(19) $\frac{d}{dx}(\text{constant}) = 0$

(20) $\frac{d}{dx}(c \cdot f(x)) = c \cdot \frac{d}{dx} f(x)$

(21) $\frac{d}{dx}[f(x) \pm g(x)] = \frac{d f(x)}{dx} \pm \frac{d}{dx} g(x)$

Example 5.30

Differentiate the following function w. r. to x from first principle :

(i) $\sin^2 x$ (ii) $\sqrt{\sin x}$

(iii) $\sin\sqrt{x}$

SOLUTION : (i) Let $y = \sin^2 x$

$$\Rightarrow \quad y + \delta y = \sin^2 (x + \delta x)$$

$$\Rightarrow \quad \delta y = \sin^2 (x + \delta x) - \sin^2 x$$

$$\Rightarrow \quad \frac{\delta y}{\delta x} = \frac{\sin^2 (x + \delta x) - \sin^2 x}{\delta x}$$

$$= \frac{\sin (x + \delta x + x) \cdot \sin (x + \delta x - x)}{\delta x}$$

$$[\because \ \sin^2 A - \sin^2 B = \sin (A + B) \sin (A - B)]$$

or $$\frac{dy}{dx} = \lim_{\delta x \to 0} \frac{\delta y}{\delta x} = \lim_{\delta x \to 0} \frac{\sin (2x + \delta x) \cdot \sin \delta x}{\delta x}$$

$$= \lim_{\delta x \to 0} \sin (2x + \delta x) \cdot \lim_{\delta x \to 0} \frac{\sin \delta x}{\delta x}$$

$$= \sin 2x \cdot 1$$

$$\frac{d}{dx} (\sin^2 x) = \sin 2x$$

(ii) Let $y = \sqrt{\sin x}$

$\Rightarrow$ $$y + \delta y = \sqrt{\sin(x + \delta x)}$$

$\Rightarrow$ $$\delta y = \sqrt{\sin(x + \delta x)} - \sqrt{\sin x}$$

$\Rightarrow$ $$\frac{\delta y}{\delta x} = \frac{\sqrt{\sin(x + \delta x)} - \sqrt{\sin x}}{\delta x}$$

or $$\frac{dy}{dx} = \lim_{\delta x \to 0} \frac{\delta y}{\delta x} = \lim_{\delta x \to 0} \frac{\sqrt{\sin (x + \delta x)} - \sqrt{\sin x}}{\delta x}$$

$$= \lim_{\delta x \to 0} \frac{\sqrt{\sin(x + \delta x)} - \sqrt{\sin x}}{\delta x} \times \frac{\sqrt{\sin(x + \delta x)} + \sqrt{\sin x}}{\sqrt{\sin(x + \delta x)} + \sqrt{\sin x}}$$

$$= \lim_{\delta x \to 0} \frac{\sin(x + \delta x) - \sin x}{\delta x [\sqrt{\sin(x + \delta x)} + \sqrt{\sin x}]}$$

$$= \lim_{\delta x \to 0} \frac{2\sin \left(\frac{x + \delta x - x}{2}\right) \cos \left(\frac{x + \delta x + x}{2}\right)}{\delta x (\sqrt{\sin(x + \delta x)} + \sqrt{\sin x})}$$

$$\left[\because \ \sin C - \sin D = 2\sin \frac{C-D}{2} \cos \frac{C+D}{2}\right]$$

$$= \lim_{\delta x \to 0} \frac{2\sin \frac{\delta x}{2} \cos \left(x + \frac{\delta x}{2}\right)}{\delta x [\sqrt{\sin (x + \delta x)} + \sqrt{\sin x}]}$$

$$= \lim_{\delta x \to 0} \frac{\sin \frac{\delta x}{2}}{\frac{\delta x}{2}} \cdot \lim_{\delta x \to 0} \cos \left(x + \frac{\delta x}{2}\right) \lim_{\delta x \to 0} \frac{1}{(\sqrt{\sin (x + \delta x)} + \sqrt{\sin x})}$$

$$= 1 \cdot \cos x \cdot \frac{1}{\sqrt{\sin x} + \sqrt{\sin x}}$$

$$= \frac{\cos x}{2\sqrt{\sin x}}$$

(iii) Let $y = \sin\sqrt{x}$

$\Rightarrow \quad y + \delta y = \sin\sqrt{x + \delta x}$

$\Rightarrow \quad \delta y = \sin\sqrt{x + \delta x} - \sin\sqrt{x}$

$\Rightarrow \quad \dfrac{\delta y}{\delta x} = \dfrac{\sin\sqrt{x + \delta x} - \sin\sqrt{x}}{\delta x}$

or $\quad \dfrac{dy}{dx} = \lim\limits_{\delta x \to 0} \dfrac{\delta y}{\delta x}$

$$= \lim_{\delta x \to 0} \frac{\sin\sqrt{x + \delta x} - \sin\sqrt{x}}{\delta x}$$

$$= \lim_{\delta x \to 0} \frac{2\sin\left(\dfrac{\sqrt{x + \delta x} - \sqrt{x}}{2}\right)\cos\left(\dfrac{\sqrt{x + \delta x} + \sqrt{x}}{2}\right)}{\delta x}$$

$$= \lim_{\delta x \to 0} \frac{2\sin\left(\dfrac{\sqrt{x + \delta x} - \sqrt{x}}{2}\right)}{\delta x\left(\dfrac{\sqrt{x + \delta x} - \sqrt{x}}{2}\right)} \times \left(\frac{\sqrt{x + \delta x} - \sqrt{x}}{2}\right)\cdot\cos\left(\frac{\sqrt{x + \delta x} + \sqrt{x}}{2}\right)$$

$$= \lim_{\delta x \to 0} \frac{\sin\left(\dfrac{\sqrt{x + \delta x} - \sqrt{x}}{2}\right)}{\left(\dfrac{\sqrt{x + \delta x} - \sqrt{x}}{2}\right)} \times \frac{(\sqrt{x + \delta x} - \sqrt{x})\,(\sqrt{x + \delta x} + \sqrt{x})}{(\sqrt{x + \delta x} + \sqrt{x})\,\delta x} \times \cos\left(\frac{\sqrt{x + \delta x} + \sqrt{x}}{2}\right)$$

$$= \lim_{\delta x \to 0} \frac{\sin\left(\dfrac{\sqrt{x + \delta x} - \sqrt{x}}{2}\right)}{\dfrac{\sqrt{x + \delta x} - \sqrt{x}}{2}} \times \frac{x + \delta x - x}{\delta x\,(\sqrt{x + \delta x} + \sqrt{x})} \times \cos\left(\frac{\sqrt{x + \delta x} + \sqrt{x}}{2}\right)$$

$$= \lim_{\delta x \to 0} \frac{\sin\left(\dfrac{\sqrt{x + \delta x} - \sqrt{x}}{2}\right)}{\left(\dfrac{\sqrt{x + \delta x} - \sqrt{x}}{2}\right)} \times \lim_{\delta x \to 0} \frac{1}{\sqrt{x + \delta x} + \sqrt{x}} \lim_{\delta x \to 0} \cos\left(\frac{\sqrt{x + \delta x} + \sqrt{x}}{2}\right)$$

$$= 1 \times \frac{1}{\sqrt{x} + \sqrt{x}} \times \cos\left(\frac{\sqrt{x} + \sqrt{x}}{2}\right)$$

$$= \frac{1}{2\sqrt{x}} \cos\sqrt{x}$$

$\therefore \quad \dfrac{d}{dx}\sin\sqrt{x} = \dfrac{\cos\sqrt{x}}{2\sqrt{x}}$

PROBLEM SET 5.3

1. Differentiate the following functions w. r. to x from first principle.

(i) $\cos x$ (ii) $\tan x$ (iii) $\cos^2 x$

(iv) $\sqrt{\cos x}$ (v) $\tan \sqrt{x}$ (vi) $\cot \sqrt{x}$

2. If $y = \sqrt{\dfrac{1 - \cos 2x}{1 + \cos 2x}}$, find $\dfrac{dy}{dx}$

3. Differentiate the following functions w. r. to x :

(i) $\tan^{-1}\left(\dfrac{1 + \tan x}{1 - \tan x}\right)$ (ii) $3^x + x^3 + 3^3$

(iii) $\left(\sqrt{x} + \dfrac{1}{\sqrt{x}}\right)^3$ (iv) $2 \sec x + 3 \cot x - 4 \tan x$

(v) $\left(x + \dfrac{1}{x}\right)\left(\sqrt{x} + \dfrac{1}{\sqrt{x}}\right)^3$, $x \neq 0$ (vi) $\dfrac{(1+x)^3}{x^3}$; $x \neq 0$

4. If $y = \sqrt{\dfrac{x}{a}} + \sqrt{\dfrac{a}{x}}$, prove that $2xy\dfrac{dy}{dx} = \left(\dfrac{x}{a} - \dfrac{a}{x}\right)$

5. If $y = \dfrac{2x^9}{3} - \dfrac{5}{7}x^7 + 6x^3 - x$, find $\dfrac{dy}{dx}$ at $x = 1$

6. If $y = \sqrt{x} + \dfrac{1}{\sqrt{x}}$, prove that $2x\dfrac{dy}{dx} + y = 2\sqrt{x}$

7. If $y = 1 + x + \dfrac{x^2}{2!} + \dfrac{x^3}{3!} + \dfrac{x^4}{4!} + \ldots$, show that $\dfrac{dy}{dx} = y$

8. If $y = x + \dfrac{1}{x}$, prove that $x^2\dfrac{dy}{dx} + y = 2\sqrt{x}$

ANSWERS

1. (i) $-\sin x$ (ii) $\sec^2 x$ (iii) $-\sin 2x$ (iv) $\dfrac{-\sin x}{2\sqrt{\cos x}}$ (v) $\dfrac{\sec^2\sqrt{x}}{2\sqrt{x}}$ (vi) $\dfrac{-\operatorname{cosec}^2\sqrt{x}}{2\sqrt{x}}$

2. $\sec^2 x$

3. (i) 1 (ii) $3^x \log 3 + 3x^2$ (iii) $\dfrac{3}{2}x^{1/2} - \dfrac{3}{2}x^{-5/2} + \dfrac{3}{2}x^{-1/2} - \dfrac{3}{2}x^{-3/2}$

(iv) $2\sec x \tan x - 3\operatorname{cosec}^2 x - 4\sec^2 x$ (v) $\dfrac{3}{2}x^{1/2} + \dfrac{1}{2}x^{-1/2} - \dfrac{1}{2}x^{-3/2} - \dfrac{1}{2}x^{-5/2}$

(vi) $\dfrac{-3}{x^4} - \dfrac{6}{x^3} - \dfrac{3}{x^2}$

5. 18

5.12 PRODUCT RULE

THEOREM 5.1. The differential coefficient of the product of two functions is the sum of the products of each function with the derivative of the other, *i.e.*,

$$\frac{d}{dx}(f(x)\cdot g(x)) = f(x)\frac{d}{dx}g(x) + g(x)\frac{d}{dx}f(x)$$

where $f(x)$ and $g(x)$ are differentiable function of x.

PROOF : Let $y = f(x)\cdot g(x)$

$\Rightarrow \quad y + \delta y = f(x+\delta x)\cdot g(x+\delta x)$

$\Rightarrow \quad \delta y = f(x+\delta x)\cdot g(x+\delta x) - f(x)\cdot g(x)$

$$\Rightarrow \quad \frac{\delta y}{\delta x} = \frac{f(x+\delta x)\,g(x+\delta x) - f(x)\,g(x)}{\delta x}$$

$$= \frac{f(x+\delta x)\,g(x+\delta x) - f(x)\,g(x+\delta x) + f(x)\,g(x+\delta x) - f(x)\,g(x)}{\delta x}$$

$$= \frac{g(x+\delta x)\,[f(x+\delta x) - f(x)] + f(x)\,[g(x+\delta x) - g(x)]}{\delta x}$$

$$= g(x+\delta x)\frac{f(x+\delta x) - f(x)}{\delta x} + f(x)\cdot\frac{g(x+\delta x) - g(x)}{\delta x}$$

Now,
$$\frac{dy}{dx} = \lim_{\delta x \to 0}\frac{\delta y}{\delta x} = \lim_{\delta x \to 0}\left(g(x+\delta x)\cdot\frac{f(x+\delta x) - f(x)}{\delta x} + f(x)\cdot\frac{g(x+\delta x) - g(x)}{\delta x}\right)$$

$$= \lim_{\delta x \to 0} g(x+\delta x)\cdot\lim_{\delta x \to 0}\frac{f(x+\delta x) - f(x)}{\delta x} + \lim_{\delta x \to 0} f(x)\cdot\lim_{\delta x \to 0}\frac{g(x+\delta x) - g(x)}{\delta x}$$

$$= g(x)\cdot\frac{d}{dx}f(x) + f(x)\frac{d}{dx}g(x)$$

$$= f(x)\frac{d}{dx}g(x) + g(x)\frac{d}{dx}f(x)$$

5.12.1 Extension of the Rule

Let $f(x)$, $g(x)$, $h(x)$ are three differentiable functions. Then

$$\frac{d}{dx}[f(x)\cdot g(x)\cdot h(x)] = \left(\frac{d}{dx}f(x)\right)g(x)\,h(x) + f(x)\left(\frac{d}{dx}g(x)\right)\cdot h(x) + f(x)\,g(x)\left(\frac{d}{dx}h(x)\right)$$

■ **NOTE :** The rule can be extended to the product of any finite number of functions, the derivative being the sum of the products as under (derivative of a function x product of the remaining functions), when all the functions have been considered. ●

Example 5.31

Differentiate the following functions w. r. to x :

(i) $x^3 e^x$ (ii) $x\cdot\sin x\cdot\log x$ (iii) $(1 - 2\tan x)(5 + 4\sin x)$

SOLUTION : (i) Let $y = x^3 e^x \Rightarrow \frac{dy}{dx} = \frac{d}{dx}(x^3 e^x) = x^3\frac{d}{dx}e^x + e^x\frac{d}{dx}x^3$

$$= x^3 e^x + e^x\cdot 3x^2$$

(ii) Let $y = x\cdot\sin x\cdot\log x \Rightarrow \frac{dy}{dx} = \frac{d}{dx}(x\cdot\sin x\cdot\log x)$

$$= \left(\frac{dx}{dx}\right)\cdot \sin x \cdot \log x + x\cdot\left(\frac{d}{dx}\sin x\right)\cdot \log x + x\cdot \sin x\left(\frac{d}{dx}\log x\right)$$

$$= 1\cdot \sin x\cdot \log x + x\cdot \cos x\cdot \log x + x\cdot \sin x\cdot \frac{1}{x}$$

$$= \sin x\cdot \log x + x\cdot \cos x\cdot \log x + \sin x$$

(iii) Let $y = (1 - 2\tan x)(5 + 4\sin x)$

$$\Rightarrow \quad \frac{dy}{dx} = \frac{d}{dx}[(1 - 2\tan x)(5 + 4\sin x)]$$

$$= (1 - 2\tan x)\frac{d}{dx}(5 + 4\sin x) + (5 + 4\sin x)\frac{d}{dx}(1 - 2\tan x)$$

$$= (1 - 2\tan x)(0 + 4\cos x) + (5 + 4\sin x)(0 - 2\cdot\sec^2 x)$$

$$= 4\cos x(1 - 2\tan x) - 2\sec^2 x(5 + 4\sin x)$$

5.12.2 Quotient Rule

THEOREM 5.2. If $f(x)$ and $g(x)$ are functions of x and $g(x) \neq 0$, then

$$\frac{d}{dx}\left(\frac{f(x)}{g(x)}\right) = \frac{g(x)\frac{d}{dx}\{f(x)\} - f(x)\frac{d}{dx}\{g(x)\}}{\{g(x)\}^2}$$

PROOF : Let $\quad y = \frac{f(x)}{g(x)}; g(x) \neq 0$

$$\Rightarrow \quad y + \delta y = \frac{f(x + \delta x)}{g(x + \delta x)}$$

$$\Rightarrow \quad \delta y = \frac{f(x + \delta x)}{g(x + \delta x)} - \frac{f(x)}{g(x)} = \frac{f(x + \delta x)\, g(x) - f(x)\, g(x + \delta x)}{g(x)\, g(x + \delta x)}$$

$$\Rightarrow \quad \frac{\delta y}{\delta x} = \frac{f(x + \delta x)\, g(x) - f(x)\, g(x + \delta x)}{\delta x\, g(x)\, g(x + \delta x)}$$

or $$\frac{dy}{dx} = \lim_{\delta x \to 0}\frac{\delta y}{\delta x} = \lim_{\delta x \to 0}\frac{f(x + \delta x)\, g(x) - f(x)\, g(x + \delta x)}{\delta x\, g(x)\, g(x + \delta x)}$$

$$= \lim_{\delta x \to 0}\frac{f(x + \delta x)\, g(x) - f(x)\, g(x) + f(x)\, g(x) - f(x)\, g(x + \delta x)}{\delta x\, g(x)\, g(x + \delta x)}$$

$$= \lim_{\delta x \to 0}\frac{g(x)\,[f(x + \delta x) - f(x)] - f(x)\,[g(x + \delta x) - g(x)]}{\delta x\, g(x)\, g(x + \delta x)}$$

$$= \left[g(x)\lim_{\delta x \to 0}\left(\frac{f(x + \delta x) - f(x)}{\delta x}\right) - f(x)\lim_{\delta x \to 0}\left(\frac{g(x + \delta x) - g(x)}{\delta x}\right)\right]\lim_{\delta x \to 0}\frac{1}{g(x)\, g(x + \delta x)}$$

$$= \frac{g(x)\frac{d}{dx}f(x) - f(x)\frac{d}{dx}g(x)}{g(x)\cdot g(x)} = \frac{g(x)\frac{d}{dx}f(x) - f(x)\frac{d}{dx}g(x)}{[g(x)]^2}$$

Example 5.32

Differentiate the following w.r. to x :

(i) $\frac{e^x}{1 + \sin x}$ (ii) $\frac{2x + 3}{x^2 - 5}$ (iii) $\frac{1}{xe^x + \cos x}$

SOLUTION : (i) Let $y = \dfrac{e^x}{1 + \sin x}$

$$\Rightarrow \quad \frac{dy}{dx} = \frac{d}{dx}\left(\frac{e^x}{1+\sin x}\right) = \frac{(1+\sin x)\dfrac{d}{dx}(e^x) - e^x \dfrac{d}{dx}(1+\sin x)}{(1+\sin x)^2}$$

$$= \frac{(1+\sin x)\, e^x - e^x\,(0+\cos x)}{(1+\sin x)^2} = \frac{(1+\sin x - \cos x)\, e^x}{(1+\sin x)^2}$$

(ii) Let $y = \dfrac{2x+3}{x^2-5}$

$$\Rightarrow \quad \frac{dy}{dx} = \frac{d}{dx}\left(\frac{2x+3}{x^2-5}\right) = \frac{(x^2-5)\dfrac{d}{dx}(2x+3) - (2x+3)\dfrac{d}{dx}(x^2-5)}{(x^2-5)^2}$$

$$= \frac{(x^2-5)(2) - (2x+3)(2x)}{(x^2-5)^2}$$

$$= \frac{2x^2 - 10 - 4x^2 - 6x}{(x^2-5)^2} = \frac{-2x^2 - 6x - 10}{(x^2-5)^2}$$

(iii) $y = \dfrac{1}{xe^x + \cos x}$

$$\Rightarrow \quad \frac{dy}{dx} = \frac{d}{dx}\left(\frac{1}{xe^x + \cos x}\right) = \frac{(xe^x + \cos x)\dfrac{d}{dx}(1) - 1\cdot\dfrac{d}{dx}(xe^x + \cos x)}{(xe^x + \cos x)^2}$$

$$= \frac{(xe^x + \cos x)(0) - \dfrac{d}{dx}(xe^x) - \dfrac{d}{dx}(\cos x)}{(xe^x + \cos x)^2}$$

$$= \frac{0 - \left(x\dfrac{d}{dx}e^x + e^x\dfrac{d}{dx}(x)\right) - \dfrac{d}{dx}(\cos x)}{(xe^x + \cos x)^2}$$

$$= \frac{-(xe^x + e^x) - (-\sin x)}{(xe^x + \cos x)^2} = \frac{-e^x(1+x) + \sin x}{(xe^x + \cos x)^2}$$

PROBLEM SET 5.4

Differentiate the following w.r.t. x :

1. $x^3 \sin x$
2. $x^2 e^x \log x$
3. $\dfrac{2^x \cot x}{\sqrt{x}}$
4. $(x \sin x + \cos x)(e^x + x^2 \log x)$

5. $\dfrac{\sin x + \cos x}{\sin x - \cos x}$

6. $\dfrac{\sec x + \tan x}{\sec x - \tan x}$

7. $\dfrac{e^x - \tan x}{\cot x - x^n}$

8. $\dfrac{\sqrt{a} + \sqrt{x}}{\sqrt{a} - \sqrt{x}}$

9. $\dfrac{1 + \log x}{1 - \log x}$

10. If $y = \dfrac{x}{x + a}$, prove that $x \dfrac{dy}{dx} = y(1 - y)$

11. If $x\sqrt{1 + y} + y\sqrt{1 + x} = 0$, prove that $\dfrac{dy}{dx} = -\dfrac{1}{(1 + x)^2}$

12. If $y = \sqrt{\dfrac{1 - x}{1 + x}}$, prove that $(1 - x^2)\dfrac{dy}{dx} + y = 0$

13. If $y = \dfrac{x - 1}{2x^2 - 7x + 5}$; find $\dfrac{dy}{dx}$ at $x = 2$

ANSWERS

1. $x^2 (x \cos x + 3 \sin x)$ **2.** $xe^x (1 + x \log x + 2\log x)$

3. $\dfrac{2^x}{\sqrt{x}}\left(\log 2 \cdot \cot x - \operatorname{cosec}^2 x - \dfrac{\cot x}{2x}\right)$

4. $x \cos x (e^x + x^2 \log x) + (x \sin x + \cos x)(e^x + x + 2x \log x)$ **5.** $\dfrac{-2}{1 - \sin 2x}$

6. $\dfrac{2 \cos x}{(1 - \sin x)^2}$ **7.** $\dfrac{(\cot x - x^n)(e^x - \sec^2 x) + (e^x - \tan x)(\operatorname{cosec}^2 x + nx^{n-1})}{(\cot x - x^n)^2}$

8. $\dfrac{\sqrt{a}}{\sqrt{x}(\sqrt{a} - \sqrt{x})^2}$ **9.** $\dfrac{2}{x(1 - \log x)^2}$ **13.** -2

5.13 DIFFERENTIATION OF FUNCTION OF A FUNCTION

5.13.1 Chain Rule

If $y = f(t)$ be a differentiable function of t and $t = g(x)$ a differentiable function of x, then $y = f(g(x))$ is a function of x.

The derivative of y w. r. to x is given by

$$\frac{dy}{dx} = \frac{dy}{dt} \cdot \frac{dt}{dx}$$

This rule is known as chain rule of differentiation.

■ **NOTE :** If y is a function of t_1, t_1 is a function of t_2, t_2 is a function of t_3 and so on. Then

$$\frac{dy}{dx} = \frac{dy}{dt_1} \cdot \frac{dt_1}{dt_2} \cdot \frac{dt_2}{dt_3} \cdots$$

Example 5.33

Differentiate the following w.r. to x :

(i) $\log(\tan e^x)$ (ii) $e^{\sin x}$ (iii) $\sin(x^2 + 1)$

SOLUTION : (i) Let $t_1 = \tan e^x$ and $t_2 = e^x$, Then $y = \log t_1$, $t_1 = \tan t_2$ and $t_2 = e^x$

$$\text{Now } \frac{dy}{dx} = \frac{dy}{dt_1} \cdot \frac{dt_1}{dt_2} \cdot \frac{dt_2}{dx} = \frac{d}{dt_1}(\log t_1) \cdot \frac{d}{dt_2}(\tan t_2) \cdot \frac{d}{dx}(e^x)$$

$$= \frac{1}{t_1} \cdot \sec^2 t_2 \cdot e^x = \frac{1}{\tan e^x} \cdot \sec^2 e^x \cdot e^x$$

(ii) Let $y = e^{\sin x}$ and $t = \sin x$

$$\therefore \quad y = e^t \text{ and } t = \sin x$$

$$\text{Now } \frac{dy}{dx} = \frac{dy}{dt} \times \frac{dt}{dx} = \frac{d}{dt}(e^t) \cdot \frac{d}{dx}(\sin x) = e^t \cdot \cos x$$

$$= e^{\sin x} \cdot \cos x$$

(iii) Let $y = \sin(x^2 + 1)$ and $t = x^2 + 1$

$$\text{Then } \quad y = \sin t \text{ and } t = x^2 + 1$$

$$\text{Now } \frac{dy}{dx} = \frac{dy}{dt} \cdot \frac{dt}{dx} = \frac{d}{dt}(\sin t)\frac{d}{dx}(x^2 + 1) = \cos t \cdot 2x$$

$$= 2x \cos(x^2 + 1)$$

Example 5.34

Differentiate the following w. r. to x :

(i) $(x^2 + x + 1)^{1/2}$ (ii) $\log(\sec x + \tan x)$ (iii) $\log \tan\left(\frac{\pi}{4} + \frac{x}{2}\right)$

SOLUTION : (i) Let $y = (x^2 + x + 1)^{1/2}$ and $t = x^2 + x + 1$

$$\text{Then } \quad y = t^{1/2} \text{ and } t = x^2 + x + 1$$

$$\text{Now } \frac{dy}{dx} = \frac{dy}{dt} \cdot \frac{dt}{dx} = \frac{d}{dt}(t^{1/2})\frac{d}{dx}(x^2 + x + 1)$$

$$= \frac{1}{2}t^{-1/2}(2x + 1) = \frac{1}{2}(x^2 + x + 1)^{-1/2}(2x + 1)$$

(ii) Let $y = \log(\sec x + \tan x)$ and $t = \sec x + \tan x$

Then $y = \log t$ and $t = \sec x + \tan x$

$$\text{Now } \frac{dy}{dx} = \frac{dy}{dt} \cdot \frac{dt}{dx} = \frac{d}{dx}(\log t) \cdot \frac{d}{dx}(\sec x + \tan x)$$

$$\frac{1}{t} \cdot (\sec x \tan x + \sec^2 x) = \frac{\sec x(\tan x + \sec x)}{\sec x + \tan x} = \sec x$$

(iii) Let $y = \log \tan\left(\frac{\pi}{4} + \frac{x}{2}\right)$

Now, let $\quad t_1 = \tan\left(\frac{\pi}{4} + \frac{x}{2}\right)$ and $t_2 = \frac{\pi}{4} + \frac{x}{2}$

$\therefore \quad y = \log t_1, \ t_1 = \tan t_2$ and $t_2 = \dfrac{\pi}{4} + \dfrac{x}{2}$

Now $\dfrac{dy}{dx} = \dfrac{dy}{dt_1} \cdot \dfrac{dt_1}{dt_2} \cdot \dfrac{dt_2}{dx} = \dfrac{d}{dt_1}(\log t_1) \cdot \dfrac{d}{dt_2}(\tan t_2) \cdot \dfrac{d}{dx}\left(\dfrac{\pi}{4} + \dfrac{x}{2}\right)$

$$= \frac{1}{t_1} \cdot \sec^2 t_2 \cdot \left(\frac{1}{2}\right) =$$

$$= \frac{1}{2} \cdot \frac{1}{\tan t_2} \cdot \sec^2 t_2 = \frac{1}{2} \frac{1}{\tan\left(\frac{\pi}{4} + \frac{x}{2}\right)} \cdot \sec^2\left(\frac{\pi}{4} + \frac{x}{2}\right)$$

$$= \frac{1}{2 \sin\left(\frac{\pi}{4} + \frac{x}{2}\right) \cos\left(\frac{\pi}{4} + \frac{x}{2}\right)}$$

$$= \frac{1}{\sin\left(\frac{\pi}{2} + x\right)} = \frac{1}{\cos x} = \sec x$$

5.13.2 Differentiation of a Function without Substitution

Example 5.35

Differentiate w. r. to x :

(i) $\log\left(x + \sqrt{a^2 + x^2}\right)$ (ii) $y = \log \sqrt{\dfrac{1 + \sin x}{1 - \sin x}}$

SOLUTION : (i) Let $y = \log\left(x + \sqrt{a^2 + x^2}\right)$

$\therefore \quad \dfrac{dy}{dx} = \dfrac{d}{dx} \log\left(x + \sqrt{a^2 + x^2}\right) = \dfrac{1}{x + \sqrt{a^2 + x^2}} \cdot \dfrac{d}{dx}\left(x + \sqrt{a^2 + x^2}\right)$

$$= \frac{1}{x + \sqrt{a^2 + x^2}}\left(1 + \frac{1}{2}(a^2 + x^2)^{-1/2} \frac{d}{dx}(a^2 + x^2)\right)$$

$$= \frac{1}{x + \sqrt{a^2 + x^2}}\left(1 + \frac{1}{2}(a^2 + x^2)^{-1/2}(0 + 2x)\right)$$

$$= \frac{1}{x + \sqrt{a^2 + x^2}}\left(1 + \frac{x}{\sqrt{a^2 + x^2}}\right) = \frac{1}{x + \sqrt{a^2 + x^2}}\left(\frac{\sqrt{a^2 + x^2} + x}{\sqrt{a^2 + x^2}}\right)$$

$$= \frac{1}{\sqrt{a^2 + x^2}}$$

(ii) Let $\quad y = \log \sqrt{\dfrac{1 + \sin x}{1 - \sin x}} = \log\left(\dfrac{1 + \sin x}{1 - \sin x}\right)^{1/2}$

$$= \frac{1}{2} \log\left(\frac{1 + \sin x}{1 - \sin x}\right) = \frac{1}{2}[\log(1 + \sin x) - \log(1 - \sin x)]$$

$\therefore \quad \dfrac{dy}{dx} = \dfrac{1}{2} \cdot \left(\dfrac{1}{1 + \sin x} \cdot \dfrac{d}{dx}(1 + \sin x) - \dfrac{1}{1 - \sin x} \cdot \dfrac{d}{dx}(1 - \sin x)\right)$

$$= \frac{1}{2}\left(\frac{1}{1 + \sin x} \cdot (0 + \cos x) - \frac{1}{1 - \sin x}(0 - \cos x)\right)$$

$$= \frac{1}{2}\left(\frac{\cos x}{1+\sin x} + \frac{\cos x}{1-\sin x}\right)$$

$$= \frac{\cos x}{2}\left(\frac{1-\sin x+1+\sin x}{1-\sin^2 x}\right) = \frac{\cos x}{2}\left(\frac{2}{\cos^2 x}\right) = \frac{1}{\cos x}$$

$$= \sec x$$

Example 5.36

Differentiate w.r.to x :

(i) $\dfrac{e^x + \log x}{\sin 3x}$ (ii) $e^{\cos^{-1}(\sqrt{1-x^2})}$

SOLUTION : (i) Let $y = \dfrac{e^x + \log x}{\sin 3x}$

$$\therefore \quad \frac{dy}{dx} = \frac{\sin 3x \dfrac{d}{dx}(e^x + \log x) - (e^x + \log x)\dfrac{d}{dx}(\sin 3x)}{\sin^2 3x}$$

$$= \frac{\sin 3x\left(e^x + \dfrac{1}{x}\right) - (e^x + \log x)\cos 3x \dfrac{d}{dx}(3x)}{\sin^2 3x}$$

$$= \frac{\sin 3x\left(e^x + \dfrac{1}{x}\right) - (e^x + \log x)\cos 3x \cdot 3}{\sin^2 3x}$$

$$= \frac{\sin 3x\left(e^x + \dfrac{1}{x}\right) - 3(e^x + \log x)\cos 3x}{\sin^2 3x}$$

(ii) Let $y = e^{\cos^{-1}\sqrt{1-x^2}}$

$$\Rightarrow \quad \frac{dy}{dx} = e^{\cos^{-1}\sqrt{1-x^2}}\frac{d}{dx}\cos^{-1}\sqrt{1-x^2}$$

$$= e^{\cos^{-1}\sqrt{1-x^2}}\frac{-1}{\sqrt{1-(\sqrt{1-x^2})^2}}\cdot\frac{d}{dx}(\sqrt{1-x^2})$$

$$= \frac{-e^{\cos^{-1}\sqrt{1-x^2}}}{\sqrt{1-(1-x^2)}}\cdot\frac{1}{2}(1-x^2)^{-1/2}\frac{d}{dx}(1-x^2)$$

$$= \frac{-e^{\cos^{-1}\sqrt{1-x^2}}}{\sqrt{x^2}}\cdot\frac{1}{2\sqrt{1-x^2}}(-2x) = \frac{x\,e^{\cos^{-1}\sqrt{1-x^2}}}{x\sqrt{1-x^2}}$$

$$= \frac{e^{\cos^{-1}\sqrt{1-x^2}}}{\sqrt{1-x^2)}}$$

Example 5.37

Find the differential coefficient of the function $\log\left(x + \dfrac{1}{x}\right)$. *[UPTU 2001]*

SOLUTION : Let $y = \log\left(x + \dfrac{1}{x}\right)$

$$\therefore \quad \frac{dy}{dx} = \frac{1}{x+\frac{1}{x}} \cdot \frac{d}{dx}\left(x + \frac{1}{x}\right) = \frac{x}{x^2+1} \cdot \left(1 - \frac{1}{x^2}\right)$$

$$= \frac{x}{x^2+1}\left(\frac{x^2-1}{x^2}\right) = \frac{x^2-1}{x\,(x^2+1}$$

Example 5.38

(i) Find $\frac{dy}{dx}$, it $y = \sin[\cos(\tan x)]$ *[UPTU 2005]*

(ii) If $y = e^x + \sin x$, find $\frac{d^2y}{dx^2}$. *[UPTU 2005]*

SOLUTION : (i) Given $y = \sin[\cos(\tan x)]$

$$\therefore \quad \frac{dy}{dx} = \cos[\cos(\tan x)]\frac{d}{dx}\cos(\tan x) = \cos[\cos(\tan x)]\,[-\sin(\tan x)]\frac{d}{dx}\tan x$$

$$= -\cos[\cos(\tan x)]\sin(\tan x)\sec^2 x$$

(ii) Given $y = e^x + \sin x \Rightarrow \frac{dy}{dx} = \frac{d}{dx}e^x + \frac{d}{dx}\sin x = e^x + \cos x$

Also, $\frac{d^2y}{dx^2} = \frac{d}{dx}\left(\frac{dy}{dx}\right) = \frac{d}{dx}(e^x + \cos x) = \frac{d}{dx}(e^x) + \frac{d}{dx}(\cos x) = e^x - \sin x$

Example 5.39

If $y = \sin(e^{\tan^{-1}x})$, find $\frac{dy}{dx}$. *[UPTU 2004]*

SOLUTION : Given $y = \sin(e^{\tan^{-1}x})$

Differentiate w.r.to x, we get

$$\frac{dy}{dx} = \cos(e^{\tan^{-1}x})\frac{d}{dx}(e^{\tan^{-1}x})$$

$$= \cos(e^{\tan^{-1}x})(e^{\tan^{-1}x})\frac{d}{dx}(\tan^{-1}x)$$

$$= \cos(e^{\tan^{-1}x})\,e^{\tan^{-1}x} \cdot \frac{1}{1+x^2}$$

Example 5.40

Differentiate $\log\sin(x^2+1)$ w.r. to x. *[UPTU 2003]*

SOLUTION : Given $y = \log\sin(x^2+1)$

Differentiating both sides w.r. to x,

$$\frac{dy}{dx} = \frac{1}{\sin(x^2+1)}\frac{d}{dx}\sin(x^2+1) = \frac{\cos(x^2+1)}{\sin(x^2+1)} \cdot \frac{d}{dx}(x^2+1)$$

$$= 2x \cdot \frac{\cos(x^2+1)}{\sin(x^2+1)}$$

Example 5.41

If $y = e^{a \sin^{-1} bx}$, find $\frac{dy}{dx}$ and $\frac{d^2y}{dx^2}$.

SOLUTION : We have

$$y = e^{a \sin^{-1} bx}$$

$$\Rightarrow \quad \frac{dy}{dx} = e^{a \sin^{-1} bx} \cdot \frac{d}{dx}(a \sin^{-1} bx)$$

$$= e^{a \sin^{-1} bx} \cdot \frac{a}{\sqrt{1-b^2x^2}} \cdot \frac{d}{dx}(bx)$$

$$= \frac{abe^{a \sin^{-1} bx}}{\sqrt{1-b^2x^2}}$$

And
$$\frac{d^2y}{dx^2} = \frac{d}{dx}\left[\frac{dy}{dx}\right] = ab\frac{d}{dx}\left[\frac{e^{a \sin^{-1} bx}}{\sqrt{1-b^2x^2}}\right]$$

$$= ab\frac{\left[\sqrt{1-b^2x^2} \cdot e^{a \sin^{-1} bx} \cdot \frac{ab}{\sqrt{1-b^2x^2}} - e^{a \sin^{-1} bx} \cdot \frac{(-2b^2x)}{2\sqrt{1-b^2x^2}}\right]}{1-b^2x^2}$$

$$= ab\left[\frac{abe^{a \sin^{-1} bx} + \frac{b^2x\, e^{a \sin^{-1} bx}}{\sqrt{1-b^2x^2}}}{1-b^2x^2}\right]$$

$$= \frac{ab^2e^{a \sin^{-1} bx}}{(1-b^2x^2)^{3/2}}\left[a\sqrt{1-b^2x^2} + bx\right]$$

PROBLEM SET 5.5

Differentiate the following function w. r. to x :

1. $\log \sin x$
2. $e^{\sin x^2}$
3. $\frac{1}{\sqrt{a^2 - x^2}}$
4. $\sin^{-1}\left(\frac{a + b\cos x}{b + a\cos x}\right), b > a$
5. $\sec(\log x^n)$
6. $\log\left(\frac{a + b\sin x}{a - b\sin x}\right)$
7. $e^x \log(1 + x^2)$

8. $\dfrac{\sin x + x^2}{\cot 2x}$

9. $\log \sqrt{\dfrac{1 - \cos x}{1 + \cos x}}$

10. $(\log \sin x)^2$

11. $e^{\tan^{-1} \sqrt{x}}$

12. $\dfrac{\sqrt{x^2 + 1} + \sqrt{x^2 - 1}}{\sqrt{x^2 + 1} - \sqrt{x^2 - 1}}$

13. $\log(\text{cosce } x - \cot x)$

14. $\dfrac{e^x + e^{-x}}{e^x - e^{-x}}$

15. If $y = \log(\sqrt{x - 1} - \sqrt{x + 1})$. Show that $\dfrac{dy}{dx} = -\dfrac{1}{2\sqrt{x^2 - 1}}$

16. If $y = (x + \sqrt{x^2 + a^2})^n$, prove that $\dfrac{dy}{dx} = \dfrac{ny}{\sqrt{x^2 + a^2}}$

17. If $y = \dfrac{x \sin^{-1} x}{\sqrt{1 - x^2}}$ prove that $(1 - x^2)\dfrac{dy}{dx} = \sqrt{x} - \dfrac{1}{\sqrt{x}}$

18. If $y = \dfrac{1}{2}\log\left(\dfrac{1 - \cos 2x}{1 + \cos 2x}\right)$. Prove that $\dfrac{dy}{dx} = 2 \text{ cosec } 2x$

19. If $x - y = 4$, prove that $x\left(\dfrac{dy}{dx} + y^2\right) = 3y$

20. If $y = x \sin^{-1} x + \sqrt{1 - x^2}$, prove that $\dfrac{dy}{dx} = \sin^{-1} x$

ANSWERS

1. $\cot x$ **2.** $2x \cos x^2 \cdot e^{\sin x^2}$ **3.** $\dfrac{x}{(a^2 - x^2)^{3/2}}$ **4.** $\dfrac{-\sqrt{b^2 - a^2}}{b + a \cos x}$

5. $\dfrac{n}{x} \sec(\log x^n) \cdot \tan(\log x^n)$ **6.** $\dfrac{2ab \cos x}{a^2 - b^2 \sin^2 x}$ **7.** $e^x\left[\dfrac{2x}{1 + x^2} + \log(1 + x^2)\right]$

8. $2(\sin x + x^2)\sec^2 2x + (\cos x + 2x)\tan 2x$ **9.** $\text{cosec } x$

10. $2 \log \sin x \cdot \cot x$ **11.** $\dfrac{e^{\tan^{-1}\sqrt{x}}}{2\sqrt{x}\,(1 + x)}$ **12.** $2x + \dfrac{2x^3}{\sqrt{x^4 - 1}}$ **13.** $\text{cosec } x$

14. $\dfrac{-4}{(e^x - e^{-x})^2}$

5.14 DIFFERENTIATION OF FUNCTIONS USING SOME TRANSFORMATIONS

Example 5.42

Differentiate the following functions w. r. to x :

(i) $\tan^{-1}\sqrt{\dfrac{1+\cos x}{1-\cos x}}$ (ii) $\tan^{-1}\left(\sqrt{1}+x^2+x\right)$

SOLUTION : (i) Let $y \tan^{-1}\sqrt{\dfrac{1+\cos x}{1-\cos x}} = \tan^{-1}\sqrt{\dfrac{1+2\cos^2\frac{x}{2}-1}{1-1+2\sin^2\frac{x}{2}}}$

$$= \tan^{-1}\sqrt{\frac{\cos^{2} x/2}{\sin^{2} x/2}} = \tan^{-1}\sqrt{\cot^2\frac{x}{2}}$$

$$= \tan^{-1}(\cot\frac{x}{2}) = \tan^{-1}\left(\tan\left(\frac{\pi}{2}-\frac{x}{2}\right)\right) = \frac{\pi}{2}-\frac{x}{2}$$

$\therefore$ $$\frac{dy}{dx} = 0-\frac{1}{2} = -\frac{1}{2}$$

(ii) Let $y = \tan^{-1}(\sqrt{1+x^2}+x)$

Again let, $x = \tan\theta \Rightarrow \theta = \tan^{-1} x$

$\therefore$ $$y = \tan^{-1}(\sqrt{1+\tan^2\theta}+\tan\theta)$$

$$= \tan^{-1}(\sqrt{\sec^2}\,\theta + \tan\theta)$$

$$= \tan^{-1}(\sec\theta + \tan\theta)$$

$$= \tan^{-1}\left(\frac{1+\sin\theta}{\cos\theta}\right)$$

$$= \tan^{-1}\left(\frac{\sin^2\frac{\theta}{2}+\cos^2\frac{\theta}{2}+2\sin\frac{\theta}{2}\cos\frac{\theta}{2}}{\cos^2\frac{\theta}{2}-\sin^2\frac{\theta}{2}}\right)$$

$$= \tan^{-1}\left(\frac{\left(\cos\frac{\theta}{2}+\sin\frac{\theta}{2}\right)^2}{\left(\cos\frac{\theta}{2}-\sin\frac{\theta}{2}\right)\left(\cos\frac{\theta}{2}+\sin\frac{\theta}{2}\right)}\right)$$

$$= \tan^{-1}\left(\frac{\left(\cos\frac{\theta}{2}+\sin\frac{\theta}{2}\right)}{\cos\frac{\theta}{2}-\sin\frac{\theta}{2}}\right) = \tan^{-1}\left(\frac{1+\tan\frac{\theta}{2}}{1-\tan\frac{\theta}{2}}\right)$$

$$= \tan^{-1}\left(\frac{\tan\frac{\pi}{4}+\tan\frac{\theta}{2}}{\tan\frac{\pi}{4}-\tan\frac{\theta}{2}}\right) = \tan^{-1}\left(\tan\left(\frac{\pi}{4}+\frac{\theta}{2}\right)\right)$$

$$y = \frac{\pi}{4} + \frac{\theta}{2} = \frac{\pi}{4} + \frac{1}{2}\tan^{-1} x$$

$$\therefore \quad \frac{dy}{dx} = 0 + \frac{1}{2}\cdot\frac{1}{1+x^2} = \frac{1}{2\,(1+x^2)}$$

Example 5.43

Differentiate the following w. r. to x :

(i) $\sin^{-1}(2x\sqrt{1-x^2})$ (ii) $\cos^{-1}\left(\dfrac{x+\sqrt{1-x^2}}{\sqrt{2}}\right)$

SOLUTION : (i) Let $y = \sin^{-1}(2x\sqrt{1-x^2})$

Again let $x = \sin\theta \Rightarrow \theta = \sin^{-1} x$

Now
$$\begin{aligned} y &= \sin^{-1}(2\sin\theta\sqrt{1-\sin^2\theta}) \\ &= \sin^{-1}(2\sin\theta\sqrt{\cos^2\theta}) \\ &= \sin^{-1}(2\sin\theta\cos\theta) \\ &= \sin^{-1}(\sin 2\theta) \\ &= 2\,\theta = 2\sin^{-1} x \quad\Rightarrow\quad \frac{dy}{dx} = 2\cdot\frac{1}{\sqrt{1-x^2}} \end{aligned}$$

(ii) Let $y = \cos^{-1}\left(\dfrac{x+\sqrt{1-x^2}}{\sqrt{2}}\right)$

Again let $x = \sin\theta \quad\Rightarrow\quad \theta = \sin^{-1} x$

$$\begin{aligned} \therefore \quad y &= \cos^{-1}\left(\frac{\sin\theta+\sqrt{1-\sin^2\theta}}{\sqrt{2}}\right) \\ &= \cos^{-1}\left(\frac{\sin\theta+\cos\theta}{\sqrt{2}}\right) \\ &= \cos^{-1}\left(\frac{1}{\sqrt{2}}\sin\theta + \frac{1}{\sqrt{2}}\cos\theta\right) = \cos^{-1}\left(\sin\frac{\pi}{4}\sin\theta + \cos\frac{\pi}{4}\cos\theta\right) \\ &= \cos^{-1}\left(\cos\left(\frac{\pi}{4}-\theta\right)\right) = \frac{\pi}{4} - \theta = \frac{\pi}{4} - \sin^{-1} x \end{aligned}$$

$$\therefore \quad y = \frac{\pi}{4} - \sin^{-1} x \quad\Rightarrow\quad \frac{dy}{dx} = 0 - \frac{1}{\sqrt{1-x^2}} = \frac{-1}{\sqrt{1-x^2}}$$

Example 5.44

If $x = a\,(t - \sin t)$, $y = a\,(1 - \cos t)$ find $\dfrac{dy}{dx}$

SOLUTION : Given, $x = a\,(t - \sin t)$, $y = a\,(1 - \cos t)$

$$\therefore \quad \frac{dx}{dt} = a\,(1-\cos t) \text{ and } \frac{dy}{dt} = a\,(0 + \sin t) = a\sin t$$

Now
$$\frac{dy}{dx} = \frac{dy}{dt}\cdot\frac{dt}{dx} = \frac{\sin t}{1-\cos t}$$

$$= \frac{2\sin\frac{t}{2}\cos\frac{t}{2}}{1-1+2\sin^2\frac{t}{2}} = \frac{\cos\frac{t}{2}}{\sin\frac{t}{2}} = \cot\frac{t}{2}$$

Example 5.45

If $x = t\cdot\log t,\ y = \frac{\log t}{t}$. Find $\frac{dx}{dt}$, $\frac{dy}{dt}$ and $\frac{dy}{dx}$

SOLUTION : Given $x = t\cdot\log t$

$$\Rightarrow \frac{dx}{dt} = t\cdot\frac{d}{dt}(\log t) + \log t\cdot\frac{d}{dt}(t)$$

$$= t\cdot\frac{1}{t} + \log t\cdot 1 = 1 + \log t$$

Also $$y = \frac{\log t}{t}$$

$$\Rightarrow \frac{dy}{dt} = \frac{t\cdot\frac{d}{dt}(\log t) - \log t\,\frac{d}{dt}(t)}{t^2}$$

$$= \frac{t\cdot\frac{1}{t} - \log t\cdot 1}{t^2} = \frac{1-\log t}{t^2}$$

Again , $$\frac{dy}{dx} = \frac{dy}{dt}\cdot\frac{dt}{dx}$$

or $$\frac{dy}{dx} = \frac{\frac{dy}{dt}}{\frac{dx}{dt}}$$

$$= \frac{(1-\log t)}{t^2}\times\frac{1}{1+\log t}$$

$$= \frac{1-\log t}{t^2(1+\log t)}$$

Example 5.46

If $x = at^2$ and $y = 2at$, find $\frac{dy}{dx}$. *[UPTU 2004]*

SOLUTION : Given $x = at^2$

$$\Rightarrow \frac{dx}{dt} = 2at$$

Also $$y = 2at$$

$$\Rightarrow \frac{dy}{dt} = 2a$$

Now $$\frac{dy}{dx} = \frac{dy}{dt}\times\frac{dt}{dx} = \frac{2a}{2at} = \frac{1}{t}$$

PROBLEM SET 5.6

Differentiate the following functions w.r. to x :

1. $\cos^{-1}\left(\dfrac{1-x^2}{1+x^2}\right)$

2. $\tan^{-1}\left(\dfrac{2x}{1-x^2}\right)$

3. $\tan^{-1}\left(\dfrac{1-\cos x}{\sin x}\right)$

4. $\tan^{-1}\left(\sqrt{\dfrac{1-\cos x}{1+\cos x}}\right)$

5. $\cot^{-1}(\sqrt{1+x^2}+x)$

6. If $x = \log t + \sin t, y = e^t + \cos t$. find $\dfrac{dy}{dx}$

7. If $x = \dfrac{3at}{(1+t)^3}, y = \dfrac{3at^2}{1+t^3}$. Find $\dfrac{dy}{dx}$

8. If $x = a\,(t \cos t - \sin t)$, $y = a\,(t \sin t + \cos t)$. Find $\dfrac{dy}{dx}$

9. If $y = \cos^{-1} 2x + 2\cos^{-1}\sqrt{1-4x^2}$. Find $\dfrac{dy}{dx}$

10. If $y = \sin^{-1}\left(\dfrac{1}{\sqrt{1+x^2}}\right)$. Find $\dfrac{dy}{dx}$

ANSWERS

1. $\dfrac{2}{1+x^2}$ **2.** $\dfrac{2}{1+x^2}$ **3.** $\dfrac{1}{2}$ **4.** $\dfrac{1}{2}$ **5.** $\dfrac{-1}{2}(1+x^2)$ **6.** $\dfrac{(e^t - \sin t)t}{(1+\cos t)}$

7. $\dfrac{t(2-t^3)(1+t)^4}{(1+t^3)^2(1-2t)}$ **8.** $\cot t$ **9.** $\dfrac{2}{\sqrt{1-4x^2}}$ **10.** $-\dfrac{1}{1+x^2}$

5.15 DERIVATIVE OF IMPLICIT FUNCTIONS

A function y is said to be an implicit function if it is not possible to express y in terms of x *i.e.*, $y = f(x)$. Such functions can be solved for y in terms of x by defining $\phi(x, y) = 0$ as $\phi(x : f(x)) = 0$. To find $\dfrac{dy}{dx}$ we differentiate each term on both sides of the equation w. r. to x and then solve for $\dfrac{dy}{dx}$.

Example 5.47

If $x^2 + 2xy + by^2 + 2gx + 2fy + c = 0$, find $\frac{dy}{dx}$

SOLUTION : We have

$$x^2 + 2xy + by^2 + 2gx + 2fy + c = 0$$

Differentiate both sides w.r. to x :

$$2x + 2\left(x\frac{dy}{dx} + y\right) + 2by\frac{dy}{dx} + 2g + 2f\frac{dy}{dx} = 0$$

$$2x + 2x\frac{dy}{dx} + 2y + 2by\frac{dy}{dx} + 2g + 2f\frac{dy}{dx} = 0$$

$$(2x + 2y + 2g) + \frac{dy}{dx}(2x + 2by + 2f) = 0$$

$$\Rightarrow \quad \frac{dy}{dx} = -\frac{2x + 2y + 2g}{2x + 2by + 2f} = -\frac{x + y + g}{x + by + f}$$

Example 5.48

If $x\sqrt{1+y} + y\sqrt{1+x} = 0$, prove that $\frac{dy}{dx} = -\frac{1}{(1+x)^2}$

SOLUTION : Given $x\sqrt{1+y} + y\sqrt{1+x} = 0$

$\Rightarrow \quad x\sqrt{1+y} = -y\sqrt{1+x}$

Squaring both sides, we get

$$x^2(1+y) = y^2(1+x)$$

$$\Rightarrow \quad x^2 + x^2y - y^2 - y^2x = 0$$

$$\Rightarrow \quad (x^2 - y^{2)} + (x^2y - y^2x) = 0$$

$$\Rightarrow \quad (x-y)(x+y) + xy(x-y) = 0$$

$$\Rightarrow \quad (x-y)(x+y+xy) = 0$$

But $x \neq y$ (y is a function of x)

$$\therefore \quad x - y \neq 0$$

$$\therefore \quad x + y + xy = 0 \Rightarrow y = -\frac{x}{1+x}$$

Diferentiate both sides, we obtain.

$$\frac{dy}{dx} = -\left(\frac{(1+x)\cdot 1 - x(0+1)}{(1+x)^2}\right) = -\left(\frac{1+x-x}{(1+x)^2}\right) = -\frac{1}{(1+x)^2}$$

Example 5.49

Differentiate the following w.r. to x :

(i) x^x (ii) $(\sin x)^{\cos^{-1} x}$

SOLUTION : (i) Let $y = x^x$

Taking log on both sides, we get

$$\log y = \log x^x \Rightarrow \log y = x \cdot \log x$$

Diferentiate both sides w. r. to x :

$$\frac{1}{y} \cdot \frac{dy}{dx} = x \cdot \frac{1}{x} + 1 \cdot \log x = 1 + \log x$$

$$\Rightarrow \qquad \frac{dy}{dx} = y\ (1 + \log x) = x^x\ (1 + \log x)$$

(ii) Let $y = (\sin x)^{\cos^{-1} x}$

Taking log on both sides, we get

$$\log y = \cos^{-1} x \log \sin x$$

Differentiate both sides w. r. to x :

$$\frac{1}{y}\frac{dy}{dx} = \frac{\cos x}{\sin x} \cos^{-1} x - \frac{\log \sin x}{\sqrt{1 - x^2}}$$

$$= \cot x \cos^{-1} x - \frac{\lambda o\gamma\ \sigma\iota\nu\ x}{\sqrt{1 - x^2}}$$

$$\Rightarrow \qquad \frac{dy}{dx} = y\left(\cos^{-1} x \cot x - \frac{\log \sin x}{\sqrt{1 - x^2}}\right)$$

$$= (\sin x)^{\cos^{-1} x}\left(\cos^{-1} x \cdot \cot x - \frac{\log \sin x}{\sqrt{1 - x^2}}\right)$$

Example 5.50

If $x^y + y^x = a^b$, find $\frac{dy}{dx}$

SOLUTION : Let $y_1 + y_2 = a^b$ ...(1)

where $y_1 = x^y$ ...(2)

and $y_2 = y^x$ ...(3)

Differentiate (1) w. r. to x :

$$\frac{dy_1}{dx} + \frac{dy_2}{dx} = 0 \qquad ...(4)$$

Now, to find $\frac{dy_1}{dx}$ and $\frac{dy_2}{dx}$, we proceed as

$$y_1 = x^y$$

Taking log on both sides, we get

$$\log y_1 = y \log x$$

Differentiate both sides w.r. to x :

$$\frac{1}{y_1}\frac{dy_1}{dx} = \frac{y}{x} + \frac{dy}{dx} \log x$$

$$\Rightarrow \qquad \frac{dy_1}{dx} = y_1\left(\frac{y}{x} + \frac{dy}{dx} \log x\right)$$

$$= x^y\left(\frac{y}{x} + \frac{dy}{dx} \log x\right)$$

Also $y_2 = y^x$

Taking log on both sides we get

$$\log y_2 = x \log y$$

Differentiate both sides w. r. to x :

$$\frac{1}{y_2}\cdot\frac{dy_2}{dx} = \frac{x}{y}\cdot\frac{dy}{dx} + \log y$$

$$\frac{dy_2}{dx} = y_2\left(\frac{x}{y}\cdot\frac{dy}{dx} + \log y\right)$$

$$= y^x\left(\frac{x}{y}\cdot\frac{dy}{dx} + \log y\right)$$

From (4), we get

$$x^y\cdot\frac{y}{x} + x^y\cdot\frac{dy}{dx}\log x + y^x\cdot\frac{x}{y}\frac{dy}{dx} + y^x\cdot\log y = 0$$

$$(y\cdot x^{y-1} + y^x\cdot\log y) + \frac{dy}{dx}(x^y\log x + x\cdot y^{x-1}) = 0$$

$$\Rightarrow \quad \frac{dy}{dx} = -\frac{yx^{y-1} + y^x\log y}{x^y\log x + x\cdot y^{x-1}}$$

Example 5.51

If $x^y = e^{x-y}$, prove that $\dfrac{dy}{dx} = \dfrac{\log x}{(1+\log x)^2}$

SOLUTION : Given $x^y = e^{x-y}$

$$\Rightarrow \quad e^{y\log x} = e^{x-y} \Rightarrow y\log x = x - y$$

$$\Rightarrow \quad y(\log x + 1) = x \ \Rightarrow\ y = \frac{x}{1+\log x}$$

Differentiate both sides w. r. to x, we get

$$\frac{dy}{dx} = \frac{(1+\log x)\cdot 1 - x\left(0 + \frac{1}{x}\right)}{(1+\log x)^2}$$

$$= \frac{1+\log x - 1}{(1+\log x)^2} = \frac{\log x}{(1+\log x)^2}$$

Example 5.52

If $x^y = y^x$, find $\dfrac{dy}{dx}$

SOLUTION : Given $x^y = y^x$.

Taking log on both sides, we get

$$y\log x = x\log y$$

Differentiate both sides w. r. to x, we get

$$y\cdot\frac{1}{x} + \frac{dy}{dx}\log x = x\cdot\frac{1}{y}\frac{dy}{dx} + \log y$$

$$\Rightarrow \quad \frac{dy}{dx}\left(\log x - \frac{x}{y}\right) = \log y - \frac{y}{x}$$

$$\Rightarrow \quad \frac{dy}{dx} = \frac{\log y - \frac{y}{x}}{\log x - \frac{x}{y}} = \frac{\frac{x \log y - y}{x}}{\frac{y \log x - x}{y}}$$

$$= \frac{y}{x}\left(\frac{x \log y - y}{y \log x - x}\right)$$

Example 5.53

If $y = x^{x^{x^{\cdot^{\cdot^{\cdot}}}}}$, find $\frac{dy}{dx}$

SOLUTION : Given $y = x^{x^{x^{\cdot^{\cdot^{\cdot}}}}}$

The given series can be written as

$$y = x^y$$

[Because deleting a single term from an infinite series does not change the series]

$$\Rightarrow \quad \log y = y \log x$$

Differentiate both sides w. r. to x, we get

$$\frac{1}{y}\frac{dy}{dx} = \frac{y}{x} + \frac{dy}{dx}\log x \quad \Rightarrow \quad \left(\frac{1}{y} - \log x\right)\frac{dy}{dx} = \frac{y}{x}$$

$$\Rightarrow \quad \left(\frac{1 - y \log x}{y}\right)\frac{dy}{dx} = \frac{y}{x} \quad \Rightarrow \quad \frac{dy}{dx} = \frac{y}{x}.\frac{y}{(1 - y \log x)}$$

$$\Rightarrow \quad \frac{dy}{dx} = \frac{y^2}{x\,(1 - y \log x)}$$

Example 5.54

If $y = \sqrt{\cos x + \sqrt{\cos x + \sqrt{\cos x + \ldots}}}$, prove that

$$\frac{dy}{dx} = \frac{\sin x}{1 - 2y}$$

SOLUTION : Given $y = \sqrt{\cos x + \sqrt{\cos x + \sqrt{\cos x + \ldots}}}$

The given series can be written as

$$y = \sqrt{\cos x + y}$$

$$\text{or} \quad y^2 = \cos x + y$$

Differentiate both sides w. r. to x, we get

$$2y\frac{dy}{dx} = -\sin x + \frac{dy}{dx}$$

$$(2y - 1)\frac{dy}{dx} = -\sin x \quad \Rightarrow \quad \frac{dy}{dx} = \frac{-\sin x}{2y - 1}$$

$$\Rightarrow \quad \frac{dy}{dx} = \frac{\sin x}{1 - 2y}$$

Example 5.55

If $y = e^{x + e^{x + \cdots}}$, then prove that $\dfrac{dy}{dx} = \dfrac{y}{1-y}$ *[UP TU 2002]*

SOLUTION : Given $y = e^{x + e^{x + \cdots}}$

The given series can be written as

$$y = e^{x + y}$$

differentiating both sides w.r. to x, we get

$$\frac{dy}{dx} = e^{x+y}\left(1 + \frac{dy}{dx}\right)$$

$$= y\left(1 + \frac{dy}{dx}\right) \qquad [\because y = e^{x+y}]$$

$$\frac{dy}{dx} = y + y\frac{dy}{dx}$$

or $$\frac{dy}{dx} - y\frac{dy}{dx} = y \Rightarrow (1-y)\frac{dy}{dx} = y \Rightarrow \frac{dy}{dx} = \frac{y}{1-y}$$

Example 5.56

If $y = e^x \cos x + x^{\sin x}$, find $\dfrac{dy}{dx}$. *[UPTU 2003]*

SOLUTION : Given $y = e^x \cos x + x^{\sin x}$

Let $y = y_1 + y_2$

where $y_1 = e^x \cos x$ and $y_2 = x^{\sin x}$

then $$\frac{dy}{dx} = \frac{dy_1}{dx} + \frac{dy_2}{dx} \qquad \ldots(1)$$

Now, $$y_1 = e^x \cos x$$

$$\Rightarrow \quad \frac{dy_1}{dx} = e^x \cos x - e^x \sin x \qquad \ldots(2)$$

Also $$y_2 = x^{\sin x}$$

Taking log of both sides

$$\log y_2 = \sin x \log x$$

Differentiate w.r.to x

$$\frac{1}{y_2}\frac{dy_2}{dx} = \sin x \cdot \frac{1}{x} + \cos x \log x$$

$$\frac{dy_2}{dx} = y_2\left[\frac{\sin x}{x} + \cos x \log x\right]$$

$$= x^{\sin x}\left[\frac{\sin x}{x} + \cos x \cdot \log x\right] \qquad \ldots(3)$$

Substituting the values of $\frac{dy_1}{dx}$ from (2) and (3) to (1), we get

$$\frac{dy}{dx} = e^x \cos x - e^x \sin x + x^{\sin x}\left(\frac{\sin x}{x} + \cos x \log x\right)$$

Example 5.57

If $\sin y = x \sin(a + y)$, prove that

$$\frac{dy}{dx} = \frac{\sin^2(a + y)}{\sin a}$$ *[UPTU (C.O.) 2006]*

SOLUTION : Given $\sin y = x \sin(a + y)$

Differentiate both sides w.r. to x

$$\cos y \cdot \frac{dy}{dx} = \sin(a + y) + x \cos(a + y) \cdot \frac{dy}{dx}$$

$$\Rightarrow \quad [\cos y - x \cos(a + y)] \frac{dy}{dx} = \sin(a + y)$$

$$\Rightarrow \quad \left[\cos y - \frac{\sin y}{\sin(a + y)} \cdot \cos(a + y)\right] \frac{dy}{dx} = \sin(a + y) \qquad \left[\because\ x = \frac{\sin y}{\sin(a + y)}\right]$$

$$\Rightarrow \quad [\cos y \sin(a + y) - \sin y \cos(a + y)] \frac{dy}{dx} = \sin^2(a + y)$$

$$\Rightarrow \quad \sin(a + y - y) \frac{dy}{dx} = \sin^2(a + y) \qquad [\because\ \sin A \cos B - \cos A \sin B = \sin(A - B)]$$

$$\Rightarrow \quad \frac{dy}{dx} = \frac{\sin^2(a + y)}{\sin a}$$

PROBLEM SET 5.7

Find $\frac{dy}{dx}$ *in each of the following*

1. $x^2 + 3xy + 3x^2 - 2x + y - 12 = 0$
2. $\frac{x}{x - y} = \log \frac{a}{x - y}$
3. $\sqrt{1 - x^2} + \sqrt{1 - y^2} = a(x - y)$
4. $x^y = e^{x + y}$
5. $e^{x - y} = \log\left(\frac{x}{y}\right)$
6. $y = x \sin(a + y)$
7. $x^{\sin x}$
8. $(\log x)^{\sin x}$
9. $(\sin x)^{\tan x} + (\cos x)^{\sec x}$
10. $(\cos x)^y = (\sin y)^x$

11. If $x^m y^n = (x+y)^{m+n}$, prove that $\dfrac{dy}{dx} = \dfrac{y}{x}$

12. If $y \cdot e^y = x$, prove that $\dfrac{dy}{dx} = \dfrac{y}{x(1+y)}$

13. If $x^y + y^x = 2$, find $\dfrac{dy}{dx}$

14. If $y = (\cos x)^{(\cos x)^{(\cos x)^{\cdot^{\cdot^{\cdot}}}}}$, prove that $\dfrac{dy}{dx} = -\dfrac{y^2 \tan x}{1 - y \log \cos x}$

15. If $y = (\sin x)^{(\sin x)^{(\sin x)^{\cdot^{\cdot^{\cdot}}}}}$, prove that $\dfrac{dy}{dx} = \dfrac{y^2 \cot x}{(1 - y \log \sin x)}$

16. If $y = (\sqrt{x})^{(\sqrt{x})^{(\sqrt{x})^{\cdot^{\cdot^{\cdot}}}}}$ prove that $\dfrac{dy}{dx} = \dfrac{y^2}{x(2 - y \log x)}$

17. If $y = \sqrt{\log x + \sqrt{\log x + \ldots}}$ prove that $(2y-1)\dfrac{dy}{dx} = \dfrac{1}{x}$

18. If $y = a^{x^{a^{x^{\cdot^{\cdot^{\cdot}}}}}}$, prove that $\dfrac{dy}{dx} = \dfrac{y^2 \log y}{x(1 - y \log x \cdot \log y)}$

19. If $y = x^{\cos x} + (\cos x)^x$, prove that

$$\frac{dy}{dx} = x^{\cos x}\left[\frac{\cos x}{x} - \sin x \cdot \log x\right] + (\cos x)^x\,[\log \cos x - x \tan x]$$

20. If $y = x^{\sin^{-1} x}$, prove that $\dfrac{dy}{dx} = x^{\sin^{-1} x}\left[\dfrac{\sin^{-1} x}{x} + \dfrac{\log x}{\sqrt{1-x^2}}\right]$

21. If $x = y + \cfrac{1}{y + \cfrac{1}{y + \ldots}}$, prove that $\dfrac{dy}{dx} = \dfrac{2x - y}{x}$

ANSWERS

1. $\dfrac{dy}{dx} = \left(\dfrac{2 - 8x - 3y}{3x+1}\right)$ **2.** $2 - \dfrac{x}{y}$ **3.** $\dfrac{\sqrt{1-y^2}}{\sqrt{1-x^2}}$ **4.** $\dfrac{2 - \log x}{(\log x - 1)^2}$ **5.** $\dfrac{y}{x}\left(\dfrac{x\,e^{x-y} - 1}{y\,e^{x-y} - 1}\right)$

6. $\dfrac{\sin^2(a+y)}{\sin(a+y) - y\cos(a+y)}$ **7.** $x^{\sin x}\left(\cos x \log x + \dfrac{\sin x}{x}\right)$

8. $(\log x)^{\sin x}\left(\log(\log x)\cdot\cos x + \dfrac{\sin x}{x \log x}\right)$

9. $(\sin x)^{\tan x}\,[\sec^2 x \log \sin x + 1] + (\cos x)^{\sec x}\,[\sec x \tan x \cdot \log \cos x - \sec x \tan x]$

10. $\dfrac{\log \sin y + y \tan x}{\log \cos x - x \cot y}$ **13.** $-\left(\dfrac{y\,x^{y-1} + y^x \log y}{x^y \log x + x\,y^{x-1}}\right)$

5.16 INTEGRATION

The inverse process of differentiation is known as integration. Consider two functions $f(x)$ and $F(x)$ such that $\frac{dF(x)}{dx} = f(x)$. Then $F(x)$ is called as integral or antiderivative of $f(x)$ with respect to x.

Mathematically, if $\frac{dF(x)}{dx} = f(x)$, then $\int f(x)dx = F(x) + C$. Here the sumbol '$\int$' denotes integration and C is the constant of integration. The function $F(x)$ is known as integrand and $f(x)dx$ is element of integration.

5.17 INDEFINITE AND DEFINITE INTEGRAL

If the integration of $f(x)$ with respect of x has no definite numerical value then the integral is known as *indefinite integral*. So, the integral $\int f(x)dx = F(x) + C$ is indefinite integral also known as *general integral*. If we substitute the value of C, then this is called *particular integral*.

Let $f(x)$ be continuous function defined on closed interval $[a_1, a_2]$ where $a_1 \le x \le a_2$, then the integration of $f(x)$ with respect to x is known as *definite integral* of $f(x)$ and is denoted by

$$\int_{a_1}^{a_2} f(x)dx = [F(x)]_{a_1}^{a_2} = F(a_2) - F(a_1)$$

Here, $\int_{a_1}^{a_2} f(x)dx$ is integral fo $f(x)$ with respect to x from a_1 to a_2 and a_1 is lower limit, a_2 is an upper limit of integration. From the definition of definite integration, we can see that there is no 'C', *i.e.*, no constant of integration. To eliminate C we proceed as follows:

$$\int_{a_1}^{a_2} f(x)dx = [F(x) + C]_{a_1}^{a_2} = [F(a_2) + C] - [F(a_1) + C]$$

$$= F(a_2) + C - F(a_1) - C = F(a_2) - F(a_1)$$

So, the value of definite integral is the difference of the values of integral $F(x)$ at a_2 and a_1 or at upper limit and lower limit of integration.

5.18 RULES OF INTEGRATION

As we know that integration is inverse of differentiation. So the rules of integration are related to the rules of differentiation. Some important rules are given below:

(1) $\frac{d}{dx}\left(\frac{x^{n+1}}{n+1}\right) = x^n, n \ne -1$ $\quad\therefore \int x^n\,dx = \frac{x^{n+1}}{n+1} + c, n \ne -1$

(2) $\frac{d}{dx}(\log x) = \frac{1}{x}$ $\quad\therefore \int \frac{1}{x}\,dx = \log|x| + c$

(3) $\frac{d}{dx}\left(e^x\right) = e^x$ $\quad\therefore \int e^x\,dx = e^x + c$

(4) $\frac{d}{dx}\left(\frac{a^x}{\log_e a}\right) = a^x,\ a > 0,\ a \neq 1$ $\quad\therefore \int a^x\,dx = \frac{a^x}{\log_e a} + c$

where a is constant

(5) $\frac{d}{dx}(-\cos x) = \sin x$ $\quad\therefore \int \sin x\,dx = -\cos x + c$

(6) $\frac{d}{dx}(\sin x) = \cos x$ $\quad\therefore \int \cos x\,dx = \sin x + c$

(7) $\frac{d}{dx}(\tan x) = \sec^2 x$ $\quad\therefore \int \sec^2 x\,dx = \tan x + c$

(8) $\frac{d}{dx}(-\cot x) = \text{cosec}^2 x$ $\quad\therefore \int \text{cosec}^2 x\,dx = -\cot x + c$

(9) $\frac{d}{dx}(\sec x) = \sec x \tan x$ $\quad\therefore \int \sec x \tan x\,dx = \sec x + c$

(10) $\frac{d}{dx}(-\text{cosec}\,x) = \text{cosec}\,x \cot x$ $\quad\therefore \int \text{cosec}\,x \cot x\,dx = -\text{cosec}\,x + c$

(11) $\frac{d}{dx}(\log \sin x) = \cot x$ $\quad\therefore \int \cot x\,dx = \log|\sin x| + c$

(12) $\frac{d}{dx}(-\log \cos x) = \tan x$ $\quad\therefore \int \tan x\,dx = -\log|\cos x| + c$

(13) $\frac{d}{dx}\{\log(\sec x + \tan x)\} = \sec x$ $\quad\therefore \int \sec x\,dx = \log|\sec x + \tan x| + c$

(14) $\frac{d}{dx}\{\log(\text{cosec}\,x - \cot x)\} = \text{cosec}\,x$ $\quad\therefore \int \text{cosec}\,x\,dx = \log|\text{cosec}\,x - \cot x| + c$

(15) $\frac{d}{dx}\left(\sin^{-1}\frac{x}{a}\right) = \frac{1}{\sqrt{a^2 - x^2}}$ $\quad\therefore \int \frac{dx}{\sqrt{a^2 - x^2}} = \sin^{-1}\frac{x}{a} + c$

(16) $\frac{d}{dx}\left(\cos^{-1}\frac{x}{a}\right) = -\frac{1}{\sqrt{a^2 - x^2}}$ $\quad\therefore \int \frac{dx}{\sqrt{a^2 - x^2}} = -\cos^{-1}\frac{x}{a} + c$

(17) $\frac{d}{dx}\left(\frac{1}{a}\tan^{-1}\frac{x}{a}\right) = \frac{1}{x^2 + a^2}$ $\quad\therefore \int \frac{dx}{x^2 + a^2} = \frac{1}{a}\tan^{-1}\frac{x}{a} + c$

(18) $\frac{d}{dx}\left(\frac{1}{a}\cot^{-1}\frac{x}{a}\right) = -\frac{1}{x^2 + a^2}$ $\quad\therefore \int \frac{dx}{x^2 + a^2} = -\frac{1}{a}\cot^{-1}\frac{x}{a} + c$

(19) $\frac{d}{dx}\left(\frac{1}{a}\sec^{-1}\frac{x}{a}\right) = \frac{1}{x\sqrt{x^2 - a^2}}$ $\quad\therefore \int \frac{dx}{x\sqrt{x^2 - a^2}} = \frac{1}{a}\sec^{-1}\left(\frac{x}{a}\right) + c$

(20) $\frac{d}{dx}\left(\frac{1}{a}\operatorname{cosec}^{-1}\frac{x}{a}\right) = -\frac{1}{\sqrt{x^2-a^2}}$ $\therefore \int \frac{dx}{x\sqrt{x^2-a^2}} = -\frac{1}{a}\operatorname{cosec}^{-1}\frac{x}{a} + c$

(21) $\frac{d}{dx}\left[\int f(x)dx\right] = f(x)$

(22) $\int k\,f(x)dx = k\int f(x)dx$, where k is constant

(23) $\int [f(x) \pm g(x)]\,dx = \int f(x)dx \pm \int g(x)\,dx$

Example 5.58

Integrate the following with respect to x.

(i) $x^3 - 3x^2 + 5x + 4$ (ii) $\frac{1}{x^5} + \frac{1}{x^2} + 2x + 3$ (iii) $(3x-3)(4x+5)$ (iv) $\frac{\cos x}{\sin^2 x}$

SOLUTION :

(i) $\int (x^3 - 3x^2 + 5x + 4)\,dx = \int x^3\,dx - \int 3x^2\,dx + \int 5x\,dx + \int 4dx$

$$= \int x^3\,dx - 3\int x^2\,dx + 5\int x\,dx + 4\int dx$$

$$= \frac{x^{3+1}}{3+1} - 3\frac{x^{2+1}}{2+1} + 5\frac{x^{1+1}}{1+1} + 4x + c$$

$$= \frac{x^4}{4} - \frac{3x^3}{3} - \frac{5x^2}{2} + 4x + c = \frac{x^4}{4} - x^3 - \frac{5}{2}x^2 + 4x + c$$

(ii) $\int\left(\frac{1}{x^5} + \frac{1}{x^2} + 2x + 3\right)dx = \int \frac{1}{x^5}\,dx + \int \frac{1}{x^2}\,dx + \int 2x\,dx + \int 3dx$

$$= \int x^{-5}\,dx + \int x^{-2}\,dx + 2\int x\,dx + 3\int dx$$

$$= \frac{x^{-5+1}}{-5+1} + \frac{x^{-2+1}}{-2+1} + \frac{2x^{1+1}}{1+1} + 3x + c$$

$$= \frac{x^{-4}}{-4} + \frac{x^{-1}}{-1} + \frac{2x^2}{2} + 3x + c$$

$$= -\frac{1}{4x^4} - \frac{1}{x} + x^2 + 3x + c$$

(iii) $\int (3x-3)(4x+5)\,dx = \int (12x^2 + 15x - 12x - 15)\,dx$

$$= \int (12x^2 + 3x - 15)\,dx$$

$$= 12\int x^2\,dx + 3\int x\,dx - 15\int dx$$

$$= \frac{12\,x^3}{3} + \frac{3x^2}{2} - 15x + c$$

$$= 4x^3 + \frac{3}{2}x^2 - 15x + c$$

(iv) $$\int \frac{\cos x}{\sin^2 x}\,dx = \int \frac{\cos x}{\sin x}\cdot\frac{1}{\sin x}\,dx = \int \cot x \cdot \text{cosec}\, x\,dx$$

$$= -\,\text{cosec}\, x + c$$

Example 5.59

Integrate with respect to x.

(i) $3e^x - \frac{4}{x^3} + 4.2^x$ (ii) $(2x^2 + 1)^2 \cdot 4x\,dx$

(iii) $\sqrt[4]{x} + \sqrt[3]{x} - x^{-2/5}$ (iv) $\frac{3x^3 + 4x^2 + 5}{x^2}$

SOLUTION : (i) $$\int\left(3e^x - \frac{4}{x^3} + 4.2^x\right)dx = 3\int e^x\,dx - 4\int x^{-3}\,dx + 4\int 2^x\,dx$$

$$= 3e^x - 4\,\frac{x^{-2}}{-2} + 4.2^x \log 2 + c$$

$$= 3e^x + \frac{2}{x^2} + 4.2^x \log 2 + c$$

(ii) $$\int (2x^2 + 1)^2\,4x\,dx = \int (4x^4 + 4x^2 + 1)4x\,dx$$

$$= \int (16x^5 + 16x^3 + 4x)\,dx$$

$$= 16\int x^5\,dx + 16\int x^3\,dx + 4\int x\,dx$$

$$= 16\,\frac{x^6}{6} + 16\,\frac{x^4}{4} + 4\frac{x^2}{2} + c$$

$$= \frac{8}{3}x^6 + 4x^4 + 2x^2 + c$$

(iii) $$\int (\sqrt[4]{x} + \sqrt[3]{x} - x^{-2/5})\,dx = \int (x^{1/4} + x^{1/3} - x^{-2/5})\,dx$$

$$= \int x^{1/4}\,dx + \int x^{1/3}\,dx - \int x^{-2/5}\,dx$$

$$= \frac{x^{\frac{1}{4}+1}}{\frac{1}{4}+1} + \frac{x^{\frac{1}{3}+1}}{\frac{1}{3}+1} - \frac{x^{-\frac{2}{5}+1}}{\frac{-2}{5}+1} + c$$

$$= \frac{4}{5}x^{5/4} + \frac{3}{4}x^{4/3} - \frac{5}{3}x^{3/5} + c$$

(iv) $$\int \frac{3x^3 + 4x^2 + 5}{x^2}\,dx = \int \frac{3x^3}{x^2}\,dx + \int \frac{4x^2}{x^2}\,dx + \int \frac{5}{x^2}\,dx$$

$$= 3\int x\,dx + 4\int dx + 5\int x^{-2}\,dx$$

$$= \frac{3x^2}{2} + 4x - 5x^{-1} + c$$

$$= \frac{3}{2}x^2 + 4x - \frac{5}{x} + c$$

Example 5.60

Evaluate : (i) $\int \frac{1}{1-\sin x}\,dx$ (ii) $\int (\operatorname{cosec} x \cot x - 3\sec^2 x)\,dx$

(iii) $\int (\sqrt{1+\cos 2x})\,dx$ (iv) $\int \frac{4 + 3\sin x}{\cos^2 x}\,dx$

SOLUTION :

(i) $$\int \frac{1}{1-\sin x}\,dx = \int \frac{1+\sin x}{(1-\sin x)(1+\sin x)}\,dx = \int \frac{1+\sin x}{1-\sin^2 x}\,dx$$

$$= \int \frac{1+\sin x}{\cos^2 x}\,dx = \int \frac{1}{\cos^2 x}\,dx + \int \frac{\sin x}{\cos^2 x}\,dx$$

$$= \int \sec^2 x\,dx + \int \tan x \cdot \sec x\,dx$$

$$= \tan x + \sec x + c$$

(ii) $$\int (\operatorname{cosec} x \cot x - 3\sec^2 x)\,dx = \int \operatorname{cosec} x \cot x\,dx - 3\int \sec^2 x\,dx$$

$$= -\operatorname{cosec} x - 3\tan x + c$$

(iii) $$\int (\sqrt{1+\cos 2x})\,dx = \int (\sqrt{1 + 2\cos^2 x - 1})\,dx$$

$$= \int \sqrt{2\cos^2 x}\,dx = \int \sqrt{2}\cos x\,dx$$

$$= \sqrt{2} \cdot \int \cos x\,dx = \sqrt{2}\sin x + c$$

(iv) $\int \frac{4 + 3\sin x}{\cos^2 x} dx = \int \left(\frac{4}{\cos^2 x} + \frac{3 \sin x}{\cos^2 x} \right) dx = \int (4 \sec^2 x + 3 \tan x \sec x)\, dx$

$$= 4\int \sec^2 x\, dx + 3\int \tan x \sec x\, dx = 4 \tan x + \sec x + c$$

Example 5.61

Evaluate the following :

(i) $\int \frac{x^2 + 1}{x + 1} dx$ (ii) $\int \left(\sqrt{x} + \frac{1}{\sqrt{x}} \right)^3 dx$ (iii) $\int \frac{x^6}{x^2 + 1} dx$ (iv) $\int \frac{(x + a)^3}{2\sqrt{x}} dx$

SOLUTION : (i) $\int \frac{x^2 + 1}{x + 1} dx$

Here, the degree of polynomial in numerator is more than the degree of polynomial in denominator. We first devide the numerator to make the degree less than the degree of denominator.

$$\begin{array}{r} x + 1\,)\, x^2 + 1\, (\, x - 1 \\ \underline{-x^2 \pm x} \\ -x + 1 \\ -x - 1 \\ \underline{+ \quad +} \\ 2 \end{array}$$

$\therefore \qquad \frac{x^2 + 1}{x + 1} = x - 1 + \frac{2}{x + 1}$

Now, $\int \frac{x^2 + 1}{x + 1} dx = \int \left(x - 1 + \frac{2}{x + 1} \right) dx = \int x\, dx - \int dx + 2\int \frac{1}{x + 1} dx$

$$= \frac{x^2}{2} - x + 2 \log (x + 1) + c$$

(ii) $\int \left(\sqrt{x} + \frac{1}{\sqrt{x}} \right)^3 dx = \int \left(x^{3/2} + \frac{1}{x^{3/2}} + 3x \frac{1}{(x)^{1/2}} + 3x^{1/2} \frac{1}{x} \right) dx$

$$= \int \left(x^{3/2} + x^{-3/2} + 3x^{1/2} + 3x^{-1/2} \right) dx$$

$$= \int x^{3/2}\, dx + \int x^{-3/2}\, dx + 3\int x^{1/2}\, dx + 3\int x^{-1/2}\, dx$$

$$= \frac{x^{\frac{3}{2} + 1}}{\frac{3}{2} + 1} + \frac{x^{\frac{-3}{2} + 1}}{\frac{-3}{2} + 1} + 3\, \frac{x^{\frac{1}{2} + 1}}{\frac{1}{2} + 1} + 3\, \frac{x^{\frac{-1}{2} + 1}}{\frac{-1}{2} + 1} + c$$

$$= \frac{2}{5} x^{5/2} - 2\, x^{-1/2} + 2\, x^{3/2} + 6\, x^{1/2} + c$$

(iii) $\int \frac{x^6}{x^2+1}\,dx$

First, we devide x^6 by $x^2 + 1$

$$
\begin{array}{r l}
x^2+1\;)\;x^6 & (\,x^4 - x^2 + 1 \\
\underline{-x^6 \; \underline{+}\, x^4} & \\
-x^4 & \\
-x^4 - x^2 & \\
\underline{+ \qquad +} & \\
x^2 & \\
\underline{\underline{-}x^2 \;\; \underline{+}1} & \\
-1 &
\end{array}
$$

$$\therefore \quad \int \frac{x^6}{x^2+1}\,dx = \int\left(x^4 - x^2 + 1 - \frac{1}{x^2+1}\right)dx$$

$$= \int x^4\,dx - \int x^2\,dx + \int dx - \int \frac{1}{x^2+1}\,dx$$

$$= \frac{x^5}{5} - \frac{x^3}{3} + x - \tan^{-1} x + c$$

(iv) $$\int \frac{(x+a)^3}{2\sqrt{x}}\,dx = \frac{1}{2}\int \frac{x^3 + a^3 + 3x\,a^2 + 3x^2\,a}{\sqrt{x}}\,dx$$

$$= \frac{1}{2}\int\left(\frac{x^3}{\sqrt{x}} + \frac{a^3}{\sqrt{x}} + \frac{3x\,a^2}{\sqrt{x}} + \frac{3x^2\,a}{\sqrt{x}}\right)dx$$

$$= \frac{1}{2}\left\{\int x^{5/2}\,dx + a^3\int x^{-1/2}\,dx + 3a^2\int x^{1/2}\,dx + 3a\int x^{3/2}\,dx\right)$$

$$= \frac{1}{2}\left(\frac{2}{7}x^{7/2} + 2a^3\,x^{1/2} + 2a^2\,x^{3/2} + \frac{6}{5}\,a\,x^{5/2}\right) + c$$

$$= \frac{1}{7}x^{7/2} + a^3\,x^{1/2} + a^2\,x^{3/2} + \frac{3}{5}\,a\,x^{5/2} + c$$

Example 5.62

Find y, if (i) $\frac{dy}{dx} = 2x^3 + 4x + 5$

(ii) $\frac{dy}{dx} = e^x + a^x + 5\sin x + 2$

(iii) $\frac{dy}{dx} = \frac{1}{x} + \sec^2 x + \cos x$

(iv) $\frac{dy}{dx} = x^4 + 5x^2 + 7$

SOLUTION : (i) We have

$$\frac{dy}{dx} = 2x^3 + 4x + 5$$

Integrating both sides w.r. to x.

$$\int \frac{dy}{dx}\, dx = \int (2x^3 + 4x + 5)\, dx$$

$$\int dy = 2\int x^3\, dx + 4\int x\, dx + 5\int dx$$

$$y = \frac{2x^4}{4} + \frac{4x^2}{2} + 5x + c$$

or
$$y = \frac{1}{2}x^4 + 2x^2 + 5x + c$$

(ii) We have
$$\frac{dy}{dx} = e^x + a^x + 5\sin x + 2$$

Integrating both sides w.r. to x

$$\int \frac{dy}{dx}\, dx = \int (e^x + a^x + 5\sin x + 2)\, dx$$

$$\int dy = \int e^x\, dx + \int a^x\, dx + 5\int \sin x\, dx + 2\int dx$$

$$y = e^x + a^x \log a - 5\cos x + 2x + c$$

(iii) Given, $\frac{dy}{dx} = \frac{1}{x} + \sec^2 x + \cos x$

Integrating both sides w.r. to x, we have

$$\int \frac{dy}{dx}\, dx = \int \left(\frac{1}{x} + \sec^2 x + \cos x\right) dx$$

$$\int dy = \int \frac{1}{x}\, dx + \int \sec^2 x\, dx + \int \cos x\, dx$$

or
$$y = \log x + \tan x + \sin x + c$$

(iv) Given, $\frac{dy}{dx} = x^4 + 5x^2 + 7$

Integrating both sides w.r. to x, we have

$$\int \frac{dy}{dx}\, dx = \int (x^4 + 5x^2 + 7)\, dx$$

$$\int dy = \int x^4\, dx + 5\int x^2\, dx + 7\int dx$$

or
$$y = \frac{x^5}{5} + \frac{5}{3}x^3 + 7x + c$$

Example 5.63

Evaluate : (i) $\int \frac{1}{\sin^2 x \cos^2 x} dx$ (ii) $\int \sqrt{1 + \sin 2x\, dx}$

(iii) $\int \frac{\sec x}{\sec x + \tan x} dx$ (iv) $\int \tan^{-1}\left(\sqrt{\frac{1 - \cos 2x}{1 + \cos 2x}}\right) dx$

SOLUTION :

(i) $\int \frac{1}{\sin^2 x \cos^2 x} dx = \int \frac{\sin^2 x + \cos^2 x}{\sin^2 \cos^2 x} dx = \int \left(\frac{1}{\cos^2 x} + \frac{1}{\sin^2 x}\right) dx$

$= \int (\sec^2 x + \operatorname{cosec}^2 x)\, dx$

$= \int \sec^2 x\, dx + \int \operatorname{cosec}^2 x\, dx \ = \tan x - \cot x + c$

(ii) $\int \sqrt{1 + \sin 2x}\, dx = \int \sqrt{(\sin^2 x + \cos^2 x + 2\sin x \cos x)}\, dx$

$= \int \sqrt{(\sin x + \cos x)^2}\, dx = \int (\sin x + \cos x)\, dx$

$= \int \sin x\, dx + \int \cos x\, dx = -\cos x + \sin x + c$

$= \sin x - \cos x + c$

(iii) $\int \frac{\sec x}{\sec x + \tan x} dx = \int \frac{\sec x (\sec x - \tan x)}{(\sec x + \tan x)(\sec x - \tan x)} dx$

$= \int \frac{\sec x (\sec x - \tan x)}{\sec^2 x - \tan^2 x} dx$

$= \int (\sec^2 x - \sec x \tan x)\, dx$ $[\because \ \sec^2 x - \tan^2 x = 1]$

$= \int \sec^2 x\, dx - \int \sec x \tan x\, dx = \tan x - \sec x + c$

(iv) $\int \tan^{-1} \sqrt{\left(\frac{1 - \cos 2x}{1 + \cos 2x}\right)}\, dx = \int \tan^{-1} \sqrt{\left(\frac{1 - 1 + 2\sin^2 x}{1 + 2\cos^2 x - 1}\right)}\, dx$

$= \int \tan^{-1} \tan x\, dx$

$= \int x\, dx = \frac{x^2}{2} + c$

PROBLEM SET 5.8

Integrate the following w.r. to x

1. $x^3 + 3x^2 + 4$
2. $\sqrt{x} - \dfrac{1}{\sqrt{x}}$
3. $\left(ax + \dfrac{1}{bx}\right)^2$
4. $\dfrac{3x^2 + 4x + 5}{x + 1}$
5. $\dfrac{4x + 7}{2x + 3}$
6. $e^x + 5^x + 3\cos x + 7$
7. $\dfrac{1}{x^2} + \dfrac{1}{x^3} + x^{3/2} + 4$

Find the value of :

8. $\displaystyle\int \frac{x^4}{x^2 - 1}\, dx$
9. $\displaystyle\int (\cot x \text{ cosce } x + 7\text{cosce}^2 x - \sec^2 x)\, dx$
10. $\displaystyle\int \frac{(x + a)^2}{\sqrt{x}}\, dx$
11. $\displaystyle\int \frac{1}{1 + \sin x}\, dx$
12. $\displaystyle\int (e + x \log 3 + x^2)\, dx$
13. $\displaystyle\int \tan^2 x\, dx$
14. $\displaystyle\int \frac{\cos 2x - \cos 2\alpha}{\cos x - \cos \alpha}\, dx$
15. $\displaystyle\int \sqrt{1 - \cos 2x}\, dx$
16. $\displaystyle\int \frac{\sin x}{1 + \sin x}\, dx$

17. $\int \tan^{-1}(\sec x + \tan x)\, dx$

ANSWERS

1. $\frac{x^4}{4} + x^3 + 4x + c$ 2. $\frac{2}{3}x^{3/2} - 2x^{1/2} + c$ 3. $\frac{a^2 x^3}{3} - \frac{1}{b^2 x} + \frac{2a}{b}x + c$

4. $\frac{3x^2}{2} + x + 4\log(x+1) + c$ 5. $2x + \frac{1}{2}\log(2x+3) + c$

6. $e^x + 5^x \log 5 + 3\sin x + 7x + c$ 7. $\frac{-1}{x} - \frac{1}{2x^2} + \frac{2}{5}x^{5/2} + 4x + c$

8. $\frac{x^3}{3} + x + \frac{1}{2}\log\left(\frac{x-1}{x+1}\right) + c$ 9. $-\operatorname{cosec} x - \cot x - \tan x + c$

10. $\frac{2}{5}x^{5/2} + \frac{4}{3}ax^{3/2} + 2a^2x^{1/2} + c$ 11. $\tan x - \sec x + c$ 12. $x^{-e} + \frac{x^2}{2}\log 3 + \frac{x^3}{3} + c$

13. $\tan x - x + c$ 14. $2\sin x + 2x\cos x + c$ 15. $-\sqrt{2}\cos x + c$

16. $\sec x - \tan x + x + c$ 17. $\frac{\pi}{4}x + \frac{x^2}{4} + c$

5.19 INTEGRATION BY SUBSTITUTION

Earliear, we have discussed the problems on integrtion of functions either in standard forms or the problems having combinations of these functions. Sometimes it is very defficult to obtain the integrals of functions directly if they are not in standard forms. But we can reduce such type of integrals to their standard forms by suitable substitution.

Thus, integration by substitution is the method of evaluating an integral by reducing it to standard form by some proper substitution.

If $f(x)$ is a continuously differentiable function than the value of integral $\int F(f(x))\, f'(x)dx$ can be obtained by substituting $f(x) = t \Rightarrow f'(x)dx = dt$

Therefore, $\int F(f(x))\, f'(x)\, dx = \int F(t)\, dt$ which is in the standard form. After evaluating this we substitute the value of $t = f(x)$ and get the required result.

Example 5.64

Integrate w.r. to x, of the following :

(i) e^{5x+4} (ii) $7x\, e^{x^2+5}$

(iii) $\frac{4x-6}{x^2-3x+7}$ (iv) $x\cos x^2$

SOLUTION :

(i) Given, $\int e^{5x+4}\,dx$

Let $5x + 4 = t \Rightarrow 5dx = dt \Rightarrow dx = \frac{1}{5}dt$

$\therefore$ $\int e^{5x+4}\,dx = \int e^t \frac{1}{5}dt = \frac{1}{5}\int e^t\,dt = \frac{1}{5}e^t + c$

or $\int e^{5x+4}\,dx = \frac{1}{5}e^{5x+4} + c$

(ii) Given, $\int 7x\,e^{x^2+5}\,dx$

Let $x^2 + 5 = t \Rightarrow 2x\,dx = dt \Rightarrow x\,dx = \frac{1}{2}dt$

$\therefore$ $\int 7x\,e^{x^2+5}\,dx = 7\int e^t \frac{1}{2}dt = \frac{7}{2}\int e^t\,dt = \frac{7}{2}e^t + c$

or $\int 7x\,e^{x^2+5}\,dx = \frac{7}{2}e^{x^2+5} + c$

(iii) Given, $\int \frac{4x-6}{x^2-3x+7}\,dx$

Let $x^2 - 3x + 7 = t \Rightarrow (2x-3)\,dx = dt$

$\therefore$ $\int \frac{4x-6}{x^2-3x+7}\,dx = 2\int \frac{2x-3}{x^2-3x+7}\,dx = 2\int \frac{dt}{t} = 2\log t + c$

or $\int \frac{4x-6}{x^2-3x+7}\,dx = 2\log(x^2-3x+7) + c$

(iv) Given, $\int x\cos x^2\,dx$

Let $x^2 = t \Rightarrow 2x\,dx = dt$

$\Rightarrow$ $x\,dx = \frac{1}{2}dt$

$\therefore$ $\int x\cos x^2\,dx = \frac{1}{2}\int \cos t\,dt = \frac{1}{2}\sin t + c$

or $\int x\cos x^2\,dx = \frac{1}{2}\sin x^2 + c$

Example 5.65

Find the integral with respect to x of the following functions

(i) $2x^2\sqrt{x^3+9}$ (ii) $\dfrac{6x^2+7}{2x^3+7x}$

(iii) $\dfrac{e^x+e^{-x}}{e^x-e^{-x}}$ (iv) $\dfrac{1}{x}\log a\,x$

SOLUTION :

(i) Given, $\int 2x^2\sqrt{x^3+9}\,dx$

Let $x^3+9=t^2 \Rightarrow 3x^2\,dx=2t\,dt \Rightarrow x^2\,dx=\frac{2}{3}t\,dt$

or $t=(x^3+9)^{1/2}$

$$\therefore \quad \int 2x^2\sqrt{x^3+9}\,dx = 2\int\frac{2}{3}t\sqrt{t^2}\,dt = \frac{4}{3}\int t\cdot t\,dt = \frac{4}{3}\int t^2\,dt$$

$$= \frac{4}{3} + \frac{t^3}{3} + c = \frac{4}{9}t^3 + c$$

or $$\int 2x^2\sqrt{x^3+9}\,dx = \frac{4}{9}(x^3+9)^{3/2}+c$$

(ii) Given, $\int\frac{6x^2+7}{2x^3+7x}\,dx$

Let $2x^3+7x=t \Rightarrow (6x^2+7)\,dx=dt$

$$\therefore \quad \int\frac{6x^2+7}{2x^3+7x}\,dx = \int\frac{dt}{t} = \log t + c$$

or $$\int\frac{6x^2+7}{2x^3+7x}\,dx = \log(2x^3+7x)+c$$

(iii) Given, $\int\frac{e^x+e^{-x}}{e^x-e^{-x}}\,dx$

Let $e^x-e^{-x}=t \Rightarrow (e^x+e^{-x})\,dx=dt$

$$\therefore \quad \int\frac{e^x+e^{-x}}{e^x-e^{-x}}\,dx = \int\frac{dt}{t} = \log t + c$$

or $$\int\frac{e^x+e^{-x}}{e^x+e^{-x}}\,dx = \log(e^x-e^{-x})+c$$

(iv) Given, $\int\frac{1}{x}\log a\,x\,dx$

Let $\log ax = t \Rightarrow \frac{1}{ax}\cdot a\,dx = dt$ or $\frac{1}{x}\,dx = dt$

$\therefore \quad \int \frac{1}{x} \log ax \, dx = \int t \, dt = \frac{t^2}{2} + c$

or $\quad \int \frac{1}{x} \log ax \, dx = \frac{1}{2} (\log ax)^2 + c$

Example 5.66

Evaluate the following :

(i) $\int x e^{-x^2} dx$ (ii) $\int \frac{e^{2x} + e^{3x}}{e^x} dx$

(iii) $\int \frac{1}{x} \cdot \sec^2 (\log x) \, dx$ (iv) $\int \operatorname{cosec} (e^x) \cdot \cot (e^x) \cdot e^x \, dx$

(v) $\int \frac{(x+1)(x + \log x)^3}{3x} dx$ (vi) $\int \frac{x^4 + x^3 + 1}{x^4 + 1} dx$

SOLUTION :

(i) $\int x e^{-x^2} dx$

Let $\quad -x^2 = t \Rightarrow -2x \, dx = dt \Rightarrow x \, dx = -\frac{1}{2} dt$

$\therefore \quad \int x e^{-x^2} dx = \frac{1}{2} \int e^t \, dt = -\frac{1}{2} e^t + c$

or $\quad \int x e^{-x^2} dx = -\frac{1}{2} e^{-x^2} + c$

(ii) $\quad \int \frac{e^{2x} + e^{3x}}{e^x} dx = \int \frac{e^{2x}}{e^x} dx + \int \frac{e^{3x}}{e^x} dx = \int e^x \, dx + \int e^{2x} dx$

$= e^x + \frac{1}{2} e^{2x} + c$

(iii) $\int \frac{1}{x} \sec^2 (\log x) \, dx$

Let $\quad \log x = t \Rightarrow \frac{1}{x} dx = dt$

$\therefore \quad \int \frac{1}{x} \sec^2 (\log x) \, dx = \int \sec^2 t \, dt = \tan t + c$

or $\quad \int \frac{1}{x} \sec^2 (\log x) \, dx = \tan (\log x) + c$

(iv) $\int \operatorname{cosec}(e^x)\cot(e^x)\,e^x\,dx$

Let $$e^x = t \Rightarrow e^x\,dx = dt$$

$$\therefore \quad \int \operatorname{cosec}(e^x)\cdot\cot(e^x)\,e^x\,dx = \int \operatorname{cosec} t \cot t\,dt = -\operatorname{cosec} t + c$$

or $$\int \operatorname{cosec}(e^x)\cot(e^x)\,e^x\,dx = \operatorname{cosec}(e^x) + c$$

(v) $$\int \frac{(x+1)(x+\log x)^3}{3x}\,dx = \frac{1}{3}\int\left(1+\frac{1}{x}\right)(x+\log x)^3\,dx$$

Let $$x + \log x = t \Rightarrow \left(1+\frac{1}{x}\right)dx = dt$$

$$\therefore \quad \int \frac{(x+1)(x+\log x)^3}{3x}\,dx = \frac{1}{3}\int t^3\,dt = \frac{t^4}{12} + c$$

or $$\int \frac{(x+1)(x+\log x)^3}{3x}\,dx = \frac{1}{12}(x+\log x)^4 + c$$

(vi) $$\int \frac{x^4+x^3+1}{x^4+1}\,dx = \int \frac{x^4+1}{x^4+1}\,dx + \int \frac{x^3}{x^4+1}\,dx$$

$$= \int dx + \int \frac{x^3}{x^4+1}\,dx$$

Let $$x^4 + 1 = t \Rightarrow 4x^3\,dx = dt \Rightarrow x^3\,dx = \frac{1}{4}\,dt$$

$$\therefore \quad \int \frac{x^4+x^3+1}{x^4+1}\,dx = x + \frac{1}{4}\int \frac{dt}{t} = x + \frac{1}{4}\log t + c$$

or $$\int \frac{x^4+x^3+1}{x^4+1}\,dx = x + \frac{1}{4}\log(x^4+1) + c$$

Example 5.67

Find the value of $\int \frac{\sin x \cos x}{1+\sin^2 x}\,dx$

SOLUTION :

Let $1 + \sin^2 x = t \Rightarrow 2\sin x \cos x\,dx = dt$

$$\therefore \quad \int \frac{\sin x \cos x}{1+\sin^2 x}\,dx = \frac{1}{2}\int \frac{dt}{t}$$

$$= \frac{1}{2}\log t + c = \frac{1}{2}\log(1+\sin^2 x) + c$$

Example 5.68

Evaluate :

(i) $\int \frac{1}{\sqrt{2x+1}-\sqrt{2x-1}}\,dx$ (ii) $\int \frac{x+1}{\sqrt{2x-1}}\,dx$

SOLUTION :

(i) $$\int \frac{1}{\sqrt{2x+1}-\sqrt{2x-1}}\,dx = \int \frac{\sqrt{2x+1}+\sqrt{2x-1}}{(\sqrt{2x+1}-\sqrt{2x-1})(\sqrt{2x+1}+\sqrt{2x-1})}\,dx$$

$$= \int \frac{\sqrt{2x+1}+\sqrt{2x-1}}{(\sqrt{2x+1})^2-(\sqrt{2x-1})^2}\,dx = \int \frac{\sqrt{2x+1}+\sqrt{2x-1}}{(2x+1)-(2x-1)}\,dx$$

$$= \int \frac{\sqrt{2x+1}+\sqrt{2x-1}}{2}\,dx = \frac{1}{2}\int \sqrt{2x+1}\,dx + \int \sqrt{2x-1}\,dx$$

$$= \frac{1}{2}\left(\frac{(2x+1)^{3/2}}{2\cdot\frac{3}{2}}\right) + \frac{1}{2}\left(\frac{(2x-1)^{3/2}}{2\cdot\frac{3}{2}}\right) + c$$

$$= \frac{1}{6}\left((2x+1)^{3/2} + (2x-1)^{3/2}\right) + c$$

(ii) $$\int \frac{x+1}{\sqrt{2x+1}}\,dx = \frac{1}{2}\int \frac{2x+2}{\sqrt{2x+1}}\,dx = \frac{1}{2}\int \frac{2x+1+1}{\sqrt{2x+1}}\,dx$$

$$= \frac{1}{2}\left(\int \frac{2x+1}{\sqrt{2x+1}}\,dx + \int \frac{1}{\sqrt{2x+1}}\,dx\right) = \frac{1}{2}\left(\int \sqrt{2x+1}\,dx + \int \frac{1}{\sqrt{2x+1}}\,dx\right)$$

$$= \frac{1}{2}\left(\frac{(2x+1)^{3/2}}{2\cdot\frac{3}{2}} + \frac{(2x+1)^{1/2}}{2\cdot\frac{1}{2}}\right) + c = \frac{1}{2}\left((2x+1)^{3/2} + (2x+1)^{1/2}\right) + c$$

Example 5.69

Evaluate (i) $\int \frac{(1+\log x)^2}{x}\,dx$ (ii) $\int \tan x\,dx$

[UPTU 2001]

SOLUTION : (i) Given $\int \frac{(1+\log x)^2}{x}\,dx$

Let $1+\log x = t \Rightarrow \frac{1}{x}\,dx = dt$

$$\therefore \quad \int \frac{(1+\log x)^2}{x}\,dx = \int t^2\,dt = \frac{t^3}{3} + c$$

Substituting the value of t; we get

$$\int \frac{(1+\log x)^2}{x}\,dx = \frac{1}{3}(1+\log x)^3 + c$$

(ii) Given $\int \tan x\, dx = \int \frac{\sin x}{\cos x} dx$

Let $\cos x = t \Rightarrow -\sin x\, dx = dt$

$\therefore$ $\int \tan x\, dx = -\int \frac{1}{t} dt = -\log t + c$

Substituting the value of t, we get

$$\int \tan x\, dx = -\log \cos x + c = \log \cos^{-1} x + c = \log \sec x + c$$

Example 5.70

Evaluate $\int \frac{\cos x}{\sin^2 x} dx$ *[UPTU 2000]*

SOLUTION : Given $\int \frac{\cos x}{\sin^2 x} dx$

Put $\sin x = t \Rightarrow \cos x\, dx = dt$

$\therefore$ $\int \frac{\cos x}{\sin^2 x} dx = \int \frac{dt}{t^2} = \int t^{-2}\, dt = \frac{t^{-1}}{-1} + c$

$$= -\frac{1}{t} + c = -\frac{1}{\cos x} + c = -\sec x + c$$

Example 5.71

Evaluate $\int \frac{\tan x}{\log \sec x} dx.$ *[UPTU 2002]*

SOLUTION : Given $\int \frac{\tan x}{\log \sec x} dx$

Let $\log \sec x = t \Rightarrow \frac{1}{\sec x} \cdot \sec x \tan x\, dx = dt$

$\Rightarrow$ $\tan x\, dx = dt$

$\therefore$ $\int \frac{\tan x}{\log \sec x} dx = \int \frac{dt}{t} = \log t + c$

Substituting the value of t, we get

$$\int \frac{\tan x}{\log \sec x} dx = \log (\log \sec x) + c$$

PROBLEM SET 5.9

Evaluate the following integrals w.r. to x

1. $\displaystyle\int \frac{2ax+b}{(ax^2+bx+c)^3}\,dx$

2. $\displaystyle\int 4x\, e^{2x^2+5}\,dx$

3. $\displaystyle\int x^2 \cdot \sin^3 dx$

4. $\displaystyle\int 3x\sqrt{2x^2+4}\,dx$

5. $\displaystyle\int \frac{e^x}{e^x+1}\,dx$

6. $\displaystyle\int \frac{x+5}{(x+2)^2}\,dx$

7. $\displaystyle\int 2^{3x+4}\,dx$

8. $\displaystyle\int (7x-2)\sqrt{3x+2}\,dx$

9. $\displaystyle\int \frac{e^x-e^{-x}}{e^x+e^{-x}}\,dx$

10. $\displaystyle\int x^3\sqrt{x^2+1}\,dx$

11. $\displaystyle\int \left(e^x+\frac{1}{e^x}\right)^2 dx$

12. $\displaystyle\int \frac{1}{\sqrt{x+a}+\sqrt{x+b}}\,dx$

13. $\displaystyle\int \frac{x^2-1}{x^3-3x+2}\,dx$

14. $\displaystyle\int \tan^5 x \cdot \sec^2 x\,dx$

15. $\displaystyle\int \frac{1}{x}(\log x)^2\, e^{(\log x)^3}\,dx$

16. $\displaystyle\int e^x \cdot \sin e^x\,dx$

17. $\displaystyle\int \frac{(\sin^{-1}x)^3}{\sqrt{1-x^2}}\,dx$

ANSWERS

1. $-\frac{1}{2}(ax^2+bx+c)^{-2}+c$ 2. $e^{2x^2+5}+c$ 3. $-\frac{1}{3}\cos x^3+c$ 4. $\frac{1}{2}(2x^2+4)^{3/2}+c$

5. $\log(e^x+1)+c$ 6. $\log(x+2)-\frac{3}{x+2}+c$ 7. $\frac{2^{3x+4}}{3\log_e 2}+c$

8. $\frac{14}{15}(3x+2)^{5/2}-\frac{40}{27}(3x+2)^{3/2}+c$ 9. $\log(e^x+e^{-x})+c$

10. $\frac{1}{5}(x^2+1)^{5/2}-\frac{1}{3}(x^2+1)^{3/2}+c$ 11. $\frac{e^{2x}}{2}-\frac{e^{-2x}}{2}+2x+c$

12. $\frac{2}{3(a-b)}\left[(x+a)^{3/2}-(x+b)^{3/2}\right]+c$ 13. $\frac{1}{3}\log(x^3-3x+2)+c$

14. $\frac{1}{6}\tan^6 x+c$ 15. $\frac{1}{3}e^{(\log x)^3}+c$ 16. $-\cos e^x+c$ 17. $\frac{1}{4}(\sin^{-1}x)^4+c$

5.20 SOME IMPORTANT FORMULAE

5.20.1 Type $\int \sin^m x\, dx$ or $\int \cos^m x\, dx$

To find the values of integrals of the type $\int \sin^m x\, dx$ or $\int \cos^m x\, dx$, we use the following trigonometrical identities.

(i) $\sin^2 x = \frac{1 - \cos 2x}{2}$ (ii) $\cos^2 x = \frac{1 + \cos 2x}{2}$

(iii) $\sin^3 x = \frac{1}{4}(3\sin x - \sin 3x)$ (iv) $\cos^3 x = \frac{1}{4}(\cos 3x + 3\cos x)$

(v) $\sin^4 x = (\sin^2 x)^2$ (vi) $\cos^4 x = (\cos^2 x)^2$ etc.

5.20.2 Type $\int \sin px \cos qx\, dx$, $\int \sin px \sin qx\, dx$, $\int \cos px \cos qx\, dx$ and $\int \cos px \sin qx\, dx$

To find the values of integrals of the type $\int \sin px \cos qx\, dx$, $\int \sin px \sin qx\, dx$, $\int \cos px \cos qx\, dx$ and $\int \cos px \sin qx\, dx$

(i) $\sin A \cos B = \frac{1}{2}[\sin(A + B) + \sin(A - B)]$

(ii) $\cos A \sin B = \frac{1}{2}[\sin(A + B) - \sin(A - B)]$

(iii) $\cos A \cos B = \frac{1}{2}[\cos(A + B) + \cos(A - B)]$

(iv) $\sin A \sin B = \frac{1}{2}[\cos(A - B) - \cos(A + B)]$

Example 5.72

Evaluate :

(i) $\int \sin^2 x\, dx$ (ii) $\int \cos^3 x\, dx$

SOLUTION :

(i) $\int \sin^2 x\, dx = \frac{1}{2}\int 2\sin^2 x\, dx = \frac{1}{2}\int (1 - \cos 2x)\, dx = \frac{1}{2}\left(x - \frac{\sin 2x}{2}\right) + c$

(ii) $\int \cos^3 x\, dx = \int \frac{\cos 3x + 3\cos x}{4} dx = \frac{1}{4}\int (\cos 3x + 3\cos x)\, dx$

$= \frac{1}{4}\left(\frac{\sin 3x}{3} + 3\sin x\right) + c$

Example 5.73

Evaluate:

(i) $\int \cos^4 x\, dx$

(ii) $\int \sin^4 x \cos^4 x\, dx$

(iii) $\int \sin 3x \cos 4x\, dx$

(iv) $\int \cos x \cos 2x \sin 3x\, dx$

SOLUTION :

(i) $\int \cos^4 x\, dx = \int (\cos^2 x)^2\, dx = \int \left(\frac{1 + \cos 2x}{2}\right)^2 dx$

$= \frac{1}{4}\int \left(1 + 2\cos 2x + \cos^2 2x\right) dx = \frac{1}{4}\int \left(1 + 2\cos 2x + \frac{1 + \cos 4x}{2}\right) dx$

$= \frac{1}{8}\int (3 + 4\cos 2x + \cos 4x)\, dx$

$= \frac{1}{8}\left(3x + 2\sin 2x + \frac{\sin 4x}{4}\right) + c$

(ii) $\int \sin^4 x \cos^4 x\, dx = \int (\sin x \cos x)^4\, dx = \int \left(\frac{2\sin x \cos x}{2}\right)^4 dx$

$= \frac{1}{2^4}\int (\sin 2x)^4\, dx = \frac{1}{16}\int (\sin^2 2x)^2\, dx$

$= \frac{1}{16}\int \left(\frac{1 - \cos 4x}{2}\right)^2 dx = \frac{1}{64}\int (1 - 2\cos 4x + \cos^2 4x)\, dx$

$= \frac{1}{64}\int \left(1 - 2\cos 4x + \frac{1 + \cos 8x}{2}\right) dx$

$= \frac{1}{128}\int (3 - 4\cos 4x + \cos 8x)\, dx$

$= \frac{1}{128}\left(3x - \sin 4x + \frac{\sin 8x}{8}\right) + c$

(iii) $\int \sin 3x \cos 4x\, dx = \frac{1}{2}\int [\sin (3x + 4x) + \sin (3x - 4x)]\, dx$

$= \frac{1}{2}\int [\sin 7x + \sin (-x)]\, dx = \frac{1}{2}\int (\sin 7x - \sin x)\, dx$

$$= \frac{1}{2}\left(-\frac{\cos 7x}{7} + \cos x\right) + c$$

(iv) $\int \cos x \cos 2x \sin 3x \, dx = \int (\cos x \cos 2x) \sin 3x \, dx$

$$= \frac{1}{2}\int [\cos (x + 2x) + \cos (x - 2x)] \sin 3x \, dx$$

$$= \frac{1}{2}\int [\cos 3x + \cos (-x)] \sin 3x \, dx$$

$$= \frac{1}{2}\int (\cos 3x \sin 3x + \cos x \sin 3x) \, dx$$

$$= \frac{1}{2}\int \cos 3x \sin 3x \, dx + \frac{1}{2}\int \cos x \sin 3x \, dx$$

$$= \frac{1}{4}\int \sin 6x \, dx + \frac{1}{4}\int [\sin (x + 3x) - \sin (x - 3x)] \, dx$$

$$= \frac{1}{4}\int \sin 6x \, dx + \frac{1}{4}\int \sin 4x \, dx + \frac{1}{4}\int \sin 2x \, dx$$

$$= \frac{1}{4}\left(-\frac{\cos 6x}{6}\right) + \frac{1}{4}\left(\frac{-\cos 4x}{4}\right) + \frac{1}{4}\left(\frac{-\cos 2x}{2}\right) + c$$

$$= -\frac{\cos 6x}{24} - \frac{\cos 4x}{16} - \frac{\cos 2x}{8} + c$$

Example 5.74

Evaluate :

(i) $\int \sqrt{1 - \sin x} \, dx$ (ii) $\int \sin x \sqrt{1 - \cos 2x} \, dx$

SOLUTION :

(i) $\int \sqrt{1 - \sin x} \, dx = \int \sqrt{\sin^2 \frac{x}{2} + \cos^2 \frac{x}{2} - 2\sin \frac{x}{2} \cos \frac{x}{2}} \, dx$

$$= \int \sqrt{\left(\sin \frac{x}{2} - \cos \frac{x}{2}\right)^2} \, dx = \int \left(\sin \frac{x}{2} - \cos \frac{x}{2}\right) dx$$

$$= \int \sin \frac{x}{2} \, dx - \int \cos \frac{x}{2} \, dx = -2 \cos \frac{x}{2} - 2\sin \frac{x}{2} + c$$

(ii) $\int \sin x \sqrt{1 - \cos 2x} \, dx = \int \sin x \sqrt{1 - 1 + 2 \sin^2 x} \, dx = \sqrt{2} \int \sin^2 x \, dx$

$$= \frac{\sqrt{2}}{2}\int 2\sin^2 x\,dx = \frac{\sqrt{2}}{2}\int (1-\cos 2x)\,dx = \frac{1}{\sqrt{2}}\left(x - \frac{\sin 2x}{2}\right) + c$$

5.20.3 Integrals of the Type $\int \sin^p x \cos^q x\,dx$

$\int \sin^p x \cos^q x\,dx$, where p and q are positive integers.

In this type of integrals following substitutions are very useful :

(i) If p is odd and q has any value i.e either even or odd then put $\cos x = t$

(ii) If p has any value *i.e.*, either even or odd but q is odd then put $\sin x = t$

(iii) If both p and q are even, then we will use De' Moivre's theorem.

(iv) If both p and q are odd, then we can put either $\cos x = t$ or $\sin x = t$.

Example 5.75

Evaluate :

(i) $\int \sin^3 x \cos^4 x\,dx$ (ii) $\int \sin^4 x \cos^3 x\,dx$ (iii) $\int \sin^3 x \cos^5 x\,dx$

SOLUTION :

(i) Let $I = \int \sin^3 x \cos^4 x\,dx = \int \sin x \cdot \sin^2 x \cos^4 x\,dx$

$$= \int \sin x(1-\cos^2 x)\cos^4 x\,dx$$

Let $\cos x = t \Rightarrow -\sin x\,dx = dt$

$\therefore$ $I = -\int (1-t^2)\,t^4\,dt = -\int (t^4 - t^6)\,dt$

$$= -\left(\frac{t^5}{5} - \frac{t^7}{7}\right) + c$$

$$= -\left(\frac{1}{5}\cos^5 x - \frac{1}{7}\cos^7 x\right) + c$$

$$= \frac{1}{7}\cos^7 x - \frac{1}{5}\cos^5 x + c$$

(ii) Let $I = \int \sin^4 x \cos^3 x\,dx$

$$= \int \sin^4 x \cos^2 x \cdot \cos x\,dx$$

$$= \int \sin^4 x\,(1-\sin^2 x)\cos x\,dx$$

Let $\sin x = t$

$\Rightarrow \quad \cos x\, dx = dt$

$$I = \int t^4 (1 - t^2)\, dt = \int (t^4 - t^6)\, dt$$

$$= \frac{t^5}{5} - \frac{t^7}{7} + c$$

or $$I = \frac{1}{5} \sin^5 x - \frac{1}{7} \sin^7 x + c$$

(iii) Let $I = \int \sin^3 x \cos^5 x\, dx$

$$= \int \sin x \cdot \sin^2 x \cos^5 x\, dx$$

$$= \int \sin (1 - \cos^2 x) \cos^5 x\, dx$$

Let $\cos x = t \Rightarrow -\sin x\, dx = dt$

$\therefore$ $$I = -\int (1 - t^2)\, t^5\, dt = -\int (t^5 - t^7)\, dt$$

$$= -\frac{t^6}{6} + \frac{t^8}{8} + c$$

$$= \frac{1}{8} \cos^8 x - \frac{1}{6} \cos^6 x + c$$

Example 5.76

Prove that $\int \frac{dx}{x^2 + a^2} = \frac{1}{a} \tan^{-1}\left(\frac{x}{a}\right)$ ***[UPTU 2002]***

SOLUTION : Given $\int \frac{dx}{x^2 + a^2}$

Put $x = a \tan \theta \Rightarrow dx = a \sec^2 \theta\, d\theta$

$\therefore$ $$\int \frac{dx}{x^2 + a^2} = \int \frac{a \sec^2 \theta\, d\theta}{a^2 \tan^2 \theta + a^2} = \frac{1}{a} \int \frac{\sec^2 \theta\, d\theta}{\tan^2 \theta + 1}$$

$$= \frac{1}{a} \int \frac{\sec^2 \theta}{\sec^2 \theta} d\theta \qquad [\because 1 + \tan^2 \theta = \sec^2 \theta]$$

$$= \frac{1}{a} \int d\theta = \frac{1}{a} \theta = \frac{1}{a} \tan^{-1}\left(\frac{x}{a}\right)$$

PROBLEM SET 5.10

Evaluate the following integrals

1. $\int \cos^3 x \sin^3 x \, dx$

2. $\int \sin^4 x \, dx$

3. $\int \cos 2x \cos 4x \, dx$

4. $\int \cos x \sin 3x \sin 2x \, dx$

5. $\int \sqrt{(1 + \sin x)} \, dx$

6. $\int \sin 4x \cos 7x \, dx$

7. $\int \sin mx \cos nx \, dx$ if $m \neq n$

8. $\int \frac{1}{\sqrt{x+3} - \sqrt{x+2}} \, dx$

9. $\int \tan x \, dx$

10. $\int \sec x \, dx$

11. $\int \text{cosec}\, x \, dx$

12. $\int \frac{1 - \tan x}{1 + \tan x} \, dx$

13. $\int \frac{\sin 2x}{a^2 \sin^2 x + b^2 \cos^2 x} \, dx$

14. $\int \frac{\sin x}{\sin (x - a)} \, dx$

15. $\int \tan x \tan 2x \tan 3x \, dx$

16. $\int \frac{1 - \sin x}{x + \cos x} \, dx$

17. $\int \frac{\sec x}{\log (\sec x + \tan x)} \, dx$

18. $\int \frac{1}{x \log x \log (\log x)} \, dx$

19. $\int \frac{1 + \cot x}{x + \log \sin x} \, dx$

20. $\int \tan^3 x \cdot \sec^2 x \, dx$

21. $\int \frac{\log \left(\tan \frac{x}{2} \right)}{\sin x} \, dx$

22. $\int \frac{\sec^2 (2\tan^{-1} x)}{1 + x^2} \, dx$

23. $\int \sin^4 x \cos^3 x \, dx$

24. $\int \sin^5 x \, dx$

25. $\int x \cos^3 x^2 \sin x^2 \, dx$

26. $\int \frac{x^2 \tan^{-1} x^3}{1 + x^6} \, dx$

27. $\int \tan^3 x \, dx$

28. $\int \left(\frac{x + 1}{x} \right) (x + \log x)^2 \, dx$

29. $\int \tan^5 x \sin^2 x\, dx$ **30.** $\int \frac{e^{m \tan^{-1} x}}{1+x^2} dx$

ANSWERS

1. $\frac{1}{32}\left(\frac{-3}{2}\cos 2x + \frac{1}{6}\cos 6x\right) + c$ **2.** $\frac{1}{8}\left(3x - 2\sin 2x + \sin\frac{4x}{4}\right) + c$

3. $\frac{\sin 6x}{12} + \frac{\sin 2x}{4} + c$ **4.** $\frac{x}{4} + \frac{\sin 2x}{8} - \frac{\sin 6x}{12} - \frac{\sin 4x}{16} + c$ **5.** $2\left(\sin\frac{x}{2} - \cos\frac{x}{2}\right) + c$

6. $\frac{-1}{22}\cos 11x + \frac{1}{6}\cos 3x + c$ **7.** $\frac{1}{2}\left(\frac{-\cos(m+n)x}{m+n} - \frac{\cos(m-n)x}{m-n}\right) + c$

8. $\frac{2}{3}\left((x+3)^{3/2} + (x+2)^{3/2}\right) + c$ **9.** $\log|\sec x| + c$ **10.** $\log|\sec x + \tan x| + c$

11. $\log|\operatorname{cosec} x - \cot x| + c$ **12.** $\log|\cos x + \sin x| + c$

13. $\frac{1}{a^2 - b^2}\log\left(a^2 \sin^2 x + b^2 \cos^2 x\right) + c$ **14.** $\sin a \log(\sin(x-a)) + (x-a)\cos a + c$

15. $-\frac{1}{3}\log(\cos^3 x) + \frac{1}{2}\log(\cos 2x) + \log(\cos x) + c$ **16.** $\log(x + \cos x) + c$

17. $\log(\log(\sec x + \tan x)) + c$ **18.** $\log(\log(\log x)) + c$ **19.** $\log(x + \log \sin x) + c$

20. $\frac{\tan^4 x}{4} + c$ **21.** $\frac{1}{2}\left(\log\tan\frac{x}{2}\right)^2 + c$ **22.** $\frac{1}{2}\tan(2\tan^{-1} x) + c$

23. $\frac{1}{5}\sin^5 x - \frac{1}{7}\sin^7 x + c$ **24.** $-\cos x + \frac{2}{3}\cos^3 x - \frac{1}{5}\cos^5 x + c$

25. $-\frac{1}{8}\cos^4 x^2 + c$ **26.** $\frac{1}{6}\left(\tan^{-1} x^3\right)^2 + c$ **27.** $\frac{1}{2}\tan^2 x - \log(\sec x) + c$

28. $\frac{1}{3}(x + \log x)^3 + c$ **29.** $\frac{1}{6}\tan^6 x + c$ **30.** $\frac{e^{m \tan^{-1} x}}{m} + c$

5.21 INTEGRATION BY PARTS

Let P and Q be any two functions of x, then

$$\int P \cdot Q \cdot dx = P\int Q\, dx - \int \frac{dp}{dx}\left(\int Q\, dx\right) dx$$

i.e., the integration of product of two functions = first function × integration of second function
− integration of [derivative of first function
× integration of second function]

Example 5.77

Evaluate :

(i) $\int x\, e^x\, dx$ (ii) $\int \log x\, dx$ (iii) $\int x \cdot \sec^2 x\, dx$ (iv) $\int x \cdot \log x^2\, dx$

SOLUTION :

(i) Let $f(x) = x \cdot e^x$

Here $P = x$ and $Q = e^x$

Using formula, $$\int PQ\, dx = P \int Q\, dx - \int \frac{dP}{dx}\left(\int Q\, dx\right) dx$$

$$\therefore \quad \int x\, e^x\, dx = x \int e^x\, dx - \int \frac{dx}{dx}\left(\int e^x\, dx\right) dx$$

$$= xe^x - \int e^x\, dx = xe^x - e^x + c$$

(ii) $$\int \log x\, dx = \int \log x \cdot 1\, dx$$

$$= \log x \int 1dx - \int \frac{d\,(\log x)}{dx}\left(\int 1dx\right) dx$$

$$= \log x \cdot x - \int \frac{1}{x} \cdot x\, dx = x \cdot \log x - \int dx$$

$$= x \log x - x + c$$

(iii) $$\int x \cdot \sec^2 x\, dx = x \int \sec^2 x\, dx - \int \frac{dx}{dx}\left(\int \sec^2 x\, dx\right) dx$$

$$= x \cdot \tan x - \int \tan x\, dx = x \cdot \tan x + \log(\cos x) + c$$

(iv) $$\int x \log x^2\, dx = \log x^2 \cdot \int x\, dx - \int \frac{d}{dx}(\log x^2)\left(\int x\, dx\right) dx$$

$$= \log x^2 \cdot \frac{x^2}{2} - \int \frac{1}{x^2} \cdot 2x \cdot \frac{x^2}{2}\, dx$$

$$= \frac{x^2}{2} \cdot \log x^2 - \int x\, dx = \frac{x^2}{2} \log x^2 - \frac{x^2}{2} + c$$

Example 5.78

Evaluate :

(i) $\int x^2 \cdot \sin x\, dx$ (ii) $\int \sin^{-1} x\, dx$ (iii) $\int x \cdot \tan^{-1} x\, dx$ (iv) $\int \sec^3 x\, dx$

SOLUTION :

(i) $\int x^2 \cdot \sin x\, dx = x^2 \int \sin x\, dx - \int \frac{d}{dx}(x^2)\left(\int \sin x\, dx\right) dx$

$$= x^2(-\cos x) - \int 2x(-\cos x)\, dx$$

$$= -x^2 \cos x + 2\int x \cdot \cos x\, dx$$

$$= -x^2 \cos x + 2\left(x\int \cos x\, dx - \int \frac{dx}{dx}\left(\int \cos x\, dx\right) dx\right)$$

$$= -x^2 \cos x + 2x \sin x - 2\int \sin x\, dx$$

$$= -x^2 \cos x + 2x \sin x + 2\cos x + c$$

(ii) $\int \sin^{-1} x\, dx = \int 1 \cdot \sin^{-1} x\, dx = \sin^{-1} x \int dx - \int \frac{d}{dx}(\sin^{-1} x)\left(\int dx\right) dx$

$$= \sin^{-1} x \cdot x - \int \frac{1}{\sqrt{1-x^2}} \cdot x\, dx$$

$$= x \cdot \sin^{-1} x - \int \frac{x}{\sqrt{1-x^2}}\, dx \qquad ...(1)$$

Now $\int \frac{x}{\sqrt{1-x^2}}\, dx$

Let $1 - x^2 = t^2 \Rightarrow -2x\, dx = 2t\, dt$

$\Rightarrow$ $x\, dx = -t\, dt$

$\therefore$ $\int \frac{x}{\sqrt{1-x^2}}\, dx = -\int \frac{t\, dt}{t} = -\int dt = -t = -\sqrt{1-x^2}$

From (1),

$$\int \sin^{-1} x\, dx = x \cdot \sin^{-1} x + \sqrt{1-x^2} + c$$

(iii) $\int x \cdot \tan^{-1} x\, dx = \tan^{-1} x \cdot \int x\, dx - \int \frac{d}{dx}(\tan^{-1} x)\left(\int x\, dx\right) dx$

$$= \tan^{-1} x \cdot \frac{x^2}{2} - \int \frac{1}{1+x^2} \cdot x^2\, dx$$

$$= \tan^{-1} x \cdot \frac{x^2}{2} - \int \frac{x^2+1-1}{1+x^2}\, dx$$

$$= \tan^{-1} x \cdot \frac{x^2}{2} - \int \frac{1 + x^2}{1 + x^2} dx + \int \frac{1}{1 + x^2} dx$$

$$= \tan^{-1} x \cdot \frac{x^2}{2} - \int dx + \int \frac{1}{1 + x^2} dx$$

$$= \tan^{-1} x \cdot \frac{x^2}{2} - x + \tan^{-1} x + c$$

(iv) $$\int \sec^3 x \, dx = \int \sec^2 x \cdot \sec x \, dx$$

$$= \sec x \cdot \int \sec^2 x \, dx - \int \frac{d}{dx} (\sec x) \left(\int \sec^2 x \, dx \right) dx$$

$$= \sec x \cdot \tan x - \int \sec x \tan x \cdot \tan x \, dx$$

$$= \sec x \tan x - \int \sec x \tan^2 x \, dx$$

$$= \sec x \tan x - \int \sec x (\sec^2 x - 1) \, dx$$

$$= \sec x \tan x - \int \sec^3 x \, dx + \int \sec x \, dx$$

$$= \sec x \tan x - \int \sec^3 x \, dx + \log (\sec x + \tan x)$$

$\Rightarrow$ $$2 \int \sec^3 x \, dx = \sec x \tan x + \log (\sec x + \tan x)$$

or $$\int \sec^3 x \, dx = \frac{1}{2} \sec x \tan x + \frac{1}{2} \log (\sec x + \tan x) + c$$

Example 5.79

Evaluate : (i) $\int \frac{x \cdot \sin^{-1} x}{\sqrt{1 - x^2}} dx$ (ii) $\int \cos^{-1} \left(\frac{1 - x^2}{1 + x^2} \right) dx$

SOLUTION :

(i) $\int \frac{x \cdot \sin^{-1} x}{\sqrt{1 - x^2}} dx$

Let $\sin^{-1} x = t$ or $x = \sin t \Rightarrow \frac{1}{\sqrt{1 - x^2}} dx = dt$

$\therefore$ $$\int \frac{x \cdot \sin^{-1} x}{\sqrt{1 - x^2}} dx = \int t \cdot \sin t \, dt$$

$$= t\cdot\int \sin t\,dt - \int \frac{dt}{dt}\left(\int \sin t\,dt\right)dt$$

$$= t\,(-\cos t) - \int -\cos t\,dt$$

$$= -t\cos t + \sin t + c = -\sin^{-1} x\cdot\sqrt{1-x^2} + x + c$$

(ii) $\int \cos^{-1}\left(\frac{1-x^2}{1+x^2}\right)dx$

Let $\qquad x = \tan\theta \Rightarrow dx = \sec^2\theta\,d\theta$

$\therefore$
$$\int \cos^{-1}\left(\frac{1-x^2}{1+x^2}\right)dx = \int \cos^{-1}\left(\frac{1-\tan^2\theta}{1+\tan^2\theta}\right)\cdot\sec^2\theta\,d\theta$$

$$= \int \cos^{-1}(\cos 2\theta)\sec^2\theta\,d\theta = \int 2\theta\cdot\sec^2\theta\,d\theta$$

$$= 2\int \theta\cdot\sec^2\theta\,d\theta = 2\left[\theta\cdot\int \sec^2\theta\,d\theta - \int \frac{d\theta}{d\theta}\left(\int \sec^2\theta\,d\theta\right)d\theta\right]$$

$$= 2\theta\cdot\tan\theta - 2\int \tan\theta\,d\theta$$

$$= 2\theta\tan\theta - 2\cdot\log(\sec\theta) + c$$
$$= 2x\tan^{-1} x - 2\log\left(\sqrt{1+x^2}\right) + c$$

Example 5.80

Evaluate (i) $\int \frac{x^2\tan^{-1} x}{1+x^2}\,dx$

[UPTU 2003]

(ii) $\int x^3 \sin x^2\,dx$

[UPTU 2003]

SOLUTION : (i) Given $\int \frac{x^2\tan^{-1} x}{1+x^2}\,dx$

$$= \int \frac{(1+x^2-1)\tan^{-1} x}{1+x^2}\,dx$$

$$= \int \frac{(1+x^2)}{1+x^2}\tan^{-1} x\,dx - \int \frac{\tan^{-1} x}{1+x^2}\,dx$$

$$= \int \tan^{-1} x\,dx - \int \frac{\tan^{-1} x}{1+x^2}\,dx \qquad \text{...(1)}$$

Now, $\int \tan^{-1} x\, dx = \int \underset{II}{1} \cdot \underset{I}{\tan^{-1} x}\, dx$

$$= \tan^{-1} x \cdot x - \int \frac{1}{1+x^2} \cdot x\, dx \qquad \text{[integration by parts]}$$

$$= x \tan^{-1} x - \int \frac{x}{1+x^2}\, dx$$

$$= x \tan^{-1} x - \frac{1}{2} \log (1+x^2)$$

Again, $\int \frac{\tan^{-1} x}{1+x^2}\, dx = \int t\, dt$ [Put $t = \tan^{-1} x$]

$$= \frac{t^2}{2} = \frac{1}{2} (\tan^{-1} x)^2$$

From (1),

$$\int \frac{x^2 \tan^{-1} x}{1+x^2}\, dx = x \tan^{-1} x - \frac{1}{2} \log (1+x^2) + \frac{1}{2} (\tan^{-1} x)^2 + c$$

(ii) Given $\int x^3 \sin x^2\, dx = \int x^2 \cdot x \sin x^2\, dx$

Put $x^2 = t \Rightarrow 2x\, dx = dt$

$$\therefore \quad \int x^3 \sin x^2\, dx = \frac{1}{2} \int \underset{I}{t}\, \underset{II}{\sin t}\, dt$$

$$= \frac{1}{2} [t\,(-\cos t) - \int 1 \cdot (-\cos t)\, dt] \qquad \text{[integration by parts]}$$

$$= -\frac{1}{2} + \cos t + \sin t + c$$

$$= -\frac{1}{2} x^2 \cos x^2 + \sin x^2 + c$$

Example 5.81

Evaluate (i) $\int \frac{x \sin^{-1} x}{\sqrt{1-x^2}}\, dx$ ***[UPTU 2005]***

(ii) $\int \frac{x^3\, dx}{(x-1)(x-)(x-3)}$ ***[UPTU 2005]***

SOLUTION : (i) Given $\int \frac{x \sin^{-1} x}{\sqrt{1-x^2}}\, dx$

Let $\sin^{-1} x = t \Rightarrow x = \sin t$

$$\Rightarrow \quad \frac{1}{\sqrt{1-x^2}}\, dx = dt$$

$$\therefore \quad \int \frac{x \sin^{-1} x}{\sqrt{1-x^2}}\, dx = \int t \cdot \sin t\, dt$$

$$= t(-\cos t) - \int 1 \cdot (-\cos t)\, dt \qquad \text{[Integration by parts]}$$

$$= -t\cos t + \sin t + c$$

$$= -\sin^{-1} x \cdot \sqrt{1-x^2} + x + c$$

(ii) Given $\int \frac{x^3\, dx}{(x-1)\,(x-)\,(x-3)}$

$$\frac{x^3}{(x-1)(x-2)(x-3)} = \frac{A}{x-1} + \frac{B}{x-2} + \frac{C}{x-3} \qquad \text{...(1)}$$

$$\Rightarrow \quad x^3 = A(x-2)(x-3) + B(x-1)(x-3) + C(x-1)(x-2) \qquad \text{...(2)}$$

Putting $x = 1$ in (2), we get $A = \frac{1}{2}$

Putting $x = 2$ in (2), we get $B = -8$

Putting $x = 3$ in (2), we get $C = \frac{27}{2}$

From (1), we obtain

$$\frac{x^3}{(x-1)(x-2)(x-3)} = \frac{1}{2(x-1)} - \frac{8}{x-2} + \frac{27}{2(x-3)}$$

$$\therefore \quad \int \frac{x^3}{(x-1)(x-2)(x-3)} = \int \frac{dx}{2(x-1)} - \int \frac{8dx}{x-2} + \int \frac{27dx}{2(x-3)}$$

$$= \frac{1}{2}\log(x-1) - 8\log(x-2) + \frac{27}{2}\log(x-3) + c$$

PROBLEM SET 5.11

Evaluate the following integrals using the method of integration by parts.

1. $\int x^3 \log x\, dx$ **2.** $\int x^3 e^x\, dx$

3. $\int x^2 e^{-x}\, dx$ **4.** $\int \log x^x\, dx$

5. $\int \frac{e^x(1 + x\log x)}{x}\, dx$ **6.** $\int x \cdot \operatorname{cosec}^2 x\, dx$

7. $\int x^3 \cdot \cos x^2 \, dx$

8. $\int \sin x \cdot \log(\cos x) \, dx$

9. $\int \frac{x + \sin x}{1 + \cos x} dx$

10. $\int \tan^{-1} \sqrt{\frac{1-x}{1+x}} \, dx$

11. $\int \frac{x \tan^{-1} x}{(1+x^2)^{3/2}} dx$

12. $\int x \sin x \cos 2x \, dx$

13. $\int \sin^{-1} \sqrt{x} \, dx$

14. $\int (x+1) \log x \, dx$

Prove that :

15. $\int \tan^3 x \, dx = \frac{1}{2} \tan^2 x - \log \sec x + c$

16. $\int \sqrt{\frac{1 + \cos 2\theta}{1 - \cos 2\theta}} \, d\theta = \log \sin \theta + c$

17. $\int \cos^{-1} \left(\frac{1 - x^2}{1 + x^2} \right) dx = 2x \tan^{-1} x + c$

18. $\int \frac{\sqrt{x}}{\sqrt{x} + 2} dx = x - 4\sqrt{x} + 8 \log (\sqrt{x} + 2) + c$

19. $\int \frac{\sqrt{2 + \log x}}{x} dx = \frac{2}{3} (2 + \log x)^{3/2} + c$

20. $\int \left(\frac{4 \sin x}{7 \cos^2 x} + \frac{5}{3 \sin^2 x} \right) dx = \frac{4}{7} \sec x - \frac{5}{3} \cot x + c$

ANSWERS

1. $\frac{x^4}{4} \left(\log x - \frac{1}{4} \right) + c$ 2. $e^x (x^3 - 3x^2 + 6x - 6) + c$ 3. $-e^{-x} (x^2 + 2x + 2) + c$

4. $\frac{1}{2} x^2 \log x - \frac{1}{4} x^2 + c$ 5. $e^x \log x + c$ 6. $-x \cot x + \log (\sin x) + c$

7. $\frac{1}{2} x^2 \sin x^2 + \frac{1}{2} \cos x^2 + c$ 8. $\cos x (1 - \log \cos x) + c$ 9. $x \cdot \tan \frac{x}{2} + c$

10. $\frac{1}{2} x (\cos^{-1} x) - \frac{1}{2} \sqrt{1 - x^2} + c$ 11. $-\frac{\tan^{-1} x}{\sqrt{1 + x^2}} + \frac{x}{\sqrt{1 + x^2}} + c$

12. $\frac{1}{2} \left(-x \frac{\cos 3x}{3} + \frac{\sin 3x}{9} + x \cos x - \sin x \right) + c$

13. $\frac{1}{2} (2x^2 - 1) \sin^{-1} \sqrt{x} + \frac{1}{2} \sqrt{x - x^2} + c$ 14. $\left(x + \frac{x^2}{2} \right) \log x - \left(x + \frac{x^2}{4} \right) + c$

5.22. SOME IMPORTANT EXAMPLES

Example 5.82

Evaluate : (i) $\int e^x \left(\frac{1-\sin x}{1-\cos x}\right) dx$ (ii) $\int \frac{\log x}{(1+\log x)^2}\, dx$

(iii) $\int e^x (\tan x - \log \cos x)\, dx$ (iv) $\int e^x \left(\log x + \frac{1}{x}\right) dx$

SOLUTION :

(i) $$\int e^x \left(\frac{1-\sin x}{1-\cos x}\right) dx = \int e^x \left(\frac{1-2\sin\frac{x}{2}\cos\frac{x}{2}}{1-1+2\sin^2\frac{x}{2}}\right) dx$$

$$= \int e^x \left(\frac{1}{2\sin^2\frac{x}{2}} - \frac{\cos\frac{x}{2}}{\sin\frac{x}{2}}\right) dx = \int e^x \left(\frac{1}{2}\text{csoec}^2\frac{x}{2} - \cot\frac{x}{2}\right) dx$$

$$= \frac{1}{2}\int e^x \operatorname{cosec}^2\frac{x}{2}\, dx - \int e^x \cot\frac{x}{2}\, dx$$

$$= \frac{1}{2}\int e^x \operatorname{cosec}^2\frac{x}{2}\, dx - \left(\cot\frac{x}{2}\, e^x - \int -\operatorname{cosec}^2\frac{x}{2}\cdot\frac{1}{2}\cdot e^x\, dx\right)$$

$$= \frac{1}{2}\int e^x \operatorname{cosec}^2\frac{x}{2}\, dx - \cot\frac{x}{2}\cdot e^x - \frac{1}{2}\int e^x\cdot\operatorname{cosec}^2\frac{x}{2}\, dx$$

$$= -e^x\cdot\cot\frac{x}{2} + c$$

(ii) Let $$I = \int \frac{\log x}{(1+\log x)^2}\, dx$$

Let $\log x = t \Rightarrow \frac{1}{x}\, dx = dt \Rightarrow x = e^t$

$$\therefore \quad I = \int \frac{t\, e^t\, dt}{(1+t)^2} = \int \frac{(t+1-1)\, e^t}{(t+1)^2}\, dt$$

$$= \int \frac{e^t}{t+1}\, dt - \int \frac{e^t}{(t+1)^2}\, dt$$

$$= \int e^t\cdot\frac{1}{t+1}\, dt - \int \frac{e^t}{(t+1)^2}\, dt$$

$$= \frac{1}{t+1}\, e^t - \int -\frac{1}{(t+1)^2}\, e^t\, dt - \int \frac{e^t}{(t+1)^2}\, dt$$

$$= \frac{t}{t+1} + \int \frac{e^t}{(t+1)^2}\, dt - \int \frac{e^t}{(t+1)^2}\, dt + c = \frac{e^t}{t+1} + c$$

or $$I = \frac{x}{1 + \log x} + c$$

(iii) $$\int e^x (\tan x - \log \cos x)\, dx = \int \underset{\text{I}}{e^x} \underset{\text{II}}{\tan x}\, dx - \int e^x \log \cos x\, dx$$

$$= e^x \cdot \log \sec x - \int e^x (-\log \cos x)\, dx - \int e^x \log \cos x\, dx$$

$$= e^x \cdot \log \sec x + \int e^x \log \cos x\, dx - \int e^x \log \cos x\, dx$$

$$= e^x \cdot \log \sec x + c$$

(iv) $$\int e^x \left(\log x + \frac{1}{x}\right) dx = \int e^x \log x\, dx + \int e^x \frac{1}{x}\, dx$$

$$= \log x \cdot e^x - \int \frac{1}{x} \cdot e^x\, dx + \int e^x \cdot \frac{1}{x}\, dx = e^x \cdot \log x + c$$

Example 5.83

Evaluate : (i) $\int e^{2x} \cdot \sin 3x\, dx$ (ii) $\int \sin(\log x)\, dx$

SOLUTION :

(i) Let $$\text{I} = \int e^{2x} \cdot \sin 3x\, dx = e^{2x} \left(\frac{-\cos 3x}{3}\right) - \int 2e^{2x} \left(\frac{-\cos 3x}{3}\right) dx$$

$$= \frac{-e^{2x}}{3} \cos 3x + \frac{2}{3} \int e^{2x} \cos 3x\, dx$$

$$= -\frac{1}{3} e^{2x} \cos 3x + \frac{2}{3} \left[e^{2x} \frac{\sin 3x}{3} - \int 2e^{2x} \cdot \frac{\sin 3x}{3}\, dx\right]$$

$$= -\frac{1}{3} e^{2x} \cos 3x + \frac{2}{9} e^{2x} \sin 3x - \frac{4}{9} \int e^{2x} \sin 3x\, dx$$

$\therefore$ $$\text{I} = -\frac{1}{3} e^{2x} \cos 3x + \frac{2}{9} e^{2x} \sin 3x - \frac{4}{9} \text{I} + c$$

or $$\text{I} + \frac{4}{9} \text{I} = -\frac{1}{3} e^{2x} \cos 3x + \frac{2}{9} e^{2x} \sin 3x + c$$

$$\frac{13}{9} \text{I} = \frac{1}{9} e^{2x} (2\sin 3x - 3\cos 3x) + c$$

$$I = \frac{1}{13} e^{2x} (2\sin 3x - 3\cos 3x) + c$$

(ii) Let $\quad I = \int \sin (\log x)\, dx$

Put $\quad \log x = t \Rightarrow \frac{1}{x} dx = dt$

$\Rightarrow \quad x = e^t$

$\therefore \quad I = \int \sin t \cdot e^t\, dt = - e^t \cos t - \int e^t (-\cos t)\, dt$

$$I = - e^t \cos t + \int e^t \cos t\, dt = - e^t \cos t + e^t \sin t - \int e^t \sin t\, dt$$

$$2I = e^t (\sin t - \cos t)$$

$$I = \frac{1}{2} e^t (\sin t - \cos t) + c$$

or $\quad I = \frac{1}{2} x\, [(\sin (\log x) - \cos (\log x)] + c$

PROBLEM SET 5.12

Evaluate :

1. $\int e^x (\sin x + \cos x)\, dx$

2. $\int e^x \left(\frac{x-1}{2x^2}\right) dx$

3. $\int e^x \frac{\sqrt{1-x^2} \sin^{-1} x + 1}{\sqrt{1-x^2}}\, dx$

4. $\int \left(\tan (\log x) + \sec^2 (\log x)\right) dx$

5. $\int e^{-x} \cos x\, dx$

6. $\int e^x \cos^2 x\, dx$

7. $\int e^{ax} \cos bx\, dx$

8. $\int e^{2x} \cos (3x + 4)\, dx$

9. $\int \sqrt{a^2 - x^2}\, dx$

10. $\int \sqrt{x^2 - a^2}\, dx$

11. $\int \frac{xe^x}{(1+x)^2}\, dx$

12. $\int \frac{e^x (x-1)}{x^2}\, dx$

ANSWERS

1. $e^x \sin x + c$ **2.** $\frac{e^x}{2x} + c$ **3.** $e^x \sin^{-1} x + c$ **4.** $x \tan (\log x) + c$

5. $\frac{e^{-x}}{2} (\sin x - \cos x) + c$ **6.** $\frac{e^x}{2} + \frac{e^x}{10} (\cos 2x + 2\sin 2x) + c$

7. $\frac{e^{ax}}{a^2 + b^2} (a \cos bx + b \sin bx) + c$ **8.** $\frac{e^{2x}}{13} (2\cos (3x + 4)) + 3\sin (3x + 4) + c$

9. $\frac{x}{2} \sqrt{a^2 - x^2} + \frac{a^2}{2} \sin^{-1}\left(\frac{x}{a}\right) + c$ **10.** $\frac{x}{2} \sqrt{x^2 - a^2} - \frac{a^2}{2} \log (x + \sqrt{x^2 - a^2}) + c$

11. $\frac{-x e^x}{1 + x} + e^x + c$ **12.** $\frac{e^x}{x} + c$

5.23 INTEGRATION USING PARTIAL FRACTIONS

5.23.1 Rational Function

Let $f(x)$ and $g(x)$ be two Polynomials of the variable x of degree m and n respectively. Then the function $\frac{f(x)}{g(x)}$ is known as rational function of x.

5.23.2 Proper Rational Function

If degree of $f(x)$ is less than the degree of $g(x)$ *i.e.*, if $m < n$, then $\frac{f(x)}{g(x)}$ is proper rational function of x.

5.23.3 Improper Rational Function

If degree of $f(x)$ is greater than equal to the degree of $g(x)$ *i.e.*, if $m \geq n$, then $\frac{f(x)}{g(x)}$ is improper rational function of x.

Example 5.84

(i) The rational function $\frac{3x + 5}{2x^2 + 3x + 4}$ is a proper rational function because the degree of $3x + 5$ is less than the degree of $2x^2 + 3x + 4$.

(ii) The rational function $\frac{2x^2 + 4}{x^2 + 3x + 2}$ is improper rational function because the degree of $2x^2 + 4$ and the degree of $x^2 + 3x + 2$ are equal.

■ **NOTE :** 1. If $\frac{f(x)}{g(x)}$ is an improper rational function it is to be first converted into proper rational function by dividing $f(x)$ by $g(x)$. For example,

$$\frac{x^2}{x^2 - 3x + 4} \text{ is Improper rational function.}$$

Hence, divide x^2 by $x^2 - 3x + 4$, we get

$$\frac{x^2}{x^2 - 3x + 4} = 1 + \frac{3x - 4}{x^2 - 3x + 4}$$

2. If the denominator $g(x)$ can be factorized as $(ax + b)^n$, $(ax^2 + bx + c)^n$ *etc.* $n \geq 1$ are integers than proper rational function $\frac{f(x)}{g(x)}$ will be a sum of a number of proper fractions. These proper fractions are known as **Partial fractions.**

5.24 TYPES OF PARTIAL FRACTIONS

The resolution of proper rational function $\frac{f(x)}{g(x)}$ into partial fractions mainly depends upon the nature of the factors of denominater $g(x)$.

CASE 1 : If $g(x)$ is the product of non-repeating linear factors, *i.e.*, if $g(x) = (x - x_1)(x - x_2) \dots (x - x_n)$. Then corresponding partial fractions will be

$$\frac{f(x)}{g(x)} = \frac{A_1}{x - x_1} + \frac{A_2}{x - x_2} + \dots + \frac{An}{x - x_n}$$

where $A_1, A_2, \dots, A_n$ are constants to be determined.

Example 5.85

Resolve : $\frac{6x + 5}{x^3 + 6x^2 + 11x + 6}$ into partial fractions.

SOLUTION : Let $\frac{6x + 5}{x^3 + 6x^2 + 11x + 6} = \frac{f(x)}{g(x)}$

Here $f(x) = 6x + 5$ and $g(x) = x^2 + 6x^2 + 11x + 6$

Clearly, the degree of $f(x)$ is less than the degree of $g(x)$. Hence, $\frac{f(x)}{g(x)}$ is a proper rational finction.

Now the factors of $x^3 + 6x^2 + 11x + 6 = (x + 1)(x + 2)(x + 3)$

These factors are linear and non-repeated.

$$\therefore \quad \frac{6x + 5}{x^3 + 6x^2\ 11x + 6} = \frac{6x + 5}{(x + 1)(x + 2)(x + 3)}$$

Let $$\frac{6x + 5}{(x + 1)(x + 2)(x + 3)} = \frac{A_1}{x + 1} + \frac{A_2}{x + 2} + \frac{A_3}{x + 3} \quad \text{...(1)}$$

$$= \frac{(x + 2)(x + 3)A_1 + (x + 1)(x + 3)A_2 + (x + 1)(x + 2)A_3}{(x + 1)(x + 2)(x + 3)}$$

or $$6x + 5 = (x + 2)(x + 3)A_1 + (x + 1)(x + 3)A_2 + (x + 1)(x + 2)A_3 \quad \text{...(2)}$$

Putting $x = -1$ in (2), we get $-1 = 2A_1 \Rightarrow A_1 = -\frac{1}{2}$

Putting $x = -2$ in (2), we get $-7 = -A_2 \Rightarrow A_2 = 7$

Putting $x = -3$ in (2), we get $-13 = 2A_3 \Rightarrow A_3 = \frac{-13}{2}$

$$\therefore \quad \frac{6x+5}{x^3+6x^2+11x+6} = \frac{6x+5}{(x+1)(x+2)(x+3)} = -\frac{1}{2(x+1)} + \frac{7}{x+2} - \frac{13}{2(x+3)}$$

Example 5.86

Resolve $\frac{x^3}{x^2-3x+2}$ into partial fractions.

SOLUTION : Let $\frac{f(x)}{g(x)} = \frac{x^3}{x^2-3x+2}\cdot$ Here degree of $f(x)$ is greater than that of $g(x)$. So we divide $f(x)$ by $g(x)$ and obtain

$$\frac{x^3}{x^2-3x+2} = x+3+\frac{7x-6}{x^2-3x+2} \quad ...(1)$$

Now $$\frac{7x-6}{x^2-3x+2} = \frac{7x-6}{(x-1)(x-2)}$$

Here, $g(x)$ have linear non-repeated factors.

Let $$\frac{7x-6}{(x-1)(x-2)} = \frac{A_1}{x-1} + \frac{A_2}{x-2} \quad ...(2)$$

$$= \frac{A_1(x-2)+A_2(x-1)}{(x-1)(x-2)}$$

$$\Rightarrow \quad 7x-6 = A_1(x-2)+A_2(x-1) \quad ...(3)$$

Putting $x = 1$ in (3), we get $1 = -A_1 \Rightarrow A_1 = -1$

Putting $x = 2$ in (3), we get $8 = A_2 \Rightarrow A_2 = 8$

From (2), we obtain,

$$\frac{7x-6}{(x-1)(x-2)} = -\frac{1}{x-1} + \frac{8}{x-2}$$

From (1), we get

$$\frac{x^3}{x^2-3x+2} = x+3-\frac{1}{x-1}+\frac{8}{x-2}$$

CAES 2 : If $g(x)$ is the product of repeating linear factors, *i.e.*, if $g(x) = (x-x_1)^m (x-x_2)\ldots\ldots(x-x_n)$, then corresponding to non-repeating linear factors we assume as in caes 1 and for each repeating factor $(x-x_1)^m$, the partial fractions are of the form.

$$\frac{A_1}{x-x_1} + \frac{A_2}{(x-x_1)^2} + \ldots. + \frac{A_m}{(x-x_1)^m}$$

where $A_1, A_2, \ldots, A_m$ are constants.

Example 5.87

Resolve $\frac{3x+1}{(x-2)^2(x+2)}$ into partial fractions.

SOLUTION : Let $\frac{3x+1}{(x-2)^2(x+2)} = \frac{A_1}{x+2} + \frac{A_2}{x-2} + \frac{A_3}{(x-2)^2}$ ...(1)

$$= \frac{(x-2)^2A_1 + (x-2)(x+2)A_2 + (x+2)A_3}{(x-2)^2(x+2)}$$

$\Rightarrow$ $3x + 1 = (x-2)^2A_1 + (x-2)(x+2)A_2 + (x+2)A_3$ ...(2)

Putting $x = 2$ in (2), we get $7 = 4A_3 \Rightarrow A_3 = \frac{7}{4}$

Putting $x = -2$ in (2), we get $-5 = 16A_1 \Rightarrow A_1 = \frac{-5}{16}$

Putting $x = 0$ in (2), we get $1 = 4A_1 - 4A_2 + 2A_3 = \frac{-5}{4} - 4A_2 + \frac{7}{2} \Rightarrow A_2 = \frac{5}{16}$

from (1),

$$\frac{3x+1}{(x-2)^2(x+2)} = \frac{-5}{16(x-2)} + \frac{5}{16(x-2)} + \frac{7}{4(x-2)^2}$$

CAES 3 : If $g(x)$ is the product of some non-repeating quadratic factors and some repeating or non-repeating linear factors *i.e.*, $g(x) = (ax^2 + bx + c)(x - x_1)^m(x - x_2)\ldots(x - x_n)$

Then corresponding to each quadratic factor $ax^2 + bx + c$, partial fraction will be $\frac{Ax+B}{ax^2+bx+c}$, where A and B are constants to be determined.

Corresponding to repeating and non-repeating linear factors we assume as in case 2.

Example 5.88

Reaolve : $\frac{x}{(x-1)(x^2+4)}$ into partial fractions.

SOLUTION : Let $\frac{f(x)}{g(x)} = \frac{x}{(x-1)(x^2+4)} = \frac{A_1}{x-1} + \frac{A_2x + A_3}{x^2+4}$ ...(1)

$$= \frac{A_1(x^2+4) + (x-1)(A_2x + A_3)}{(x-1)(x^2+4)}$$

$\Rightarrow$ $x = A_1(x^2+4) + (x-1)(A_2x + A_3)$ ...(2)

Putting $x = 1$ in (2), we get $1 = 5A_1 \Rightarrow A_1 = \frac{1}{5}$

Putting $x = 0$ in (2), we get $0 = 4A_1 - A_3 \Rightarrow A_3 = \frac{4}{5}$

Putting $x = -1$ in (2), we get $-1 = 5A_1 + 2A_2 - 2A_3 \Rightarrow A_2 = \frac{-1}{5}$

Now from (1), we obtion

$$\frac{x}{(x-1)(x^2+4)} = \frac{1}{5(x-1)} + \frac{-\frac{1}{5}x + \frac{4}{5}}{x^2+4} = \frac{1}{5(x-1)} - \frac{x-4}{5(x^2+4)}$$

CASE 4 : If $g(x)$ is the product of some repeating quadratic factors and some repeating or non-repeating linear factors.

Example 5.89

Resolve $\frac{2x-3}{(x-1)(x^2+1)^2}$ into partial fractions.

SOLUTION : Let $$\frac{2x-3}{(x-1)(x^2+1)^2} = \frac{A}{x-1} + \frac{Bx+C}{x^2+1} + \frac{Dx+E}{(x^2+1)^2} \quad ...(1)$$

$$\Rightarrow \quad 2x-3 = A(x^2+1)^2 + (Bx+C)(x-1)(x^2+1) + (Dx+E)(x-1) \quad ...(2)$$

$$2x-3 = A(x^4+2x^2+1) + (Bx+C)(x^3-x^2+x-1) + (Dx+E)(x-1)$$
$$= Ax^4 + 2Ax^2 + A + Bx^4 - Bx^3 + Bx^2 - Bx + Cx^3 - Cx^2 + Cx - C + Dx^2 - Dx + Ex - E$$
$$= (A+B)x^4 + (-B+C)x^3 + (2A+B-C+D)x^2 + (-B+C-D+E)x + (A-C-E)$$

Equating the coefficients of like powers of x, we get

Coefficient of x^4, $0 = A + B$

Coefficient of x^3, $0 = -B + C$

Coefficient of x^2, $0 = 2A + B - C + D$

Coefficient of x, $2 = -B + C - D + E$

Coefficient of constant term, $-3 = A - C - E$

Solving these, we get $A = -\frac{1}{4}$, $B = \frac{1}{4}$, $C = \frac{1}{4}$, $D = \frac{1}{2}$, $E = \frac{5}{2}$

From (1), we obtain.

$$\frac{2x-3}{(x-1)(x^2+1)^2} = \frac{-1}{4(x-1)} + \frac{x+1}{4(x^2+1)} + \frac{x+5}{2(x^2+1)^2}$$

Example 5.90

Evaluate the following integrals :

(i) $\int \frac{dx}{x^2-a^2}$ (ii) $\int \frac{7x}{(x+2)(2x-3)}dx$

SOLUTION : (i) Let $\frac{1}{x^2-a^2} = \frac{1}{(x-a)(x+a)} = \frac{A}{x-a} + \frac{B}{x+a}$

$$\Rightarrow \quad 1 = A(x+a) + B(x-a) \quad ...(1)$$

Putting $x = a$ in (1), we get $1 = 2aA \Rightarrow A = \frac{1}{2a}$

Putting $x = -a$ in (1), we get $1 = -2aB \Rightarrow B = \frac{-1}{2a}$

$$\therefore \quad \frac{1}{x^2-a^2} = \frac{1}{2a(x-a)} - \frac{1}{2a(x+a)}$$

Hence, $$\int \frac{1}{x^2-a^2}dx = \frac{1}{2a}\int \frac{dx}{(x-a)} - \frac{1}{2a}\int \frac{dx}{x+a}$$

$$= \frac{1}{2a}\log(x-a) - \frac{1}{2a}\log(x+a) + c$$

$$= \frac{1}{2a} \log\left(\frac{x-a}{x+a}\right) + c$$

(ii) Let $\qquad \dfrac{7x}{(x+2)(2x-3)} = \dfrac{A}{x+2} + \dfrac{B}{2x-3}$ ...(1)

$\Rightarrow \qquad 7x = A(2x-3) + B(x+2)$ ...(2)

Putting $x = -2$ in (2), we get $-14 = -7A \Rightarrow A = 2$

Putting $x = \frac{3}{2}$ in (2), we get $\frac{21}{2} = \frac{7}{2}B \Rightarrow B = 3$

From (1), we obtain

$$\frac{7x}{(x+2)(2x-3)} dx = \frac{2}{x+2} + \frac{3}{2x-3}$$

So, $$\int \frac{7x}{(x+2)(2x-3)} dx = 2\int \frac{dx}{x+2} + 3\int \frac{dx}{2x-3}$$

$$= 2\log(x+2) + \frac{3}{2}\log(2x-3) + c$$

Example 5.91

Evaluate : (i) $\displaystyle\int \frac{x^3}{(x-1)(x-2)} dx$ (ii) $\displaystyle\int \frac{2x}{(x^2+1)(x^2+2)} dx$

SOLUTION : (i) Here the degree of numerator is higher than that of denominator. So, first we divide the numerator by denominator.

$$\begin{array}{r|l} x^2 - 3x + 2 \,)\, x^3 & (x+3 \\ \underline{-x^3 \mp 3x^2 \pm 2x} & \\ 3x^2 - 2x & \\ \underline{-3x^2 \mp 9x \pm 6} & \\ 7x - 6 & \end{array}$$

$\therefore \qquad \dfrac{x^3}{(x-1)(x-2)} = x + 3 + \dfrac{7x-6}{(x-1)(x-2)}$ ...(1)

Again let $\qquad \dfrac{7x-6}{(x-1)(x-2)} = \dfrac{A}{x-1} + \dfrac{B}{x-2}$ ...(2)

$\Rightarrow \qquad 7x - 6 = A(x-2) + B(x-1)$ ...(3)

Putting $x = 1$ in (3), we get $A = -1$

Putting $x = 2$ in (3), we get $B = 8$

From (2), $\qquad \dfrac{7x-6}{(x-1)(x-2)} = \dfrac{8}{x-2} - \dfrac{1}{x-1}$

From (1), $\qquad \dfrac{x^3}{(x-1)(x-2)} = x + 3 + \dfrac{8}{x-2} - \dfrac{1}{x-1}$

$\therefore \qquad \displaystyle\int \frac{x^3}{(x-1)(x-2)} dx = \int\left(x + 3 + \frac{8}{x-2} - \frac{1}{x-1}\right)dx$

$$= \frac{x^2}{2} + 3x + 8\log(x-2) - \log(x-1) + c$$

(ii) Let $$\frac{2x}{(x^2+1)(x^2+2)} = \frac{Ax+B}{x^2+1} + \frac{Cx+D}{x^2+2} \quad ...(1)$$

$\Rightarrow$ $$2x = (Ax+B)(x^2+2) + (Cx+D)(x^2+1)$$
$$= (A+C)x^3 + (B+D)x^2 + (2A+C)x + (2B+D)$$

Equating to the coefficients of different powers of x

Coefficient of x^3, $0 = A + C$

Coefficient of x^2, $0 = B + D$

Coefficient of x, $2 = 2A + C$

Constant term, $0 = 2B + D$

Solving these, we get $A = 2, B = 0, C = -2, D = 0$

From (1), we have

$$\frac{2x}{(x^2+1)(x^2+2)} = \frac{2x}{x^2+1} - \frac{2x}{x^2+2}$$

So, $$\int \frac{2x}{(x^2+1)(x^2+2)}dx = \int \frac{2x\,dx}{x^2+1} - \int \frac{2x\,dx}{x^2+2}$$
$$= \log(x^2+1) - \log(x^2+2) + c$$
$$= \log\left(\frac{x^2+1}{x^2+2}\right) + c$$

Example 5.92

Evaluate :

(i) $\int \frac{3x+1}{(x-2)^2(x+2)}dx$ (ii) $\int \frac{x}{(x-1)(x^2+4)}dx$

SOLUTION : (i) Let $$\frac{3x+1}{(x-2)^2(x+2)} = \frac{A}{x+2} + \frac{B}{x-2} + \frac{C}{(x-2)^2} \quad ...(1)$$

$\Rightarrow$ $$3x + 1 = A(x-2)^2 + B(x+2)(x-2) + C(x+2) \quad ...(2)$$

Putting $x = 2$ in (2), we get $7 = 4C \Rightarrow C = \frac{7}{4}$

Putting $x = -2$ in (2), we get $-5 = 16A \Rightarrow A = \frac{-5}{16}$

Putting $x = 0$ in (2), we get $1 = 4A - 4B + 2C \Rightarrow B = \frac{5}{16}$

From (1), we obtain

$$\frac{3x+1}{(x+2)(x-2)^2} = \frac{-5}{16(x+2)} + \frac{5}{16(x-2)} + \frac{7}{4(x-2)^2}$$

So, $$\int \frac{3x+1}{(x-2)^2(x+2)}dx = \frac{-5}{16}\int \frac{dx}{x+2} + \frac{5}{16}\int \frac{dx}{x-2} + \frac{7}{4}\int \frac{dx}{(x-2)^2}$$
$$= \frac{-5}{16}\log(x+2) + \frac{5}{16}\log(x-2) - \frac{7}{4(x-2)} + c$$
$$= \frac{5}{16}\log\left(\frac{x-2}{x+2}\right) - \frac{7}{4(x-2)} + c$$

(ii) Let $\dfrac{x}{(x-1)(x^2+4)} = \dfrac{A}{x-1} + \dfrac{Bx+C}{x^2+4}$

$$\Rightarrow \quad x = A(x^2+4) + (Bx+C)(x-1)$$

$$= (A+B)x^2 + (-B+C)x + (4A-C)$$

Now equating the coefficients of different powers of x, we obtain

$$A+B=0,\ C-B=1,\ 4A-C=0$$

Solving these, we get

$$A = \frac{1}{5},\ B = -\frac{1}{5},\ C = \frac{4}{5}$$

$$\therefore \quad \frac{x}{(x-1)(x^2+4)} = \frac{1}{5(x-1)} + \frac{\frac{-1}{5}x + \frac{4}{5}}{x^2+4}$$

So, $$\int \frac{x}{(x-1)(x^2+4)}\,dx = \frac{1}{5}\int \frac{dx}{x-1} - \frac{1}{5}\int \frac{x-4}{x^2+4}\,dx$$

$$= \frac{1}{5}\int \frac{dx}{x-1} - \frac{1}{5}\int \frac{x}{x^2+4}\,dx + \frac{4}{5}\int \frac{dx}{x^2+4}$$

$$= \frac{1}{5}\log(x-1) - \frac{1}{10}\log(x^2+4) + \frac{4}{5}\cdot\frac{1}{2}\tan^{-1}\frac{x}{2} + c$$

$$= \frac{1}{5}\log(x-1) - \frac{1}{10}\log(x^2+4) + \frac{2}{5}\tan^{-1}\frac{x}{2} + c$$

Example 5.93

Evaluate :

(i) $\displaystyle\int \frac{1-\cos\theta}{\cos\theta\,(1+\cos\theta)}\,d\theta$ (ii) $\displaystyle\int \frac{\tan\theta + \tan^3\theta}{1+\tan^3\theta}$

SOLUTION : (i) Let $\cos\theta = x$

$$\therefore \quad \frac{1-\cos\theta}{\cos\theta\,(1+\cos\theta)} = \frac{1-x}{x(1+x)} \qquad \text{...(1)}$$

Let $$\frac{1-x}{x(1+x)} = \frac{A}{x} + \frac{B}{1+x} \qquad \text{...(2)}$$

$$\Rightarrow \quad 1-x = (1+x)A + xB \qquad \text{...(3)}$$

Putting $x = 0$ in (3), we get $1 = A$

Putting $x = -1$ in (3), we get $2 = -B \Rightarrow B = -2$

from (2), we obtain,

$$\frac{1-x}{x(1+x)} = \frac{1}{x} - \frac{2}{1+x}$$

$$\therefore \quad \frac{1-\cos\theta}{\cos\theta\,(1+\cos\theta)} = \frac{1}{x} - \frac{2}{1+x} = \frac{1}{\cos\theta} - \frac{2}{1+\cos\theta}$$

So, $$\int \frac{1-\cos\theta}{\cos\theta\,(1+\cos\theta)}\,d\theta = \int \frac{1}{\cos\theta}\,d\theta - 2\int \frac{1}{1+\cos\theta}\,d\theta$$

$$= \int \sec\theta \, d\theta - \int \frac{2}{2\cos^2\theta/2} d\theta$$

$$= \int \sec\theta \, d\theta - \int \sec^2\theta/2 \, d\theta$$

$$= \log(\sec\theta - \tan\theta) - 2\tan\theta/2 + c$$

(ii) $$\int \frac{\tan\theta + \tan^3\theta}{1+\tan^3\theta} d\theta = \int \frac{\tan\theta\,(1+\tan^2\theta)}{1+\tan^3\theta} d\theta = \int \frac{\tan\theta \sec^2\theta}{1+\tan^3\theta} d\theta$$

Let $\tan\theta = x \Rightarrow \sec^2\theta \, d\theta = dx$

$$\therefore \quad \int \frac{\tan\theta + \tan^3\theta}{1+\tan^3\theta} d\theta = \int \frac{x\,dx}{1+x^3}$$

Let $$\frac{x}{1+x^3} = \frac{x}{(1+x)(1-x+x^2)} = \frac{A}{1+x} + \frac{Bx+C}{1-x+x^2} \quad ...(1)$$

$$\Rightarrow \quad x = A(1-x+x^2) + (1+x)(Bx+C) \quad ...(2)$$

Putting $x = -1$ in (2), we get $-1 = 3A \Rightarrow A = \frac{-1}{3}$

Putting $x = 0$ in (2), we get $0 = A + C \Rightarrow C = \frac{1}{3}$

Putting $x = 1$ in (2), we get $1 = A + 2B + 2C \Rightarrow B = \frac{1}{3}$

from (1), we obtain

$$\frac{x}{1+x^3} = -\frac{1}{3(1+x)} + \frac{\frac{1}{3}x + \frac{1}{3}}{1-x+x^2}$$

$$= -\frac{1}{3(1+x)} + \frac{1}{3}\frac{(1+x)}{1-x+x^2}$$

$$\therefore \quad \int \frac{xdx}{1+x^3} = -\frac{1}{3}\int \frac{dx}{1+x} + \frac{1}{3}\int \frac{1+x}{1-x+x^2} dx$$

$$= -\frac{1}{3}\int \frac{dx}{1+x} + \frac{1}{3}\int \frac{x+1}{x^2-x+1} dx$$

$$= -\frac{1}{3}\int \frac{dx}{1+x} + \frac{1}{6}\int \frac{2x-1+3}{x^2-x+1} dx$$

$$= \frac{-1}{3}\int \frac{dx}{1+x} + \frac{1}{6}\int \frac{2x-1}{x^2-x+1} dx + \frac{1}{2}\int \frac{dx}{x^2-x+1}$$

$$= \frac{-1}{3}\int \frac{dx}{1+x} + \frac{1}{6}\int \frac{2x-1}{x^2-x+1} dx + \frac{1}{2}\int \frac{dx}{\left(x-\frac{1}{2}\right)^2 + \left(\frac{\sqrt{3}}{2}\right)^2}$$

$$= \frac{-1}{3}\log(1+x) + \frac{1}{6}\log(x^2 - x + 1) + \frac{1}{2}\cdot\frac{2}{\sqrt{3}}\tan^{-1}\left(\frac{x - \frac{1}{2}}{\frac{\sqrt{3}}{2}}\right) + c$$

$$= \frac{-1}{3}\log(1+\tan\theta) + \frac{1}{6}\log(\tan^2\theta - \tan\theta + 1) + \frac{1}{\sqrt{3}}\tan^{-1}\left(\frac{2\tan\theta - 1}{\sqrt{3}}\right) + c$$

Example 5.94

Evaluate $\displaystyle\int \frac{x^2}{(x^2+1)(x^2+4)}\,dx$

SOLUTION : Given $\displaystyle\int \frac{x^2}{(x^2+1)(x^2+4)}\,dx$

Let $x^2 = y$

Then $$\frac{x^2}{(x^2+1)(x^2+4)} = \frac{y}{(y+1)(y+4)}$$

$$= \frac{A}{y+1} + \frac{B}{y+4}$$

$\Rightarrow$ $$y = A(y+4) + B(y+1)$$

Putting $y = -1, A = -\frac{1}{3}$

Putting $y = -4, B = \frac{4}{3}$

$\therefore$ $$\frac{y}{(y+1)(y+4)} = -\frac{1}{3(y+1)} + \frac{4}{3(y+4)}$$

$\Rightarrow$ $$\frac{x^2}{(x^2+1)(x^2+4)} = -\frac{1}{3(x^2+1)} + \frac{4}{3(x^2+4)}$$

So, $$\int \frac{x^2\,dx}{(x^2+1)(x^2+4)} = -\frac{1}{3}\int\frac{dx}{x^2+1} + \frac{4}{3}\int\frac{dx}{x^2+4}$$

$$= -\frac{1}{3}\tan^{-1}x + \frac{4}{3}\cdot\frac{1}{2}\tan^{-1}\frac{x}{2} + c$$

$$= -\frac{1}{3}\tan^{-1}x + \frac{2}{3}\tan^{-1}\frac{x}{2} + c$$

Example 5.95

Evaluate : (i) $\displaystyle\int \frac{dx}{x - x^3}$ *[UPTU 2003]*

(ii) $\displaystyle\int \frac{dx}{x + x^2}$ *[UPTU 2004]*

SOLUTION : (i) $\displaystyle\int \frac{dx}{x - x^3} = \int \frac{dx}{x(1-x)(1+x)}$

Let $$\frac{1}{x(1-x)(1+x)} = \frac{A}{x} + \frac{B}{1-x} + \frac{C}{1+x} \qquad ...(1)$$

$\Rightarrow$ $$1 = A(1-x)(1+x) + Bx(1+x) + Cx(1-x) \qquad ...(2)$$

Putting $x = 0$ in (2), we get $1 = A$

Putting $x = 1$ in (2), we get $1 = 2B \Rightarrow B = \frac{1}{2}$

Putting $x = -1$ in (2), we get $1 = 2C \Rightarrow C = \frac{1}{2}$

From (1), we obtain

$$\frac{1}{x(1-x)(1+x)} = \frac{1}{x} + \frac{1}{2(1-x)} + \frac{1}{2(1+x)}$$

$\therefore$ $$\int \frac{dx}{x(1-x)(1+x)} = \int \frac{1}{x}\,dx + \frac{1}{2}\int \frac{1}{1-x}\,dx + \frac{1}{2}\int \frac{1}{1+x}\,dx$$

$$= \log x - \frac{1}{2}\log(1-x) + \frac{1}{2}\log(1+x) + c$$

(ii) Let $$\frac{1}{x+x^2} = \frac{1}{x(1+x)} = \frac{A}{x} + \frac{B}{1+x} \qquad ...(1)$$

$\Rightarrow$ $$1 = A(1+x) + Bx \qquad ...(2)$$

Putting $x = 0$ in (2), we get $1 = A$

Putting $x = -1$ in (2), we get $1 = -B \Rightarrow B = -1$

From (1), we obtain

$$\frac{1}{x+x^2} = \frac{1}{x} - \frac{1}{1+x}$$

$\therefore$ $$\int \frac{dx}{x+x^2} = \int \frac{1}{x}\,dx - \int \frac{1}{1+x}\,dx$$

$$= \log x - \log(1+x) + c$$

Example 5.96

Evaluate $\int \frac{3x\,dx}{(x-1)(x-2)}$

SOLUTION : Let $$\frac{3x}{(x-1)(x-2)} = \frac{A}{x-1} + \frac{B}{x-2} \qquad ...(1)$$

$\therefore$ $$3x = A(x-2) + B(x-1) \qquad ...(2)$$

Putting $x = 1$ in (2), we get $3 = -A \Rightarrow A = -3$

Putting $x = 2$ in (2), we get $6 = B \Rightarrow B = 6$

From (1), we obtain

$$\frac{3x}{(x-1)(x-2)} = \frac{-3}{x-1} + \frac{6}{x-2}$$

So $$\int \frac{3x}{(x-1)(x-2)}\,dx = -3\int \frac{dx}{x-1} + 6\int \frac{dx}{x-2}$$

$$= -3\log(x-1) + 6\log(x-2) + c$$

PROBLEM SET 5.13

Using method of partial fractions evaluate the following integrals:

1. $\int \frac{3x}{(x-2)(x+1)} dx$

2. $\int \frac{1}{x(x-2)(x-4)} dx$

3. $\int \frac{dx}{x^2 - 3x + 2}$

4. $\int \frac{3 + 4x - x^2}{(x+2)(x-1)} dx$

5. $\int \frac{3x - 20}{x^2 + 3x - 10} dx$

6. $\int \frac{2x}{(x^2+1)(x^2+3)} dx$

7. $\int \frac{\cos\theta}{(1+\sin\theta)(2+\sin\theta)} d\theta$

8. $\int \frac{23x - 11x^2}{(2x-1)(9-x^2)} dx$

9. $\int \frac{1}{x \log x (2 + \log x)} dx$

10. $\int \frac{x^2 + 5x - 1}{x^3 + x^2 - 6x} dx$

11. $\int \frac{x}{(x^2 - a^{2)}(x^2 - b^{2)}} dx$

12. $\int \frac{e^x \, dx}{3e^{2x} + 2e^x - 1}$

13. $\int \frac{x^2}{(x^2+1)(3x^2+4)} dx$

14. $\int \frac{x^2+1}{(x-2)^2 (x+3)} dx$

15. $\int \frac{d\theta}{\sin\theta (2 + \cos\theta)} d\theta$

16. $\int \frac{dx}{a^2 - x^2}$

17. $\int \frac{x^3}{(x-1)(x-2)(x-3)} dx$

18. $\int \frac{3x+1}{(x-1)^3 (x+1)} dx$

19. $\int \frac{dx}{5 + 4\cos x}$

20. $\int \frac{dx}{4 + 5\sin x}$

21. $\int \frac{\sin x}{1 + \sin x} dx$

ANSWERS

1. $2\log(x-2) + \log(x+1) + c$ **2.** $\frac{1}{8}\log\left(\frac{x(x-4)}{(x-2)^2}\right) + c$ **3.** $\log\left(\frac{x-2}{x-1}\right) + c$

4. $-x + 3\log(x+2) + 2\log(x-1) + c$ **5.** $5\log(x+5) - 2\log(x-2) + c$

6. $\frac{1}{2}\log\left(\frac{x^2+1}{x^2+3}\right) + c$ **7.** $\log\left(\frac{1+\sin\theta}{2+\sin\theta}\right) + c$

8. $\frac{1}{2}\log(2x-1)+4\log(3+x)+\log(3-x)+c$ 9. $\frac{1}{2}\log\left(\frac{\log x}{\log(x+2)}\right)+c$

10. $\frac{1}{6}\log x+\frac{13}{10}\log(x-2)-\frac{7}{15}\log(x+3)+c$

11. $\frac{1}{2(a^2-b^2)}\log\left(\frac{x^2-a^2}{x^2-b^2}\right)+c$ 12. $\frac{1}{4}\log\left(\frac{3e^x-1}{e^x+1}\right)+c$

13. $\frac{2}{\sqrt{3}}\tan^{-1}\left(\frac{\sqrt{3}x}{2}\right)-\tan^{-1}x+c$ 14. $\frac{2}{5}\log(x+3)+\frac{3}{5}\log(x-2)-\frac{1}{x-2}+c$

15. $\frac{1}{6}\log(1-\cos\theta)-\frac{1}{2}\log(1+\cos\theta)-\frac{1}{3}\log(2+\cos\theta)+c$

16. $\frac{1}{2a}\log\left(\frac{a+x}{a-x}\right)$ 17. $x+\frac{1}{2}\log(x-1)-8\log(x-2)+\frac{27}{2}\log(x-3)$

18. $\frac{2}{(x-1)^3}+\frac{1}{2(x-1)^2}-\frac{1}{4(x-1)}+\frac{1}{4(x+1)}$ 19. $\frac{2}{3}\tan^{-1}\left(\frac{1}{3}\tan\frac{x}{2}\right)+c$

20. $\frac{1}{3}\log\left[\frac{1+2\tan\frac{x}{2}}{4+2\tan\frac{x}{2}}\right]+c$ 21. $\sec x-\tan x+x+c$

5.25 DEFINITE INTEGRAL

We know if $\int f(x)\,dx=\varphi(x)+C$ then $\varphi(x)$ is said to be the indfinite integral of $f(x)$. The definite integral of $f(x)$ over closed interval $[a, b]$ is defined by $\varphi(b)-\varphi(a)$ denoted by $\int_a^b f(x)\,dx$ and read as "the integral of $f(x)$ from a to b".

Thus, $$\int_a^b f(x)\,dx=[\theta(x)+C]_a^b=[\theta(b)+C]-[\theta(a)+C]$$
$$=\varphi(b)+C-\varphi(a)-C$$
$$=\varphi(b)-\varphi(a)$$

The numbers a and b are called limits of integration, a is the lower limit and b is the upper limit of integration.

Example 5.97

Evaluate $\int_1^2 x^2\,dx$

SOLUTION : $\int_1^2 x^2\,dx=\left[\frac{x^3}{3}\right]_1^2=\frac{1}{3}[(2)^3-(1)^3]=\frac{1}{3}[8-1]=\frac{7}{3}$

5.26 PROPERTIES OF DEFINITE INTEGRALS

(1) To prove that $\int_a^b f(x)\,dx = \int_a^b f(y)\,dy$

i.e., **integration is independent of the change of variable.**

PROOF :

By definition, $$\int_a^b f(x)\,dx = [\phi(x)]_a^b = \phi(b) - \phi(a) \quad ...(1)$$

Also $$\int_a^b f(y)\,dy = [\phi(y)]_a^b = \phi(b) - \phi(a) \quad ...(2)$$

From (1) and (2), we get

$$\int_a^b f(x)\,dx = \int_a^b f(y)\,dy$$

(2) To prove that

$$\int_a^b f(x)\,dx = -\int_b^a f(x)\,dx$$

i.e., **if the limits of the integration are interchanged then the sign of the definite integral is changed.**

PROOF : By definition, $$\int_a^b f(x)\,dx = [\phi(x)]_a^b = \phi(b) - \phi(a)$$

$$= -[\phi(a) - \phi(b)] = -\int_b^a f(x)\,dx$$

(3) To prove that

$$\int_a^b f(x)\,dx = \int_a^c f(x)\,dx + \int_c^b f(x)\,dx\,,\ a < c < b$$

i.e., **a definite integral can be expressed as the sum of a number of finite definite integrals by dividing the range of integration.**

PROOF : By definition, $$\int_a^b f(x)\,dx = \phi(b) - \phi(a) = \phi(b) - \phi(c) + \phi(c) - \phi(a)$$

$$= \int_c^b f(x)\,dx + \int_a^c f(x)\,dx = \int_a^c f(x)\,dx + \int_c^b f(x)\,dx$$

(4) To prove that $\int_0^a f(x)\,dx = \int_0^a f(a-x)\,dx$

i.e., **if the lower limit of definite integral is zero and upper limit is a then, we can change the integrand $f(x)$ to $f(a-x)$.**

PROOF : RHS $= \int_0^a f(a-x)dx$

Let $a - x = t \Rightarrow -dx = dt$

when $x = 0, t = a$

when $x = a, t = 0$

$$\therefore \quad \text{RHS} = \int_a^0 f(t)\,(-dt) = -\int_a^0 f(t)dt$$

$$= \int_0^a f(t)\,dt \qquad \text{[using prop. 2]}$$

$$= \int_0^a f(x)\,dx \qquad \text{[using prop. 1]}$$

$$= \text{LHS}$$

(5) To prove that $\int_{-a}^{a} f(x)dx = \begin{cases} 2\int_0^a f(x)dx, & \text{if } f(x) \text{ is an even function} \\ 0 & \text{, if } f(x) \text{ is an odd function} \end{cases}$

PROOF : (i) A function $f(x)$ is said to be an even function if $f(-x) = f(x)$, for example $\cos x$, x^2 etc.

(ii) A function $f(x)$ is said to be an odd function if $f(-x) = -f(x)$, for example, $\sin x$, x , x^3 etc.

$$\text{LHS} = \int_{-a}^{a} f(x)\,dx = \int_{-a}^{0} f(x)\,dx + \int_0^a f(x)\,dx \qquad \text{[using prop. 3]} \qquad ...(1)$$

Now let $I_1 = \int_{-a}^{0} f(x)\,dx$

Put $x = -t \Rightarrow dx = -dt$

when $x = -a, t = a$

when $x = 0, t = 0$

$$\therefore \quad I_1 = -\int_a^0 f(-t)\,dt = \int_0^a f(-t)\,dt \qquad \text{[prop. 2]}$$

$$= \int_0^a f(-x)\,dx \qquad \text{[prop. 1]}$$

from (1),

$$\text{LHS} = \int_{-a}^{a} f(x)\,dx = \int_0^a f(-x)\,dx + \int_0^a f(x)\,dx \qquad ...(2)$$

Now, if function $f(x)$ is even then $f(-x) = f(x)$

$$\therefore \quad \text{LHS} = \int_0^a f(x)\,dx + \int_0^a f(x)\,dx = 2\int_0^a f(x)\,dx \qquad ...(3)$$

and if function $f(x)$ is odd then $f(-x) = -f(x)$

$\therefore \qquad \text{LHS} = -\int_0^a f(x)\,dx + \int_0^a f(x)\,dx = 0 \qquad ...(4)$

from (3) and (4), we get

$$\int_{-a}^{a} f(x)dx = \begin{cases} 2\int_0^a f(x)dx, \text{ if } f(x) \text{ is even function} \\ 0 \quad , \text{ if } f(x) \text{ is an odd function} \end{cases}$$

(6) To prove that

$$\int_0^{2a} f(x)dx = \begin{cases} 2\int_0^a f(x)dx, \text{ if } f(2a-x) = f(x) \\ 0 \quad , \text{ if } f(2a-x) = -f(x) \end{cases}$$

PROOF : Let $I = \int_0^{2a} f(x)\,dx = \int_0^a f(x)\,dx + \int_a^{2a} f(x)\,dx$ [using prop. 3] ...(1)

Consider $\quad I_1 = \int_a^{2a} f(x)\,dx$

Let $x = 2a - t \Rightarrow dx = -dt$

when $x = a, t = a$

when $x = 2a, t = 0$

$\therefore \qquad I_1 = -\int_a^0 f(2a-t)\,dt = \int_0^a f(2a-t)\,dt$ [using prop. 2]

$= \int_0^a f(2a-x)\,dx$ [using prop. 1]

From (1), we get

$$I = \int_0^a f(x)\,dx + \int_0^a f(2a-x)\,dx \qquad ...(2)$$

Now, if $f(2a-x) = f(x)$, then from (2),

$$I = \int_0^a f(x)\,dx + \int_0^a f(x)\,dx = 2\int_0^a f(x)\,dx$$

and if $f(2a-x) = -f(x)$, then from (2)

$$I = \int_0^a f(x)\,dx - \int_0^a f(x)\,dx = 0$$

$$\therefore \qquad \int_0^{2a} f(x)dx = \begin{cases} 2\int_0^a f(x)dx, \text{ if } f(2a-x) = f(x) \\ 0 \quad , \text{ if } f(2a-x) = -f(x) \end{cases}$$

Example 5.98

Evaluate : (i) $\int_1^2 \frac{dx}{3x+5}$ (ii) $\int_0^{\pi/2} \sin^2 x\, dx$ (iii) $\int_0^{\pi/2} \cos^3 x\, dx$

SOLUTION : (i) $\int_1^2 \frac{dx}{3x+5} = \frac{1}{3}\left[\log(3x+5)\right]_1^2 = \frac{1}{3}[\log 11 - \log 8]$

$$= \frac{1}{3}\log\frac{11}{8}$$

(ii) $$\int_0^{\pi/2} \sin^2 x\, dx = \frac{1}{2}\int_0^{\pi/2} 2\sin^2 x\, dx = \frac{1}{2}\int_0^{\pi/2}(1-\cos 2x)dx$$

$$= \frac{1}{2}\left[x - \frac{\sin 2x}{2}\right]_0^{\pi/2} = \frac{1}{2}\left[\left(\frac{\pi}{2} - \frac{\sin\pi}{2}\right) - \left(0 - \frac{\sin 0}{2}\right)\right]$$

$$= \frac{1}{2}\left[\frac{\pi}{2}\right] = \frac{\pi}{4}$$

(iii) $$\int_0^{\pi/2} \cos^3 x\, dx = \int_0^{\pi/2} \cos^2 x \cdot \cos x\, dx = \int_0^{\pi/2}(1-\sin^2 x)\cos x\, dx$$

Let $\sin x = t \Rightarrow \cos x\, dx = dt$

when $x = 0, t = \sin 0 = 0$

when $x = \frac{\pi}{2}, \; t = \sin\frac{\pi}{2} = 1$

$\therefore$ $$\int_0^{\pi/2} \cos^3 x\, dx = \int_0^1 (1-t^2)dt = \left[t - \frac{t^3}{3}\right]_0^1 = 1 - \frac{1}{3} = \frac{2}{3}$$

Example 5.99

(i) $\int_0^{\pi/4} \sqrt{1+\sin 2x}\, dx$ (ii) $\int_1^3 \frac{\log x}{(1+x)^2} dx$ (iii) $\int_0^1 x e^x\, dx$

SOLUTION : (i) $\int_0^{\pi/4} \sqrt{1+\sin 2x}\, dx = \int_0^{\pi/4} \sqrt{\sin^2 x + \cos^2 x + 2\sin\cos x}\, dx$

$$= \int_0^{\pi/4} \sqrt{(\sin x + \cos x)^2}\, dx = \int_0^{\pi/4} (\sin x + \cos x)\, dx$$

$$= [-\cos x + \sin x]_0^{\pi/4}$$

$$= \left[-\cos\frac{\pi}{4} + \sin\frac{\pi}{4} + \cos 0 - \sin 0\right]$$

$$= -\frac{1}{\sqrt{2}} + \frac{1}{\sqrt{2}} + 1 - 0 = 1$$

(ii) $$\int_1^3 \frac{\log x}{(1+x)^2} dx = \int_1^3 \log x\,(1+x)^{-2} dx = \left[\log x\,\frac{(1+x)^{-1}}{-1}\right]_1^3 - \int_1^3 \frac{1}{x}\,\frac{(1+x)^{-1}}{-1}\, dx$$

$$= \left[-\frac{\log x}{1+x}\right]_1^3 + \int_1^3 \frac{1}{x(1+x)}\,dx$$

$$= -\frac{\log 3}{4} + \frac{\log 1}{2} + \int_1^3 \left(\frac{1}{x} - \frac{1}{1+x}\right) dx \qquad \text{[Using partial fraction]}$$

$$= -\frac{\log 3}{4} + [\log x - \log(1+x)]_1^3$$

$$= -\frac{\log 3}{4} + \log 3 - \log 4 - \log 1 + \log 2$$

$$= \frac{-\log 3}{4} + \log 3 - \log 4 + \log 2$$

$$= \frac{3 \log 3}{4} - 2 \log 2 + \log 2$$

$$= \frac{3 \log 3}{4} - \log 2$$

(iii) $\displaystyle \int_0^1 x\, e^x\, dx = \left[xe^x\right]_0^1 - \int_0^1 1 \cdot e^x\, dx = 1 \cdot e^1 - 0 \cdot e^0 - [e^x]_0^1$

$$= e - (e^1 - e^0) = e - e + 1 = 1$$

Example 5.100

Evaluate:

(i) $\displaystyle \int_0^{\pi} \frac{1}{5 + 4\cos x}\,dx$ (ii) $\displaystyle \int_0^{\pi/2} \frac{1}{2\cos x + 4\sin x}\,dx$

SOLUTION :

(i) Let $\displaystyle I = \int_0^{\pi} \frac{1}{5 + 4\cos x}\,dx = \int_0^{\pi} \frac{dx}{5 + 4\left(\frac{1-\tan^2 \frac{x}{2}}{1+\tan^2 \frac{x}{2}}\right)}$

$$= \int_0^{\pi} \frac{\left(1 + \tan^2 \frac{x}{2}\right)}{5 + 5\tan^2 \frac{x}{2} + 4 - 4\tan^2 \frac{x}{2}}\,dx$$

$$= \int_0^{\pi} \frac{\sec^2 \frac{x}{2}}{9 + \tan^2 \frac{x}{2}}\,dx$$

Let $\tan \frac{x}{2} = t \Rightarrow \sec^2 \frac{x}{2} \cdot \frac{1}{2}\,dx = dt$

when $x = 0$, $t = \tan 0 = 0$

when $x = \pi$, $t = \tan \frac{\pi}{2} = \infty$

$\therefore \qquad I = \int_0^{\infty} \frac{2dt}{9+t^2} = \frac{2}{3}\left[\tan^{-1}\frac{t}{3}\right]_0^{\infty} = \frac{2}{3}\,[\tan^{-1}\infty - \tan^{-1}0]$

$$= \frac{2}{3}\left(\frac{\pi}{2} - 0\right) = \frac{\pi}{3}$$

(ii) Let $\qquad I = \int_0^{\pi/2} \frac{1}{2\cos x + 4\sin x}\,dx$

$$= \int_0^{\pi/2} \frac{1}{2\left(\frac{1-\tan^2\frac{x}{2}}{1+\tan^2\frac{x}{2}}\right) + 4\left(\frac{2\tan\frac{x}{2}}{1+\tan^2\frac{x}{2}}\right)}\,dx$$

$$= \int_0^{\pi/2} \frac{1+\tan^2\frac{x}{2}}{2 - 2\tan^2\frac{x}{2} + 8\tan\frac{x}{2}}\,dx$$

$$= \frac{1}{2}\int_0^{\pi/2} \frac{sec^2\frac{x}{2}}{1 - \tan^2\frac{x}{2} + 4\tan\frac{x}{2}}\,dx$$

Let $\tan\frac{x}{2} = t \;\Rightarrow\; \sec^2\frac{x}{2}.\frac{1}{2}\,dx = dt$

when $x = 0,\ t = \tan 0 = 0$

when $x = \pi/2,\ t = \tan\frac{\pi}{4} = 1$

$\therefore \qquad I = \frac{1}{2}\int_0^1 \frac{2\,dt}{1-t^2+4t} = \int_0^1 \frac{dt}{1+4t-t^2} = -\int_0^1 \frac{dt}{t^2-4t-1}$

$$= -\int_0^1 \frac{dt}{(t-2)^2-5} = -\int_0^1 \frac{dt}{(t-2)^2-(\sqrt{5})^2}$$

$$= \int_0^1 \frac{dt}{(\sqrt{5})^2-(t-2)^2}$$

$$= \frac{1}{2\sqrt{5}}\left[\log\left(\frac{\sqrt{5}+t-2}{\sqrt{5}-t+2}\right)\right]_0^1$$

$$= \frac{1}{2\sqrt{5}}\left[\log\left(\frac{\sqrt{5}-1}{\sqrt{5}+1}\right) - \log\left(\frac{\sqrt{5}-2}{\sqrt{5}+2}\right)\right]$$

$$= \frac{1}{2\sqrt{5}}\left[\frac{(\sqrt{5}-1)(\sqrt{5}+2)}{(\sqrt{5}+1)(\sqrt{5}-2)}\right] = \frac{1}{2\sqrt{5}}\log\left[\frac{5-\sqrt{5}+2\sqrt{5}-2}{5+\sqrt{5}-2\sqrt{5}-2}\right]$$

$$= \frac{1}{2\sqrt{5}}\log\left(\frac{3+\sqrt{5}}{3-\sqrt{5}}\right)$$

Example 5.101

Evaluate : (i) $\int_0^{\pi} |\cos x|\, dx$ (ii) $\int_0^{\pi/2} \frac{\sin x}{\sin x + \cos x}\, dx = \frac{\pi}{4}$

SOLUTION : (i) We know that $|\cos x| = \begin{cases} \cos x & \text{if } 0 \le x \le \frac{\pi}{2} \\ -\cos x & \text{if } \frac{\pi}{2} \le x \le \pi \end{cases}$

$$\therefore \quad \int_0^{\pi} |\cos x|\, dx = \int_0^{\pi/2} \cos x\, dx - \int_{\pi/2}^{\pi} \cos x\, dx$$

$$= [\sin x]_0^{\pi/2} - [\sin x]_{\pi/2}^{\pi}$$

$$= \left(\sin\frac{\pi}{2} - \sin 0\right) - \left(\sin \pi - \sin\frac{\pi}{2}\right)$$

$$= 1 - 0 - 0 + 1 = 2$$

(ii) Let $I = \int_0^{\pi/2} \frac{\sin x}{\sin x + \cos x}\, dx$ ...(1)

Using $\int_0^a f(x)\, dx = \int_0^a f(a-x)\, dx$ ...[Prop 4]

$$I = \int_0^{\pi/2} \frac{\sin(\pi/2 - x)}{\sin(\pi/2 - x) + \cos(\pi/2 - x)}\, dx$$

$$= \int_0^{\pi/2} \frac{\cos x}{\cos x + \sin x}\, dx \qquad \text{...(2)}$$

On adding 1 and 2, we get

$$2I = \int_0^{\pi/2} \frac{\sin x + \cos x}{\sin x + \cos x}\, dx = \int_0^{\pi/2} dx = [x]_0^{\pi/2} = \frac{\pi}{2} - 0$$

$$I = \frac{\pi}{4}$$

Example 5.102

Evaluate : (i) $\int_0^{\pi/2} \frac{\sqrt{\sin x}}{\sqrt{\sin x} + \sqrt{\cos x}}\, dx$, (ii) $\int_0^{\pi/2} \log \tan x\, dx$

SOLUTION : (i) Let $I = \int_0^{\pi/2} \frac{\sqrt{\sin x}}{\sqrt{\sin x} + \sqrt{\cos x}}\, dx$ ...(1)

$$= \int_0^{\pi/2} \frac{\sqrt{\sin(\pi/2 - x)}}{\sqrt{\sin(\pi/2 - x)} + \sqrt{\cos(\pi/2 - x)}}\, dx \qquad \text{[using prop.4]}$$

$$I = \int_0^{\pi/2} \frac{\sqrt{\cos x}}{\sqrt{\cos x} + \sqrt{\sin x}}\, dx \qquad \text{...(2)}$$

Adding (1) and (2), we get

$$2I = \int_0^{\pi/2} \frac{\sqrt{\sin x} + \sqrt{\cos x}}{\sqrt{\sin x} + \sqrt{\cos x}}\, dx$$

$$= \int_0^{\pi/2} dx = [x]_0^{\pi/2} = \frac{\pi}{2} - 0 = \frac{\pi}{2}$$

$\therefore$ $$I = \frac{\pi}{4}$$

(ii) Let $$I = \int_0^{\pi/2} \log \tan x \, dx \qquad \text{...(i)}$$

$$= \int_0^{\pi/2} \log \tan\left(\frac{\pi}{2} - x\right) dx \qquad \text{[using prop.4]}$$

$$I = \int_0^{\pi/2} \log \cot x \qquad \text{...(ii)}$$

Adding (i) and (ii), we get

$$2I = \int_0^{\pi/2} (\log \tan x + \log \cot x)\, dx$$

$$= \int_0^{\pi/2} \log(\tan x \cdot \cot x)\, dx$$

$$= \int_0^{\pi/2} \log 1 \, dx = \int_0^{\pi/2} 0 \, dx = 0$$

$\therefore$ $$I = 0$$

Example 5.103

Evaluate :

(i) $\displaystyle\int_0^{\pi/2} \frac{\sin x - \cos x}{1 + \sin x \cos x}\, dx$ (ii) $\displaystyle\int_0^{\pi} \frac{x \sin x}{1 + \cos^2 x}\, dx$

SOLUTION :

(i) Let $$I = \int_0^{\pi/2} \frac{\sin x - \cos x}{1 + \sin x \cos x}\, dx \qquad \text{...(1)}$$

$$= \int_0^{\pi/2} \frac{\sin\left(\frac{\pi}{2} - x\right) - \cos\left(\frac{\pi}{2} - x\right)}{1 + \sin\left(\frac{\pi}{2} - x\right)\cos\left(\frac{\pi}{2} - x\right)}\, dx \qquad \text{[using prop. 4]}$$

$$I = \int_0^{\pi/2} \frac{\cos x - \sin x}{1 + \cos x \sin x} \qquad \text{...(2)}$$

Adding (1) and (2), we get

$$2I = \int_0^{\pi/2} \frac{\sin x - \cos x + \cos x - \sin x}{1 + \sin x \cos x}\, dx$$

$$= \int_0^{\pi/2} 0\, dx = 0$$

$$\therefore \qquad I = 0$$

(ii) Let $\qquad I = \int_0^{\pi} \frac{x \sin x}{1 + \cos^2 x}$ ...(1)

$$= \int_0^{\pi} \frac{(\pi - x) \sin (\pi - x)}{1 + \cos^2 (\pi - x)}\, dx \qquad \text{[using prop. 4]}$$

$$I = \int_0^{\pi} \frac{(\pi - x) \sin x}{1 + \cos^2 x} \qquad \text{...(2)}$$

On adding (1) and (2), we get

$$2I = \int_0^{\pi} \frac{(\pi - x + x) \sin x}{1 + \cos^2 x}\, dx$$

$$= \pi \int_0^{\pi} \frac{\sin x}{1 + \cos^2 x}\, dx$$

$$\therefore \qquad I = \frac{\pi}{2} \int_0^{\pi} \frac{\sin x}{1 + \cos^2 x}\, dx$$

Let $\qquad \cos x = t \Rightarrow -\sin x\, dx = dt$

when $x = 0$, then $t = \cos 0 = 1$, when $x = \pi$, then $t = \cos \pi = -1$

$$\therefore \qquad I = \frac{\pi}{2} \int_1^{-1} \frac{-dt}{1 + t^2} = -\frac{\pi}{2} (\tan^{-1} t)_1^{-1}$$

$$= -\frac{\pi}{2}\left(\tan^{-1}(-1) - \tan^{-1}(1)\right) = -\frac{\pi}{2}\left(-\frac{\pi}{4} - \frac{\pi}{4}\right) = \frac{\pi^2}{4}$$

Example 5.104

Evaluate:

(i) $\int_0^{\pi/2} \log \sin x\, dx$ $\qquad$ (ii) $\int_{-\pi/2}^{\pi/2} \sin^2 x\, dx$

SOLUTION : (i) Let $\quad I = \int_0^{\pi/2} \log \sin dx$ ...(1)

$$= \int_0^{\pi/2} \log \sin\left(\frac{\pi}{2} - x\right) dx \qquad \left[\text{using } \int_0^a f(x)dx = \int_0^a f(a-x)dx\right]$$

$$I = \int_0^{\pi/2} \log \cos x \, dx \qquad \text{...(2)}$$

Adding (1) and (2), we get

$$2I = \int_0^{\pi/2} (\log \sin x + \log \cos x)\, dx$$

$$= \int_0^{\pi/2} \log(\sin x \cos x)\, dx$$

$$= \int_0^{\pi/2} \log\left(\frac{2\sin x \cos x}{2}\right) dx$$

$$= \int_0^{\pi/2} \log\left(\frac{\sin 2x}{2}\right) dx$$

$$= \int_0^{\pi/2} \log \sin 2x \, dx - \int_0^{\pi/2} \log 2 \, dx \qquad \text{...(3)}$$

consider, $\int_0^{\pi/2} \log \sin 2x \, dx$

Let $2x = t \Rightarrow 2dx = dt$

when $x = 0, t = 0$, when $x = \pi/2$, $t = \pi$

$$\therefore \quad \int_0^{\pi/2} \log \sin 2x \, dx = \frac{1}{2}\int_0^{\pi} \log \sin t \, dt \qquad [\because \log \sin(\pi - t) = \log \sin t]$$

$$\because \quad \frac{1}{2}\int_0^{\pi} \log \sin t \, dt = \int_0^{\pi/2} \log \sin t \, dt \qquad \text{[Prop. 6]}$$

Thus $\int_0^{\pi/2} \log \sin 2x \, dx = \int_0^{\pi/2} \log \sin x \, dx$

From (3),

$$2I = \int_0^{\pi/2} \log \sin x \, dx - \int_0^{\pi/2} \log 2 \, dx$$

$$= I - \log 2 \, [x]_0^{\pi/2}$$

$$\therefore \quad I = -\frac{\pi}{2} \log 2$$

(ii) Here $f(x) = \sin^2 x \Rightarrow f(-x) = \sin^2(-x) = (-\sin x)^2 = \sin^2 x = f(x)$

$\therefore \quad f(x) = \sin^2 x$ is an even function.

Thus,

$$\int_{-\pi/2}^{\pi/2} \sin 2x\, dx = 2\int_{0}^{\pi/2} \sin^2 x\, dx = 2I \qquad \text{[using prop. 5]}$$

where,

$$I = \int_{0}^{\pi/2} \sin^2 x\, dx \qquad ...(1)$$

$$= \int_{0}^{\pi/2} \sin^2 (\pi/2 - x) dx$$

$$= \int_{0}^{\pi/2} \cos^2 x\, dx \qquad ...(2)$$

Adding (1) and (2), we get

$$2I = \int_{0}^{\pi/2} (\sin^2 + \cos^2 x)\, dx = \int_{0}^{\pi/2} dx = [x]_0^{\pi/2} = \frac{\pi}{2}$$

$$\therefore \qquad I = \frac{\pi}{4}$$

Hence, $\int_{-\pi/2}^{\pi/2} \sin^2 x\, dx = \frac{\pi}{2}$

Example 5.105

Evaluate

$$\int_{0}^{\pi/2} \frac{\sin x \cos x}{\cos^2 x + 3\cos x + 2} dx \qquad \textit{[UPTU 2003]}$$

SOLUTION : We have

$$\int_{0}^{\pi/2} \frac{\sin x \cos x}{\cos^2 x + 3\cos x + 2} dx$$

Let $\cos x = t \Rightarrow -\sin x\, dx = dt$

At $x = 0$, $t = 1$ at $x = \frac{\pi}{2}$, $t = 0$

$$\therefore \quad \int_{0}^{\pi/2} \frac{\sin x \cos x}{\cos^2 x + 3\cos x + 2} dx = -\int_{1}^{0} \frac{t\,dt}{t^2 + 3t + 2} = \int_{0}^{1} \frac{t}{(t+2)(t+1)} dt$$

Now, $$\frac{t}{(t+2)(t+1)} = \frac{A}{(t+2)} + \frac{B}{t+1} \qquad ...(1)$$

$$\Rightarrow \qquad t = A(t+1) + B(t+2) \qquad ...(2)$$

Putting $t = -2$ in (2), we get $-2 = -A \Rightarrow A = 2$

Putting $t = -1$ in (2), we get $-1 = B \Rightarrow B = -1$

From Eq. (1), we obtain

$$\frac{t}{(t+2)(t+1)} = \frac{2}{t+2} - \frac{1}{t+1}$$

$$\therefore \quad \int_0^1 \frac{t}{(t+2)(t+1)} dt = 2\int_0^1 \frac{dt}{t+2} - \int_0^1 \frac{dt}{t+1}$$

$$= 2\,[\log(t+2)]_0^1 - [\log(t+1)]_0^1$$

$$= 2\,[\log 3 - \log 2] - [\log 2 - \log 1]$$

$$= 2\log 3 - 2\log 2 - \log 2 \qquad [\because \log 1 = 0]$$

$$= 2\log 3 - 3\log 2$$

Example 5.106

Evaluate $\int_{\pi/4}^{\pi/2} \cos\theta \operatorname{cosec}^2\theta \, d\theta$

[UPTU 2001]

SOLUTION : Given

$$\int_{\pi/4}^{\pi/2} \cos\theta \operatorname{cosec}^2\theta \, d\theta = \int_{\pi/4}^{\pi/2} \frac{\cos\theta}{\sin^2\theta} d\theta$$

Put $\sin\theta = t \Rightarrow \cos\theta \, d\theta = dt$

At $\theta = \frac{\pi}{4},\ t = \sin\frac{\pi}{4} = \frac{1}{\sqrt{2}}$

$\theta = \frac{\pi}{2},\ t = \sin\frac{\pi}{2} = 1$

$$\therefore \quad \int_{\pi/4}^{\pi/2} \frac{\cos\theta}{\sin^2\theta} d\theta = \int_{1/\sqrt{2}}^{1} \frac{dt}{t^2} = \left[-\frac{1}{t}\right]_{1/\sqrt{2}}^{1} = -[1-\sqrt{2}] = \sqrt{2} - 1$$

Example 5.107

Evaluate $\int_0^a \frac{\sqrt{x}\,dx}{\sqrt{x} + \sqrt{a-x}}$

SOLUTION : We know that $\int_0^a f(x) = \int_0^a f(a-x)\,dx$

Let $$I = \int_0^a \frac{\sqrt{x}\,dx}{\sqrt{x} + \sqrt{a-x}} \qquad \text{...(1)}$$

$$\Rightarrow \qquad I = \int_0^a \frac{\sqrt{a-x}}{\sqrt{a-x}+\sqrt{x}}\,dx \qquad ...(2)$$

Adding (1) and (2), we get

$$2I = \int_0^a \frac{\sqrt{x}+\sqrt{a-x}}{\sqrt{x}+\sqrt{a-x}}\,dx = \int_0^a dx = [x]_0^a = a$$

$$\therefore \qquad I = \frac{a}{2}$$

PROBLEM SET 5.14

Evaluate :

1. $\int_2^3 (e^{2x} + e^x + 2x)\,dx$

2. $\int_0^{\pi/4} \tan^2 x\,dx$

3. $\int_0^{\pi} \sin^3 x\,dx$

4. $\int_0^{\pi/4} \sqrt{1-\sin 2x}\,dx$

5. $\int_0^{\pi/2} \sqrt{1-\cos 2x}\,dx$

6. $\int_0^1 \frac{e^{-x}}{1+e^x}\,dx$

7. $\int_1^2 \frac{\log x}{x^2}\,dx$

8. $\int_0^{\pi/2} x \sin x\,dx$

9. $\int_1^3 \frac{1}{x^2(x+1)}\,dx$

10. $\int_0^{\sqrt{2}} \sqrt{2-x^2}\,dx$

11. $\int_0^{\pi/6} \cos x \cos 2x\,dx$

12. $\int_0^{\pi/2} x^2 \cos^2 x\,dx$

13. $\int_0^{\pi/2} \frac{\cos x}{(1+\sin x)(2+\sin x)}\,dx$

14. $\int_0^{\frac{1}{\sqrt{2}}} \frac{\sin^{-1} x}{(1-x^2)^{3/2}}\,dx$

15. $\int_0^{\pi/4} \tan^3 x\,dx$

16. $\int_0^{\pi/2} \frac{1}{3+2\cos x}\,dx$

17. $\int_0^{\pi/2} \frac{\sin x}{1+\cos^2 x}\,dx$

18. $\int_0^{\pi/2} \frac{\cos x}{1+\cos x+\sin x}\,dx$

19. $\int_0^{\pi} \frac{1}{5+3\cos x}\,dx$

20. $\int_{-1}^{1} e^{|x|}\,dx$

21. $\int_0^{\pi/4} \log(1 + \tan x)\, dx$

22. $\int_0^{\pi/2} \frac{\sin x - \cos x}{1 + \sin x \cos x}\, dx$

23. $\int_0^{\pi/2} \sin 2x \log \tan x\, dx$

24. $\int_0^{\pi} \frac{x}{1 + \sin x}\, dx$

25. $\int_{-\pi/2}^{\pi/2} \sin^2 x\, dx$

26. $\int_0^{\pi/2} \log \cos x\, dx$

27. $\int_0^{\pi/2} \frac{dx}{1 + \tan x}$

28. $\int_0^{\pi/2} \frac{\sqrt{\cot x}}{\sqrt{\cot x} + \sqrt{\tan x}}\, dx$

29. $\int_0^{\pi/2} \frac{\sin^n x}{\sin^n x + \cos^n x}\, dx$

30. $\int_0^{\pi} \frac{x \tan x}{\sec x + \tan x}\, dx$

31. $\int_0^{\pi} \frac{x \sin x}{1 + \sin x}\, dx$

32. $\int_0^{\pi} x \log \sin x\, dx$

33. $\int_0^{\pi} \frac{x}{a^2 \cos^2 x + b^2 \sin^2 x}\, dx$

34. $\int_0^{\pi/2} (\sqrt{\tan x} + \sqrt{\cot x})\, dx$

35. $\int_0^{\pi/2} \log \cot x\, dx$

36. $\int_0^{\pi/2} \left(\sin \frac{x}{2} + \cos \frac{x}{2}\right)^2 dx$

37. $\int_0^{\pi/2} \frac{dx}{1 + \sqrt{\tan x}}$

ANSWERS

1. $\frac{e^6 - e^4}{2} + e^3 - e^2 + 5$ 2. $1 - \frac{\pi}{4}$ 3. $\frac{4}{3}$ 4. $\sqrt{2} - 1$ 5. $\sqrt{2}$ 6. $\log\left(\frac{1+e}{2e}\right) - \frac{1}{e} + 1$

7. $\frac{1}{2}\log\frac{e}{2}$ 8. 1 9. $\log\frac{2}{3} + \frac{2}{3}$ 10. $\frac{\pi}{2}$ 11. $\frac{5}{12}$ 12. $\frac{\pi^3}{48} - \frac{\pi}{8}$ 13. $\log\frac{4}{3}$

14. $\frac{\pi}{4} - \frac{1}{2}\log 2$ 15. $\frac{1}{2}(1 - \log 2)$ 16. $\frac{2}{\sqrt{5}}\tan^{-1}\frac{1}{\sqrt{5}}$ 17. $\frac{\pi}{4}$ 18. $\frac{1}{2}\left(\frac{\pi}{2} - \log 2\right)$

19. $\frac{\pi}{4}$ 20. $2e - 2$ 21. $\frac{\pi}{8}\log 2$ 22. 0 23. 0 24. π 25. $\frac{\pi}{2}$ 26. $-\frac{\pi}{2}\log 2$

27. $\frac{\pi}{4}$ 28. $\frac{\pi}{4}$ 29. $\frac{\pi}{4}$ 30. $\frac{\pi}{2}(\pi - 2)$ 31. $\pi\left(\frac{\pi}{2} - 1\right)$ 32. $-\frac{\pi^2}{2}\log 2$ 33. $\frac{\pi^2}{2ab}$

34. $\sqrt{2}\,\pi$ 35. 0 36. $\frac{\pi}{2} + 1$ 37. $\frac{\pi}{4}$

MULTIPLE CHOICE QUESTIONS

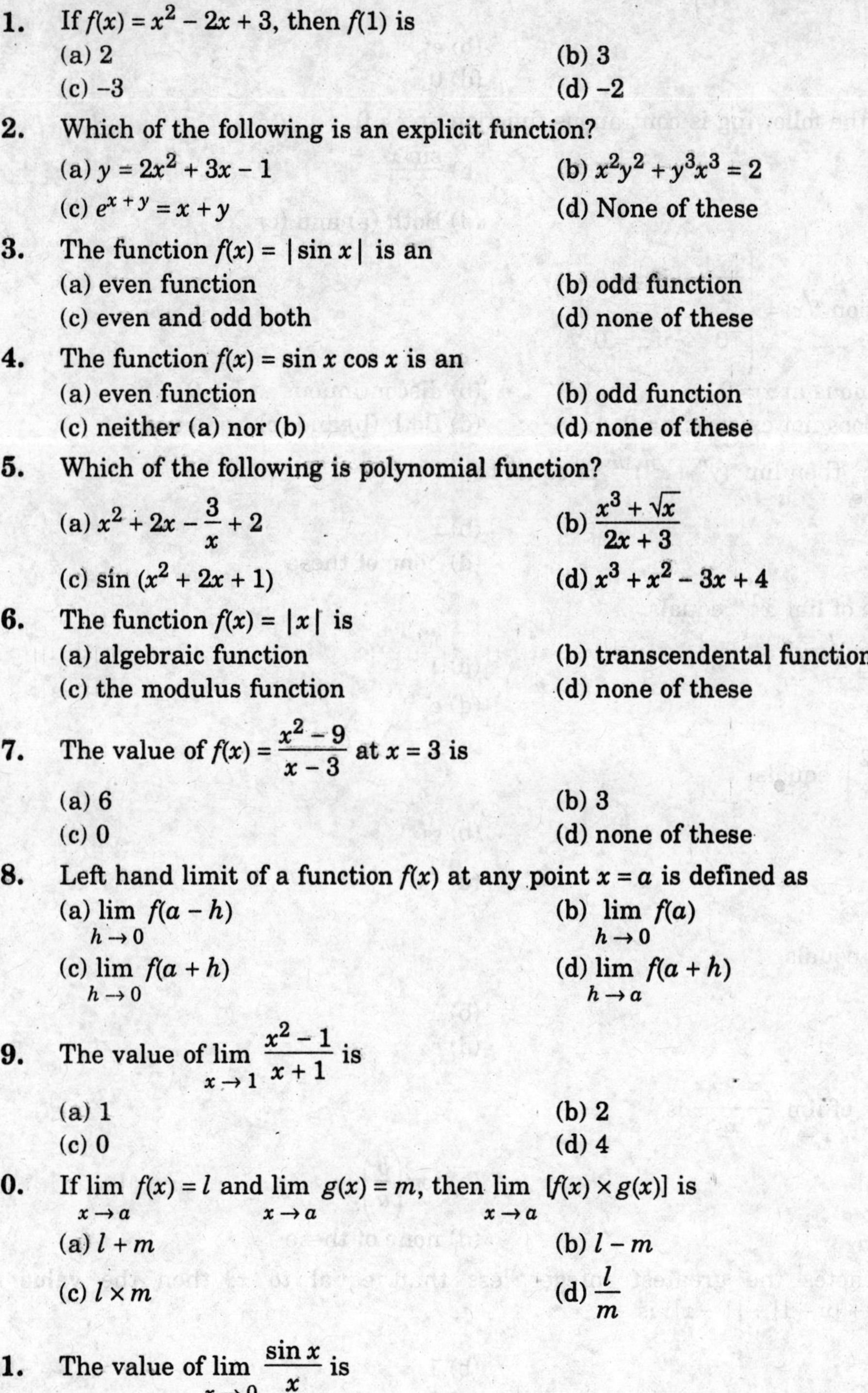

1. If $f(x) = x^2 - 2x + 3$, then $f(1)$ is

(a) 2 (b) 3
(c) –3 (d) –2

2. Which of the following is an explicit function?

(a) $y = 2x^2 + 3x - 1$ (b) $x^2y^2 + y^3x^3 = 2$
(c) $e^{x+y} = x + y$ (d) None of these

3. The function $f(x) = |\sin x|$ is an

(a) even function (b) odd function
(c) even and odd both (d) none of these

4. The function $f(x) = \sin x \cos x$ is an

(a) even function (b) odd function
(c) neither (a) nor (b) (d) none of these

5. Which of the following is polynomial function?

(a) $x^2 + 2x - \frac{3}{x} + 2$ (b) $\frac{x^3 + \sqrt{x}}{2x + 3}$
(c) $\sin(x^2 + 2x + 1)$ (d) $x^3 + x^2 - 3x + 4$

6. The function $f(x) = |x|$ is

(a) algebraic function (b) transcendental function
(c) the modulus function (d) none of these

7. The value of $f(x) = \frac{x^2 - 9}{x - 3}$ at $x = 3$ is

(a) 6 (b) 3
(c) 0 (d) none of these

8. Left hand limit of a function $f(x)$ at any point $x = a$ is defined as

(a) $\lim_{h \to 0} f(a - h)$ (b) $\lim_{h \to 0} f(a)$
(c) $\lim_{h \to 0} f(a + h)$ (d) $\lim_{h \to a} f(a + h)$

9. The value of $\lim_{x \to 1} \frac{x^2 - 1}{x + 1}$ is

(a) 1 (b) 2
(c) 0 (d) 4

10. If $\lim_{x \to a} f(x) = l$ and $\lim_{x \to a} g(x) = m$, then $\lim_{x \to a} [f(x) \times g(x)]$ is

(a) $l + m$ (b) $l - m$
(c) $l \times m$ (d) $\frac{l}{m}$

11. The value of $\lim_{x \to 0} \frac{\sin x}{x}$ is

(a) 1 (b) ∞
(c) 0 (d) 2

12. The value of $\lim_{x \to \infty} \left(1 + \frac{1}{x}\right)^x$ is

(a) e^x (b) e^2
(c) e (d) 0

13. Which of the following is continuous function at $x = 0$

(a) $|x|$ (b) $\frac{\sin x}{x}$
(c) $\sin x$ (d) Both (a) and (c)

14. The function $f(x) = \begin{cases} \frac{|x|}{x} & \text{if } x \neq 0 \\ 0 & \text{if } x = 0 \end{cases}$ is

(a) continuous at $x = 0$ (b) discontinuous at $x = 0$
(c) limit does not exist at $x = 0$ (d) Both (b) and (c)

15. If $0 < x < y$, then $\lim_{n \to \infty} (y^n + x^n)^{1/n}$ is equal to

(a) e (b) x
(c) y (d) none of these

16. The value of $\lim_{n \to \infty} x^{1/n}$ equals

(a) 0 (b) 1
(c) e (d) e^{-1}

17. $\lim_{n \to \infty} \left(1 + \frac{2}{x}\right)^x$ equals

(a) e (b) ∞
(c) e^2 (d) $\frac{1}{e}$

18. $\lim_{x \to \infty} \frac{\sin x}{x}$ equals

(a) 0 (b) 1
(c) ∞ (d) e

19. The value of $\lim_{x \to 0} \frac{a^x - b^x}{x}$ is

(a) $\log\left(\frac{a}{b}\right)$ (b) $\log\left(\frac{b}{a}\right)$
(c) $\log ab$ (d) none of these

20. If $[x]$ denotes the greatest integer less than equal to x; then the value of $\lim_{x \to 1} (1 - x + [x - 1] + [1 - x])$ is

(a) 0 (b) 1
(c) -1 (d) 2

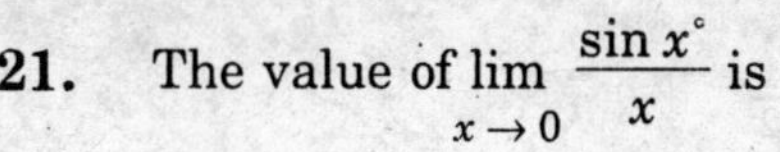

21. The value of $\lim\limits_{x \to 0} \dfrac{\sin x^\circ}{x}$ is

(a) 1 (b) π

(c) 0 (d) $\dfrac{\pi}{180}$

22. Let $f(x) = |x| + |x - 1|$, then
(a) $f(x)$ is continuous at $x = 0$, as well as at $x = 1$
(b) $f(x)$ is continuous at $x = 0$, but not at $x = 1$
(c) $f(x)$ is continuous at $x = 1$, but not at $x = 0$
(d) none of these

23. If $f(x) = (x + 1)^{\cot x}$ is continuous at $x = 0$, then $f(0)$ is equal to
(a) 0 (b) e
(c) $\dfrac{1}{e}$ (d) none of these

24. If $\lim\limits_{x \to c} \dfrac{f(x) - f(c)}{x - c}$ exists finitely, then
(a) $\lim\limits_{x \to c} f(x) = f(c)$ (b) $\lim\limits_{x \to c} f'(c) = f'(c)$
(c) $\lim\limits_{x \to c} f(x)$ does not exist (d) $\lim\limits_{x \to c} f(x)$ may or may not exist

25. Let $f(x)$ be a differentiable function at $x = c$, then $\lim\limits_{x \to c} f(x)$ equals
(a) $f'(c)$ (b) $f''(c)$
(c) $\dfrac{1}{f(c)}$ (d) none of these

26. The function $f(x) = |x| + |x - 1|$ is
(a) continuous at $x = 1$, but not differentiable
(b) both continuous and differentiable at $x = 1$
(c) not continuous at $x = 1$
(d) none of these

27. The function $f(x) = \begin{cases} x^n \sin \dfrac{1}{x} & x \neq 0 \\ 0 & x = 0 \end{cases}$

is continuous but not differentiable at $x = 0$ if
(a) $0 < n \leq 1$ (b) $1 \leq n < \infty$
(c) $-\infty < n \leq 0$ (d) $n = 0$

28. The value of derivative $|x - 1| + |x - 3|$ at $x = 2$ is
(a) -2 (b) 0
(c) 2 (d) does not exist

29. If $y = x + e^x$, then $\dfrac{dx}{dy}$ is

(a) e^x (b) $-\dfrac{e^x}{1 + e^x}$

(c) $\frac{1}{1+e^x}$ (d) $\frac{1}{(1+e^x)^2}$

30. If $2^x + 2^y = 2^{x+y}$, then the value of $\frac{dy}{dx}$ at $x = y = 1$

(a) 0 (b) –1

(c) 1 (d) 2

31. The expression of $\frac{dy}{dx}$ of the function $y = a^{x^{a^{x \dots \infty}}}$

(a) $\frac{y^2}{x(1 - y \log x)}$ (b) $\frac{y^2 \log y}{x(1 - y \log x)}$

(c) $\frac{y^2 \log y}{x(1 - y \log x \log y)}$ (d) $\frac{y^2 \log y}{x(1 + y \log x \log y)}$

32. If $x^y = e^{x-y}$, the $\frac{dy}{dx}$ is equal to

(a) $(1 + \log x)^{-1}$ (b) $(1 + \log x)^{-2}$

(c) $\log x\, (1 + \log x)^{-2}$ (d) None of these

33. The differential coefficient of $y = \log |x|$ is

(a) $\frac{1}{x}, x > 0$ (b) $\frac{1}{|x|}\, x \neq 0$

(c) $\frac{1}{x}, x \neq 0$ (d) None of these

34. If $x = a \cos \theta, y = b \sin \theta$, then $\frac{d^3y}{dx^3}$ equals

(a) $-\frac{3b}{a^3} \operatorname{cosec}^4 \theta \cot^4 \theta$ (b) $\frac{3b}{a^3} \operatorname{cosec}^4 \theta \cot \theta$

(c) $-\frac{3b}{a^3} \operatorname{cosec}^4 \theta \cot \theta$ (d) None of these

35. If $y = \sin x + e^x$, then $\frac{d^2y}{dx^2}$ equals

(a) $(-\sin x + e^x)^{-1}$ (b) $\frac{\sin x - e^x}{(\cos x + e^x)^2}$

(c) $\frac{\sin x - e^x}{(\cos x + e^x)^3}$ (d) $\frac{\sin x + e^x}{(\cos x + e^x)^3}$

36. If $\sqrt{x+y} + \sqrt{y-x} = c$, then $\frac{d^2y}{dx^2}$ is equal to

(a) $\frac{2}{c}$ (b) $\frac{2}{c^2}$

(c) $-\frac{2}{c^2}$ (d) $\frac{2}{c^3}$

37. $\int a^x\, dx$ is equal to

(a) $a^x + c$ (b) $a^x \log_e a + c$

(c) $\dfrac{a^x}{\log_e a}$ (d) $a^x + \log_e a + c$

38. $\int \frac{1}{x}\, dx$ is equal to

(a) $\log x + c$ (b) $e^x + c$

(c) $x \log x + c$ (d) $x^2 \log x + c$

39. $\int \tan x\, dx$ is equal to

(a) $\log \cos x$ (b) $\log \sin x$

(c) $\log \sec x$ (d) None of these

40. The value of the $\int \sec x$ is

(a) $\log |\sec x + \tan x| - c$ (b) $\log \sec x + c$

(c) $\log \sec x \tan x + c$ (d) None of these

41. $\int \dfrac{1}{x^2(x^4+1)^{3/4}}\, dx$ is equal to

(a) $\left(1 + \dfrac{1}{x^4}\right)^{1/4} + c$ (b) $(x^4 + 1)^{1/4} + c$

(c) $\left(1 - \dfrac{1}{x^4}\right)^{1/4} + c$ (d) $-\left(1 + \dfrac{1}{x^4}\right)^{1/4} + c$

42. $\int \dfrac{x + \sin x}{1 + \cos x}\, dx$ equals

(a) $x \tan \dfrac{x}{2} + c$ (b) $\cot \dfrac{x}{2} + c$

(c) $\log (1 + \cos x)$ (d) $\log (x + \sin x) + c$

43. $\int \dfrac{\sqrt{\tan x}}{\sin x \cos x}\, dx$ is equal to

(a) $2\sqrt{\tan x} + c$ (b) $2\sqrt{\cot x} + c$

(c) $\dfrac{\sqrt{\tan x}}{2} + c$ (d) None of these

44. $\int 5^{5^{5^x}} . 5^{5^x} . 5^x\, dx$ is equal to

(a) $\dfrac{5^{5^x}}{(\log 5)^3} + c$ (b) $5^{5^{5^x}} (\log 5)^3 + c$

(c) $\dfrac{5^{5^{5^x}}}{(\log 5)^3}$ (d) None of these

45. $\int [f(x)\, g''(x) - f''(x)\, g(x)]\, dx$ is equal to

(a) $\dfrac{f(x)}{g'(x)}$ (b) $f'(x)\, g(x) - f(x)\, g'(x)$

(c) $f(x)\, g'(x) - f'(x)\, g(x)$ (d) $f(x)\, g'(x) + f'(x)\, g(x)$

46. $\int \dfrac{1}{\sqrt{\sin^3 x \cos x}}\, dx$ is equal to

(a) $\dfrac{-2}{\sqrt{\tan x}} + c$ (b) $2\sqrt{\tan x} + c$

(c) $\dfrac{2}{\sqrt{\tan x}} + c$ (d) $-2\sqrt{\tan x} + c$

47. $\int \sqrt{\dfrac{\cos x - \cos^3 x}{1 - \cos^3 x}}\, dx$ is equal to

(a) $\dfrac{2}{3} \sin^{-1} (\cos^{3/2} x) + c$ (b) $\dfrac{3}{2} \sin^{-1} (\cos^{3/2} x) + c$

(c) $\dfrac{2}{3} \cos^{-1} (\cos^{3/2} x) + c$ (d) None of these

48. $\int \dfrac{x + 2}{(x^2 + 3x + 3)\sqrt{x + 1}}$ is equal to

(a) $\dfrac{1}{\sqrt{3}} \tan^{-1}\left(\dfrac{x}{\sqrt{3(x + 1)}}\right)$ (b) $\dfrac{2}{\sqrt{3}} \tan^{-1}\left(\dfrac{x}{\sqrt{3(x + 1)}}\right)$

(c) $\dfrac{2}{\sqrt{3}} \tan^{-1}\left(\dfrac{x}{\sqrt{x + 1}}\right)$ (d) None of these

45. $\int \dfrac{(x^4 - x)^{1/4}}{x^5}\, dx$ is equal to

(a) $\dfrac{4}{15}\left(1 - \dfrac{1}{x^3}\right)^{5/4} + c$ (b) $\dfrac{4}{5}\left(1 - \dfrac{1}{x^3}\right)^{5/4} + c$

(c) $\dfrac{4}{15}\left(1 + \dfrac{1}{x^3}\right)^{5/4} + c$ (d) None of these

50. $\int \dfrac{1}{x(x^n + 1)}\, dx$ equals

(a) $\dfrac{1}{n} \log\left(\dfrac{x^n}{x^n + 1}\right) + c$ (b) $\dfrac{1}{n} \log\left(\dfrac{x^n + 1}{x^n}\right)$

(c) $\log\left(\dfrac{x^n}{x^n + 1}\right) + c$ (d) None of these

51. The value if $\int_{-1/2}^{1/2} \cos x \log\left(\frac{1+x}{1-x}\right) dx$ is equal to

(a) 0 (b) $\frac{1}{2}$

(c) $-\frac{1}{2}$ (d) 1

52. The value of the integral $\int_0^{\pi/2} \log \tan x \, dx$ is

(a) $\frac{\pi}{4}$ (b) 0

(c) $\frac{\pi}{2}$ (d) π

53. The value of $\int_0^{\pi/2} \frac{\sqrt{\sin x}}{\sqrt{\sin x}+\sqrt{\cos x}} dx$ is

(a) $\frac{\pi}{2}$ (b) 0

(c) $\frac{\pi}{4}$ (d) 1

54. The value of the integral $\int_{-1}^{1} |x| \, dx$ is

(a) 1 (b) 0
(c) –1 (d) 2

55. The value of the integral $\int_{-\pi/4}^{\pi/4} \sin^4 x \, dx$ is

(a) $-\frac{8}{3}$ (b) $\frac{3}{2}$

(c) $\frac{8}{3}$ (d) None of these

56. The value of the integral $\int_0^{\pi/2} |\sin x - \cos x| \, dx$ is

(a) 0 (b) $2(\sqrt{2}-1)$
(c) $2\sqrt{2}$ (d) $2(\sqrt{2}+1)$

57. $\frac{d}{dx}\left(\int_{f(x)}^{g(x)} \phi(t)\, dt\right)$ is equal to

(a) $\phi\, g(x) - \phi f f(x))$ (b) $\frac{1}{2}[\phi(g(x))]^2 - \frac{1}{2}[\phi(f(x))]^2$

(c) $g'(x)\,\phi(g(x) - f'(x)\,\phi(f(x))$ (d) $\phi'(g(x))\,g'(x) - \phi'(f(x))\,f'(x)$

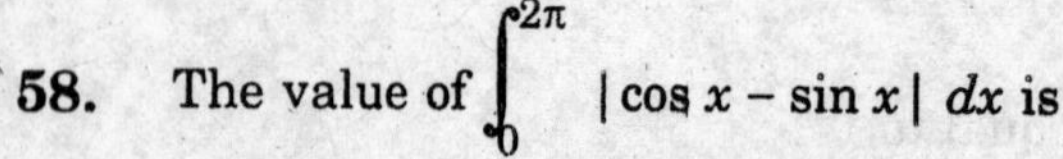

58. The value of $\int_0^{2\pi} |\cos x - \sin x|\, dx$ is

(a) $\frac{4}{\sqrt{2}}$ (b) $2\sqrt{2}$

(c) $\frac{2}{\sqrt{2}}$ (d) $4\sqrt{2}$

59. The value of $\int_0^{2\pi} \cos^{99} x\, dx$ is

(a) 1 (b) –1

(c) 99 (d) 0

60. The value of the integral $\int_{-2}^{4} x\,[x]\, dx$ is

(a) $\frac{41}{2}$ (b) 20

(c) $\frac{21}{2}$ (d) None of these

61. The value of the integral $\int_{-1}^{2} \frac{|x|}{x}\, dx$ is

(a) 0 (b) 1

(c) 3 (d) None of these

62. The value of the integral $\int_0^{1} \log \sin\left(\frac{\pi x}{2}\right) dx$ is

(a) log 2 (b) – log 2

(c) $\frac{\pi}{2} \log 2$ (d) $-\frac{\pi}{2} \log 2$

63. The value of the integral $\int_0^{\pi} x \log \sin x\, dx$ is

(a) $\frac{\pi}{2} \log 2$ (b) $\frac{\pi^2}{2} \log 2$

(c) $-\frac{\pi^2}{2} \log 2$ (d) None of these

64. The value of the integral $\int_0^{\pi/2} \sin^6 x\, dx$ is

(a) $\frac{3\pi}{4}$ (b) $\frac{5}{32}\pi$

(c) $\frac{\pi}{2\sqrt{2}}$ (d) None of these

66. The value of the integral $\int_1^e (\log x)^3\, dx$ is

(a) $6 + 2e$ (b) $6 - 2e$
(c) $2e - 6$ (d) None of these

67. The integral $\int_{-1}^{1} \frac{|x+2|}{x+2}\, dx$ is equal to

(a) 1 (b) 2
(c) 0 (d) -1

68. If $I = \int_0^1 \frac{e^x}{x+1}\, dx$, then the value of the integral $\int_0^1 \frac{x\, e^{x^2}}{x^2+1}\, dx$ is

(a) I^2 (b) $\frac{I}{2}$
(c) $2I$ (d) $\frac{I^2}{2}$

69. If $\int_0^\infty e^{-x} \sin^n x\, dx = \frac{24}{125}$, then the value of n is

(a) 3 (b) 4
(c) 5 (d) 6

70. The value of integral $\int_{-1}^{1} |1 - x|\, dx$ is equal

(a) -2 (b) 0
(c) 2 (d) 4

Fill in the blanks :

71. If $\lim_{x \to a} f(x) = l$ and k is a constant, then $\lim_{x \to a} k\, f(x) =$ ....

72. $\lim_{x \to 0} \frac{1}{x} \sin^{-1}\left(\frac{2x}{1+x^2}\right)$ is equal to ....

73. $\lim_{x \to 0} (\cos x)^{\cot^2 x}$ is equal to ....

74. If $f(x) = x \sin \frac{1}{x}$ is continuous, at $x = 0$, then $f(0)$ is equal to ....

75. The value of $f(0)$ for which the function $f(x) = \frac{2 - \sqrt{x+4}}{\sin 2x}$ is continuous at $x = 0$ is ....

76. The value of $\lim_{x \to 0} x^x$ is equal to ....

77. The point of discontinuity of the function $f(x) = |x| + |x - 1|$ is ....

78. For the function $f(x) = |\log x|$, the left hand derivative and right hand derivative of $f(x)$ at $x = 1$ are .... and ....

79. If $y = \log(x + \sqrt{1 + x^2})$, then $y_2(0)$ is equal to ....

80. If $y = \sin^{-1} x$ satisfies the equation $(1 - x^2)\, y_2 = f(x)\, y_1$, then $f(x)$ is equal to ....

81. If $f(x) = \sin(\log x)$ and $y = f\left(\frac{2x + 3}{3 - 2x}\right)$, then $\frac{dy}{dx} =$ ....

82. The derivative of the function $\sin^{-1}\left(\frac{2x}{1 + x^2}\right)$ does not exist for ....

83. The integral $\int \sqrt{\frac{x}{1 + x^3}}\, dx$ equals ....

84. $\int \frac{e^x}{\sqrt{1 - e^2}}\, dx =$ ....

85. If $\int \frac{1}{x^2 - 13x + 42}\, dx = \log\left|\frac{x - A}{x - B}\right|$, then $A =$ .... and $B =$ ....

86. The value of $\int \frac{e^{2x} - 1}{e^{2x} + 1}\, dx$ is ....

87. The value of $\int_{-1}^{1} (|x - 1| + |x|)\, dx$ is equal to ....

88. The value of $\int_{0}^{1.5} x\, [x^2]\, dx$ ....

89. The value of $\int_{0}^{1} \frac{\tan^{-1} x}{x}\, dx = \lambda \int_{0}^{\pi/2} \frac{t}{\sin t}$, then $\lambda =$ ....

90. The value of $\int_{-\infty}^{\infty} x\, e^{-x^2/2}\, dx$ is ....

True and False :

91. The function $f(x) = |\sin x|$ is an even function.

92. $f(x) = \frac{2x + 3}{4x + 7}$ is an algebraic function.

93. $\lim_{x \to a} f(x) \Leftrightarrow \lim_{x \to a^-} f(x) = \lim_{x \to a^+} f(x)$.

94. If $\lim_{x \to a} f(x)$ exists finitely, then $f(x)$ is continuous at $x = a$.

95. Every continuous function is differentiable.

96. Continuity is the necessary condition for differentiability.

97. $y^x = x^y$ is an implicit function.

98. If $f(x) = [x] - 5$, the $f'(4.5) = 1$.

99. If $f(x) = x - \frac{1}{x}$, $f(x^2) = f(x)\, f(-x)$.

100. The function $f(x) = |x| + |x - 4|$ is continuous at $x = 0$ and $x = 4$.

101. The function $f(x) = |x + 1| + |x - 2|$ is differentiable everywhere except $x = 2$.

102. If $u = ax + b$, then $\frac{d^2}{dx^2}[f(ax + b)]$ is equal to $a^n \frac{d^n}{dx^n}[f(u)]$.

103. The function $f(x) = \frac{x}{1 + |x|}$ is differentiable everywhere except $x = 0$.

104. Indefinite integration is a reverse operator of differentiation.

105. The value of $\int_a^b f(a + b - x)\, dx = \int_a^b f(x)\, dx$.

106. The value of $\int_0^{2a} \frac{f(x)}{f(x) + f(2a - x)}\, dx$ is equal to a.

107. If $f(x)$ is an odd function, then $\int_a^x f(t)\, dt$ is an even function.

108. $\int uv\, dx = u \int v\, dx - \int \left(\frac{du}{dx}\right) \int v\, dxdx + C.$

109. The value of integral $\int_0^{\pi/4} \frac{\sqrt{\tan x}}{\sin x \cos x}\, dx$ is 2.

110. The value of integral $\int_{-1}^{1} \sin^{11} x\, dx$ is 10.

Answers

Objective type questions :

1. a	**2.** a	**3.** a	**4.** b	**5.** d	**6.** c	**7.** d	**8.** c	**9.** c	**10.** c
11. a	**12.** c	**13.** d	**14.** d	**15.** c	**16.** b	**17.** c	**18.** a	**19.** a	**20.** c
21. d	**22.** a	**23.** c	**24.** a	**25.** d	**26.** a	**27.** a	**28.** b	**29.** c	**30.** b
31. c	**32.** c	**33.** c	**34.** c	**35.** c	**36.** b	**37.** c	**38.** a	**39.** b	**40.** a
41. d	**42.** a	**43.** a	**44.** c	**45.** c	**46.** b	**47.** c	**48.** b	**49.** a	**50.** a
51. a	**52.** b	**53.** c	**54.** a	**55.** a	**56.** b	**57.** c	**58.** d	**59.** d	**60.** a
61. b	**62.** b	**63.** c	**64.** b	**65.** c	**66.** b	**67.** b	**68.** b	**69.** b	**70.** c

Fill in the blanks :

71. ke, **72.** 2, **73.** $e^{-1/2}$, **74.** 0, **75.** $-\frac{1}{8}$, **76.** 1,

77. continuous everywhere, **78.** LHD = –1, RHD = 1, **79.** 0, **80.** x,

81. $\frac{12}{9-4x^2}\cos\left[\log\left(\frac{2x+3}{3-2x}\right)\right]$, **82.** $\frac{2}{\sqrt{1-x^2}}$, **83.** $\frac{2}{2}\log\left[x^{3/2}+\sqrt{1+x^3}\right]$,

84. sin–1 $(e^x)+c$, **85.** $A = 7, B = 6$, **86.** $\log(e^x + e^{-x})$, **87.** 3, **88.** $\frac{5}{8}$,

89. $\frac{1}{2}$, **90.** 0

True and False :

91. T, **92.** T, **93.** T, **94.** F, **95.** F, **96.** T, **97.** T, **98.** F, **99.** F,
100. T, **101.** F, **102.** T, **103.** F, **104.** T, **105.** T, **106.** T, **107.** T,
108. T, **109.** T, **110.** F

□ ■ □

B. PHARM

FIRST SEMESTER SPECIAL EXAMINATION, 2000—2001

REMEDIAL MATHEMATICS

Time : 3 Hours **Total Marks : 80**

Note : (1) Attempt **ALL** the questions.

(2) All questions carry equal marks.

1. Answer any FOUR of the following :

(a) Solve the following system of equations by Cramer's Rule :

$$x - 2y = 4$$
$$-3x + 5y = -7$$

(b) Find A^{-1}, if $A = \begin{bmatrix} 1 & 2 & 5 \\ 1 & -1 & -1 \\ 2 & 3 & -1 \end{bmatrix}$

(c) Prove that $\begin{vmatrix} 1 & 1 & 1 \\ a & b & c \\ a^2 & b^2 & c^2 \end{vmatrix} = (a-b)(b-c)(c-a)$

(d) Divide 33 into two parts whose product is 242.

(e) The sum of the present ages of Kamal and his mother is 89 years. After 11 years, mother's age will be twice Kamal's age. Find their present ages.

2. Answer any TWO of the following :

(a) Calculate the median for the following statistical distribution :

Value	Less than 100	100–200	200–300	300–400	400 and above
Frequency	50	90	158	68	134

(b) Give merits and demerits of Median.

(c) The arithmetic mean of the daily wages of two pharmaceutical concerns A and B are Rs. 30 and Rs. 40 respectively. If the number of workers employed in A and B were 2000 and 4000 respectively, find the average daily wage of all the workers.

3. Answer any FOUR of the following :

(a) Prove that $\dfrac{\sin A + \sin B}{\cos A + \cos B} = \tan\left(\dfrac{A+B}{2}\right)$

(b) Prove that $\sin^2 72^\circ - \sin^2 60^\circ = \dfrac{\sqrt{5}-1}{8}$.

(c) Find the seventh root of 2.965.

(d) Prove that $\dfrac{\cos\theta}{1+\sin\theta} = \tan\left(\dfrac{\pi}{4} - \dfrac{\theta}{2}\right)$.

(e) Prove that the points $(a, b+c)$, $(b, c+a)$, $(c, a+b)$ are collinear.

4. (a) Prove that $\tan^2\theta + \cot^2\theta + 2 = \sec^2\theta . \text{cosec}^2\theta$

OR

Differentiate $y = \log_3(x^2 + 3x)$ w.r.t. x.

(b) Evalute $\int \frac{\cos x}{\sin^2 x} dx$.

(c) Solve that in ΔABC,

$$\tan A + \tan B + \tan C = \tan A \tan B \tan C$$

5. Answer any TWO of the following :

(a) Find the equation of a line passing through (1, 1) and perpendicular to the straight line $3x - 4y = 6$.

(b) Find the equation of a line passing through the two points (1, 2) and (7, 8).

(c) Find the angle in between the lines

$$x - y\sqrt{3} - 5 = 0 \text{ and } \sqrt{3}x + y - 7 = 0$$

B. PHARM

FIRST SEMESTER EXAMINATION 2001-2002

REMEDIAL MATHEMATICS

Time : 3 Hours **Total Marks : 80**

NOTE : (1) Answer **ALL** the questions.
(2) All questions carry equal marks.

1. Answer any *Four* of the following : (4 × 4)

(a) If α be a root of the equation, $4x^2 + 2x - 1 = 0$, prove that $4\alpha^3 - 3\alpha$ is the other root.

(b) Without expanding the determinant, show that :

$$\begin{vmatrix} 0 & b & -c \\ -b & 0 & a \\ c & -a & 0 \end{vmatrix} = 0$$

(c) If $A = \begin{bmatrix} 4 & 1 \\ 3 & 2 \end{bmatrix}$, then prove that :

$$A^2 - 6A + 5I = \begin{bmatrix} 0 & 0 \\ 0 & 0 \end{bmatrix}, \text{ when } I = \begin{bmatrix} 1 & 0 \\ 0 & 1 \end{bmatrix}$$

(d) Find the inverse of the matrix $A = \begin{bmatrix} 2 & 3 & 4 \\ 4 & 3 & 1 \\ 1 & 2 & 4 \end{bmatrix}$

(e) Solve the following by using Cramer's Rule :

$$x - 2y + 3 = 2$$
$$2x - 3z = 3$$
$$x + y + z = 6$$

2. Answer any *Two* of the following : (8 × 2)

(a) Compute the mean of the following frequency distribution :

Class	0–11	11–22	22–33	33–44	44–55	55–66
Frequency	9	17	28	26	15	8

(b) Find the median of the following distribution :

Wages (in Rs.)	0–10	10–20	20–30	30–40	40–50
No. of Workers	22	38	46	35	20

(c) Compute the mode of the following distribution :

Class-interval	0–7	7–14	14–21	21–28	28–35	35–42	42–49
Frequency	19	25	36	72	51	43	28

3. Answer any *Four* of the following : (4 × 4)

(a) Prove that : $\sec\alpha + \tan\alpha = \tan\left(\frac{\pi}{4} + \frac{\alpha}{2}\right)$

(b) If $\sin 2A = \frac{4}{5}$, find the value of $\tan A$. $\left(0 \le A \le \frac{\pi}{4}\right)$

(c) Find the value of $\tan 22\frac{1}{2}^{\circ}$.

(d) Show that $\frac{\cos 8^{\circ} + \sin 8^{\circ}}{\cos 8^{\circ} - \sin 8^{\circ}} = \cot 37^{\circ}$.

(e) Calculate cot 75° by using cotangent formula.

4. Answer any *Two* of the following : (8 × 2)

(a) Find the equation of straight line which makes equal intercepts on the axes and passes through the point (3, –5).

(b) Find the slope and the equation of the straight line joining the points (2, –5) and (4, 1).

(c) Find the area of a triangle whose vertices are (0, 5), (2, 3) and (4, 5).

5. Answer the following : (8 × 2)

(a) Evalute : $\int \frac{(1 + \log x)^2}{x} dx$

OR

Prove that : $\int \tan x \, dx = \log \sec x$

(b) Find the Differential Coefficient of the function $\log\left(x + \frac{1}{x}\right)$.

OR

Find the value of $\int_{\pi/4}^{\pi/2} \cos\theta \operatorname{cosec}^2\theta \, d\theta$

B. PHARM

FIRST SEMESTER EXAMINATION 2002-2003

REMEDIAL MATHEMATICS

Time : 3 Hours **Total Marks : 80**

Note : (1) Answer **ALL** the questions.

(2) All questions carry equal marks.

1. Attempt any FOUR of the following : (4 × 4)

(a) Solve the equation:

$$[\sqrt{5+2\sqrt{6}}]^x + [\sqrt{6-2\sqrt{6}}]^x = 10$$

(b) For what value of k, are the points (1, 5), (k, 1) and (11, 7) collinear ? (Use determinants).

(c) Using properties of determinant prove that :

$$\begin{vmatrix} a & b & c \\ b & c & a \\ c & a & b \end{vmatrix} = (a+b+c)(ab+bc+ca-a^2-b^2-c^2)$$

(d) If $A = \begin{vmatrix} 1 & 2 & 2 \\ 2 & 1 & 2 \\ 2 & 2 & 1 \end{vmatrix}$, using $A^2 - 4A - 5I = 0$, find A^{-1}.

(e) Solve the following equations :

$$2x - y + 3z = 1$$
$$x + 2y - z = 2$$
$$5y - 5z = 3$$

Using matrix method.

2. Attempt any TWO of the following : (8 × 2)

(a) From the following particulars relating to employees of a factory, determine the number of male and female emplyees :

Total number of employee = 100
Average salary of all employees = Rs. 5,000
Average salary of male employees = Rs. 5,200
Average salary of female employees = Rs. 4,200

(b) Following is the distribution of marks obtained by 50 students in Mathematics. Calculate the median marks :

Marks more than	0	10	20	30	40	50
No. of Students	50	46	40	20	10	3

(c) Compute Mode of the following series :

Age group	0–5	5–10	10–15	15–20	20–25	25–30	30–35	35–40	40–45
Frequency	20	24	32	28	20	16	34	10	8

3. Attempt any FOUR of the following : (4 × 4)

(a) Evaluate : $\cos \frac{\pi}{5}$

(b) If $\cos \alpha + \sin \alpha = \sqrt{2} \cos \alpha$, prove that

$$\cos \alpha - \sin \alpha = \sqrt{2} \sin \alpha$$

(c) Prove that : $\sin 10^\circ \sin 30^\circ \sin 50^\circ \sin 70^\circ = \frac{1}{16}$

(d) Prove that : $\sin^2\left(\frac{\pi}{8} + \frac{\alpha}{2}\right) - \sin^2\left(\frac{\pi}{8} - \frac{\alpha}{2}\right) = \frac{1}{\sqrt{2}} \sin \alpha$

(e) Prove that : $\sin \frac{5\pi}{18} - \cos \frac{4\pi}{9} = \sqrt{3} \sin \frac{\pi}{9}$

(f) Prove that : $\tan\left(\frac{\pi}{4} - \frac{A}{2}\right) = \sqrt{\frac{1 - \sin A}{1 + \sin A}}$

4. Attempt any TWO of the following : (8×2)

(a) (i) Find the value of a when distance bettween the points $(3, a)$ and $(4, 1)$ is $\sqrt{10}$.

(ii) The line joining $(-5, 7)$ and $(0, -2)$ is perpendicular to the line joining $(1, 3)$ and $(4, x)$, then find x.

(b) Find the equation of the locus of a point which moves so that sum of its distances from $(3, 0)$ and $(-3, 0)$ is less than 9.

(c) A straight line, drawn through the point A (2, 1), makes an angle $\frac{\pi}{4}$ with +ve x-axis and intersects another line $x + 2y + 1 = 0$ at point B. Find the length AB.

5. Attempt any TWO of the following : (8×2)

(a) (i) If $y = e^{x + e^{x} + \ldots \text{ to } \infty}$, then prove that : $\frac{dy}{dx} = \frac{y}{1 - y}$

(ii) Evaluate : $\int \frac{\tan x}{\log \sec x} dx$

(b) (i) Differentiate w.r.t. x : $x^{\sin^{-1} x}$

(ii) Prove that : $\int \frac{dx}{x^2 + a^2} = \frac{1}{a} \tan^{-1}\left(\frac{x}{a}\right)$

(c) Evaluate : $\int_0^{\frac{\pi}{2}} \frac{\sin x \cos x}{\cos^2 x + 3 \cos x + 2} dx$

B. PHARM

FIRST SEMESTER EXAMINATION, 2003-2004

REMEDIAL MATHS

Time : 3 Hours **Total Marks : 80**

Note : (1) **ALL** the questions are compulsory.
(2) **ALL** the questions carry equal marks.

1. Attempt any FOUR of the following questions : **(4 × 4)**

(a) Solve $\left(x+\frac{1}{x}\right)^2-\frac{3}{2}\left(x-\frac{1}{x}\right)=4, x \neq 0$

(b) By using Cramer's rule, solve the following system of equations :

$$3x - 4y + 5z = -6$$
$$x + y - 2z = -1$$
$$2x + 3y + z = 5$$

(c) Prove that :

$$\begin{vmatrix} 1+a & 1 & 1 \\ 1 & 1+b & 1 \\ 1 & 1 & 1+c \end{vmatrix} = (abc)\left(\frac{1}{a}+\frac{1}{b}+\frac{1}{c}+1\right)$$

(d) If $[1\ 1\ x]\begin{bmatrix} 1 & 0 & 2 \\ 0 & 2 & 1 \\ 2 & 1 & 0 \end{bmatrix}\begin{bmatrix} 1 \\ 1 \\ 1 \end{bmatrix} = 0$, find x.

(e) Find the inverse of matrix A, if

$$A = \begin{bmatrix} 2 & -1 & 1 \\ -1 & 2 & -1 \\ 1 & -1 & 2 \end{bmatrix}$$

2. Attempt any FOUR of the following parts : **(4 × 4)**

(a) Prove that $\frac{\tan\theta + \sec\theta - 1}{\tan\theta - \sec\theta + 1} = \frac{1+\sin\theta}{\cos\theta}$

(b) Find the value of sin 15°.

(c) Prove that $\sin 20° \sin 40° \sin 60° \sin 80° = \frac{3}{16}$.

(d) If $A + B + C = \pi/2$, prove that

$$\cos^2 A + \cos^2 B + \cos^2 C = 2 + 2 \sin A \sin B \sin C.$$

(e) pH of a solution is given as $pH = \frac{1}{[H^+]}$, if $[H^+] = 2.57 \times 10^{-4}$ g-ion/L, find pH of solution.

3. Attempt any FOUR of the following parts : **(4 × 4)**

(a) A quadrilateral has the vertices at the points (–4, 2), (2, 6), (8, 5) and (9, –7). Show that the mid-points of the sides of the quadrilateral are the vertices of a parallelogram.

(b) Find the equation of perpendicular bisector of the line segment joining the points A (2, 3) and B (6, –5).

(c) Find the locus of a point which moves so that the sum of its distances from (3, 0) and (–3, 0) is less than 9.

(d) Find the distance between the parallel lines $3x + 4y = 12$ and $3x + 4y = 3$.

(e) Reduce $4x + 3y - 9 = 0$ to the normal form and find the distance (perpendicular distance p) from origin.

4. Attempt any FOUR of the following parts : (4 × 4)

(a) Evaluate : $\lim_{x \to a} \dfrac{\cos x - \cos a}{x - a}$

(b) If $y = e^x \cos x + (x)^{\sin x}$, find $\dfrac{dy}{dx}$.

(c) Evaluate : $\displaystyle\int \frac{x^2 \tan^{-1} x}{1 + x^2}$.

(d) Evaluate : $\displaystyle\int x^3 \sin x^2 \, dx$.

(e) Evaluate : $\displaystyle\int \frac{dx}{x - x^3}$.

5. Attempt any FOUR of the following parts : (4 × 4)

(a) The mean age (in years) of a group of 100 mem, was found to be 35. Later, it was found that during calculation, 77 years are misread as 27. Calculate the correct Arithmetic Mean of the given group.

(b) Calculate the Median of the following data :

Marks :	0–10	10–20	20–30	30–40	40–50	50–60	60–70	70–80	80–90	90–100
No. of Students :	4	9	15	20	16	14	10	6	4	2

(c) Find the Mode from the following data :

Size :	1	2	3	4	5	6	7	8	9	10
Frequency :	2	7	13	17	16	26	28	32	18	20

(d) Differentiate $\log \sin (x^2 + 1)$ with respect to x.

(e) Evaluate : $\displaystyle\int e^{2x} (-\sin x + 2 \cos x) \, dx$.

B. PHARM

FIRST SEMESTER EXAMINATION, 2004-2005

REMEDIAL MATHEMATICS

Time : 3 Hours **Total Marks : 80**

Note : (1) Attempt **ALL** questions.

(2) **ALL** questions carry equal marks.

1. Attempt *any four* of the following : (4 × 4 = 16)

(a) Solve the following equation :

$$\left(\frac{2x+1}{x-1}\right)^4 - 10\left(\frac{2x+1}{x-1}\right)^2 + 9 = 0$$

(b) A straight line passes through the point (–2, 6) and the portion of the line intercepted between the axes is divided at this point in the ratio 3 : 2. Find the equation of the line.

(c) Show that

$$\begin{vmatrix} 1 & x & x^3 \\ 1 & y & y^3 \\ 1 & z & z^3 \end{vmatrix} = (x-y)(y-z)(z-x)(x+y+z)$$

(d) If $A = \begin{bmatrix} -1 & 1 & 0 \\ 3 & -3 & 3 \\ 5 & -5 & 5 \end{bmatrix}$, $B = \begin{bmatrix} 0 & 4 & 3 \\ 2 & 4 & 6 \\ -1 & 4 & 4 \end{bmatrix}$, show that $A^2B^2 = A^2$.

(e) Solve using Matrix method, the following equations :

$$x + y = 0, y + z = 1, z + x = 3$$

2. Attempt *any two* of the following : (8 × 2 = 16)

(a) A candidate obtained the following marks in an examination in a paper of 100 marks each :

Subject	English	Maths	Physics	Chemistry	Biology
Marks (Out of 100)	48	82	70	64	60

It is agreed to give double weight to Physics and Maths as compared to other subjects. What is the Arithmetic Mean ?

(b) From the following table, calculate the Median of the cost of living index :

Cost of Living Index (Rs.)	140–150	150–160	160–170	170–180	180–190	190–200
No. of weeks	5	10	20	9	6	2

(c) Calculate the Mode of the following data :

Wages (Rs.)	below 100	100–200	200–300	300–400	400–500	500 & above
No. of weeks	8	12	25	15	10	5

3. Attempt *any four* of the following : $(4 \times 4 = 16)$

(a) Find the number of degrees in the angle subtended at the centre of radius 10 ft. by an arc of length 20 feet.

(b) Prove that the expression :

$$2(\sin^6\theta + \cos^6\theta) - 3(\sin^4\theta + \cos^4\theta)$$

is independent of the angle θ.

(c) If $A + B = \pi/4$, prove that

$$(1 + \tan A)(1 + \tan B) = 2$$

(d) Show that $\cos 20^\circ \cos 40^\circ \cos 80^\circ = \frac{1}{8}$.

(e) Prove that $\tan 13A - \tan 9A - \tan 4A = \tan 13A \tan 9A \tan 4A$.

4. Attempt *any two* of the following : $(8 \times 2 = 16)$

(a) Find the distance between the parallel lines $2x - 3y + 9 = 0$ and $4x - 6y + 1 = 0$.

(b) If $A(-1, 3)$, $B(1, -1)$ and $C(5, 1)$ are vertices of a triangle, find the length of the median through the vertex $(-1, 3)$.

(c) A point moves so that its distance from (3, 0) is twice the distance from (–3, 0). Find the equation of the locus.

5. Attempt *any two* of the following : $(8 \times 2 = 16)$

(a) (i) Evaluate $\lim\limits_{x \to 0} \frac{\sqrt{1+x} - 1}{x}$

(ii) If $f(x) = \log\left(\frac{1+x}{1-x}\right)$, show that $f(x) + f(y) = f\left(\frac{x+y}{1+xy}\right)$.

(b) (i) If $y = \sin(e \tan^{-1} x)$, find $\frac{dy}{dx}$.

(ii) If $x = at^2$ and $y = 2at$, find $\frac{dy}{dx}$.

(c) (i) Evaluate $\int \frac{dx}{x + x^2}$.

(ii) Evaluate $\int_0^{\frac{\pi}{2}} \frac{\sqrt{\sin x}}{\sqrt{\sin x} + \sqrt{\cos x}}\, dx$.

B. PHARM

FIRST SEMESTER EXAMINATION, 2005—2006

REMEDIAL MATHEMATICS

Time : 3 Hours **Total Marks : 80**

Note : (1) Answer **ALL** the questions.

(2) All questions carry equal marks.

(3) In case of numerical problems assume data wherever not provided.

1. Answer *any four* parts of the following: **(4 × 4 = 16)**

(a) Solve the following equation

$$(x+1)(x+2)(x+3)(x+4) = 120$$

(b) Prove that

$$\begin{vmatrix} 1+a & 1 & 1 \\ 1 & 1+b & 1 \\ 1 & 1 & 1+c \end{vmatrix} = abc\left(\frac{1}{a}+\frac{1}{b}+\frac{1}{c}+1\right)$$

(c) Solve the following equations (use matrix method).

$$2x - y + 3z = 1$$
$$x + 2y - z = 2$$
$$5y - 5z = 3$$

(d) Find the inverse

$$A = \begin{bmatrix} 2 & 3 & 4 \\ 4 & 3 & 1 \\ 1 & 2 & 4 \end{bmatrix}$$

(e) If the points (a, b), (a', b') and $(a - a', b - b')$ are collinear, prove that $ab' = a'b$.

(f) If $[1 \quad 1 \quad x] \begin{bmatrix} 1 & 0 & 2 \\ 0 & 2 & 1 \\ 2 & 1 & 0 \end{bmatrix} = 0$

Find the value of x.

2. Attempt *any two* parts of the following: **(8 × 2 = 16)**

(a) Compute the mean of the following frequency distribution

Class	0–10	10–20	20–30	30–40	40–50	50–60	60–70
Frequency	3	7	10	15	20	12	8

(b) Find the median of the following distribution:

Class	5–10	10–15	15–20	20–25	25–30	30–35	35–40
Frequency	12	15	18	20	12	10-	6

(c) Compute the mode of the following distribution

Class	10–20	20–30	30–40	40–50	50–60	60–70	70–80
Frequency	22	35	40	62	50	45	40

3. Attempt *any four* parts of the following: **(4 × 4 = 16)**

(a) Prove that $\sec\alpha + \tan\alpha = \tan\left(\frac{\pi}{4} + \frac{\alpha}{2}\right)$

(b) Evaluate : $\sin \frac{\pi}{5}$

(c) Prove that : $\sin 20^\circ \sin 40^\circ \sin 60^\circ \sin 80^\circ = \frac{3}{16}$

(d) If $A + B + C = \frac{\pi}{2}$, prove that

$$\cos^2 A + \cos^2 B + \cos^2 C = 2 + 2 \sin A \sin B \sin C$$

(e) Find the value of tan 15°

(f) Prove that : $\tan\left(\frac{\pi}{4} - \frac{A}{2}\right) = \sqrt{\frac{1 - \sin A}{1 + \sin A}}$

4. Attempt *any four* parts of the following: **(4 × 4 = 16)**

(a) Show that the three points (–2, 3, 5), (1, 2, 3) and (7, 0, –1) are collinear

(b) Four points $A(6, 3)$, $B(-3, 5)$, $C(4, -2)$ and $D(x, 3x)$ are given in such a way that $\frac{\Delta DBC}{\Delta ABC} = \frac{1}{2}$, find x.

(c) Find the locus of a point such that the line segments having end points (2, 0) and (–2, 0) subtend a right angle at that point.

(d) What is the value of x so that the line through $(3, x)$ and (2, 7) is parallel to the line through (–1, 4) and (0, 6).

(e) Find the equation of a line at a distance of 3 units from the origitn such that the perpendicular from the origin to the line makes an angle $\tan^{-1}\left(\frac{3}{4}\right)$ with the positive direction of x-axis.

(f) Find the equation of the straight line which divides the line joining points (2, 3) and (–5, 8) in the ratio 3 : 4 and is also perpendicular to it.

5. Attempt *any two* parts of the following: **(8 × 2 = 16)**

(a) (i) Evaluate $\lim_{x \to 0} \frac{a^x - 1}{x}$.

(ii) $f(x) = \frac{e^{\sin x} - 1}{x}$ for $x \neq 0$

$= 0$ for $x = 0$

Determine whether the function is continuous at $x = 0$.

(b) (i) Find $\frac{dy}{dx}$, if $y = \sin [\cos (\tan x)]$

(ii) Find $y = e^x + \sin x$, find $\frac{d^2y}{dx^2}$

(c) Evaluate : (i) $\int \frac{x \sin^{-1} x}{\sqrt{1 - x^2}} dx$

(ii) $\int \frac{x^3 \, dx}{(x - 1)(x - 2)(x - 3)}$

B. PHARM

CARRY OVER PAPER EXAMINATION, 2006

REMEDIAL MATHEMATICS

Time : 3 Hours **Total Marks : 80**

Note : (1) Attempt **ALL** the questions.
(2) All questions carry equal marks.
(3) Assume missing data suitably and state assumption made.
(4) Be precise in your answer.

1. Answer *any four* parts of the following: **(4 × 4 = 16)**

(a) Solve $(x^2 - 5x)^2 - 30(x^2 - 5x) - 216 = 0$

(b) Show that $\begin{vmatrix} a & a+b & a+b+c \\ 2a & 3a+2b & 4a+3b+2c \\ 3a & 6a+3b & 10a+6b+3c \end{vmatrix} = a^3$

(c) Using Cramer's rule, solve the system of equations

$$5x - y + 4z = 5$$
$$2x + 3y + 5z = 2$$
$$5x - 2y + 6z = -1$$

(d) If $A = \begin{bmatrix} \cos\theta & \sin\theta \\ -\sin\theta & \cos\theta \end{bmatrix}$, prove that $A^n = \begin{bmatrix} \cos n\theta & \sin n\theta \\ -\sin n\theta & \cos n\theta \end{bmatrix}$,

for every positive integer n.

(e) Solve the following system of equations by matrix method :

$$3x - 2y + 3z = 8$$
$$2x + y - z = 1$$
$$4x - 3y + 2z = 4$$

2. Attempt any two parts of the following : **(8 × 2 = 16)**

(a) Define and find arithmetic mean and mean deviation of the following data :

12, 3, 18, 17, 4, 9, 17, 19, 20, 15, 8, 17, 2, 3, 16, 11, 3, 1, 0, 5.

(b) Define variance and standard deviation for a set of data. Find the variance and standard deviation of the following data :

16, 58, 68, 44, 48, 45, 60, 62, 60, 50

(c) The variance of 20 observations is 5. If each observation is multiplied by a fixed number P, find the new variance of the resulting observations.

3. Attempt any four parts of the following : **(4 × 4 = 16)**

(a) Prove that $\sin(45^\circ + A)\sin(45^\circ - A) = \frac{1}{2}\cos 2A$.

(b) Prove that

$$2\cos\frac{\pi}{13}\cos\frac{9\pi}{13} + \cos\frac{3\pi}{13} + \cos\frac{5\pi}{13} = 0$$

(c) Find tan 75° and tan 15°.

(d) Prove that $\cos\frac{\pi}{9}\cos\frac{2\pi}{9}\cos\frac{3\pi}{9}\cos\frac{4\pi}{9}=\frac{1}{16}$.

(e) The population of a certain bacteria increases at the rate of 27% per week. Find the total population of the bacteria at the end of 10 week assume that initially the population of the bacteria was 10^5.

4. Attempt **any two** parts of the following : **(8 × 2 = 16)**

(a) Show that the product of the perpendicular drawn from the points $(\sqrt{a^2-b^2}, 0)$, $(-\sqrt{a^2-b^2}, 0)$ upon the line $\frac{x}{a}\cos\alpha+\frac{y}{b}\sin\alpha=1$ is b^2.

(b) Show that the three straight lines $2x-3y+5=0$, $3x+4y-7=0$ and $9x-5y+8=0$ meet in a point.

(c) Find the locus of a point moving in such a way that the sum of the squares its distances from the points $(-a, b)$ adn (a, b) is always constant.

5. Attempt **any two** parts of the following : **(8 × 2 = 16)**

(a) (i) Evaluate $\lim\limits_{x\to 0}\frac{\cos x-1+x^2/2}{x^4}$

(ii) If $f(x)=\log\frac{1+x}{1-x}$, show that $f\left(\frac{2x}{1+x^2}\right)=2f(x)$

(b) (i) If $\sin y=x\sin(a+y)$, prove that $\frac{dy}{dx}=\frac{\sin^2(a+y)}{\sin a}$

(ii) If $y=e^{a\sin^{-1}bx}$. Find $\frac{dy}{dx}$ and $\frac{d^2y}{dx^2}$.

(c) (i) Evaluate $\int_0^a\frac{\sqrt{x}\,dx}{\sqrt{x}+\sqrt{a-x}}$

(ii) Evaluate $\int\frac{3x\,dx}{(x-1)(x-2)}$.

B. PHARM

FIRST SEMESTER EXAMINATION, 2006–07

REMEDIAL MATHEMATICS

Time : 3 Hours **Total Marks : 80**

Note : (1) Answer **ALL** the questions.
(2) All questions carry equal marks.
(3) Be precise in your answer.

1. Answer *any four* parts of the following: **(4 × 4 = 16)**

(a) In a given function, if the numerator is multiplied by 3 and the denominator is reduced by 1, we get $\frac{3}{2}$. But if the numerator is increased by 12 and the denominator is multiplied by 3, we get $\frac{1}{3}$. Find the fraction.

(b) Solve $7^{x+1} + 7^{1-x} = 50$.

(c) Solve $\begin{vmatrix} x-2 & 2x-3 & 3x-4 \\ x-4 & 2x-9 & 3x-16 \\ x-8 & 2x-27 & 3x-64 \end{vmatrix} = 0$.

(d) If $A = \begin{bmatrix} 1 & 1 & 1 \\ 1 & -1 & 1 \\ 2 & 1 & -1 \end{bmatrix}$, find A^{-1}. Hence solve the equations $x + y + z = 6$, $x - y + z = 2$, $2x + y - z = 1$.

(e) Solve using Cramer's rule $x + y = 5$, $y + z = 3$, $z + x = 4$.

(f) Show that the three points $(a, b + c)$,$(b, c + a)$ and $(c, a + b)$ are colinear.

2. Attempt any two parts of the following : **(8 × 2 = 16)**

(a) From the following data find the missing frequency when mean is 15.38 :

Size	10	12	14	16	18	20
f	3	7	?	20	8	5

(b) Find the missing frequencies in the following distributions if median = 50 and $n = 100$:

Marks	0–20	20–40	40–60	60–80	80–100
No. of students	14	?	27	?	15

(c) Calculate the mode of the following distribution :

Value less than	10	20	30	40	50	60
Frequency	4	20	35	55	62	67

3. Attempt any four parts of the following : **(4 × 4 = 16)**

(a) Find the length of an arc of a circle of radius 5 cms, subtending an angle of 15° at the centre.

(b) Prove that $\sin^6 \alpha + \cos^6 \alpha = 1 - 3 \sin^2 \alpha \cos^2 \alpha$.

(c) If $\tan A = \frac{m}{m+1}$ and $\tan B = \frac{1}{2m+1}$, find the value of $(A + B)$.

(d) Prove that $\left(2 \cos \frac{\pi}{13} \cos \frac{9\pi}{13} + \cos \frac{3\pi}{13} + \cos \frac{5\pi}{13}\right)$ is equal to zero.

(e) If $A + B + C = \pi$, prove that $\tan A + \tan B + \tan C = \tan A \tan B \tan C$.

(f) If $\tan \frac{\theta}{2} = \frac{p}{q}$, find the vaule of $(p \sin \theta + q \cos \theta)$.

4. Attempt any four parts of the following : **(4 × 4 = 16)**

(a) Show that (2, –2), (8, 4), (5, 7) and (–1, 1) are the vertices of a rectangle.

(b) The mid-points of the sides of a triangle are (1, 2) (0, 1) and (2, –1). Find the coordinates of the vertices of the triangle.

(c) Find the equation of the locus of a point which moves so that its distance from $(a, 0)$ is equal to its distance from the y-axis.

(d) Find the equation of the line passing through (–2, –4) and perpendicular to the line $3x - y + 5 = 0$.

(e) Find the angle made by the line $x \cos 30° + y \sin 30° + \sin 120° = 0$ with the positive direction of x-axis.

(f) Show that the equation of a line passing through $(a \cos^3 \theta, a \sin^3 \theta)$ and perpendicular to the line :

$$x \sec \theta + y \operatorname{cosec} \theta = p \text{ is}$$

$$x \cos \theta - y \sin \theta = p \cos 2\theta.$$

5. Attempt any two parts of the following : **(8 × 2 = 16)**

(a) (i) State the domain of the function $f(x) = \frac{x^2 + 3x + 1}{x^2 - 5x + 6}$

(ii) Evaluate $\lim_{x \to 0} \left(\frac{e^x + e^{-x} - 2}{x^2}\right)$

(b) (i) If $y = (e^x + e^x \log x)$, find $\frac{dy}{dx}$.

(ii) If $y = \log \log \log x^3$, fidn $\frac{dy}{dx}$.

(c) (i) Evaluate $\int \frac{\sin x}{\sin (x - \alpha)} dx$.

(ii) Evaluate $\int \frac{x e^x}{(1 + x)^2} dx$.

B. PHARM

(SEM. I) EXAMINATION, 2007–08

REMEDIAL MATHEMATICS

Time : 3 Hours **Total Marks : 80**

Note : (1) Answer **ALL** the questions.
(2) All questions carry equal marks.

1. Attempt any four of the following : 4×4

(a) Solve the equations

$$(x+y)(z+x) = 30$$
$$(y+z)(x+y) = 15$$
$$(z+x)(y+z) = 18$$

(b) Show that $\begin{vmatrix} a+b & b+c & c+a \\ b+c & c+a & a+b \\ c+a & a+b & b+c \end{vmatrix} = 2\begin{vmatrix} a & b & c \\ b & c & a \\ c & a & b \end{vmatrix}$

(c) By using determinant solve the following system of equations :

$$x + y + z = 1$$
$$x + 2y + 3z = 1$$
$$x + 3y + 5z = 7$$

(d) If $A = \begin{bmatrix} -4 & -3 & -3 \\ 1 & 0 & 1 \\ 4 & 4 & 3 \end{bmatrix}$. Show that adj $A = A$.

(e) Find the inverse of matrix : $A = \begin{bmatrix} 2 & 3 & 1 \\ 2 & 4 & 1 \\ 3 & 7 & 2 \end{bmatrix}$

using elementary row transformation.

(f) The sum of three numbers is 2. If twice the second number is added to the sum of first and third, the sum is 1. By adding second and third number to five times the first number, we get 6. Find the three numbers, using matrix.

2. Attempt any two of the following : 8×2

(a) Income of employees in an industrial concern are given below : The total income of the 10 employees in the class over Rs. 250 is Rs. 3000. Compute the mean income.

Income (Rs.)	0–50	50–100	100–150	150–200	200–250	250 and over
Frequency	90	150	100	80	70	10

(b) Find the median of the following data :

x :	Above 48	Above 40	32–40	16–32	8–24	Below 18	Below 8
f:	5	15	20	45	32	20	8

(c) Calculate mode of the following :

Size :	10–20	10–30	10–40	10–50	10–60	10–70	10–80	10–90
Frequency :	4	16	56	97	124	137	146	150

3. Attempt any four of the following : **4 × 4**

(a) pH of a solution is given as $pH = \log \frac{1}{(H^+)}$. If $[H^+] = 2.57 \times 10^{-4}$ g-ion/l.

Find pH of solution. (Given log 257 = 2.4099).

(b) Prove that :

$$\frac{\cos 3\alpha + 2\cos 5\alpha + \cos 7\alpha}{\cos \alpha + 2\cos 3\alpha + \cos 5\alpha} = \frac{\cos 5\alpha}{\cos 3\alpha}$$

(c) Prove that :

$$\sin\frac{\pi}{5}\sin\frac{2\pi}{5}\sin\frac{3\pi}{5}\sin\frac{4\pi}{5} = \frac{5}{16}.$$

(d) Prove that

$$\tan 70^\circ = \tan 20^\circ + 2\tan 50^\circ$$

(e) Prove that :

$$\sin^2 B = \sin^2 A + \sin^2 (A - B) - 2\sin A\cos B\sin(A - B)$$

(f) Prove that :

$$(\cos A + \cos B)^2 + (\sin A - \sin B)^2 = 4\cos^2\left(\frac{A+B}{2}\right).$$

4. Attempt any two of the following : **8 × 2**

(a) (i) Find the equation of a locus of a point equidistant from the points $A(1, 3)$ and $B(-2, 1)$.

(ii) Find the equation of line that has y-intercept 4 and is perpendicular to line joining (2, –3) and (4, 2).

(b) Find the distance of a point (3, 5) from the line $3x + y + 4 = 0$ measured parallel to a line having slope $\frac{3}{4}$.

(c) Find the bisector of acute angle between the lines $3x - 4y + 11 = 0$ and $12x - 5y - 2 = 0$.

5. Attempt any two of the following : **8 × 2**

(a) (i) If $y = \sin e^{\tan^{-1} x}$, find $\frac{dy}{dx}$.

(ii) Prove that : $\int \frac{dx}{x\sqrt{x^2 - a^2}} = \frac{1}{a}\sec^{-1}\left(\frac{x}{a}\right)$.

(b) (i) If $y = \sqrt{x + \sqrt{x + \sqrt{x + \ldots \infty}}}$. Prove that : $(2y - 1)\frac{dy}{dx} = 1$.

(ii) Evaluate : $\int \frac{x + 1}{x^2 + 4x + 5}\,dx$

(c) Evaluate : $\int_0^\pi \frac{x}{1 + \cos^2 x}\,dx$.

B. PHARM

(SEM. I) EXAMINATION, 2008–09

REMEDIAL MATHEMATICS

Time : 3 Hours **Total Marks : 80**

Note : (1) Answer **all** questions in Section-A, any six questions in Section-B and any **four** questions in Section-C.

SECTION-A

$1 \times 16 = 16$

1. State whether true or false for questions (i) to (iv) :

(i) The roots of equation $x - \frac{2}{x} = 1$ and 1 and 3.

(ii) The order and value of $\begin{vmatrix} 4 & -3 \\ 2 & -1 \end{vmatrix}$ are 2×2 and 2 respectively.

(iii) If $\begin{bmatrix} a+b & 2 \\ 5 & a-b \end{bmatrix} = \begin{bmatrix} 6 & 2 \\ 5 & 8 \end{bmatrix}$, the values of a and b are 7 and -1 respectively.

(iv) The median and mode of the data

4, 10, 7, 15, 7, 3, 5, 3, 7 are 5 and 3 respectively.

Choose the correct answer for questions (v) to (viii)

(v) The gain in weights of 5 rats over a period of 5 days are 5, 6, 4, 8, 7. The arithmetic mean is

(a) 6 (b) 4 (c) 8 (d) None of these

(vi) The median of the data 25, 1275, 748, 162, 967, 162 is

(a) 162 (b) 455 (c) 560 (d) None of these

(vii) The angle $\frac{3\pi}{5}$ radian is equal to

(a) 100° (b) 120° (c) 108° (d) None of these

(viii) If $\sin \theta = -\frac{3}{5}$ and $\tan \theta$ is positive, the value of $\cos \theta$ is

(a) $\frac{3}{5}$ (b) $\frac{2}{5}$ (c) $-\frac{4}{5}$ (d) None of these

Fill in the blanks for questions (ix) to (xvi)

(ix) The value of sin 75° is ....

(x) If $(3, a)$ is the mid point of the line joining the points (4, 3) and (2, 5), the value of a is ....

(xi) The area of the triangle of vertices (0, 5), (2, 3) and (4, 5) is ....

(xii) The slope of the line joining the points (2, –5) and (4, 1) is ....

(xiii) The value of $\lim_{x \to 1} \frac{x^2 - 1}{x - 1}$ is ....

(xiv) If $y = x^2 + 2^x$, then $\frac{dy}{dx}$ is ....

(xv) $\int \sin 2x \, dx =$ ....

(xvi) $\int \log x \, dx =$ ....

SECTION-B

2. Answer any four parts of the following : **4 × 6 = 24**

(a) Solve $7^{x+1} + 7^{1-x} = 50$.

(b) Prove that

$$\begin{vmatrix} 1 & a & a^2 \\ 1 & b & b^2 \\ 1 & c & c^2 \end{vmatrix} = (a-b)(b-c)(c-a)$$

(c) Find the median of the following data :

Daily wages (Rs.)	22	24.50	28	31.50	34	36.50
No. of workers	10	23	32	28	12	5

(d) Prove that $\cos \frac{\pi}{5} - \cos \frac{2\pi}{5} = \frac{1}{2}$.

(e) Find the value of sec $(-1500)^\circ$ sin 390°.

(f) Find the equation of the line which is perpendicular to the line $3x - 2y + 4 = 0$ and passes through the point of intersection of the lines $x + 2y + 1 = 0$ and $y = x + 7$.

(g) If $x = a(\cos t + t \sin t)$, $y = a(\sin t - t \cos t)$, find $\frac{dy}{dx}$.

(h) Evaluate $\int \frac{1}{x^2} \log x \, dx$.

SECTION-C

Answer any four questions of the following : **10 × 4 = 40**

3. If $\frac{4x+1}{x+1} + \frac{x+1}{4x+1} - \frac{5}{2} = 0$, find the values of x.

4. If $A = \begin{bmatrix} 1 & 2 & 2 \\ 2 & 1 & 2 \\ 2 & 2 & 1 \end{bmatrix}$, find A^{-1}.

5. Three teachers of statistics reported mean marks of their classes, consisting of 69, 64 and 71 students as 30, 26 and 18 respectively. Determine the mean mark of all the three classes.

6. A line passes through (3, 4) and sum of its intercepts on the axes is 14. Find the equation of the straight line.

7. If $(\tan x)^{\cot x} + (\cot x)^{\tan x}$, find $\frac{dy}{dx}$.

LOGARITHMS

	0	1	2	3	4	5	6	7	8	9	Mean Differences								
											1	2	3	4	5	6	7	8	9
10	0000	0043	0086	0128	0170	0212	0253	0294	0334	0374	4	8	12	17	21	25	29	33	37
11	0414	0453	0492	0531	0569	0607	0645	0682	0719	0755	4	8	11	15	19	23	26	30	34
12	0792	0828	0864	0899	0934	0969	1004	1038	1072	1106	3	7	10	14	17	21	24	28	31
13	1139	1173	1206	1239	1271	1303	1335	1367	1399	1430	3	6	10	13	16	19	23	26	29
14	1461	1492	1523	1553	1584	1614	1644	1673	1703	1732	3	6	9	12	15	18	21	24	27
15	1761	1790	1818	1847	1875	1903	1931	1959	1987	2014	3	6	8	11	14	17	20	22	25
16	2041	2068	2095	2122	2148	2175	2201	2227	2253	2279	3	5	8	11	13	16	18	21	24
17	2304	2330	2355	2380	2405	2430	2455	2480	2504	2529	2	5	7	10	12	15	17	20	22
18	2553	2577	2601	2625	2648	2672	2695	2718	2742	2765	2	5	7	9	12	14	16	19	21
19	2788	2810	2833	2856	2878	2900	2923	2945	2967	2989	2	4	7	9	11	13	16	18	20
20	3010	3032	3054	3075	3096	3118	3139	3160	3181	3201	2	4	6	8	11	13	15	17	19
21	3222	3243	3263	3284	3304	3324	3345	3365	3385	3404	2	4	6	8	10	12	14	16	18
22	3424	3444	3464	3483	3502	3522	3541	3560	3579	3598	2	4	6	8	10	12	14	15	17
23	3617	3636	3655	3674	3692	3711	3729	3747	3766	3784	2	4	6	7	9	11	13	15	17
24	3802	3820	3838	3856	3874	3892	3909	3927	3945	3962	2	4	5	7	9	11	12	14	16
25	3979	3997	4014	4031	4048	4065	4082	4099	4116	4133	2	3	5	7	9	10	12	14	15
26	4150	4166	4183	4200	4216	4232	4249	4265	4281	4298	2	3	5	7	8	10	11	13	15
27	4314	4330	4346	4362	4378	4393	4409	4425	4440	4456	2	3	5	6	8	9	11	13	14
28	4472	4487	4502	4518	4533	4548	4564	4579	4594	4609	2	3	5	6	8	9	11	12	14
29	4624	4639	4654	4669	4683	4698	4713	4728	4742	4757	1	3	4	6	7	9	10	12	13
30	4771	4786	4800	4814	4829	4843	4857	4871	4886	4900	1	3	4	6	7	9	10	11	13
31	4914	4928	4942	4955	4969	4983	4997	5011	5024	5038	1	3	4	6	7	8	10	11	12
32	5051	5065	5079	5092	5105	5119	5132	5145	5159	5172	1	3	4	5	7	8	9	11	12
33	5185	5198	5211	5224	5237	5250	5263	5276	5289	5302	1	3	4	5	6	8	9	10	12
34	5315	5328	5340	5353	5366	5378	5391	5403	5416	5428	1	3	4	5	6	8	9	10	11
35	5441	5453	5465	5478	5490	5502	5514	5527	5539	5551	1	2	4	5	6	7	9	10	11
36	5563	5575	5587	5599	5611	5623	5635	5647	5658	5670	1	2	4	5	6	7	8	10	11
37	5682	5694	5705	5717	5729	5740	5752	5763	5775	5786	1	2	3	5	6	7	8	9	10
38	5798	5809	5821	5832	5843	5855	5866	5877	5888	5899	1	2	3	5	6	7	8	9	10
39	5911	5922	5933	5944	5955	5966	5977	5988	5999	6010	1	2	3	4	5	7	8	9	10
40	6021	6031	6042	6053	6064	6075	6085	6096	6107	6117	1	2	3	4	5	6	8	9	10
41	6128	6138	6149	6160	6170	6180	6191	6201	6212	6222	1	2	3	4	5	6	7	8	9
42	6232	6243	6253	6263	6274	6284	6294	6304	6314	6325	1	2	3	4	5	6	7	8	9
43	6335	6345	6355	6365	6375	6385	6395	6405	6415	6425	1	2	3	4	5	6	7	8	9
44	6435	6444	6454	6464	6474	6484	6493	6503	6513	6522	1	2	3	4	5	6	7	8	9
45	6532	6542	6551	6561	6571	6580	6590	6599	6609	6618	1	2	3	4	5	6	7	8	9
46	6628	6637	6646	6656	6665	6675	6684	6693	6702	6712	1	2	3	4	5	6	7	7	8
47	6721	6730	6739	6749	6758	6767	6776	6785	6794	6803	1	2	3	4	5	5	6	7	8
48	6812	6821	6830	6839	6848	6857	6866	6875	6884	6893	1	2	3	4	4	5	6	7	8
49	6902	6911	6920	6928	6937	6946	6955	6964	6972	6981	1	2	3	4	4	5	6	7	8
50	6990	6998	7007	7016	7024	7033	7042	7050	7059	7067	1	2	3	3	4	5	6	7	8
51	7076	7084	7093	7101	7110	7118	7126	7135	7143	7152	1	2	3	3	4	5	6	7	8
52	7160	7168	7177	7185	7193	7202	7210	7218	7226	7235	1	2	2	3	4	5	6	7	7
53	7243	7251	7259	7267	7275	7284	7292	7300	7308	7316	1	2	2	3	4	5	6	6	7
54	7324	7332	7340	7348	7356	7364	7372	7380	7388	7396	1	2	2	3	4	5	6	6	7

LOGARITHMS

	0	1	2	3	4	5	6	7	8	9	Mean Differences								
											1	2	3	4	5	6	7	8	9
55	7404	7412	7419	7427	7435	7443	7451	7459	7466	7474	1	2	2	3	4	5	5	6	7
56	7482	7490	7497	7505	7513	7520	7528	7536	7543	7551	1	2	2	3	4	5	5	6	7
57	7559	7566	7574	7582	7589	7597	7604	7612	7619	7627	1	2	2	3	4	5	5	6	7
58	7634	7642	7649	7657	7664	7672	7679	7686	7694	7701	1	1	2	3	4	4	5	6	7
59	7709	7716	7723	7731	7738	7745	7752	7760	7767	7774	1	1	2	3	4	4	5	6	7
60	7782	7789	7796	7803	7810	7818	7825	7832	7839	7846	1	1	2	3	4	4	5	6	6
61	7853	7860	7868	7875	7882	7889	7896	7903	7910	7917	1	1	2	3	4	4	5	6	6
62	7924	7931	7938	7945	7952	7959	7966	7973	7980	7987	1	1	2	3	3	4	5	6	6
63	7993	8000	8007	8014	8021	8028	8035	8041	8048	8055	1	1	2	3	3	4	5	5	6
64	8062	8069	8075	8082	8089	8096	8102	8109	8116	8122	1	1	2	3	3	4	5	5	6
65	8129	8136	8142	8149	8156	8162	8169	8176	8182	8189	1	1	2	3	3	4	5	5	6
66	8195	8202	8209	8215	8222	8228	8235	8241	8248	8254	1	1	2	3	3	4	5	5	6
67	8261	8267	8274	8280	8287	8293	8299	8306	8312	8319	1	1	2	3	3	4	5	5	6
68	8325	8331	8338	8344	8351	8357	8363	8370	8376	8382	1	1	2	3	3	4	4	5	6
69	8388	8395	8401	8407	8414	8420	8426	8432	8439	8445	1	1	2	2	3	4	4	5	6
70	8451	8457	8463	8470	8476	8482	8488	8494	8500	8506	1	1	2	2	3	4	4	5	6
71	8513	8519	8525	8531	8537	8543	8549	8555	8561	8567	1	1	2	2	3	4	4	5	5
72	8573	8579	8585	8591	8597	8603	8609	8615	8621	8627	1	1	2	2	3	4	4	5	5
73	8633	8639	8645	8651	8657	8663	8669	8675	8681	8686	1	1	2	2	3	4	4	5	5
74	8692	8698	8704	8710	8716	8722	8727	8733	8739	8745	1	1	2	2	3	4	4	5	5
75	8751	8756	8762	8768	8774	8779	8785	8791	8797	8802	1	1	2	2	3	3	4	5	5
76	8808	8814	8820	8825	8831	8837	8842	8848	8854	8859	1	1	2	2	3	3	4	5	5
77	8865	8871	8876	8882	8887	8893	8899	8904	8910	8915	1	1	2	2	3	3	4	4	5
78	8921	8927	8932	8938	8943	8949	8954	8960	8965	8971	1	1	2	2	3	3	4	4	5
79	8976	8982	8987	8993	8998	9004	9009	9015	9020	9025	1	1	2	2	3	3	4	4	5
80	9031	9036	9042	9047	9053	9058	9063	9069	9074	9079	1	1	2	2	3	3	4	4	5
81	9085	9090	9096	9101	9106	9112	9117	9122	9128	9133	1	1	2	2	3	3	4	4	5
82	9138	9143	9149	9154	9159	9165	9170	9175	9180	9186	1	1	2	2	3	3	4	4	5
83	9191	9196	9201	9206	9212	9217	9222	9227	9232	9238	1	1	2	2	3	3	4	4	5
84	9243	9248	9253	9258	9263	9269	9274	9279	9284	9289	1	1	2	2	3	3	4	4	5
85	9294	9299	9304	9309	9315	9320	9325	9330	9335	9340	1	1	2	2	3	3	4	4	5
86	9345	9350	9355	9360	9365	9370	9375	9380	9385	9390	1	1	2	2	3	3	4	4	5
87	9395	9400	9405	9410	9415	9420	9425	9430	9435	9440	0	1	1	2	2	3	3	4	4
88	9445	9450	9455	9460	9465	9469	9474	9479	9484	9489	0	1	1	2	2	3	3	4	4
89	9494	9499	9504	9509	9513	9518	9523	9528	9533	9538	0	1	1	2	2	3	3	4	4
90	9542	9547	9552	9557	9562	9566	9571	9576	9581	9586	0	1	1	2	2	3	3	4	4
91	9590	9595	9600	9605	9609	9614	9619	9624	9628	9633	0	1	1	2	2	3	3	4	4
92	9638	9643	9647	9652	9657	9661	9666	9671	9675	9680	0	1	1	2	2	3	3	4	4
93	9685	9689	9694	9699	9703	9708	9713	9717	9722	9727	0	1	1	2	2	3	3	4	4
94	9731	9736	9741	9745	9750	9754	9759	9763	9768	9773	0	1	1	2	2	3	3	4	4
95	9777	9782	9786	9791	9795	9800	9805	9809	9814	9818	0	1	1	2	2	3	3	4	4
96	9823	9827	9832	9836	9841	9845	9850	9854	9859	9863	0	1	1	2	2	3	3	4	4
97	9868	9872	9877	9881	9886	9890	9894	9899	9903	9908	0	1	1	2	2	3	3	4	4
98	9912	9917	9921	9926	9930	9934	9939	9943	9948	9952	0	1	1	2	2	3	3	4	4
99	9956	9961	9965	9969	9974	9978	9983	9987	9991	9996	0	1	1	2	2	3	3	3	4

ANTILOGARITHMS

	0	1	2	3	4	5	6	7	8	9	Mean Differences								
											1	2	3	4	5	6	7	8	9
.00	1000	1002	1005	1007	1009	1012	1014	1016	1019	1021	0	0	1	1	1	1	2	2	2
.01	1023	1026	1028	1030	1033	1035	1038	1040	1042	1045	0	0	1	1	1	1	2	2	2
.02	1047	1050	1052	1054	1057	1059	1062	1064	1067	1069	0	0	1	1	1	1	2	2	2
.03	1072	1074	1076	1079	1081	1084	1086	1089	1091	1094	0	0	1	1	1	1	2	2	2
.04	1096	1099	1102	1104	1107	1109	1112	1114	1117	1119	0	1	1	1	1	2	2	2	2
.05	1122	1125	1127	1130	1132	1135	1138	1140	1143	1146	0	1	1	1	1	2	2	2	2
.06	1148	1151	1153	1156	1159	1161	1164	1167	1169	1172	0	1	1	1	1	2	2	2	2
.07	1175	1178	1180	1183	1186	1189	1191	1194	1197	1199	0	1	1	1	1	2	2	2	2
.08	1202	1205	1208	1211	1213	1216	1219	1222	1225	1227	0	1	1	1	1	2	2	2	3
.09	1230	1233	1236	1239	1242	1245	1247	1250	1253	1256	0	1	1	1	1	2	2	2	3
.10	1259	1262	1265	1268	1271	1274	1276	1279	1282	1285	0	1	1	1	1	2	2	2	3
.11	1288	1291	1294	1297	1300	1303	1306	1309	1312	1315	0	1	1	1	2	2	2	2	3
.12	1318	1321	1324	1327	1330	1334	1337	1340	1343	1346	0	1	1	1	2	2	2	2	3
.13	1349	1352	1355	1358	1361	1365	1368	1371	1374	1377	0	1	1	1	2	2	2	3	3
.14	1380	1384	1387	1390	1393	1396	1400	1403	1406	1409	0	1	1	1	2	2	2	3	3
.15	1413	1416	1419	1422	1426	1429	1432	1435	1439	1442	0	1	1	1	2	2	2	3	3
.16	1445	1449	1452	1455	1459	1462	1466	1469	1472	1476	0	1	1	1	2	2	2	3	3
.17	1479	1483	1486	1489	1493	1496	1500	1503	1507	1510	0	1	1	1	2	2	2	3	3
.18	1514	1517	1521	1524	1528	1531	1535	1538	1542	1545	0	1	1	1	2	2	2	3	3
.19	1549	1552	1556	1560	1563	1567	1570	1574	1578	1581	0	1	1	1	2	2	3	3	3
.20	1585	1589	1592	1596	1600	1603	1607	1611	1614	1618	0	1	1	1	2	2	3	3	3
.21	1622	1626	1629	1633	1637	1641	1644	1648	1652	1656	0	1	1	2	2	2	3	3	3
.22	1660	1663	1667	1671	1675	1679	1683	1687	1690	1694	0	1	1	2	2	2	3	3	3
.23	1698	1702	1706	1710	1714	1718	1722	1726	1730	1734	0	1	1	2	2	2	3	3	4
.24	1738	1742	1746	1750	1754	1758	1762	1766	1770	1774	0	1	1	2	2	2	3	3	4
.25	1778	1782	1786	1791	1795	1799	1803	1807	1811	1816	0	1	1	2	2	2	3	3	4
.26	1820	1824	1828	1832	1837	1841	1845	1849	1854	1858	0	1	1	2	2	3	3	3	4
.27	1862	1866	1871	1875	1879	1884	1888	1892	1897	1901	0	1	1	2	2	3	3	3	4
.28	1905	1910	1914	1919	1923	1928	1932	1936	1941	1945	0	1	1	2	2	3	3	4	4
.29	1950	1954	1959	1963	1968	1972	1977	1982	1986	1991	0	1	1	2	2	3	3	4	4
.30	1995	2000	2004	2009	2014	2018	2023	2028	2032	2037	0	1	1	2	2	3	3	4	4
.31	2042	2046	2051	2056	2061	2065	2070	2075	2080	2084	0	1	1	2	2	3	3	4	4
.32	2089	2094	2099	2104	2109	2113	2118	2123	2128	2133	0	1	1	2	2	3	3	4	4
.33	2138	2143	2148	2153	2158	2163	2168	2173	2178	2183	0	1	1	2	2	3	3	4	4
.34	2188	2193	2198	2203	2208	2213	2218	2223	2228	2234	1	1	2	2	3	3	4	4	5
.35	2239	2244	2249	2254	2259	2265	2270	2275	2280	2286	1	1	2	2	3	3	4	4	5
.36	2291	2296	2301	2307	2312	2317	2323	2328	2333	2339	1	1	2	2	3	3	4	4	5
.37	2344	2350	2355	2360	2366	2371	2377	2382	2388	2393	1	1	2	2	3	3	4	4	5
.38	2399	2404	2410	2415	2421	2427	2432	2438	2443	2449	1	1	2	2	3	3	4	4	5
.39	2455	2460	2466	2472	2477	2483	2489	2495	2500	2506	1	1	2	2	3	3	4	5	5
.40	2512	2518	2523	2529	2535	2541	2547	2553	2559	2564	1	1	2	2	3	4	4	5	5
.41	2570	2576	2582	2588	2594	2600	2606	2612	2618	2624	1	1	2	2	3	4	4	5	5
.42	2630	2636	2642	2649	2655	2661	2667	2673	2679	2685	1	1	2	2	3	4	4	5	6
.43	2692	2698	2704	2710	2716	2723	2729	2735	2742	2748	1	1	2	3	3	4	4	5	6
.44	2754	2761	2767	2773	2780	2786	2793	2799	2805	2812	1	1	2	3	3	4	4	5	6
.45	2818	2825	2831	2838	2844	2851	2858	2864	2871	2877	1	1	2	3	3	4	5	5	6
.46	2884	2891	2897	2904	2911	2917	2924	2931	2938	2944	1	1	2	3	3	4	5	5	6
.47	2951	2958	2965	2972	2979	2985	2992	2999	3006	3013	1	1	2	3	3	4	5	5	6
.48	3020	3027	3034	3041	3048	3055	3062	3069	3076	3083	1	1	2	3	4	4	5	6	6
.49	3090	3097	3105	3112	3119	3126	3133	3141	3148	3155	1	1	2	3	4	4	5	6	6

ANTILOGARITHMS

	0	1	2	3	4	5	6	7	8	9	Mean Differences								
											1	2	3	4	5	6	7	8	9
.50	3162	3170	3177	3184	3192	3199	3206	3214	3221	3228	1	1	2	3	4	4	5	6	7
.51	3236	3243	3251	3258	3266	3273	3281	3289	3296	3304	1	2	2	3	4	5	5	6	7
.52	3311	3319	3327	3334	3342	3350	3357	3365	3373	3381	1	2	2	3	4	5	5	6	7
.53	3388	3396	3404	3412	3420	3428	3436	3443	3451	3459	1	2	2	3	4	5	6	6	7
.54	3467	3475	3483	3491	3499	3508	3516	3524	3532	3540	1	2	2	3	4	5	6	6	7
.55	3548	3556	3565	3573	3581	3589	3597	3606	3614	3622	1	2	2	3	4	5	6	7	7
.56	3631	3639	3648	3656	3664	3673	3681	3690	3698	3707	1	2	3	3	4	5	6	7	8
.57	3715	3724	3733	3741	3750	3758	3767	3776	3784	3793	1	2	3	3	4	5	6	7	8
.58	3802	3811	3819	3828	3837	3846	3855	3864	3873	3882	1	2	3	4	4	5	6	7	8
.59	3890	3899	3908	3917	3926	3936	3945	3954	3963	3972	1	2	3	4	5	5	6	7	8
.60	3981	3990	3999	4009	4018	4027	4036	4046	4055	4064	1	2	3	4	5	6	6	7	8
.61	4074	4083	4093	4102	4111	4121	4130	4140	4150	4159	1	2	3	4	5	6	7	8	9
.62	4169	4178	4188	4198	4207	4217	4227	4236	4256	4256	1	2	3	4	5	6	7	8	9
.63	4266	4276	4285	4295	4305	4315	4325	4335	4345	4355	1	2	3	4	5	6	7	8	9
.64	4365	4375	4385	4395	4406	4416	4426	4436	4446	4457	1	2	3	4	5	6	7	8	9
.65	4467	4477	4487	4498	4508	4519	4529	4539	4550	4560	1	2	3	4	5	6	7	8	9
.66	4571	4581	4592	4603	4613	4624	4634	4645	4656	4667	1	2	3	4	5	6	7	9	10
.67	4677	4688	4699	4710	4721	4732	4742	4753	4764	4775	1	2	3	4	5	7	8	9	10
.68	4786	4797	4808	4819	4831	4842	4853	4864	4875	4887	1	2	3	4	6	7	8	9	10
.69	4898	4909	4920	4932	4943	4955	4966	4977	4989	5000	1	2	3	5	6	7	8	9	10
.70	5012	5023	5035	5047	5058	5070	5082	5093	5105	5117	1	2	4	5	6	7	8	9	11
.71	5129	5140	5152	5164	5176	5188	5200	5212	5224	5236	1	2	4	5	6	7	8	10	11
.72	5248	5260	5272	5284	5297	5309	5321	5333	5346	5358	1	2	4	5	6	7	9	10	11
.73	5370	5383	5395	5408	5420	5433	5445	5458	5470	5483	1	3	4	5	6	8	9	10	11
.74	5495	5508	5521	5534	5546	5559	5572	5585	5598	5610	1	3	4	5	6	8	9	10	12
.75	5623	5636	5649	5662	5675	5689	5702	5715	5728	5741	1	3	4	5	7	8	9	10	12
.76	5754	5768	5781	5794	5808	5821	5834	5848	5861	5875	1	3	4	5	7	8	9	11	12
.77	5888	5902	5916	5929	5943	5957	5970	5984	5998	6012	1	3	4	5	7	8	10	11	12
.78	6026	6039	6053	6067	6081	6095	6109	6124	6138	6152	1	3	4	6	7	8	10	11	13
.79	6166	6180	6194	6209	6223	6237	6252	6266	6281	6295	1	3	4	6	7	9	10	11	13
.80	6310	6324	6339	6353	6368	6383	6397	6412	6427	6442	1	3	4	6	7	9	10	12	13
.81	6457	6471	6486	6501	6516	6531	6546	6561	6577	6592	2	3	5	6	8	9	11	12	14
.82	6607	6622	6637	6653	6668	6683	6699	6714	6730	6745	2	3	5	6	8	9	11	12	14
.83	6761	6776	6792	6808	6823	6839	6855	6871	6887	6902	2	3	5	6	8	9	11	13	14
.84	6918	6934	6950	6966	6982	6998	7015	7031	7047	7063	2	3	5	6	8	10	11	13	15
.85	7079	7096	7112	7129	7145	7161	7178	7194	7211	7228	2	3	5	7	8	10	12	13	15
.86	7244	7261	7278	7295	7311	7328	7345	7362	7379	7396	2	3	5	7	8	10	12	13	15
.87	7413	7430	7447	7464	7482	7499	7516	7534	7551	7568	2	3	5	7	9	10	12	14	16
.88	7586	7603	7621	7638	7656	7674	7691	7709	7727	7745	2	4	5	7	9	11	12	14	16
.89	7762	7780	7798	7816	7834	7852	7870	7889	7907	7925	2	4	5	7	9	11	13	14	16
.90	7943	7962	7980	7998	8017	8035	8054	8072	8091	8110	2	4	6	7	9	11	13	15	17
.91	8128	8147	8166	8185	8204	8222	8241	8260	8279	8299	2	4	6	8	9	11	13	15	17
.92	8318	8337	8356	8375	8395	8414	8433	8453	8472	8492	2	4	6	8	10	12	14	15	17
.93	8511	8531	8551	8570	8590	8610	8630	8650	8670	8690	2	4	6	8	10	12	14	16	18
.94	8710	8730	8750	8770	8790	8810	8831	8851	8872	8892	2	4	6	8	10	12	14	16	18
.95	8913	8933	8954	8974	8995	9016	9036	9057	9078	9099	2	4	6	8	10	12	15	17	19
.96	9120	9141	9162	9183	9204	9226	9247	9268	9290	9311	2	4	6	8	11	13	15	17	19
.97	9333	9354	9376	9397	9419	9441	9462	9484	9506	9528	2	4	7	9	11	13	15	17	20
.98	9550	9572	9594	9616	9638	9661	9683	9705	9727	9750	2	4	7	9	11	13	16	18	20
.99	9772	9795	9817	9840	9863	9886	9908	9931	9954	9977	2	5	7	9	11	14	16	18	20

NATURAL SINES

Degree	0′ 0°.0	6′ 0°.1	12′ 0°.2	18′ 0°.3	24′ 0°.4	30′ 0°.5	36′ 0°.6	42′ 0°.7	48′ 0°.8	54′ 0°.9	Mean differences 1′	2′	3′	4′	5′
0	.0000	0017	0035	0052	0070	0087	0105	0122	0140	0157	3	6	9	12	15
1	.0175	0192	0209	0227	0244	0262	0279	0297	0314	0332	3	6	9	12	15
2	.0349	0366	0384	0401	0419	0436	0454	0471	0488	0506	3	6	9	12	15
3	.0523	0541	0558	0576	0593	0610	0628	0645	0663	0680	3	6	9	12	15
4	.0698	0715	0732	0750	0767	0785	0802	0819	0837	0854	3	6	9	12	15
5	.0872	0889	0906	0924	0941	0958	0976	0993	1011	1028	3	6	9	12	14
6	.1045	1063	1080	1097	1115	1132	1149	1167	1184	1201	3	6	9	12	14
7	.1219	1236	1253	1271	1288	1305	1323	1340	1357	1374	3	6	9	12	14
8	.1392	1409	1426	1444	1461	1478	1495	1513	1530	1547	3	6	9	12	14
9	.1564	1582	1599	1616	1633	1650	1668	1685	1702	1719	3	6	9	12	14
10	.1736	1754	1771	1788	1805	1822	1840	1857	1874	1891	3	6	9	11	14
11	.1908	1925	1942	1959	1977	1994	2011	2028	2045	2062	3	6	9	11	14
12	.2079	2096	2113	2130	2147	2164	2181	2198	2215	2233	3	6	9	11	14
13	.2250	2267	2284	2300	2317	2334	2351	2368	2385	2402	3	6	8	11	14
14	.2419	2436	2453	2470	2487	2504	2521	2538	2554	2571	3	6	8	11	14
15	.2588	2605	2622	2639	2656	2672	2689	2706	2723	2740	3	6	8	11	14
16	.2756	2773	2790	2807	2823	2840	2857	2874	2890	2907	3	6	8	11	14
17	.2924	2940	2957	2974	2990	3007	3024	3040	3057	3074	3	6	8	11	14
18	.3090	3107	3123	3140	3156	3173	3190	3206	3223	3239	3	6	8	11	14
19	.3256	3272	3289	3305	3322	3338	3355	3371	3387	3404	3	5	8	11	14
20	.3420	3437	3453	3469	3486	3502	3518	3535	3551	3567	3	5	8	11	14
21	.3584	3600	3616	3633	3649	3665	3681	3697	3714	3730	3	5	8	11	14
22	.3746	3762	3778	3795	3811	3827	3843	3859	3875	3891	3	5	8	11	14
23	.3907	3923	3939	3955	3971	3987	4003	4019	4035	4051	3	5	8	11	14
24	.4067	4083	4099	4115	4131	4147	4163	4179	4195	4210	3	5	8	11	13
25	.4226	4242	4258	4274	4289	4305	4321	4337	4352	4368	3	5	8	11	13
26	.4384	4399	4415	4431	4446	4462	4478	4493	4509	4524	3	5	8	10	13
27	.4540	4555	4571	4586	4602	4617	4633	4648	4664	4679	3	5	8	10	13
28	.4695	4710	4726	4741	4756	4772	4787	4802	4818	4833	3	5	8	10	13
29	.4848	4863	4879	4894	4909	4924	4939	4955	4970	4985	3	5	8	10	13
30	.5000	5015	5030	5045	5060	5075	5090	5105	5120	5135	3	5	8	10	13
31	.5150	5165	5180	5195	5210	5225	5240	5255	5270	5284	2	5	7	10	12
32	.5299	5314	5329	5344	5358	5373	5388	5402	5417	5432	2	5	7	10	12
33	.5446	5461	5476	5490	5505	5519	5534	5548	5563	5577	2	5	7	10	12
34	.5592	5606	5621	5635	5650	5664	5678	5693	5707	5721	2	5	7	10	12
35	.5736	5750	5764	5779	5793	5807	5821	5835	5850	5864	2	5	7	9	12
36	.5878	5892	5906	5920	5934	5948	5962	5976	5990	6004	2	5	7	9	12
37	.6018	6032	6046	6060	6074	6088	6101	6115	6129	6143	2	5	7	9	12
38	.6157	6170	6184	6198	6211	6225	6239	6252	6266	6280	2	5	7	9	11
39	.6293	6307	6320	6334	6347	6361	6374	6388	6401	6414	2	4	7	9	11
40	.6428	6441	6455	6468	6481	6494	6508	6521	6534	6547	2	4	7	9	11
41	.6561	6574	6587	6600	6613	6626	6639	6652	6665	6678	2	4	7	9	11
42	.6691	6704	6717	6730	6743	6756	6769	6782	6794	6807	2	4	6	9	11
43	.6820	6833	6845	6858	6871	6884	6896	6909	6921	6934	2	4	6	8	11
44	.6947	6959	6972	6984	6997	7009	7022	7034	7046	7059	2	4	6	8	10

NATURAL SINES

Degree	0′ 0°.0	6′ 0°.1	12′ 0°.2	18′ 0°.3	24′ 0°.4	30′ 0°.5	36′ 0°.6	42′ 0°.7	48′ 0°.8	54′ 0°.9	Mean Differences 1′	2′	3′	4′	5′
45	.7071	7083	7096	7108	7120	7133	7145	7157	7169	7181	2	4	6	8	10
46	.7193	7206	7218	7230	7242	7254	7266	7278	7290	7302	2	4	6	8	10
47	.7314	7325	7337	7349	7361	7373	7385	7396	7408	7420	2	4	6	8	10
48	.7431	7443	7455	7466	7478	7490	7501	7513	7524	7536	2	4	6	8	10
49	.7547	7559	7570	7581	7593	7604	7615	7627	7638	7649	2	4	6	8	9
50	.7660	7672	7683	7694	7705	7716	7727	7738	7749	7760	2	4	6	7	9
51	.7771	7782	7793	7804	7815	7826	7837	7848	7859	7869	2	4	5	7	9
52	.7880	7891	7902	7912	7923	7934	7944	7955	7965	7976	2	4	5	7	9
53	.7986	7997	8007	8018	8028	8039	8049	8059	8070	8080	2	3	5	7	9
54	.8090	8100	8111	8121	8131	8141	8151	8161	8171	8181	2	3	5	7	8
55	.8192	8202	8211	8221	8231	8241	8251	8261	8271	8281	2	3	5	7	8
56	.8290	8300	8310	8320	8329	8339	8348	8358	8368	8377	2	3	5	6	8
57	.8387	8396	8406	8415	8425	8434	8443	8453	8462	8471	2	3	5	6	8
58	.8480	8490	8499	8508	8517	8526	8536	8545	8554	8563	2	3	5	6	8
59	.8572	8581	8590	8599	8607	8616	8625	8634	8643	8652	1	3	4	6	7
60	.8660	8669	8678	8686	8695	8704	8712	8721	8729	8738	1	3	4	6	7
61	.8746	8755	8763	8771	8780	8788	8796	8805	8813	8821	1	3	4	6	7
62	.8829	8838	8846	8854	8862	8870	8878	8886	8894	8902	1	3	4	5	7
63	.8910	8918	8926	8934	8942	8949	8957	8965	8973	8980	1	3	4	5	6
64	.8988	8996	9003	9011	9018	9026	9033	9041	9048	9056	1	3	4	5	6
65	.9063	9070	9078	9085	9092	9100	9107	9114	9121	9128	1	2	4	5	6
66	.9135	9143	9150	9157	9164	9171	9178	9184	9191	9198	1	2	3	5	6
67	.9205	9212	9219	9225	9232	9239	9245	9252	9259	9265	1	2	3	4	6
68	.9272	9278	9285	9291	9298	9304	9311	9317	9323	9330	1	2	3	4	5
69	.9336	9342	9348	9354	9361	9367	9373	9379	9385	9391	1	2	3	4	5
70	.9397	9403	9409	9415	9421	9426	9432	9438	9444	9449	1	2	3	4	5
71	.9455	9461	9466	9472	9478	9483	9489	9494	9500	9505	1	2	3	4	5
72	.9511	9516	9521	9527	9532	9537	9542	9548	9553	9558	1	2	3	3	4
73	.9563	9568	9573	9578	9583	9588	9593	9598	9603	9608	1	2	2	3	4
74	.9613	9617	9622	9627	9632	9636	9641	9646	9650	9655	1	2	2	3	4
75	.9659	9664	9668	9673	9677	9681	9686	9690	9694	9699	1	1	2	3	4
76	.9703	9707	9711	9715	9720	9724	9728	9732	9736	9740	1	1	2	3	3
77	.9744	9748	9751	9755	9759	9763	9767	9770	9774	9778	1	1	2	3	3
78	.9781	9785	9789	9792	9796	9799	9803	9806	9810	9813	1	1	2	2	3
79	.9816	9820	9823	9826	9829	9833	9836	9839	9842	9845	1	1	2	2	3
80	.9848	9851	9854	9857	9860	9863	9866	9869	9871	9874	0	1	1	2	2
81	.9877	9880	9882	9885	9888	9890	9893	9895	9898	9900	0	1	1	2	2
82	.9903	9905	9907	9910	9912	9914	9917	9919	9921	9923	0	1	1	2	2
83	.9925	9928	9930	9932	9934	9936	9938	9940	9942	9943	0	1	1	1	2
84	.9945	9947	9949	9951	9952	9954	9956	9957	9959	9960	0	1	1	1	2
85	.9962	9963	9965	9966	9968	9969	9971	9972	9973	9974	0	0	1	1	1
86	.9976	9977	9978	9979	9980	9981	9982	9983	9984	9985	0	0	1	1	1
87	.9986	9987	9988	9989	9990	9990	9991	9992	9993	9993	0	0	0	1	1
88	.9994	9995	9995	9996	9996	9997	9997	9997	9998	9998	0	0	0	0	0
89	.9998	9999	9999	9999	9999	1.000	1.000	1.000	1.000	1.000	0	0	0	0	0

NATURAL COSINES

Degree	0′ 0°.0	6′ 0°.1	12′ 0°.2	18′ 0°.3	24′ 0°.4	30′ 0°.5	36′ 0ā.6	42′ 0°.7	48′ 0°.8	54′ 0°.9	Mean Differences 1′	2′	3′	4′	5′
0	1.000	1.000	1.000	1.000	1.000	1.000	.9999	9999	9999	9999	0	0	0	0	0
1	.9998	9998	9998	9997	9997	9997	9996	9996	9995	9995	0	0	0	0	0
2	.9994	9993	9993	9992	9991	9990	9990	9989	9988	9987	0	0	0	1	1
3	.9986	9985	9984	9983	9982	9981	9980	9979	9978	9977	0	0	1	1	1
4	.9976	9974	9973	9972	9971	9969	9968	9966	9965	9963	0	0	1	1	1
5	.9962	9960	9959	9957	9956	9954	9952	9951	9949	9947	0	1	1	1	2
6	.9945	9943	9942	9940	9938	9936	9934	9932	9930	9928	0	1	1	1	2
7	.9925	9923	9921	9919	9917	9914	9912	9910	9907	9905	0	1	1	2	2
8	.9903	9900	9898	9895	9893	9890	9888	9885	9882	9880	0	1	1	2	2
9	.9877	9874	9871	9869	9866	9863	9860	9857	9854	9851	0	1	1	2	2
10	.9848	9845	9842	9839	9836	9833	9829	9826	9823	9820	1	1	2	2	3
11	.9816	9813	9810	9806	9803	9799	9796	9792	9789	9785	1	1	2	2	3
12	.9781	9778	9774	9770	9767	9763	9759	9755	9751	9748	1	1	2	3	3
13	.9744	9740	9736	9732	9728	9724	9720	9715	9711	9707	1	1	2	3	3
14	.9703	9699	9694	9690	9686	9681	9677	9673	9668	9664	1	1	2	3	4
15	.9659	9655	9650	9646	9641	9636	9632	9627	9622	9617	1	2	2	3	4
16	.9613	9608	9603	9598	9593	9588	9583	9578	9573	9568	1	2	2	3	4
17	.9563	9558	9553	9548	9542	9537	9532	9527	9521	9516	1	2	3	3	4
18	.9511	9505	9500	9494	9489	9483	9478	9472	9466	9461	1	2	3	4	5
19	.9455	9449	9444	9438	9432	9426	9421	9415	9409	9403	1	2	3	4	5
20	.9397	9391	9385	9379	9373	9367	9361	9354	9348	9342	1	2	3	4	5
21	.9336	9330	9323	9317	9311	9304	9298	9291	9285	9278	1	2	3	4	5
22	.9272	9265	9259	9252	9245	9239	9232	9225	9219	9212	1	2	3	4	6
23	.9205	9198	9191	9184	9178	9171	9164	9157	9150	9143	1	2	3	5	6
24	.9135	9128	9121	9114	9107	9100	9092	9085	9078	9070	1	2	4	5	6
25	.9063	9056	9048	9041	9033	9026	9018	9011	9003	8996	1	3	4	5	6
26	.8988	8980	8973	8965	8957	8949	8942	8934	8926	8918	1	3	4	5	6
27	.8910	8902	8894	8886	8878	8870	8862	8854	8846	8838	1	3	4	5	7
28	.8829	8821	8813	8805	8796	8788	8780	8771	8763	8755	1	3	4	6	7
29	.8746	8738	8729	8721	8712	8704	8695	8686	8678	8669	1	3	4	6	7
30	.8660	8652	8643	8634	8625	8616	8607	8599	8590	8581	1	3	4	6	7
31	.8572	8563	8554	8545	8536	8526	8517	8508	8499	8490	2	3	5	6	8
32	.8480	8471	8462	8453	8443	8434	8425	8415	8406	8396	2	3	5	6	8
33	.8387	8377	8368	8358	8348	8339	8329	8320	8310	8300	2	3	5	6	8
34	.8290	8281	8271	8261	8251	8241	8231	8221	8211	8202	2	3	5	7	8
35	.8192	8181	8171	8161	8151	8141	8131	8121	8111	8100	2	3	5	7	8
36	.8090	8080	8070	8059	8049	8039	8028	8018	8007	7997	2	3	5	7	9
37	.7986	7976	7965	7955	7944	7934	7923	7912	7902	7891	2	4	5	7	9
38	.7880	7859	7869	7848	7837	7826	7815	7804	7793	7782	2	4	5	7	9
39	.7771	7760	7749	7738	7727	7716	7705	7694	7683	7672	2	4	6	7	9
40	.7660	7649	7638	7627	7615	7604	7593	7581	7570	7559	2	4	6	8	9
41	.7547	7536	7524	7513	7501	7490	7478	7466	7455	7443	2	4	6	8	10
42	.7431	7420	7408	7396	7385	7373	7361	7349	7337	7325	2	4	6	8	10
43	.7314	7302	7290	7278	7266	7254	7242	7230	7218	7206	2	4	6	8	10
44	.7193	7181	7169	7157	7145	7133	7120	7108	7096	7083	2	4	6	8	10

Note : *Subtract Mean Differences*

NATURAL COSINES

Degree	0′ 0°.0	6′ 0°.1	12′ 0°.2	18′ 0°.3	24′ 0°.4	30′ 0°.5	36′ 0°.6	42′ 0°.7	48′ 0°.8	54′ 0°.9	Mean Differences 1′	2′	3′	4′	5′
45	.7071	7059	7046	7034	7022	7009	6997	6984	6972	6959	2	4	6	8	10
46	.6947	6934	6921	6909	6896	6884	6871	6858	6845	6833	2	4	6	8	11
47	.6820	6807	6794	6782	6769	6756	6743	6730	6717	6704	2	4	6	9	11
48	.6691	6678	6665	6652	6639	6626	6613	6600	6587	6574	2	4	7	9	11
49	.6561	6547	6534	6521	6508	6494	6481	6468	6455	6441	2	4	7	9	11
50	.6428	6414	6401	6388	6374	6361	6347	6334	6320	6307	2	4	7	9	11
51	.6293	6280	6266	6252	6239	6225	6211	6198	6184	6170	2	5	7	9	11
52	.6157	6143	6129	6115	6101	6088	6074	6060	6046	6032	2	5	7	9	12
53	.6018	6004	5990	5976	5962	5948	5934	5920	5906	5892	2	5	7	9	12
54	.5878	5864	5850	5835	5821	5807	5793	5779	5764	5750	2	5	7	9	12
55	.5736	5721	5707	5693	5678	5664	5650	5635	5621	5606	2	5	7	10	12
56	.5592	5577	5563	5548	5534	5519	5505	5490	5476	5461	2	5	7	10	12
57	.5446	5432	5417	5402	5388	5373	5358	5344	5329	5314	2	5	7	10	12
58	.5299	5284	5270	5255	5240	5225	5210	5195	5180	5165	2	5	7	10	12
59	.5150	5135	5120	5105	5090	5075	5060	5045	5030	5015	3	5	8	10	13
60	.5000	4985	4970	4955	4939	4924	4909	4894	4879	4863	3	5	8	10	13
61	.4848	4833	4818	4802	4787	4772	4756	4741	4726	4710	3	5	8	10	13
62	.4695	4679	4664	4648	4633	4617	4602	4586	4571	4555	3	5	8	10	13
63	.4540	4524	4509	4493	4478	4462	4446	4431	4415	4399	3	5	8	10	13
64	.4384	4368	4352	4337	4321	4305	4289	4274	4258	4242	3	5	8	11	13
65	.4226	4210	4195	4179	4163	4147	4131	4115	4099	4083	3	5	8	11	13
66	.4067	4051	4035	4019	4003	3987	3971	3955	3939	3923	3	5	8	11	14
67	.3907	3891	3875	3859	3843	3827	3811	3795	3778	3762	3	5	8	11	14
68	.3746	3730	3714	3697	3681	3665	3649	3633	3616	3600	3	5	8	11	14
69	.3584	3567	3551	3535	3518	3502	3486	3469	3453	3437	3	5	8	11	14
70	.3420	3404	3387	3371	3355	3338	3322	3305	3289	3272	3	5	8	11	14
71	.3256	3239	3223	3206	3190	3173	3156	3140	3123	3107	3	6	8	11	14
72	.3090	3074	3057	3040	3024	3007	2990	2974	2957	2940	3	6	8	11	14
73	.2924	2907	2890	2874	2857	2840	2823	2807	2790	2773	3	6	8	11	14
74	.2756	2740	2723	2706	2689	2672	2656	2639	2622	2605	3	6	8	11	14
75	.2588	2571	2554	2538	2521	2504	2487	2470	2453	2436	3	6	8	11	14
76	.2419	2402	2385	2368	2351	2334	2317	2300	2284	2267	3	6	8	11	14
77	.2250	2233	2215	2198	2181	2164	2147	2130	2113	2096	3	6	9	11	14
78	.2079	2062	2045	2028	2011	1994	1977	1959	1942	1925	3	6	9	11	14
79	.1908	1891	1874	1857	1840	1822	1805	1788	1771	1754	3	6	9	11	14
80	.1736	1719	1702	1685	1668	1650	1633	1616	1599	1582	3	6	9	12	14
81	.1564	1547	1530	1513	1495	1478	1461	1444	1426	1409	3	6	9	12	14
82	.1392	1374	1357	1340	1323	1305	1288	1271	1253	1236	3	6	9	12	14
83	.1219	1201	1184	1167	1149	1132	1115	1097	1080	1063	3	6	9	12	14
84	.1045	1028	1011	0993	0976	0958	0941	0924	0906	0889	3	6	9	12	14
85	.0872	0854	0837	0819	0802	0785	0767	0750	0732	0715	3	6	9	12	14
86	.0698	0680	0663	0645	0623	0610	0593	0576	0558	0541	3	6	9	12	15
87	.0523	0506	0488	0471	0454	0436	0419	0401	0384	0366	3	6	9	12	15
88	.0349	0332	0314	0297	0279	0262	0244	0227	0209	0192	3	6	9	12	15
89	.0175	0157	0140	0122	0105	0087	0070	0052	0035	0017	3	6	9	12	15

NATURAL TANGENTS

Degree	0′ 0°.0	6′ 0°.1	12′ 0°.2	18′ 0°.3	24′ 0°.4	30′ 0°.5	36′ 0°.6	42′ 0°.7	48′ 0°.8	54′ 0°.9	Maen Differences				
											1′	2′	3′	4′	5′
0	.0000	0017	0035	0052	0070	0087	0105	0122	0140	0157	3	6	9	12	15
1	.0175	0192	0209	0227	0244	0262	0279	0297	0314	0332	3	6	9	12	15
2	.0349	0367	0384	0402	0419	0437	0454	0472	0489	0507	3	6	9	12	15
3	.0524	0542	0559	0577	0594	0612	0629	0647	0664	0682	3	6	9	12	15
4	.0699	0717	0734	0752	0769	0787	0805	0822	0840	0857	3	6	9	12	15
5	.0875	0892	0910	0928	0945	0963	0981	0998	1016	1033	3	6	9	12	15
6	.1051	1069	1086	1104	1122	1139	1157	1175	1192	1210	3	6	9	12	15
7	.1228	1246	1263	1281	1299	1317	1334	1352	1370	1388	3	6	9	12	15
8	.1405	1423	1441	1459	1477	1495	1512	1530	1548	1566	3	6	9	12	15
9	.1584	1602	1620	1638	1655	1673	1691	1709	1727	1745	3	6	9	12	15
10	.1763	1781	1799	1817	1835	1853	1871	1890	1908	1926	3	6	9	12	15
11	.1944	1962	1980	1998	2016	2035	2053	2071	2089	2107	3	6	9	12	15
12	.2126	2144	2162	2180	2199	2217	2235	2254	2272	2290	3	6	9	12	15
13	.2309	2327	2345	2364	2382	2401	2419	2438	2456	2475	3	6	9	12	15
14	.2493	2512	2530	2549	2568	2586	2605	2623	2642	2661	3	6	9	12	16
15	.2679	2698	2717	2736	2754	2773	2792	2811	2830	2849	3	6	9	13	16
16	.2867	2886	2905	2924	2943	2962	2981	3000	3019	3038	3	6	9	13	16
17	.3057	3076	3096	3115	3134	3153	3172	3191	3211	3230	3	6	10	13	16
18	.3249	3269	3288	3307	3327	3346	3365	3385	3404	3424	3	6	10	13	16
19	.3443	3463	3482	3502	3522	3541	3561	3581	3600	3620	3	7	10	13	16
20	.3640	3659	3679	3699	3719	3739	3759	3779	3799	3819	3	7	10	13	17
21	.3839	3859	3879	3899	3919	3939	3959	3979	4000	4020	3	7	10	13	17
22	.4040	4061	4081	4101	4122	4142	4163	4183	4204	4224	3	7	10	14	17
23	.4245	4265	4286	4307	4327	4348	4369	4390	4411	4431	3	7	10	14	17
24	.4452	4473	4494	4515	4536	4557	4578	4599	4621	4642	4	7	11	14	18
25	.4663	4684	4706	4727	4748	4770	4791	4813	4834	4856	4	7	11	14	18
26	.4877	4899	4921	4942	4964	4986	5008	5029	5051	5073	4	7	11	15	18
27	.5095	5117	5139	5161	5184	5206	5228	5250	5272	5295	4	7	11	15	18
28	.5317	5340	5362	5384	5407	5430	5452	5475	5498	5520	4	8	11	15	19
29	.5543	5566	5589	5612	5635	5658	5681	5704	5727	5750	4	8	12	15	19
30	.5774	5797	5820	5844	5867	5890	5914	5938	5961	5985	4	8	12	16	20
31	.6009	6032	6056	6080	6104	6128	6152	6176	6200	6224	4	8	12	16	20
32	.6249	6273	6297	6322	6346	6371	6395	6420	6445	6469	4	8	12	16	20
33	.6494	6519	6544	6569	6594	6619	6644	6669	6694	6720	4	8	13	17	21
34	.6745	6771	6796	6822	6874	6873	6899	6924	6950	6976	4	9	13	17	21
35	.7002	7028	7054	7080	7107	7133	7159	7186	7212	7239	4	9	13	18	22
36	.7265	7292	7319	7346	7373	7400	7427	7454	7481	7508	5	9	14	18	23
37	.7536	7563	7590	7618	7646	7673	7701	7729	7757	7785	5	9	14	18	23
38	.7813	7841	7869	7898	7926	7954	7983	8012	8040	8069	5	9	14	19	24
39	.8098	8127	8156	8185	8214	8243	8273	8302	8332	8361	5	10	15	20	24
40	.8391	8421	8451	8481	8511	8541	8571	8601	8632	8662	5	10	15	20	25
41	.8693	8724	8754	8785	8816	8847	8878	8910	8941	8972	5	10	16	21	26
42	.9004	9036	9067	9099	9131	9163	9195	9228	9260	9293	5	11	16	21	27
43	.9325	9358	9391	9424	9457	9490	9523	9556	9590	9623	6	11	17	22	28
44	.9657	9691	9725	9759	9793	9827	9861	9896	9930	9965	6	11	17	23	29

NATURAL TANGENTS

Degree	0′ 0°.0	6′ 0°.1	12′ 0°.2	18′ 0°.3	24′ 0°.4	30′ 0°.5	36′ 0°.6	42′ 0°.7	48′ 0°.8	54′ 0°.9	Mean Differences 1′	2′	3′	4′	5′
45	1.0000	0035	0070	0105	0141	0176	0212	0247	0283	0319	6	12	18	24	30
46	1.0355	0392	0428	0464	0501	0538	0575	0612	0649	0686	6	12	18	25	31
47	1.0724	0761	0799	0837	0875	0913	0951	0990	1028	1067	6	13	19	25	32
48	1.1106	1145	1184	1224	1263	1303	1343	1383	1423	1463	7	13	20	27	33
49	1.1504	1544	1585	1626	1667	1708	1750	1792	1833	1875	7	14	21	28	34
50	1.1918	1960	2002	2045	2088	2131	2174	2218	2261	2305	7	14	22	29	36
51	1.2349	2393	2437	2484	2527	2572	2617	2662	2708	2753	8	15	23	30	38
52	1.2799	2846	2892	2938	2985	3032	3079	3127	3175	3222	8	16	24	31	39
53	1.3270	3319	3367	3416	3465	3514	3564	3613	3663	3713	8	16	25	33	41
54	1.3764	3814	3865	3916	3968	4019	4071	4124	4176	4229	9	17	26	34	43
55	1.4281	4335	4388	4442	4496	4550	4605	4659	4715	4770	9	18	27	36	45
56	1.4826	4882	4938	4994	5051	5108	5166	5224	5282	5340	10	19	29	38	48
57	1.5399	5458	5517	5577	5637	5697	5757	5818	5880	5941	10	20	30	40	50
58	1.6003	6066	6128	6191	6255	6319	6383	6447	6512	6577	11	21	32	43	53
59	1.6643	6709	6775	6842	6909	6977	7045	7113	7182	7251	11	23	34	45	56
60	1.7321	7391	7461	7532	7603	7675	7747	7820	7893	7966	12	24	36	48	60
61	1.8040	8115	8190	8265	8341	8418	8495	8572	8650	8728	13	26	38	51	64
62	1.8807	8887	8967	9047	9128	9210	9292	9375	9458	9542	14	27	41	55	68
63	1.9626	9711	9797	9883	9970	2.0057	2.0145	2.0233	2.0323	2.0413	15	29	44	58	73
64	2.0503	0594	0686	0778	0872	0965	1060	1155	1251	1348	16	31	47	63	78
65	2.1445	1543	1642	1742	1842	1943	2045	2148	2251	2355	17	34	51	68	85
66	2.2460	2566	2673	2781	2889	2998	3109	3220	3332	3445	18	37	55	73	92
67	2.3559	3673	3789	3906	4023	4142	4262	4383	4504	4627	20	40	60	79	99
68	2.4751	4876	5002	5129	5257	5386	5517	5649	5782	5916	22	43	65	87	108
69	2.6051	6187	6325	6464	6605	6746	6889	7034	7179	7326	24	47	71	95	119
70	2.7475	7625	7776	7929	8083	8239	8397	8556	8716	8878	26	52	78	104	133
71	2.9042	9208	9375	9544	9714	9887	3.0061	3.0237	3.0415	3.0595	29	58	87	116	145
72	3.0777	0961	1146	1334	1524	1716	1910	2106	2305	2506	32	64	96	129	161
73	3.2709	2914	3122	3332	3544	3759	3977	4197	4420	4646	36	72	108	144	180
74	3.4874	5105	5339	5576	5816	6059	6305	6554	6806	7062	41	81	122	163	204
75	3.7321	7583	7848	8118	8391	8667	8947	9232	9520	9812	46	93	139	186	232
76	4.0108	0408	0713	1022	1335	1653	1976	2303	2635	2972	53	107	160	213	267
77	4.3315	3662	4015	4374	4737	5107	5483	5864	6252	6646					
78	4.7046	7453	7867	8288	8716	9152	9594	5.0045	5.0504	5.0970					
79	5.1446	1923	2422	2924	3435	3955	4486	5026	5578	6140					
80	5.6713	7297	7894	8502	9124	9758	6.0405	6.1066	6.1742	6.2432					
81	6.3138	3859	4596	5350	6122	6912	7720	8548	9395	7.0264					
82	7.1154	2066	3002	3962	4947	5958	6996	8062	9158	8.0285					
83	8.1443	2636	3863	5126	6427	7769	9152	9.0579	9.2052	9.3572					
84	9.5144	9.677	9.845	10.02	10.20	10.39	10.58	10.78	10.99	11.20	Mean differences no longer sufficiently accurate.				
85	11.43	11.66	11.91	12.16	12.43	12.71	13.00	13.30	13.62	13.95					
86	14.30	14.67	15.06	15.46	15.89	16.35	16.83	17.34	17.89	18.46					
87	19.08	19.74	20.45	21.20	22.02	22.90	23.86	24.90	26.03	27.27					
88	28.64	30.14	31.82	33.69	35.80	38.19	40.92	44.07	47.74	52.08					
89	57.29	63.66	71.62	81.85	95.49	114.6	143.2	191.0	286.5	573.0					

NATURAL COSECANTS

(Numbers in difference columns to be subtracted, not added.)

Degrees	0′ 0°.0	6′ 0°.1	12′ 0°.2	18′ 0°.3	24′ 0°.4	30′ 0°.5	36′ 0°.6	42′ 0°.7	48′ 0°.8	54′ 0°.9	Mean Differences 1	2	3	4	5
0	Inf.	573.0	286.5	191.0	143.2	114.6	95.49	81.85	71.62	63.66					
1	57.30	52.09	47.75	44.08	40.93	38.20	35.81	33.71	31.84	30.16					
2	28.65	27.29	26.05	24.92	23.88	22.93	22.04	21.23	20.47	19.77					
3	19.11	18.49	17.91	17.37	16.86	16.38	15.93	15.50	15.09	14.70					
4	14.34	13.99	13.65	13.34	13.03	12.75	12.47	12.20	11.95	11.71					
5	11.47	11.25	11.03	10.83	10.63	10.43	10.25	10.07	9.895	9.728					
6	9.5668	9.4105	9.2593	9.1129	8.9711	8.8337	8.7004	8.5711	8.4457	8.3238					
7	8.2055	8.0905	7.9787	7.8700	7.7642	7.6613	7.5611	7.4635	7.3684	7.2757	Mean differences are not sufficiently accurate.				
8	7.1853	7.0972	7.0112	6.9273	6.8454	6.7655	6.6874	6.6111	6.5366	6.4637					
9	6.3925	6.3228	6.2546	6.1880	6.1227	6.0589	5.9963	5.9351	5.8751	5.8164					
10	5.7588	5.7023	5.6470	5.5928	5.5396	5.4874	5.4362	5.3860	5.3367	5.2883					
11	5.2408	5.1942	5.1484	5.1034	5.0593	5.0159	4.9732	4.9313	4.8901	4.8496					
12	4.8097	4.7706	4.7321	4.6942	4.6569	4.6202	4.5841	4.5486	4.5137	4.4793					
13	4.4454	4.4121	4.3792	4.3469	4.3150	4.2837	4.2527	4.2223	4.1923	4.1627					
14	4.1336	4.1048	4.0765	4.0486	4.0211	3.9939	3.9672	3.9408	3.9147	3.8890					
15	3.8637	8387	8140	7897	7657	7420	7186	6955	6727	6502					
16	3.6280	6060	5843	5629	5418	5209	5003	4799	4598	4399					
17	3.4203	4009	3817	3628	3440	3255	3072	2891	2712	2535					
18	3.2361	2188	2017	1848	1681	1515	1352	1190	1030	0872	27	55	82	105	137
19	3.0716	0561	0407	0256	0106	2.9957	2.9811	2.9665	2.9521	2.9379	25	49	74	99	123
20	2.9238	9099	8960	8824	8688	8555	8422	8291	8161	8032	22	44	67	89	119
21	2.7904	7778	7653	7529	7407	7285	7165	7046	6927	6811	20	40	60	81	101
22	2.6695	6580	6466	6354	6242	6131	6022	5913	5805	5699	18	37	55	73	92
23	2.9593	5488	5384	5282	5180	5078	4978	4879	4780	4683	17	34	50	67	84
24	2.4586	4490	4395	4300	4207	4114	4022	3931	3841	3751	15	31	46	62	77
25	2.3662	3574	3486	3400	3314	3228	3144	3060	2976	2894	14	28	43	57	71
26	2.2812	2730	2650	2570	2490	2412	2333	2256	2179	2103	13	26	39	52	65
27	2.2027	1952	1877	1803	1730	1657	1584	1513	1441	1371	12	24	36	48	60
28	2.1301	1231	1162	1093	1015	0957	0890	0824	0757	0692	11	22	34	45	56
29	2.0627	0562	0498	0434	0371	0308	0245	0183	0122	0061	10	21	31	42	52
30	2.0000	1.9940	1.9880	1.9821	1.9762	1.9703	1.9645	1.9587	1.9530	1.9473	10	19	29	39	49
31	1.9416	9360	9304	9249	9194	9139	9084	9031	8977	8924	9	18	27	36	45
32	1.8871	8818	8766	8714	8663	8612	8561	8510	8460	8410	8	17	25	34	42
33	1.8361	8312	8263	8214	8166	8118	8070	8023	7976	7929	8	16	24	32	40
34	1.7883	7837	7791	7745	7700	7655	7610	7566	7522	7478	7	15	22	30	37
35	1.7434	7391	7348	7305	7263	7221	7179	7137	7095	7054	7	14	21	28	35
36	1.7013	6972	6932	6892	6852	6812	6772	6733	6694	6655	7	13	20	26	33
37	1.6616	6578	6540	6502	6464	6427	6390	6353	6316	6279	6	12	19	25	31
38	1.6243	6207	6171	6135	6099	6064	6029	5994	5959	5925	6	12	18	23	29
39	1.5890	5856	5822	5788	5755	5721	5688	5655	5622	5590	6	11	17	22	28
40	1.5557	5525	5493	5461	5429	5398	5366	5335	5304	5273	5	10	16	21	26
41	1.5243	5212	5182	5151	5121	5092	5062	5032	5003	4974	5	10	15	20	25
42	1.4945	4916	4887	4859	4830	4802	4774	4746	4718	4690	5	9	14	19	23
43	1.4663	4635	4608	4581	4554	4527	4501	4474	4448	4422	4	9	13	18	22
44	1.4396	4370	4344	4318	4293	4267	4242	4217	4192	4167	4	8	13	17	21

NATURAL COSECANTS

(Numbers in difference column to be subtracted, not added.)

Degrees	0′ 0°.0	6′ 0°.1	12′ 0°.2	18′ 0°.3	24′ 0°.4	30′ 0°.5	36′ 0°.6	42′ 0°.7	48′ 0°.8	54′ 0°.9	Mean Differences 1	2	3	4	5
45	1.4142	4118	4093	4069	4044	4020	3996	3972	3949	3925	4	8	12	16	20
46	1.3902	3878	3855	3832	3809	3786	3763	3741	3718	3696	4	8	11	15	19
47	1.3673	3651	3629	3607	3585	3563	3542	3520	3499	3478	4	7	11	14	18
48	1.3456	3435	3414	3393	3373	3352	3331	3311	3291	3270	3	7	10	14	17
49	1.3250	3230	3210	3190	3171	3151	3131	3112	3093	3073	3	7	10	13	16
50	1.3054	3035	3016	2997	2978	2960	2941	2923	2904	2886	3	6	9	12	15
51	1.2868	2849	2831	2813	2796	2778	2760	2742	2725	2708	3	6	9	12	15
52	1.2690	2673	2656	2639	2622	2605	2588	2571	2554	2538	3	6	8	11	14
53	1.2521	2505	2489	2472	2456	2440	2424	2408	2392	2376	3	5	8	11	13
54	1.2361	2345	2329	2314	2299	2283	2268	2253	2238	2223	3	5	8	10	13
55	1.2208	2193	2178	2163	2149	2134	2120	2105	2091	2076	2	5	7	10	12
56	1.2062	2048	2034	2020	2006	1992	1978	1964	1951	1936	2	5	7	9	12
57	1.1924	1910	1897	1883	1870	1857	1844	1831	1818	1805	2	4	7	9	11
58	1.1792	1779	1766	1753	1741	1728	1716	1703	1691	1679	2	4	6	8	10
59	1.1666	1654	1642	1630	1618	1606	1594	1582	1570	1559	2	4	6	8	10
60	1.1547	1535	1524	1512	1501	1490	1478	1467	1456	1445	2	4	6	8	9
61	1.1434	1423	1412	1401	1390	1379	1368	1357	1347	1336	2	4	5	7	9
62	1.1326	1315	1305	1294	1284	1274	1264	1253	1243	1233	2	3	5	7	9
63	1.1223	1213	1203	1194	1184	1174	1164	1155	1145	1136	2	3	5	6	8
64	1.1126	1117	1107	1098	1089	1079	1070	1061	1052	1043	2	3	5	6	8
65	1.1034	1025	1016	1007	0998	0989	0987	0972	0963	0955	1	3	4	6	7
66	1.0946	0938	0929	0921	0913	0904	0896	0888	0880	0872	1	3	4	6	7
67	1.0864	0856	0848	0840	0832	0824	0816	0808	0801	0793	1	3	4	5	6
68	1.0785	0778	0770	0763	0755	0748	0740	0733	0726	0719	1	2	4	5	6
69	1.0711	0704	0697	0690	0683	0676	0669	0662	0655	0649	1	2	3	5	6
70	1.0642	0635	0628	0622	0615	0608	0602	0595	0589	0583	1	2	3	4	5
71	1,0576	0570	0564	0557	0551	0545	0539	0533	0527	0521	1	2	3	4	5
72	1.0515	0509	0503	0497	0491	0485	0480	0474	0468	0463	1	2	3	4	5
73	1.0457	0451	0446	0440	0435	0429	0424	0419	0413	0408	1	2	3	3	4
74	1.0403	0398	0393	0388	0382	0377	0372	0367	0363	0358	1	2	2	3	4
75	1.0353	0348	0343	0338	0334	0329	0324	0320	0315	0311	1	2	2	3	4
76	1.0306	0302	0297	0293	0288	0284	0280	0276	0271	0267	1	1	2	3	4
77	1.0263	0259	0255	0251	0247	0243	0239	0235	0231	0227	1	1	2	3	3
78	1.0223	0220	0216	0212	0209	0205	0201	0198	0194	0191	1	1	2	2	3
79	1.0187	0184	0180	0177	0174	0170	0167	0164	0161	0157	1	1	2	2	3
80	1.0154	0151	0148	0145	0142	0139	0136	0133	0130	0127	0	1	1	2	2
81	1.0125	0122	0119	0116	0114	0111	0108	0106	0103	0101	0	1	1	2	2
82	1.0098	0096	0093	0091	0089	0086	0084	0082	0079	0077	0	1	1	2	2
83	1.0075	0073	0071	0069	0067	0065	0063	0061	0059	0057	0	1	1	1	2
84	1.0055	0053	0051	0050	0048	0046	0045	0043	0041	0040	0	1	1	1	1
85	1.0038	0037	0035	0034	0032	0031	0030	0028	0027	0026	0	0	1	1	1
86	1.0024	0023	0022	0021	0020	0019	0018	0017	0016	0015	0	0	0	1	1
87	1.0014	0013	0012	0011	0010	0010	0009	0008	0007	0007	0	0	0	1	1
88	1.0006	0006	0005	0004	0004	0003	0003	0003	0002	0002	0	0	0	0	0
89	1.0002	0001	0001	0001	0001	0000	0000	0000	0000	0000	0	0	0	0	0

NATURAL SECANTS

Degrees	0′ 0°.0	6′ 0°.1	12′ 0°.2	18′ 0°.3	24′ 0°.4	30′ 0°.5	36′ 0°.6	42′ 0°.7	48′ 0°.8	54′ 0°.9	Mean Differences 1	2	3	4	5
0	1.0000	0000	0000	0000	0000	0000	0001	0001	0001	0001	0	0	0	0	0
1	1.0002	0002	0002	0003	0003	0003	0004	0004	0005	0006	0	0	0	0	0
2	1.0006	0007	0007	0008	0009	0010	0010	0011	0012	0013	0	0	0	0	0
3	1.0014	0015	0016	0017	0018	0019	0020	0021	0022	0023	0	0	1	1	1
4	1.0024	0026	0027	0028	0030	0031	0032	0034	0035	0037	0	0	1	1	1
5	1.0038	0040	0041	0043	0045	0046	0048	0050	0051	0053	0	1	1	1	1
6	1.0055	0057	0059	0061	0063	0065	0067	0069	0071	0073	0	1	1	1	2
7	1.0075	0077	0079	0082	0084	0086	0089	0091	0093	0096	0	1	1	2	2
8	1.0098	0101	0103	0106	0108	0111	0114	0116	0119	0122	0	1	1	2	2
9	1.0125	0127	0130	0133	0136	0139	0142	0145	0148	0151	0	1	1	2	2
10	1.0154	0157	0161	0164	0167	0170	0174	0177	0180	0184	1	1	2	2	3
11	1.0187	0191	0194	0198	0201	0205	0209	0212	0216	0220	1	1	2	2	3
12	1.0223	0227	0231	0235	0239	0243	0247	0251	0255	0259	1	1	2	3	3
13	1.0263	0267	0271	0276	0280	0284	0288	0293	0297	0302	1	1	2	3	4
14	1.0306	0311	0315	0320	0324	0329	0334	0338	0343	0348	1	2	2	3	4
15	1.0353	0358	0363	0367	0372	0377	0382	0388	0393	0398	1	2	3	3	4
16	1.0403	0408	0413	0419	0424	0429	0435	0440	0446	0451	1	2	3	3	4
17	1.0457	0463	0468	0474	0480	0485	0491	0497	0503	0509	1	2	3	4	5
18	1.0515	0521	0527	0533	0539	0545	0551	0557	0564	0570	1	2	3	4	5
19	1.0576	0583	0589	0595	0602	0608	0615	0622	0628	0635	1	2	3	4	5
20	1.0642	0649	0655	0662	0669	0676	0683	0690	0697	0704	1	2	3	5	6
21	1.0711	0719	0726	0733	0740	0748	0755	0763	0770	0778	1	2	4	5	6
22	1.0785	0793	0801	0808	0816	0824	0832	0840	0848	0856	1	3	4	5	6
23	1.0864	0872	0880	0888	0896	0904	0913	0921	0929	0938	1	3	4	6	7
24	1.0946	0955	0963	0972	0981	0989	0998	1007	1016	1025	1	3	4	6	7
25	1.1034	1043	1052	1061	1070	1079	1089	1098	1107	1117	2	3	5	6	8
26	1.1126	1136	1145	1155	1164	1174	1184	1194	1203	1213	2	3	5	6	8
27	1.1223	1233	1243	1253	1264	1274	1284	1294	1305	1315	2	3	5	7	9
28	1.1326	1336	1347	1357	1368	1379	1390	1401	1412	1423	2	4	5	7	9
29	1.1434	1445	1456	1467	1478	1490	1501	1512	1524	1535	2	4	6	8	9
30	1.1547	1559	1570	1582	1594	1606	1618	1630	1642	1654	2	4	6	8	10
31	1.1666	1679	1691	1703	1716	1728	1741	1753	1766	1779	2	4	6	8	10
32	1.1792	1805	1818	1831	1844	1857	1870	1883	1897	1910	2	4	7	9	11
33	1.1924	1937	1951	1964	1978	1992	2006	2020	2034	2048	2	5	7	9	12
34	1.2062	2076	2091	2105	2120	2134	2149	2163	2178	2193	2	5	7	10	12
35	1.2208	2223	2238	2253	2268	2283	2299	2314	2329	2345	3	5	8	10	13
36	1.2361	2376	2392	2408	2424	2440	2456	2472	2489	2505	3	5	8	11	13
37	1.2521	2538	2554	2571	2588	2605	2622	2639	2656	2673	3	6	8	11	14
38	1.2690	2708	2725	2742	2760	2778	2796	2813	2831	2849	3	6	9	12	15
39	1.2868	2886	2904	2923	2941	2960	2978	2997	3016	3035	3	6	9	12	16
40	1.3054	3073	3093	3112	3131	3151	3171	3190	3210	3230	3	7	10	13	16
41	1.3250	3270	3291	3311	3331	3352	3373	3393	3414	3435	3	7	10	14	17
42	1.3456	3478	3499	3520	3542	3563	3585	3607	3629	3651	4	7	11	14	18
43	1.3673	3696	3718	3741	3763	3786	3809	3832	3855	3878	4	8	11	15	19
44	1.3902	3925	3949	3972	3996	4020	4044	4069	4093	4118	4	8	12	16	20

NATURAL SECANTS

Degrees	0′	6′	12′	18′	24′	30′	36′	42′	48′	54′	Mean Differences				
	0°.0	0°.1	0°.2	0°.3	0°.4	0°.5	0°.6	0°.7	0°.8	0°.9	1	2	3	4	5
45	1.4142	4167	4192	4217	4242	4267	4293	4318	4344	4370	4	8	13	17	21
46	1.4396	4422	4448	4474	4501	4527	4554	4581	4608	4635	4	9	13	18	22
47	1.4663	4690	4718	4746	4774	4802	4830	4859	4887	4916	5	9	14	19	23
48	1.4945	4974	5003	5032	5062	5092	5121	5151	5182	5212	5	10	15	20	25
49	1.5243	5273	5304	5335	5366	5398	5429	5461	5493	5525	5	10	16	21	26
50	1.5557	5590	5622	5655	5688	5721	5755	5788	5822	5856	6	11	17	22	28
51	1.5890	5925	5959	5994	6029	6064	6099	6135	6171	6207	6	12	18	24	29
52	1.6243	6279	6316	6353	6390	6427	6464	6502	6540	6578	6	12	19	25	31
53	1.6616	6655	6694	6733	6772	6812	6852	6892	6932	6972	7	13	20	26	33
54	1.7013	7054	7095	7137	7179	7221	7263	7305	7348	7391	7	14	21	28	35
55	1.7434	7478	7522	7566	7610	7655	7700	7745	7791	7837	7	15	22	30	37
56	1.7883	7929	7976	8023	8070	8118	8166	8214	8263	8312	8	16	24	32	40
57	1.8361	8410	8460	8510	8561	8612	8663	8714	8766	8818	9	17	26	34	43
58	1.8871	8924	8977	9031	9084	9139	9194	9249	9304	9360	9	18	27	36	45
59	1.9416	9473	9530	9587	9645	9703	9762	9821	9880	9940	10	19	29	39	49
60	2.0000	0061	0122	0183	0245	0308	0371	0434	0498	0562	10	21	31	42	52
61	2.0627	0692	0757	0824	0890	0957	1025	1093	1162	1231	11	22	34	45	56
62	2.1301	1371	1441	1513	1584	1657	1730	1803	1877	1952	12	24	36	48	61
63	2.2027	2103	2179	2256	2333	2412	2490	2570	2650	2730	13	26	39	52	66
64	2.2812	2894	2976	3060	3144	3228	3314	3400	3486	3574	14	28	43	57	71
65	2.3662	3751	3841	3931	4022	4114	4207	4300	4395	4490	15	31	46	62	77
66	2.4586	4683	4780	4879	4978	5078	5180	5282	5384	5488	17	34	50	67	84
67	2.5593	5699	5805	5913	6022	6131	6242	6354	6466	6580	18	37	55	73	92
68	2.6695	6811	6927	7046	7165	7285	7407	7529	7653	7778	20	40	60	81	101
69	2.7904	8032	8161	8291	8422	8555	8688	8824	8960	9099					
70	2.9238	9379	9521	9665	9811	9957	3.0106	3.0256	3.0407	3.0561					
71	3.0716	0872	1030	1190	1350	1515	1681	1848	2017	2188					
72	3.2361	2535	2712	2891	3072	3255	3440	3628	3817	4009					
73	3.4203	4399	4598	4799	5003	5209	5418	5629	5843	6060					
74	3.6280	6502	6727	6955	7186	7420	7657	7897	8140	8387					
75	3.8637	8890	9147	9408	9672	9939	4.0211	4.0486	4.0765	4.1048					
76	4.1336	1627	1923	2223	2527	2837	3150	3469	3792	4121					
77	4.4454	4793	5137	5486	5841	6202	6569	6942	7321	7706					
78	4.8097	8496	8901	0313	9732	5.0159	5.0593	5.1034	5.1484	5.1942	Mean differences cease to be sufficiently accurate.				
79	5.2408	2883	3367	3860	4362	4874	5396	5928	6470	7023					
80	5.7588	8164	8751	9351	9963	6.0589	6.1227	6.1880	6.2546	6.3228					
81	6.3925	4637	5366	6111	6874	7655	8454	9273	7.0112	7.0972					
82	7.1853	2757	3684	4635	5611	6613	7642	8700	9787	8.0905					
83	8.2055	3238	4457	5711	7004	8337	9711	9.1129	9.2593	9.4105					
84	9.5668	9.728	9.895	10.07	10.25	10.43	10.63	10.83	11.03	11.25					
85	11.47	11.71	11.95	12.20	12.47	12.75	13.03	13.34	13.65	13.99					
86	14.34	14.70	15.09	15.50	15.93	16.38	16.86	17.37	17.91	18.49					
87	19.11	19.77	20.47	21.23	22.04	22.93	23.88	24.92	26.05	27.29					
88	28.65	30.16	31.84	33.71	35.81	38.20	40.93	44.08	47.75	52.09					
89	57.30	03.66	71.62	81.85	95.49	114.6	143.2	191.0	286.5	573.0					

NATURAL COTANGENTS

(Mean Difference to be Subtracted)

Degrees	0′ 0°.0	6′ 0°.1	12′ 0°.2	18′ 0°.3	24′ 0°.4	30′ 0°.5	36′ 0°.6	42′ 0°.7	48′ 0°.8	54′ 0°.9	Mean Differences 1	2	3	4	5
0	Inf.	573.0	286.5	191.0	743.2	114.6	95.4	981.8	571.6	263.66					
1	57.29	52.08	47.74	44.07	40.92	38.19	35.80	33.69	31.82	30.14					
2	28.64	27.27	26.03	24.90	23.86	22.90	22.02	21.20	20.45	19.74					
3	19.08	18.46	17.89	17.34	16.83	16.35	15.89	15.46	15.06	14.67					
4	14.30	13.95	13.62	13.30	13.00	12.71	12.43	12.16	11.91	11.66					
5	11.43	11.20	10.99	10.78	10.58	10.39	10.20	10.02	9.845	9.677					
6	9.5144	9.3572	9.2052	9.0579	8.9152	8.7769	8.6427	8.5126	8.3863	8.2636					
7	8.1443	8.0285	7.9158	7.8062	7.6996	7.5958	7.4947	7.3962	7.3002	7.2066					
8	7.1154	7.0264	6.9395	6.8548	6.7720	6.6912	6.6122	6.5350	6.4596	6.3859					
9	6.3138	6.2432	6.1742	6.1066	6.0405	5.9758	5.9124	5.8502	5.7894	5.7297					
10	5.6713	6140	5578	5026	4486	3955	3435	2924	2422	1929					
11	5.1446	0970	0504	0045	4.9594	4.9152	4.8716	4.8288	4.7867	4.7453					
12	4.7046	6646	6252	5864	5483	5107	4737	4374	4015	3662					
13	4.3315	2972	2635	2303	1976	1653	1335	1022	0713	0408	53	107	160	213	267
14	4.0108	3.9812	3.9520	3.9232	3.8947	3.8667	3.8391	3.8118	3.7848	3.7583	46	93	139	186	232
15	3.7321	7062	6806	6554	6305	6059	5816	5576	5339	5105	41	81	122	163	204
16	3.4874	4646	4420	4197	3977	3759	3544	3332	3122	2914	36	72	108	144	180
17	3.2709	2506	2305	2106	1910	1716	1524	1334	1146	0961	32	64	96	129	161
18	3.0777	0595	0415	0237	0061	2.9887	2.9714	2.9544	2.9375	2.9208	29	58	87	116	144
19	2.9042	8878	8716	8556	8397	5239	8083	7929	7776	7625	26	52	78	104	131
20	2.7475	7326	7179	7034	6889	6746	6605	6464	6325	6187	24	47	71	95	119
21	2.6051	5916	5782	5649	5517	5386	5257	5129	5002	4876	22	43	65	87	108
22	2.4751	4627	4504	4383	4262	4142	4023	3906	3789	3673	20	40	60	79	99
23	2.3559	3445	3332	3220	3109	2998	2889	2781	2673	2566	18	37	55	73	92
24	2.2460	2355	2251	2148	2045	1943	1842	1742	1642	1543	17	34	51	68	85
25	2.1445	1348	1251	1155	1060	0965	0872	0778	0686	0594	16	31	47	63	78
26	2.0503	0413	0323	9233	0145	0057	1.9970	1.9883	1.9797	1.9711	15	29	44	58	73
27	1.9626	9542	0458	9375	9292	9210	9128	9047	9867	8887	14	27	41	55	68
28	1.8807	8728	8650	8572	8495	8418	8341	8265	8190	8115	13	26	38	51	64
29	1.8040	7966	7893	7820	7747	7675	7603	7532	7461	7391	12	24	36	48	60
30	1.7321	7251	7182	7113	7045	6977	6909	6842	6775	6709	11	23	34	45	56
31	1.6643	6577	6512	6447	6383	6319	6255	6191	6128	6066	11	21	32	43	53
32	1.6003	5941	5880	5818	5757	5697	5637	5577	5517	5458	10	20	30	40	50
33	1.5399	5340	5282	5224	5166	5108	5051	4994	4938	4882	10	19	29	38	48
34	1.4826	4770	4715	4659	4605	4550	4496	4442	4388	4335	9	18	27	36	45
35	1.4281	4229	4176	4124	4071	4019	3968	3916	3865	3814	9	17	26	34	43
36	1.2764	3713	3663	3613	3564	3514	3465	3416	3367	3319	8	16	25	33	41
37	1.3270	3222	3175	3127	3079	3032	2985	2938	2892	2846	8	16	23	31	39
38	1.2799	2753	2708	2662	2617	2572	2527	2582	2437	2393	8	15	23	30	38
39	1.2349	2305	2261	2218	2174	2131	2088	2045	2002	1960	7	14	22	29	36
40	1.1918	1875	1833	1782	1750	1708	1667	1626	1585	1544	7	14	21	28	34
41	1.1504	1463	1423	1383	1343	1303	1263	1224	1184	1145	7	13	20	26	33
42	1.1106	1067	1028	0990	0951	0913	0875	0837	0799	0761	6	13	19	25	32
43	1.0724	0686	0649	0612	0575	0538	0501	0464	0428	0392	6	12	18	25	31
44	1.0335	0319	0283	0247	0212	0176	0141	0105	0070	0035	6	12	18	24	30

NATURAL COTANGENTS

(Mean Difference to be Subtracted)

Degrees	0′ 0°.0	6′ 0°.1	12′ 0°.2	18′ 0°.3	24′ 0°.4	30′ 0°.5	36′ 0°.6	42′ 0°.7	48′ 0°.8	54′ 0°.9	Mean Differences 1	2	3	4	5
45	1.0000	9965	9930	9896	9861	9828	9793	9759	9725	9691	6	11	17	23	29
46	0.9657	9623	9540	9556	9523	9490	9457	9424	9391	9358	6	11	17	22	28
47	0.9325	9293	9260	9228	9195	9163	9131	9099	9067	9036	5	11	16	21	27
48	0.9004	8972	8941	8910	8878	8847	8816	8785	8754	8724	5	10	16	21	26
49	0.8693	8662	8632	8601	8571	8541	8511	8481	8451	8421	5	10	15	20	25
50	0.8391	8361	8332	8302	8273	8243	8214	8185	8156	8127	5	10	15	20	24
51	0.8098	8069	8040	8012	7983	7954	7926	7898	7869	7841	5	10	14	19	24
52	0.7813	7785	7757	7729	7701	7673	7646	7614	7590	7563	5	9	14	18	23
53	0.7536	7508	7481	7454	7427	7400	7373	7346	7319	7292	5	9	14	18	23
54	0.7265	7239	7212	7186	7159	7133	7107	7080	7054	7028	4	9	13	17	22
55	0.7002	6976	6950	6924	6899	6873	6847	6822	6796	6771	4	9	13	17	21
56	0.6745	6720	6694	6669	6644	6619	6594	6569	6544	6519	4	8	13	17	21
57	0.6494	6496	6445	6420	6395	6371	6346	6322	6297	6273	4	8	12	16	20
58	0.6249	6224	6200	6176	6152	6128	6104	6080	6056	6032	4	8	12	16	20
59	0.6009	5985	5961	5938	5914	5890	5867	5844	5820	5797	4	8	12	16	20
60	0.5774	5750	5727	5704	5681	5658	5635	5612	5589	5566	4	8	12	15	19
61	0.5543	5520	5498	5475	5452	5430	5407	5384	5362	5340	4	8	11	15	19
62	0.5317	5295	5272	5250	5228	5206	5184	5161	5139	5117	4	7	11	15	18
63	0.5095	5073	5051	5029	5008	4986	4964	4942	4921	4899	4	7	11	15	18
64	0.4877	4856	4834	4813	4791	4770	4748	4727	4706	4648	4	7	11	14	18
65	0.4663	4642	4621	4599	4578	4557	4536	4515	4494	4473	4	7	10	14	18
66	0.4452	4431	4411	4390	4369	4348	4327	4307	4286	4265	3	7	10	14	17
67	0.4245	4224	4204	4183	4163	4142	4122	4101	4081	4061	3	7	10	14	17
68	0.4040	4020	4000	3979	3959	3939	3919	3899	3878	3859	3	7	10	13	17
69	0.3839	3819	3799	3779	3759	3739	3719	3699	3679	3659	3	7	10	13	17
70	0.3640	3620	3600	3581	3561	3541	3522	3502	3482	3463	3	7	10	13	16
71	0.3443	3424	3404	3385	3365	3346	3327	3307	3288	3269	3	6	10	13	16
72	0.3249	3230	3211	3191	3172	3153	3134	3115	3096	3076	3	6	10	13	16
73	0.3057	3038	3019	3000	2981	2962	2943	2924	2905	2886	3	6	9	13	16
74	0.2867	2849	2830	2811	2792	2773	2754	2736	2717	2698	3	6	9	13	16
75	0.2679	2661	2642	2623	2605	2586	2568	2549	2530	2512	3	6	9	12	16
76	0.2493	2475	2456	2438	2419	2401	2382	2364	2345	2327	3	6	9	12	15
77	0.2309	2290	2272	2254	2235	2217	2199	2180	2162	2144	3	6	9	12	15
78	0.2126	2107	2089	2071	2053	2035	2016	1998	1980	1962	3	6	9	12	15
79	0.1994	1926	1908	1890	1871	1853	1835	1817	1799	1781	3	6	9	12	15
80	0.1763	1745	1727	1709	1691	1673	1655	1638	1620	1602	3	6	9	12	15
81	0.1584	1566	1548	1530	1512	1495	1477	1459	1441	1423	3	6	9	12	15
82	0.1405	1388	1370	1352	1334	1317	1299	1281	1363	1246	3	6	9	12	15
83	0.1228	1210	1192	1175	1157	1139	1122	1104	1086	1069	3	6	9	12	15
84	0.1051	1033	1016	0998	0981	0963	0945	0928	0910	0892	3	6	9	12	15
85	0.0875	0857	0840	0822	0805	0787	0769	0752	0734	0717	3	6	9	12	15
86	0.0699	0682	0664	0647	0629	0612	0549	0577	0559	0542	3	6	9	12	15
87	0.0524	0507	0489	0472	0454	0437	0419	0402	0384	0367	3	6	9	12	15
88	0.0349	0332	0314	0297	0279	0262	0244	0227	0209	0192	3	6	9	12	15
89	0.0175	0157	0140	0122	0105	0087	0070	0052	0035	0017	3	6	9	12	15